D0478660

# Exploration of the
# solar system

# Exploration of the solar system

## WILLIAM J. KAUFMANN, III

*Jet Propulsion Laboratory, California Institute of Technology, and Department of Physics, San Diego State University*

**Macmillan Publishing Co., Inc.**

*New York*

**Collier Macmillan Publishers**

*London*

Copyright © 1978, Macmillan Publishing Co., Inc.

Printed in the United States of America

All rights reserved. No part of this book may be reproduced or transmitted in
any form or by any means, electronic or mechanical, including photocopying,
recording, or any information storage and retrieval system, without permission
in writing from the Publisher.

Macmillan Publishing Co., Inc.
866 Third Avenue, New York, New York 10022

Collier Macmillan Canada, Ltd.

Library of Congress Cataloging in Publication Data

Kaufmann, William J
    Exploration of the solar system.

    Includes index.
    1. Solar system.   I. Title.
QB501.K38            523.2            77-5543
ISBN 0-02-362140-0

Printing:    3 4 5 6 7 8        Year:    9 0 1 2 3 4

*To My Mother and Father,*
*with Love*

# Preface

FROM THE pyramids to Palomar, man has looked to the skies for clues to his origin, his fate, and sometimes the meaning of existence itself. This preoccupation with the heavens spans the entire course of human history. Every civilization and every religion to appear on our planet have always had a cosmology at the core of their teachings. From a cosmology, or theory of the universe, it should be possible to understand the motions of the planets, the nature of the stars, and even creation of the cosmos. In the most ancient times, the contents of a cosmology were often based on myth, legend, and a good measure of spiritual revelation. Over the past few centuries, however, man has turned to the task of discovering basic scientific principles at work in nature from which the universe can be better understood.

Almost all of western civilization is a direct result of man's increased knowledge of the physical world. Quite often, this growth of physical insight is an immediate consequence of astronomical investigation. For example, in attempting to explain the motions of the planets, Sir Isaac Newton developed the foundations of classical mechanics, a branch of science dealing with the behavior of physical objects. In examining the motions of the planets, physical laws are revealed in their simplest and purest form, unhampered by friction and air resistance encountered in the laboratory. Using classical mechanics, engineers today not only build bridges and automobiles but send astronauts to the moon. As another example, an understanding of the nature of the sun intimately involved the development of nuclear physics and the special theory of relativity along with Einstein's famous equation $E = mc^2$. Such knowledge today plays a critical role in the balance of international power. There can be little doubt that basic astonomical research will continue to have a profound effect on the future of mankind.

Over the past several decades, an interesting trend has emerged in astronomical research. Beginning in the seventeenth century, astronomers have traditionally focused much of their attention on the planets and the solar system. With the advent of space exploration, however, most of this field of study has been taken over by physicists, geologists, meteorologists, and even biologists. Subjects like planetology, planetary physics, space science, solar physics, space medicine, exobiology, and selenology—to name a few—have sprung up in abundance. And the "pure" astronomer has turned his sights to objects far beyond the confines of our solar system. Indeed, as evidenced by the subjects discussed in the professional journals as well as topics chosen for Ph.D. theses, modern astronomy today begins at distances greater than one light year from earth.

Realizing the potential significance of planetary research, many colleges and universities have recently instituted courses dealing with the solar system. Often this course appears as one half of a two–quarter or two–semester sequence in general astronomy. Unfortunately, no adequate up–to–date texts exist and the instructor is often forced to use the available material from the beginning chapters of a traditional astronomy book.

The purpose of this text is to meet the growing need for a readable comprehensive book on the solar system sciences. It is written primarily for the liberal arts major at typical colleges and universities. Every effort has been made to include the most recent material and much of the manuscript was written as data was being transmitted back from the Martian surface. In this regard, I am deeply grateful to Rita Beebe, Louis Serrano, Bill Pickering, Ben Casados, Gary Price, and Cargill Hall for the many courtesies they showed me during my stay at JPL.

I am also grateful to many of my colleagues who advised me on topics such as geology, geophysics, and meteorology. Through their assistance I have been able to relate many of the findings of the Apollo, Mariner, and Pioneer Missions directly to our own planet. Indeed, one of the key themes running through this book is the applicability of planetary physics to improving man's understanding of the earth sciences. One of the primary goals of this book therefore has been a major synthesis of material from widely divergent and apparently unrelated sources. It is sincerely hoped that the student will appreciate how, for example, the geology of the Martian surface and meteorology of Venus' atmosphere can have a direct bearing on improving life for the future of mankind here on earth.

Finally, I am deeply grateful to Ms. Louise Nelson for typing the manuscript and for her encouragement during the course of this and similar projects.

WILLIAM J. KAUFMANN, III

# Contents

# The Dawn of Astronomy 1

## Ancient Astronomy 1.1

For thousands of years people have looked up into the star-filled night sky, seen what we see and asked many of the same questions we ask. As have countless generations before us, we feel a sense of awe and mystery as we gaze out into space and wonder about the meaning and purpose of the universe. But few people are satisfied with an endless collection of observations. It is simply not enough to go out night after night and make long lists of what you see. The rising and setting of the stars, the silver moon going through its phases and the planets wandering among the constellations of the zodiac imply order rather than chaos to the rational human mind. Man's natural curiosity therefore prompts him to ask why things are the way they are, why the universe works the way it does. An attempt to formulate answers, to understand the nature, properties, and origin of the things we observe in the sky constitutes the basis of the field of science known as *astronomy*. A theory of the universe, from which many of these astronomical questions can be answered, is called a *cosmology*.

In ancient times, man frequently turned to myth, legend, and religion to supply the answers to astronomical questions. Indeed, the resulting fanciful cosmologies were often based in large measure on spiritual revelation. As with the story in Genesis, every religion ever invented by man has a cosmology at the core of its teachings. Great works of literature, art, sculpture, architecture, and music have been inspired by these mythical cosmologies over the ages. Of course, these myths and legends usually reveal far more about the psychological orientation of ancient cultures than the true nature of physical reality.

This total dependence on myth, legend, and religion persisted up until about the fifth century, B.C. At that time, a new approach to astronomy appeared in ancient Greece—an approach that set the stage for all modern science. Rather than relying on prophets and priests, the Greeks felt that the nature of the universe could be deduced from a combination of observation, logic, and reasoning. For example, the earliest Greek astonomers realized that the phases of the moon could be explained by assuming that the moon orbits the earth and shines by reflected sunlight. As shown in Figure 1–1, when the moon is located in the same part of the

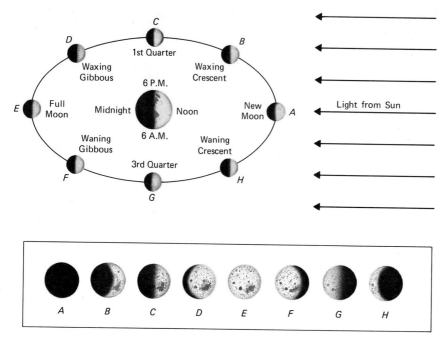

**Figure 1–1.** *Phases of the Moon.* The phases of the moon are determined by how much of the illuminated side can be seen from Earth.

sky as the sun, we on the earth see only the dark side of the moon. This phase is called *new moon.* Two weeks later, when the moon is located in the direction opposite the sun, we on the earth see the fully illuminated side. This phase is called *full moon.* Half way between new moon and full moon phases, we can see half of the illuminated side and half of the dark side. These phases are called *first quarter* and *last quarter.*

It takes the moon about four weeks to go around the earth. Therefore, approximately one week elapses between each of the major phases (new moon ⟶ first quarter ⟶ full moon ⟶ last quarter). When less than one half of the illuminated side is seen (i.e., from last quarter to first quarter), the moon is in a *crescent* phase. Conversely, when more than one half of the illuminated side is seen (i.e., from first quarter to last quarter), the moon is in a *gibbous* phase.

During the two weeks following a new moon, more and more of the illuminated side of the moon is seen from one night to the next. Astronomers therefore speak of *waxing crescent* and *waxing gibbous* phases since "waxing" means "getting bigger." Conversely, during the two weeks following a full moon, less and less of the illuminated side of the moon is seen from one night to the next. The terms *waning gibbous* and *waning crescent* are used to denote these phases since "waning" means "getting smaller."

In addition to explaining the phases of the moon, the ancient Greeks also understood why eclipses occur. Occasionally at the time of new moon, the alignment of the earth, moon, and sun is so precise that the moon's shadow falls directly on the earth, as shown in Figure 1–2. At such times of perfect alignment, the moon momentarily blocks out the sun, resulting in a *solar eclipse.* Similarly, on rare occasions, a nearly perfect alignment of the earth, moon, and sun occurs at the time of full moon. When this happens, the full moon passes the earth's shadow, resulting in a *lunar eclipse,* as shown in Figure 1–3. Such explanations seemed far more reasonable than the idea that goblins or demons swallowed and regurgitated the sun and moon during solar and lunar eclipses.

In examining lunar eclipses, the Greeks noticed that the shape of the earth's shadow is always circular. Regardless of the day of the year or hour of the night at which a lunar eclipse occurs, as the moon enters or leaves the earth's shadow, the edge of the shadow is always circular. Reasoning that only spherical objects could always cast a circular shadow, the Greeks concluded that the earth is round. From that time on, anyone familiar with the writings of Plato and Aristotle was fully aware of the fact that the earth is spherical.

In addition to arriving at some important conclusions about the sun, moon, and earth, the Greeks used these discoveries in an attempt to make astronomical measurements. For example, in the third century B.C., Eratosthenes used the angle of the sun at high noon to measure the circumference of the earth. Eratosthenes noticed that on the first day of summer, the sun is directly overhead at noon as seen

**Figure 1-2.** *Geometry of a Solar Eclipse.* When the moon passes directly between the earth and the sun, the moon casts its shadow onto the earth. People standing inside the moon's shadow see an eclipse of the sun.

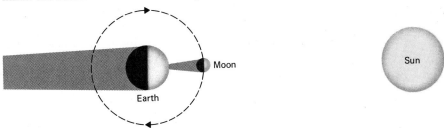

**Figure 1-3.** *Geometry of a Lunar Eclipse.* An eclipse of the moon occurs when the moon passes through the earth's shadow. A lunar eclipse is visible to anyone standing on the nighttime side of the earth.

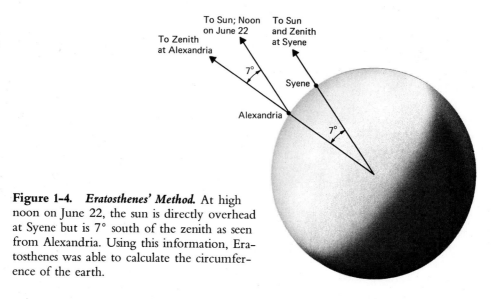

**Figure 1-4.** *Eratosthenes' Method.* At high
noon on June 22, the sun is directly overhead
at Syene but is 7° south of the zenith as seen
from Alexandria. Using this information, Eratosthenes was able to calculate the circumference of the earth.

from Syene, Egypt (near the modern Aswan). Incidentally, the "overhead point" is
called the *zenith* while the opposite point in the sky, directly below one's feet, is
called the *nadir*. Eratosthenes lived in Alexandria where the sun is never directly
overhead. In fact, on the first day of summer (June 22), the sun is 7° south of the
zenith at high noon. As shown in Figure 1–4, Eratosthenes reasoned that the angle
between Syene and Alexandria, as measured from the center of the earth, must also
be 7°. Since there are 360° in a circle, and since 7° is about ⅟₅₀ of 360°, Eratosthenes concluded that the distance between Syene and Alexandria must be ⅟₅₀ of
the circumference of the earth. Knowing the distance between these two cities, he
simply multiplied this length by 50 to arrive at the earth's circumference.

At about the same time Eratosthenes was measuring the size of the earth,

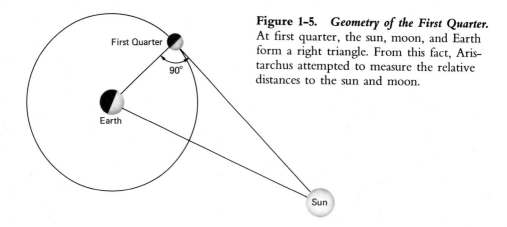

**Figure 1-5.** *Geometry of the First Quarter.*
At first quarter, the sun, moon, and Earth
form a right triangle. From this fact, Aristarchus attempted to measure the relative
distances to the sun and moon.

another Greek astronomer, Aristarchus, was trying to measure the relative distances to the moon and sun. Aristarchus realized that at the precise moment of first quarter or last quarter phases, the angle between the sun and the earth as seen from the moon must be exactly 90°. As shown in Figure 1–5, at first or last quarter the sun, moon, and earth therefore form a right triangle. Greek mathematicians were very familiar with the geometrical properties of right triangles, and using this knowledge Aristarchus attempted to deduce the relative distances to the moon and sun. Specifically, Aristarchus compared the time it takes the moon to go from first quarter to last quarter with the time it takes to go from last quarter to first quarter in order to measure the other angles in the right triangle in Figure 1–5. Unfortunately, with the naked eye it is extremely difficult to know precisely when the moon is at first or last quarter. Consequently, Aristarchus' final answers were not terribly precise. Nevertheless, Aristarchus clearly demonstrated that the sun must be many times farther away from us than the moon.

While the Greeks had a reasonably good understanding of some properties of the sun, moon, and earth, the motions of the planets posed some rather difficult problems. Looking out at the nighttime sky, we see lots of stars. The stars rise and set over the course of a day, just as the sun and moon rise and set. Yet, in spite of the daily or *diurnal* motion of the stars, the relative positions of the stars remain virtually unchanged over the ages. The Big Dipper or the constellation of Orion, for example, looks exactly the same tonight as it did thousands of years ago. Astronomers therefore speak of the *fixed stars* because it is impossible to detect any changes in the patterns of the stars with the naked eye.

In sharp contrast to the fixed stars, there are five objects (Mercury, Venus, Mars, Jupiter, and Saturn) that at first glance look like ordinary stars but that do change their locations from night to night. Careful observation over the course of several nights reveals that these objects appear to wander slowly against the background of the fixed stars. Of course, these objects are called *planets,* a term which comes from the Greek word meaning "wanderer."

The planets confine their wanderings to a band of twelve constellations around the sky called the *zodiac.* Usually the planets move eastward against the background stars. This gradual eastward movement is called *direct motion.* Occasionally, however, a planet stops and backs up for a few weeks. This occasional westward movement is called *retrograde motion.* By way of an example, Figure 1–6 shows the path of Mars through Capricornus during the summer months of 1971. Mars exhibited retrograde motion from mid-July to September.

As a result of occasional retrograde motion, the paths of the planets across the heavens are far more complicated than those of the sun or moon. The Greeks therefore spent many centuries struggling to devise a scheme to account for these complicated planetary movements. They first of all had to establish some basic ideas about how the universe is organized. For example, the Greeks felt that the earth is at rest at the center of the universe. It was simply inconceivable that our huge, massive earth could be in motion. Furthermore, the daily rising and setting of the sun, moon, planets, and stars indicated that the heavens revolve around the earth.

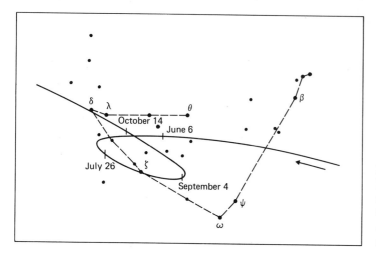

**Figure 1-6. *The Path of Mars.*** During the summer of 1971, Mars moved through the constellation of Capricornus. From mid-July to early September Mars was moving retrograde.

The Greeks therefore concluded that any reasonable cosmology must be *geocentric;* it must be centered on the earth. In addition, since God created the heavens and since God is perfect, only the most perfect geometrical forms could be used in devising a cosmology. The Greeks felt that circles are the most perfect of all geometrical shapes. Consequently, a reasonable cosmology must be geocentric and the motions of the planets must be explainable in terms of circles.

In the second century A.D., the last great Greek astronomer, Claudius Ptolemy, succeeded in devising a system which met all of these "reasonable" requirements. Ptolemy found that he could account for the motion of the planets by assuming that each planet revolves about a small circle, called an *epicycle,* which in turn revolves about a larger circle, called a *deferent,* centered approximately on the earth. Usually the motion of the planet around the epicycle adds to the motion of the epicycle around the deferent resulting in direct (eastward) movement against the

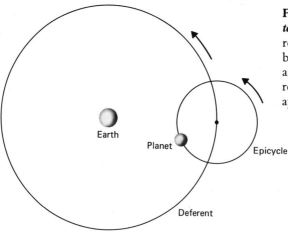

**Figure 1-7. *The Ptolemaic System.*** Ptolemy accounted for the retrograde motion of the planets by assuming that the planets go around epicycles which, in turn, revolve around deferents centered approximately on the earth.

**6**

background stars. Occasionally, however, the rapid motion of the planet around the epicycle subtracts from the slower motion of the epicycle around the deferent resulting in periods of retrograde (westward) movement. The essential idea behind Ptolemy's geocentric cosmology is shown in Figure 1–7. He spent most of his life figuring out the sizes and rotation rates of all the epicycles and deferents of all the planets. His findings were compiled in a series of 13 volumes called the *Almagest*.

For the next 1300 years the Ptolemaic system and the *Almagest* completely dominated all of astronomy. Ptolemy had performed his calculations with such precision that his geocentric theory succeeded in predicting the positions of celestial bodies for many centuries. By the time of the Renaissance, however, tiny errors which would have gone unnoticed in earlier years had compounded over the centuries to such an extent that calculations based on the *Almagest* no longer gave the anticipated accuracy. In spite of eleventh-hour attempts to save the Ptolemaic system, the time was at hand for a fundamental reorientation in man's thinking about his place in the universe.

## *The Copernican Revolution* **1.2**

ONE OF THE primary and most difficult tasks of ancient astronomy was to explain the motions of the planets. As mentioned in the previous section, the planets usually appear to move slowly eastward from night to night, as seen against the background stars. Occasionally, however, each planet stops its usual direct motion and backs up for a few weeks or months. These periods of retrograde motion resulted in the complexity of Ptolemy's geocentric cosmology with its system of circles on circles. Indeed, as long as astronomers insisted that the earth is at the center of the universe, as long as they insisted that the planets go around the earth, their theories and cosmologies were bound to be complicated.

Almost two thousand years ago, the Greek astronomer Aristarchus suggested a very simple explanation for the apparent retrograde motions of the planets. To appreciate Aristarchus' suggestion, imagine driving down a freeway at 55 miles per hour. If you pass a car going in the same direction at 45 miles per hour, it will appear that this slower vehicle is going backward relative to you. It was from this type of an analogy (probably using chariots rather than cars) that Aristarchus proposed the first *heliocentric* or sun-centered cosmology. He realized that if the earth and the planets were all orbiting the sun, then as the earth passes by a slower-moving planet, that planet will appear to go backward for a period of time.

Perhaps the best way of illustrating Aristarchus' explanation of retrograde motion is with the aid of a diagram such as Figure 1–8. In this diagram, the earth catches up with and overtakes a slowly moving planet such as Mars. While the earth and the planet are far apart, the planet appears to be moving eastward among the stars. But when the earth is overtaking this outer planet, retrograde motion

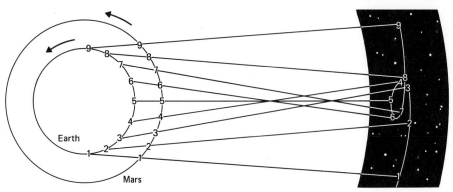

**Figure 1-8.** *Retrograde Motion.* As the earth overtakes and passes a slowly moving outer planet, that planet will appear to stop its usual eastward motion and back up for a period of time.

occurs because the earth is moving at a higher speed. Finally, when the separation between the earth and the planet is again large, the planet resumes its usual eastward course. In other words, as seen from earth, the motions of the planets look complicated only because the earth itself is moving.

Although this scheme would obviously explain retrograde motion, it is necessary to assume that the earth is moving. In addition, Aristarchus evidently did not have the data or motivation to prove in detail that this heliocentric cosmology could, in fact, be used to predict the courses of the planets. As a result, his ideas were dismissed as clever fantasy. After all, if the earth is moving, why don't we fall off?

As mentioned in the previous section, the Ptolemaic system worked so well and seemed so reasonable that it completely dominated all of astronomy for over a thousand years. But as the centuries passed by, it became clear that there were discrepancies between the observed positions of the planets and the predicted position from the *Almagest*. To correct this, astronomers added more epicycles. Planets were then assumed to travel around epicycles which in turn moved along additional epicycles. By the early 1500s the geocentric picture of the universe contained no less than 79 separate circles. This system had become artificial and had so many circles that it was in imminent danger of collapsing under the sheer weight of its own complexity. The time was at hand for a basic change in thinking.

In the early part of the sixteenth century, a young monk visited the Vatican. The Vatican Library contained copies of Aristarchus' works, but it is unknown whether or not this young man ever read about Aristarchus' heliocentric ideas. Nevertheless, this humble monk who was also a gifted artist, a distinguished physician, and a renowned economist, was destined to give to the world a new way of thinking about the nature of the universe. His name was Nicholas Copernicus.

Copernicus' main contribution is that he was the first astronomer to work out all the details of a heliocentric cosmology. Unlike Aristarchus who merely showed that it was plausible to assume that the sun was at the center of the universe,

Copernicus used mathematics to prove that the positions of the planets could be predicted with accuracy in such a system. His heliocentric cosmology was published in a book called *De Revolutionibus Orbium Celestium* in 1543, the year of his death.

It should be emphasized that Copernicus' cosmology was entirely emperical. In other words, he set up his theory in such a way that he would get the right answers for the positions of the planets. For example, Copernicus explained retrograde motion by assuming that the order of the planets, starting nearest the sun, must be: Mercury, Venus, Earth, Mars, Jupiter, Saturn. With any other ordering, he would not be able to account for their motions. Also, he assumed that the nearer a planet is to the sun, the greater its orbital speed. Thus, Mars goes around the sun faster than Jupiter. Using these two assumptions, he was able to make his heliocentric system fit the observations.

From this ordering of the solar system, it is possible to distinguish between so-called *inferior planets* and *superior planets*. An inferior planet is one which has an orbit inside the earth's orbit. There are only two such planets: Mercury and Venus.

**Figure 1-9.** *Nicholas Copernicus,* 1473–1543. (*Yerkes Observatory photograph*)

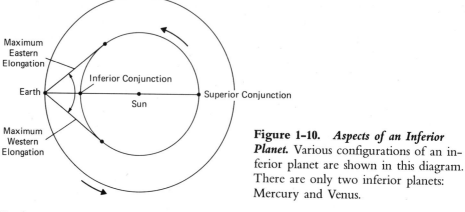

**Figure 1-10.** *Aspects of an Inferior Planet.* Various configurations of an inferior planet are shown in this diagram. There are only two inferior planets: Mercury and Venus.

Both Mercury and Venus always appear near the sun in the sky. Under favorable conditions they are seen either in the west shortly after sunset, in which case they are sometimes called "evening stars," or in the east as "morning stars" just before sunrise. Due to the brilliance of the sun, the best time to observe these planets is when they are farthest from the sun. As shown in Figure 1-10, this condition occurs when the inferior planet is at either *greatest eastern elongation* (in the evening sky) or *greatest western elongation* (in the morning sky).

From time to time, an inferior planet passes between the earth and the sun. Such an event is called *inferior conjunction*. On the other hand, when Mercury or Venus passes behind the sun, astronomers speak of *superior conjunction*. While it might seem that Mercury and Venus should be seen as little black dots against the solar disc at every inferior conjunction, this rarely occurs. Actually, the orbits of Venus and Mercury are inclined to the earth's orbit and therefore the inferior planets usually pass to the north or south of the sun. Those rare occasions when they actually cross directly in front of the sun are called *transits*. The most recent transit of Venus occurred in the nineteenth century and the next transit is not due until June 8, 2004. Transits of Mercury are far more common with about thirteen occurring each century.

Superior planets have orbits that are larger than the earth's orbit. Obviously a superior planet such as Mars, Jupiter, or Saturn can never appear at inferior conjunction. But they do, however, pass behind the sun. As shown in Figure 1-11, when the sun lies between the earth and a superior planet, we say simply that the planet is in *conjunction*. Unlike Venus and Mercury, the superior planets can appear directly opposite the sun at what we call *opposition*. When a superior planet is in opposition, the earth lies between the planet and the sun. At such times, the planet is seen high in the nighttime sky at midnight.

In setting up his heliocentric theory, Copernicus realized that he had to distinguish between the *sidereal period* and the *synodic period* of a planet. The sidereal period of a planet is its real orbital period. It is simply how long it takes for a planet to go once all the way around the sun. For example, the sidereal period of Mars is 687 days and thus you might say that a "year" on Mars lasts for 687 Earth-days.

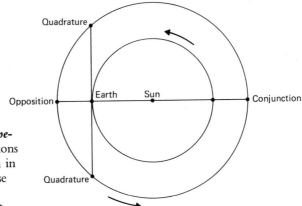

**Figure 1-11.** *Aspects of a Superior Planet.* Various configurations of a superior planet are shown in this diagram. Any planet whose orbit is larger than the earth's orbit is called a superior planet.

Unfortunately, we can never directly measure the sidereal period of a planet from the earth because the earth itself is moving. What can be measured is the synodic period, which is the time it takes the planet to go from one configuration back to the same configuration. For example, the synodic period of Mars is 780 days meaning that slightly more than two years pass between one opposition of Mars and the next.

The meaning of a synodic period is shown in Figure 1–12. In the case of an inferior planet, we could choose to measure the synodic period as the time from one inferior conjunction to the next. As shown on the left of Figure 1–12, the inferior

**Figure 1-12.** *Synodic Periods of Inferior and Superior Planets.* The synodic period of a planet is the time it takes the planet to go from one configuration back to the same configuration. For example, for an inferior planet the synodic period is the time between successive inferior conjunctions (shown on the left) while for a superior planet, it is the time between successive oppositions (shown on the right).

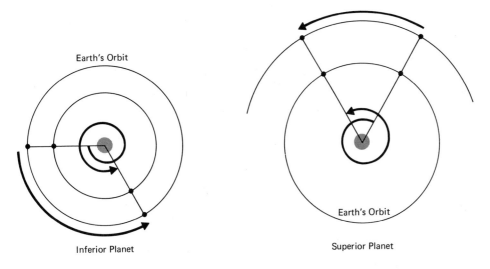

planet must make more than one complete trip around the sun between successive inferior conjunctions. Consequently, the synodic periods of the inferior planets must always be longer than their sidereal periods. In the case of a superior planet, we could choose to measure the synodic period as the time from one opposition to the next. As shown on the right of Figure 1–12, the earth must make more than one complete trip around the sun between successive oppositions. Consequently, the synodic periods of the superior planets are always longer than one Earth-year (365 days).

Copernicus was able to prove that the sidereal and synodic periods of the planets are related in a very simple mathematical fashion. Thus, if we measure the synodic period of a planet, we can always calculate its sidereal period. A table giving the sidereal and synodic periods of the planets known to Copernicus is shown as follows.

| Planet | Sidereal Period | Synodic Period |
|--------|-----------------|----------------|
| Mercury | 88 days | 116 days |
| Venus | 225 days | 584 days |
| Earth | 365 days | |
| Mars | 687 days | 780 days |
| Jupiter | 11.9 years | 390 days |
| Saturn | 29.5 years | 378 days |

Knowing the various periods of the planets, and measuring their positions in the sky, Copernicus was able to figure out the relative sizes of their orbits. He had no way of coming up with answers in miles or kilometers, so he expressed his results in terms of the size of the earth's orbit. These answers agree remarkably well with modern results as shown in Table 1–1.

The work of Nicholas Copernicus marks an important turning point in the course of human consciousness. He started a revolution that spread through Europe during the decades following the publication of his book, culminating with the

**Table 1-1**

| Planet | Distance of Planet from Sun | |
|--------|------------|--------|
| | Copernicus | Modern |
| Mercury | 0.38 | 0.39 |
| Venus | 0.72 | 0.72 |
| Earth | 1.00 | 1.00 |
| Mars | 1.52 | 1.52 |
| Jupiter | 5.22 | 5.20 |
| Saturn | 9.17 | 9.54 |

brilliant accomplishments of Sir Isaac Newton. Yet, in the interim, the Copernican cosmology met stiff opposition, primarily from the Church. *De Revolutionibus* was banned and burned, and had Copernicus lived in Rome, he would have most certainly been executed for heresy.

In spite of the successes of his heliocentric cosmology, Copernicus' system still had a few problems. Following the example of many astronomers before him, Copernicus chose to express the planetary orbits using only circles. As explained in the next section, the true orbits of planets are not precisely circular, and therefore he resorted to epicycles. Although Copernicus was able to dismiss the idea of the earth's central location in the universe, he still clung in his theory to circles and circular motion. Two generations passed before Johannes Kepler proved that ellipses, not circles, describe the true orbits of planets. Kepler succeeded in forever banishing the troublesome epicycle from the field of astronomy.

## Kepler's Cosmology 1.3

HOW COULD Copernicus possibly be right? How could something as "heavy" and "sluggish" as the earth be in motion about the sun? And to make matters worse, it was virtually impossible to reconcile a heliocentric cosmology with the Bible or with the great writings of Aristotle and Plato.

It occurred to some of Copernicus' contemporaries that observations might prove or disprove his theory. After all, if you move from one place on the earth to another, the scenery changes; the relative position of distant mountains, trees and houses appear to change as you walk down the street. If the earth is really moving, shouldn't we observe a similar effect with the stars? It seemed reasonable to suppose that if Copernicus was right, then the relative positions of the stars should move slightly as the earth allegedly goes around the sun. To decide this issue, extremely accurate measurements of the positions of the stars were needed, far more precise than anything that had ever been done before.

Three years after the death of Copernicus, Tycho Brahe was born in Denmark. Extravagant, arrogant, and often obnoxious, Tycho Brahe met the challenge of the times by establishing a superb observatory on the Danish island of Hveen. For twenty years, this colorful astronomer and his assistants carried out a program of making the most complete and precise astronomical observations of the locations of the sun, moon, stars, and planets ever produced.

Tycho Brahe assumed that the stars were only 7000 times farther away from the earth than the sun. He therefore expected to detect slight changes in the positions of the stars over the period of a year. No such changes were observed and he therefore concluded that Copernicus must have been wrong. Actually, the real distances to the stars are more like 300,000 times the distance from the earth to the sun. Thus, while slight changes in stellar positions do indeed occur, they are so small that Tycho Brahe could not have detected them with his sixteenth-century instruments.

In spite of his rejection of a heliocentric cosmology, Tycho Brahe's observations proved invaluable to astronomy in the seventeenth century. After his death in 1601, most of Brahe's notes and records were given to his gifted young assistant, Johannes Kepler. The precision of the measurements in these records allowed Kepler to proceed in establishing the groundwork for all of modern celestial mechanics.

During his youth as a theology student in Germany, Kepler became an early convert to the Copernican system. It was soon found, however, that Copernicus' theory did not accurately predict the positions of the planets as measured by Tycho Brahe. Kepler therefore began working on the details of planetary motions. He chose to concentrate his attention on the particularly troublesome planet Mars. For nearly ten years Kepler tried all kinds of epicycles, deferents, and equants in a heliocentric system in an effort to reproduce Brahe's observations. Every attempt met with failure. In desperation, it occurred to Kepler that perhaps the cause of his difficulties was the use of circles. For thousands of years, all astronomers had used circles to describe the motions of the planets. Perhaps the circle was simply the *wrong* curve. After several attempts with ovals, Kepler discovered that the orbit of Mars conformed extremely well with a curve called an *ellipse*.

**Figure 1–13.** *Johannes Kepler,* 1571–1630. ( *Yerkes Observatory photograph*)

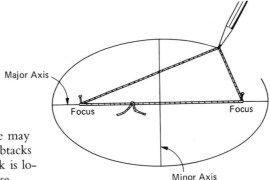

**Figure 1-14.** *The Ellipse.* An ellipse may be drawn with the aid of two thumbtacks and a loop of string. Each thumbtack is located at a focus of the resulting ellipse.

An ellipse is a basic geometrical curve that can be drawn with the aid of two thumbtacks and a loop of string, as shown in Figure 1-14. To draw an ellipse, first stick two thumbtacks into a piece of paper on a drawing board. Then place the loop of string over the thumbtacks and use a pencil to hold the string taut. As you move the pencil around the taut loop of string, an ellipse is drawn on the paper.

Each thumbtack in the above construction is located at a *focus* of the ellipse. An ellipse has two foci and the distance between the foci plays an important role in how flat or round the ellipse is. If the foci are far apart in relation to the size of the loop of string, then the ellipse will be skinny. If the foci are close together, then the ellipse will have a circular appearance. In fact, you may think of a circle as an ellipse in which one focus is on top of the other. In addition, just as we speak of the "diameter" of a circle, the size of an ellipse may be specified by giving the lengths of its *major axis* and *minor axis*. As shown in the diagram, the major axis is simply a straight line passing through both foci. It is the maximum diameter of the ellipse, and half of this length is called the *semimajor axis*. The minor axis is the perpendicular bisector of the major axis, and the *semiminor axis* is equal to one half of the minor axis.

This new curve provided Kepler with the tools for the first major breakthrough in astronomy since the publication of *De Revolutionibus*. Not only did Kepler succeed in showing that the orbit of Mars is an ellipse, he found that the orbits of *all* the planets are ellipses—provided you assume that the sun is located at one focus of each elliptical orbit. This discovery was published in 1609 in *The New Astronomy* and has come to be known as *Kepler's first law*. It may be stated concisely as follows:

> *Each planet moves about the sun in an orbit that is an ellipse,
> with the sun at one focus of the ellipse.*

In addition to establishing the true shapes of the orbits of the planets, Kepler also investigated the speeds of the planets along their orbits. He found that planets moved faster when they are near the sun and slower when they are far from the sun. After considerable research, he discovered that there was a fairly simple way of expressing how fast a planet moves along its orbit at different times. This discovery is called *the*

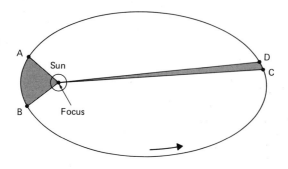

**Figure 1-15.** *The Law of Equal Areas.* If it takes a planet one month to go from "A" to "B," it will also take the planet one month to go from "C" to "D," provided the shaded areas are equal.

*law of equal areas,* or more commonly *Kepler's second law.* It was also published in *The New Astronomy* and may be stated as follows:

> *A straight line joining the sun and a planet sweeps out equal areas in space in equal intervals of time.*

To demonstrate the meaning of Kepler's second law, consider the orbit of a hypothetical planet, as shown in Figure 1–15. According to Kepler's first law, this orbit is an ellipse with the sun at one focus; the other focus is empty. Suppose it takes one month for the planet to go from point "A" to point "B." In doing so, a line joining the planet and the sun sweeps out a triangular area in space. Kepler's second law tells us that during any other month, a line joining the sun and the planet must sweep out an area equal in size to the original triangular segment. Thus, if it takes the planet one month to go from point "C" to point "D," the two shaded triangular segments in the diagram must have equal areas.

Although the above example illustrates Kepler's second law, it perhaps erroneously suggests that the orbits of the planets are flattened ellipses. Actually, all of the planetary orbits are very nearly circular. If you draw the orbit of Mars to scale on a piece of paper, it would look almost circular. It is therefore incredible that Kepler was able to discover that the orbits of planets were ellipses at all.

The final major contribution Kepler made in building the foundations of modern astronomy is stated in his third law. In general terms, this law tells us the sizes of the orbits of each of the planets. Just like Copernicus, Kepler could not figure out the sizes of the orbits in miles or kilometers. Instead, he chose to express the scale of the solar system relative to the size of the earth's orbit. For this purpose it is extremely useful to define the *astronomical unit* (often abbreviated *AU*) as the length of the semimajor axis of the earth's orbit. Since the earth's orbit is very nearly circular, we can say that the average distance from the earth to the sun is 1 AU. It was not until the twentieth century that astronomers succeeded in accurately measuring the astronomical unit, arriving at the value of 149,597,893 kilometers (92,956,000 miles) for the average Earth–sun distance.

The easiest way to express Kepler's third law is with the aid of a graph, as shown in Figure 1–16. Here is plotted the orbital periods (i.e., sidereal periods) of the planets against the lengths of the semimajor axes of their orbits (expressed in AUs).

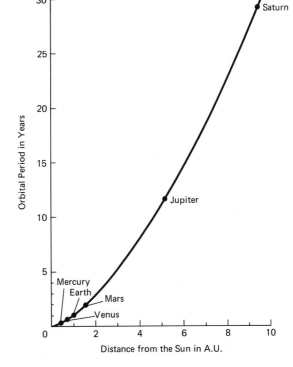

**Figure 1-16.** *Kepler's Third Law.* The sidereal periods of the planets are plotted against the sizes of their orbits. Since a smooth curve fits through all the points, there must be a simple relation between these two quantities.

The fact that a smooth curve fits through all the points on the graph proves that there must be a simple mathematical relationship between the sidereal period of a planet and the length of its semimajor axis. This relationship, published in *The Harmony of the Worlds* in 1619, is called *Kepler's third law* and may be stated as follows:

> *The squares of the sidereal periods of the planets are in direct proportion to the cubes of the semimajor axes of their orbits.*

Although this may sound complicated, consider Mars as an example. The sidereal period of Mars is 687 days, or 1.88 years. This is how long it takes Mars to go once around the sun. Thus the "square of the sidereal period" is simply 1.88 × 1.88 equals 3.54. We then realize that the length of the semimajor axis of Mars' orbit must be 1.52 AU since "the cube of the semimajor axis" is 1.52 × 1.52 × 1.52 also equals 3.54. Parenthetically note that given the modern value of nearly 150 milllion kilometers (93 million miles) for the AU, the average Mars–sun distance must be 228 million kilometers (330 million miles).

While Kepler was doing primarily theoretical work in Northern Europe, important new observations were being made in Italy by Galileo Galilei. In 1609, Galileo heard about the invention of a remarkable device called a telescope. After some experimentation with lenses, Galileo succeeded in building several telescopes,

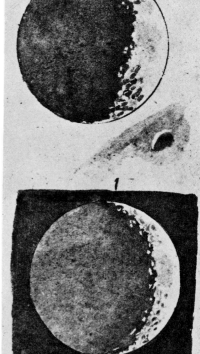

**Figure 1-18.** *Galileo's Drawings of the Moon.* When Galileo turned his telescope toward the moon, he saw mountains, valleys, craters, and plains. This was very blasphemous to many of his contemporaries who thought that the moon (up in "heaven") should be perfectly smooth since it was made by a perfect God. (*Yerkes Observatory photograph*)

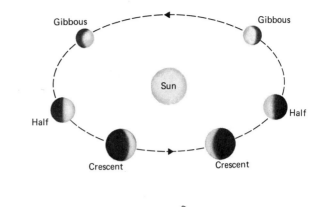

**Figure 1-19.** *The Phases of Venus.* Observations of the phases and apparent size of Venus demonstrate that Venus must be in orbit about the sun, *not* about the earth.

the best having a 30-power magnification. Galileo's major contribution to astronomy is that he was the the first person to point the telescope toward the stars. He promptly discovered that the moon was covered with craters and mountains and that the sun had "sunspots." This, of course, was considered heresy since heavenly bodies created by God must be perfect, uncorruptable, and without blemish. So Galileo found himself in big trouble with the Church. Nevertheless, he persevered and discovered that Venus goes through phases, just like our moon. Furthermore, after a couple of years he found that when Venus presents a gibbous phase, it appears small in size. However, when Venus shows a crescent phase, it appears large through the telescope. As shown in Figure 1–19, the relationship between the phases and sizes of Venus constitutes conclusive proof that Venus goes around the sun.

Galileo also noticed that there were four little "stars" near Jupiter. From night to night these "stars" seemed to move back and forth with respect to the planet. After only a few nights of observation, it was clear to Galileo that these "stars" were not stars at all, but rather they must be four moons in orbit about Jupiter. News of the discovery of four new moons in the universe spread quickly. Upon learning the details of Galileo's observations, Kepler immediately proved that his third law worked for the moons of Jupiter by demonstrating that the cubes of the distances of the Jovian satellites from Jupiter were proportional to the squares of their orbital periods around the planet.

With the brilliant work of Johannes Kepler and the important observations of Galileo Galilei, it was now totally impossible to go back to a geocentric picture of the universe. On the theoretical side, Kepler had succeeded in formulating the laws of planetary motion about the sun. Galileo had made a number of observations and discoveries that only make sense in a heliocentric cosmology. In spite of the efforts

**19**

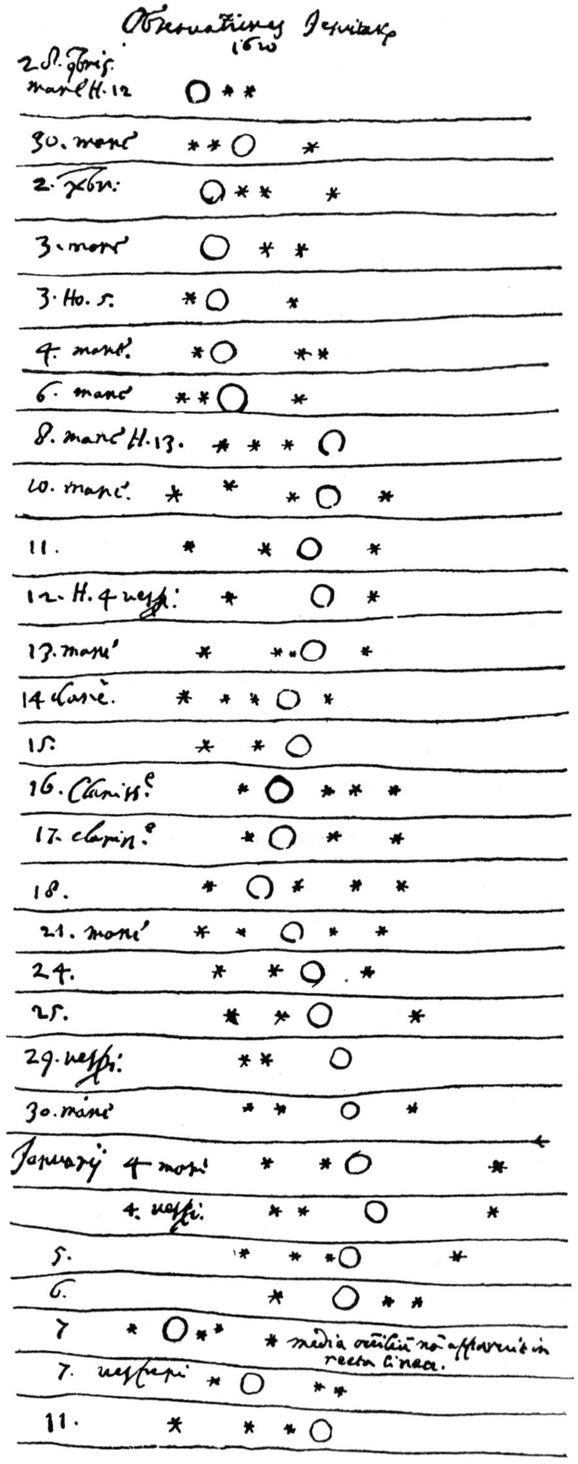

**Figure 1-20.** *Galileo's Drawings of Jupiter and Its Satellites.* Galileo noticed that there were four little "stars" near Jupiter. From night to night, these "stars" appeared to move back and forth from one side of Jupiter to the other. Galileo concluded that Jupiter must have four moons, which orbit the planet. (*Yerkes Observatory photograph*)

of the Church, the geocentric universe of Ptolemy that had dominated astronomy for twelve centuries was dead and would never again be resurrected.

Finally, it should be noted that, as was the case with Copernicus, all of Kepler's work was empirical. In other words, Kepler tried all kinds of different curves for the orbit of Mars until he finally came up with one (the ellipse) that would fit the observations of Tycho Brahe. But it would be much more elegant if we could make some fundamental assumptions about the nature of the physical world, and then by using mathematics actually prove that the orbits of the planets around the sun *must* be ellipses. Accomplishing such a feat would be a monumental achievement of the human intellect.

## *Newton and Gravity*   1.4

ONE OF THE greatest figures in the entire history of science, Sir Isaac Newton, was born in England in 1642, the same year in which Galileo died. From the time of his youth, Newton was preoccupied with trying to understand the physical reasons behind the heliocentric cosmologies of Copernicus and Kepler. What was the physical force that kept the planets going around the sun? It was apparent to Newton that a truly elegant approach to these and related questions could only be achieved by making some fundamental assumptions about the nature of the universe. Then, using mathematics he hoped to be able to prove the validity of Kepler's laws.

Newton began by making three basic assumptions about the nature of the physical world, now known as *Newton's laws of motion*. The groundwork for his first law had actually been laid by Galileo several decades earlier. This first law simply states that:

> *In the absence of outside forces, a body at rest will remain at
> rest, or a body in motion will remain in motion at a constant
> speed in a straight line.*

At first glance, this law of motion might seem unreasonable. For example, if you shove a chair and let it slide across the floor, the chair will come to rest rather than continuing "in motion at a constant speed in a straight line." But you realize that there were, in fact, "outside forces" acting on the chair. It was the force of friction between the legs of the chair and the floor that caused the chair to slow down. If the floor had been perfectly smooth and frictionless, the chair would indeed have continued unhampered until it hit the wall or some other object.

Using this first law, Newton immediately realized that some kind of force must be acting on the planets. As demonstrated by Kepler, the planets move about the sun in ellipses with the sun at one focus. The planets do not move in straight lines at

**Figure 1–21.** *Sir Isaac Newton,* 1643–1727. (*Yerkes Observatory photograph*)

constant speeds. Therefore there must be a force constantly acting on the planets causing them to move along elliptical orbits, preventing them from flying straight off into space. Using mathematics, Newton was able to prove that this force on the planets is always pointed directly at the sun, as shown schematically in Figure 1–22. It is as if the sun exerts a force on the planets directly across empty space thereby keeping them in their orbits.

With further study, Newton realized that he had to make a clear distinction between the two commonly confused concepts of *mass* and *weight*. Every material object has mass. Mass is an inherent property of matter; it is that property of matter that resists a change in its state of motion. If you push on an object, the object will begin to move or to accelerate. If the mass of the object is small, the object will give only a little resistance to the force with which you push it. The object will accelerate very easily. If, however the mass is large, the object will put up much more resistance to the force with which you push, and therefore the object will accelerate much more slowly. Thus, when your car runs out of gas, if you own a

22

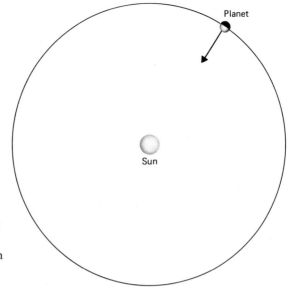

**Figure 1-22.** *An Orbit About the Sun.* Since the planets do not travel in straight lines, a force must be acting on them. Newton proved that this force is always pointed directly toward the sun.

Volkswagen you will find it much easier to push the car to a gas station than if you had been driving a Cadillac. The mass of a Cadillac is greater than the mass of a Volkswagen. This relationship between mass, force, and acceleration is stated concisely in Newton's second law as follows:

> *If a force acts on a body, it produces an acceleration that is proportional to the force and inversely proportional to the mass of the body.*

While the details of Newton's second law are not important to the basic theme of this text, we nevertheless realize that mass is a fundamental property of matter. An object has a certain mass regardless of where that object is or what it is doing. Thus, it is always more difficult to push a Cadillac than a Volkswagen, regardless of whether these cars are on the earth, on Jupiter, or floating in free space.

Weight, however, is a different subject. The weight of an object is a measure of how hard something pushes downward under the influence of gravity. Thus, a Cadillac weighs more than a Volkswagen because the Cadillac pushes down harder on the pavement of a street than a Volkswagen does. But unlike mass, the weight of an object does depend on where that object is. On Jupiter or on the moon, these cars would weigh differently than here on earth. In fact, floating in outer space, they would have no weight at all; they would be weightless.

As a final example of the difference between mass and weight, consider a man who weighs 150 pounds on the earth. It turns out that this same man would weigh only 25 pounds on the moon, while he would tip the scales at 380 pounds on Jupiter. Floating in space, he would weigh 0 pounds. Yet in all these examples, he has exactly the same mass. There are the same number of atoms in his body and his body will give the same resistance to your pushing and shoving regardless of where he is.

In discussing the distinction between mass and weight, we have "jumped the gun" by introducing the concept of *gravity*. The earth obviously exerts a force on everything around us. This force is called gravity and without it, tables, chairs, and people would go floating off into space. This fact was apparent to any intelligent person in the seventeenth century. Yet, Newton realized that while gravity exerts a force on tables, chairs, and people, these objects exert an equal and opposite force back on the earth. This idea is contained in very general terms in his third law of motion:

*For every action there is an equal and opposite reaction.*

Thus, for example, if you weigh 150 pounds, you are pushing down on the floor with a force of 150 pounds. But the floor is pushing back up against your feet with a force also equal to 150 pounds. Similarly, if the sun is exerting a force on the planets which keeps them in elliptical orbits, each planet must be exerting an equal and opposite force back on the sun.

In thinking about the concept of gravity, Newton wondered whether or not this force of gravity could be the same force which causes the planets to stay in orbit about the sun. Perhaps the same force which pulls an apple toward the earth could also be the force that pulls Jupiter toward the sun. In order to proceed, he realized that he needed a better understanding of how gravity works.

One reason why mass and weight are so easily confused is that there is an intimate relation between mass, gravity, and weight. If one object has twice the mass of another object, under the influence of gravity it will also have twice the weight. Similarly, if a Volkswagen weighs 2000 pounds and a Cadillac weighs 4000 pounds, by pushing the two cars we find that the mass of the Cadillac is twice the mass of the Volkswagen. From a clear understanding of this relationship between mass, gravity, and weight, Newton was able to conclude that the gravitational force between objects must be proportional to their masses. The bigger the mass, the bigger the force, and conversely.

But how does gravity vary with distances between objects? Is the gravitational force of the sun at the distance of Mercury larger or smaller than the gravitational force of the sun at the distance of Mars? To answer this, Newton had to realize that Kepler's laws worked. He therefore could reduce the question to: "What must I *assume* the relationship between gravitational force and distance to be in order to come out with the planets travelling along elliptical orbits according to Kepler's laws?" After some considerable mathematical work, he found a simple answer which he stated concisely in his *Universal Law of Gravitation*:

> *Two bodies attract each other with a force that is proportional to the product of their masses and inversely proportional to the square of the distance between them.*

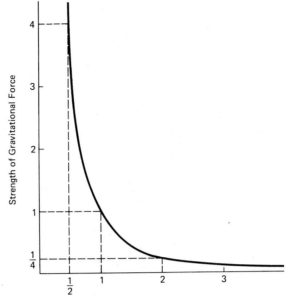

**Figure 1-23. *Newton's Law of Gravitation.*** This graph shows how the strength of the gravitational force of a body varies with distance from that body. Move twice as far away, and the force is only one quarter as great.

Before proceeding to the remarkable implications of this powerful law, perhaps we should examine a few examples of what it means. Imagine two objects in space. Newton's law says that they will attract each other; they will exert a force on each other that tries to pull them together. The strength of this force first of all depends on the masses of the objects. If you double the mass of one of the objects, the force will be twice as strong. Triple the mass, and the force goes up by a factor of three. In addition, Newton's law states that the strength of the force depends "inversely on the square of the distance." To understand what this means, suppose these two objects are 1 foot apart and exert a force of 1 pound on each other. If you were to separate them by 2 feet, the force would be reduced to $\frac{1}{4}$ pound because $2 \times 2 = 4$. If the separation were increased to 3 feet, the force would go down to $\frac{1}{9}$ pound since $3 \times 3 = 9$. This behavior of the gravitational force with distance is shown graphically in Figure 1–23.

By way of another example, consider a man weighing 150 pounds standing on the earth. If he overeats substantially, so as to double the number of atoms in his body, his mass will have doubled and he will weigh 300 pounds. On the other hand, if he had gone on a starvation diet, so as to lose half of the atoms in his body, his mass would then be one half of its original value and he would weigh only 75 pounds.

To see the dependence of gravitation on distance, let us assume that our hypothetical man neither diets nor overeats; the number of atoms in his body remain constant. In standing on the earth, this 150-pound man is 4000 miles from the center of the earth. If he climbs to the top of a 4000-mile-long ladder, he is now 8000 miles from the center of the earth. Since his distance is doubled, a bathroom scale at the top of the ladder will show that he weighs only $\frac{1}{4}$ as much as he did originally, or $37\frac{1}{2}$ pounds. It is important to note that this situation is identical to doubling the size of the earth. If we simply blew up the earth to twice its original size without adding any new matter (i.e., if we double the distances between all the atoms inside the earth), our man would again find himself 8000 miles from the earth's center and he would again weigh only $37\frac{1}{2}$ pounds. Conversely, if we squeezed the earth down to half of its original size, being sure not to gain or lose any matter in the process, our man would be only 2000 miles from the center of this compressed earth. As a result, his weight would now be four times its original value, or 600 pounds.

Armed with his laws of motion and the law of gravitation, Newton now proceeded to tackle the problem of the solar system. The careful reader will notice that Newton has cheated a little bit. He used Kepler's laws to discover the nature of

**Figure 1–24.** *A Man and the Earth.* This series of examples illustrates how the force of gravity depends on distance. (See text for discussion.)

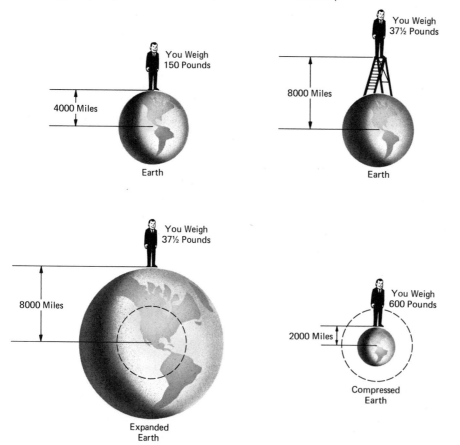

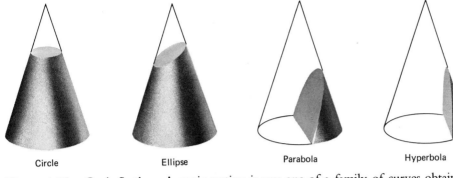

Circle    Ellipse    Parabola    Hyperbola

**Figure 1-25.** *Conic Sections.* A conic section is any one of a family of curves obtained by cutting a cone with a plane. By slicing the cone at different angles, you obtain one of four curves: circle, ellipse, parabola, hyperbola.

gravity. Rigorously speaking, however, he then assumed the validity of the laws of motion and the law of gravitation. From these assumptions, Newton found that he could easily prove Kepler's laws as well as predict a great deal more. Specifically, in calculating the orbits of planets around the sun, Newton found that allowable orbits could be any *conic section* and not just ellipses. A conic section is any curve obtained by cutting a cone with a plane, as shown in Figure 1-25. There are four types of curves that result: circle, ellipse, parabola, and hyperbola. Thus, in addition to proving Kepler's first law, Newton showed that parabolic and hyperbolic orbits were also possible. Parabolic and hyperbolic orbits are "open" while circular and elliptical orbits are "closed." If an object travels along one of these open curves, it will pass by the sun only once and return to the reaches of interstellar space from whence it came. Comets sometimes travel along such orbits.

The precise orbit that an object follows is determined by how much energy it has. To see how this is so, consider a satellite in a circular orbit about the sun. Suppose this satellite has a rocket attached to it and at some point we turn on the rocket engines for a short time. This boost of energy will enable the satellite to get a little further away from the sun, provided the thrust from the rocket is parallel to the direction in which the satellite is moving. As a result of this added speed, the satellite will go into an elliptical orbit. The longer the rocket burns, the bigger the resulting elliptical orbit. In fact, if the rockets are very powerful, the satellite could achieve a speed high enough to escape from the sun's gravitational pull altogether. Such a speed is called the *escape velocity*. An object traveling at the escape velocity moves along a parabolic orbit. And finally, if the satellite has a speed greater than the escape velocity, it will rapidly fly off into interstellar space along a hyperbolic orbit. These kinds of orbits are contrasted in Figure 1-26.

The power of Newton's work, or *Newtonian mechanics* as it is sometimes called, lies in the fact that in addition to explaining everything that had been known earlier, his methods could be used to predict new phenomena. For example, Newton's good friend, Edmond Halley, had a great interest in comets. In particular, he noticed that the bright comets seen in 1531, 1607, and 1682 seemed to have

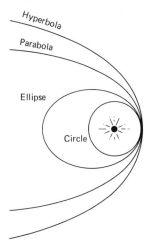

**Figure 1-26.** *Orbits About the Sun.* Newton proved that orbits about the sun can be any conic section (not just ellipses).

almost identical orbits. After some study, he concluded that these three comets were actually the *same* comet returning to the sun on three separate occasions. Using Newton's ideas about orbits, Halley predicted the return of this comet in 1758. The comet was sighted on Christmas night of that year thereby providing a dramatic verification of Newtonian mechanics. Since that time this comet has been known as *Halley's comet.* It has been observed since 240 B.C. and passes near the sun approximately every 76 years.

It is perhaps instructive to mention a few more details about Halley's comet. Actually, the orbit is *not* a perfect ellipse with the sun at one focus. This does not mean that Kepler's first law is wrong, but rather the orbit is affected by the planets, especially Jupiter and Saturn. When the comet passes near a planet, the gravitational force of the planet pulls on the comet. This causes the comet to deviate slightly

**Figure 1-27.** *Halley's Comet.* Using Newton's ideas about orbits, Edmond Halley correctly concluded that the bright comets seen in 1531, 1607, and 1682 were actually the *same* comet. This comet is due to return in the spring of 1986. (*Lick Observatory photograph*)

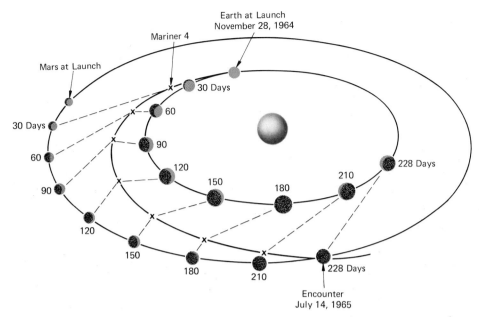

Earth at Launch
November 28, 1964

Mariner 4

Mars at Launch

30 Days

30 Days

60

60

90

90

120

150

180

210

120

228 Days

150

180

210

228 Days

Encounter
July 14, 1965

**Figure 1-28.** ***The Mariner 4 Trajectory to Mars.*** Mariner 4 took over seven months to get from Earth to Mars. Although this path may look indirect, it was chosen because it was very economical.

from its elliptical orbit. Such minor variations are called *perturbations*. In calculating the orbit of any object about the sun, astronomers must be sure to include perturbations; they can be important.

Although it might seem that perturbations by the gravitational fields of planets are simply annoying complications, in the 1960s astronomers realized that perturbations could be of great assistance in planetary exploration. Whenever a mission is planned to another planet, scientists spend a lot of time calculating the best orbit or *trajectory* for their spacecraft. Ordinary Newtonian mechanics and Newton's law of gravity are used along with computers to speed up the calculations. The path that the scientists choose may look indirect (for example, Figure 1–28 shows the trajectory of Mariner 4 during 1965 on its way to Mars). But this path is the most economical in terms of how much energy is necessary at blast-off. After all, the rockets used to launch a satellite can carry only a limited amount of fuel.

On rare occasions, the planets are positioned in such a way that the gravitational field of one planet can be used to propel a spacecraft on toward a second planet. The spacecraft is aimed to pass very near the first planet. At the time of *encounter* with the first planet, the spacecraft experiences large perturbations, which dramatically change its orbit. By aiming the spacecraft properly before the initial encounter, the vehicle can be redirected toward the second planet. In this way, the gravity of the first planet acts like a huge "gravitational slingshot" and scientists have the opportunity to view two planets by launching only one rocket.

Large perturbations have recently been used in planetary exploration with great success. For example, during the mid–1970s Mariner 10 completed a mission to both Venus and Mercury. The spacecraft was launched from Earth in November of 1973, sped past Venus in February of 1974 and arrived at Mercury seven weeks later. By using Venus' gravitational field to direct the vehicle to Mercury, two planets were examined in the same mission.

In September of 1979, Pioneer 11 is due to arrive at Saturn. This spacecraft was launched in April of 1973, and passed by Jupiter in December of 1974. Jupiter's huge gravitational field was used to redirect Pioneer 11 toward Saturn, thereby again using large gravitational pertubations to observe two planets in a single mission. All of these historic space flights will be discussed in detail in later chapters.

In conclusion, we see that from studying the motions of the planets about the sun Newton arrived at a theory of gravity which is so powerful and universal that it is still used today in computing orbits of interplanetary spacecrafts. In formulating his law of gravity, Newton developed mathematical methods and physical principles that comprise the body of *classical mechanics*. Such knowledge is commonly used in building bridges and automobiles as well as sending astronauts to the moon. It is interesting to note that man's most fundamental understanding of physical reality was destined to come from examining the heavens. In the motions of the planets we see fundamental laws of physics revealed in their simplest and purest form, unhampered by friction and air resistance encountered in the laboratory.

## Review Questions and Exercises

1. Describe the phases of the moon.
2. Explain why the moon goes through phases.
3. What is meant by waxing crescent, waning crescent, waxing gibbous, and waning gibbous phases? Relative to the positions of the earth and sun, where is the moon in its orbit when these phases are seen?
4. Explain why the full moon rises at sunset.
5. Explain why the new moon sets at sunset.
6. Explain why the first quarter moon rises at approximately high noon.
7. Explain why the last quarter moon rises at approximately midnight.
8. Briefly describe how Eratosthenes measured the circumference of the earth.
9. Who was the first person to prove that the sun is farther from the earth than the moon?
10. What is meant by direct and retrograde motion of the planets?
11. Briefly describe the key features of Ptolemy's geocentric cosmology.
12. What is the *Almagest?*
13. How did Copernicus account for the retrograde motion of the planets? And how did his explanation differ from that of Ptolemy?

14. What is *De Revolutionibus?*
15. What is the order of the planets from the sun?
16. How many inferior planets are there? What are their names?
17. Why can't Jupiter be seen at inferior conjunction?
18. What is meant by greatest eastern elongation and greatest western elongation? To which planets do such terms apply?
19. What is the difference between the sidereal period of a planet and the synodic period of a planet?
20. Who was Tycho Brahe?
21. How would you draw an ellipse?
22. How are ellipses related to the orbits of planets about the sun? Who discovered such a relationship?
23. Explain how Kepler's second law tells us that planets move faster when they are near the sun and slower when they are far away.
24. What is an AU?
25. Who wrote *The New Astronomy* and *The Harmony of the Worlds?*
26. What major contributions did Galileo make to astronomy?
27. Explain why the phases and sizes of Venus clearly demonstrate that Venus goes around the sun and *not* around the earth.
28. Explain the difference between mass and weight.
29. How did Newton conclude that a force must be acting on the planets?
30. Briefly describe the sequence of events or reasons leading up to Newton's formulation of his Universal Law of Gravitation.
31. What is a conic section?
32. What is meant by escape velocity?
33. Who was Edmond Halley?
34. How can gravitational perturbations be used in interplanetary space flight?

# 2 The Earth and the Sky

## 2.1 *The Sun, Earth, and Seasons*

STANDING underneath the star-studded nighttime sky, we can imagine ourselves at the center of a huge hollow sphere. This impression of the spherical shape of the sky is so strong that ancient Greek astronomers actually believed that we are indeed at the center of a huge crystalline sphere on which the stars are embedded like jewels. This sphere was presumed to rotate once each day carrying with it the sun, moon, planets, and stars, as shown schematically in Figure 2–1. The Greeks reasoned that the diameter of this sphere must be very large, otherwise we would notice a slight shifting of the stars as we travel from one place to another on the earth.

Today, of course, we realize that the stars in the sky are not embedded in a huge

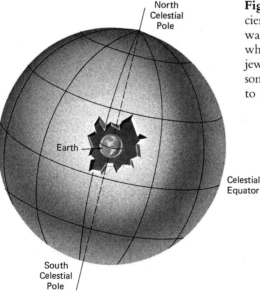

**Figure 2-1.** *The Celestial Sphere.* Ancient astronomers believed that the earth was at the center of a huge sphere on which the stars were embedded like jewels. Even today modern astronomers sometimes find it convenient and useful to refer to the celestial sphere.

sphere but rather are scattered through space. Nevertheless it is often very useful for the modern astronomer to speak of the stars *as if* they were on such a sphere with the earth at the center. This huge imaginary sphere is called the *celestial sphere*.

One way in which the concept of the celestial sphere is extremely useful is that it provides a convenient means of expressing the directions of the stars and other objects in the sky. In order to set up a system of denoting directions in space, it is first of all necessary to establish basic reference points and circles from which the angles to stars and planets can be measured. The easiest way to do this is with a straightforward analogy with the earth.

Any location on the earth can be specified by giving the *latitude* and *longitude* of that location. Latitude is measured in degrees either north or south of the earth's equator while longitude is measured in degrees along the earth's equator either east or west of the so-called *Greenwich meridian*. The Greenwich meridian is a circle which passes through both the north and south poles and the town of Greenwich, England. For historical reasons, this circle was chosen as the reference circle from which longitude could be measured. Thus, for example, Moscow in the Soviet Union is located at 37°37'E and 55°45'N, which means that Moscow lies 37°37' east of the Greenwich meridian and 55°45' north of the equator.

If we think of the stars and planets as being on the celestial sphere, it should be possible to devise a coordinate system to denote the locations of astronomical objects, just as longitude and latitude can be used to give the locations of objects on the earth. We must begin by establishing points and circles in the sky from which the angles to astronomical objects can be measured. For example, the point on the celestial sphere directly above the earth's north pole is called the *celestial north pole* while the point directly above the earth's south pole is called the *south celestial pole*. In other words, the north and south celestial poles are obtained by simply extending the earth's axis of rotation (which passes through the earth's north and south poles) out into space. Similarly we can imagine extending the plane of the earth's equator out into space. The intersection of the plane of the earth's equator and the celestial sphere denotes the location of the *celestial equator*.

While the celestial poles and celestial equator are direct analogies of the poles and equator of the earth, we cannot project the Greenwich meridian onto the celestial sphere and hope to get something useful. The reason for this is that the earth rotates once in about 24 hours and therefore the stars overhead at Greenwich are constantly changing. Rather, we must select some relatively fixed point on the celestial sphere just as the city of Greenwich is a fixed point on the earth. By mutual agreement, the point chosen on the celestial sphere for this purpose is the location of the sun on the first day of spring, usually March 21. On the first day of spring, the sun is directly on the celestial equator at a point we call the *vernal equinox*. We then can imagine drawing a circle on the celestial sphere passing through the vernal equinox and the celestial poles. This circle among the stars is our second reference circle; it plays a role similar to the meridian passing through Greenwich.

We are now in a position to set up a coordinate system for the purpose of denoting the positions of astronomical objects. This system is shown in Figure 2–2. Just as latitude is measured in degrees north or south of the earth's equator, the

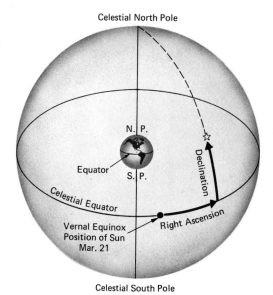

Celestial North Pole

Celestial South Pole

**Figure 2–2.** *The Equatorial System.*
In this coordinate system the position
of an astronomical object is specified
by its right ascension and declination.
Right ascension and declination on
the celestial sphere are analogous to
longitude and latitude on the earth.

astronomer defines *declination* as the angular distance (degrees, minutes, seconds) of
a star north or south of the celestial equator. If a star is north of the celestial equator
its declination is preceded with a plus sign, while a minus sign is used in connection
with declinations in the southern half of the sky. The Greek letter delta ($\delta$) is
commonly used by astronomers as an abbreviation for the word "declination."

As an analogy to longitude, the astronomer defines *right ascension* as a measure of
angular distance eastward from the vernal equinox. Unlike longitude, however,
right ascension is *not* measured in degrees. As a result of this historical motivation as
well as the fact that the earth rotates about its axis once every 24 hours, right
ascension is measured in units of time (hours, minutes, seconds) eastward from the
vernal equinox. The Greek letter alpha ($\alpha$) is commonly used by astronomers as an
abbreviation for the term "right ascension."

Since declination is measured from the celestial equator to the celestial poles, the
values of $\delta$ can range from $0°$ to $+90°$ in the northern half of the sky, or from $0°$
to $-90°$ in the southern half. Right ascension is measured in units of time and $\alpha$
has a range of $0^h$ to $24^h$.

Using right ascension and declination we can describe the location of any
astronomical object in the sky. This coordinate system is used extensively in books
such as *The American Ephemeris and Nautical Almanac* (sometimes simply called *The
Ephemeris*), published annually by the United States Naval Observatory. It contains
data such as the location of the sun, moon, and planets over the year. By way of
illustration, on page 225 of the 1974 edition we find that the location of Saturn on
August 27 of that year was

$$\alpha = 7^h\ 05^m\ 59^s$$
$$\delta = +22°03'00''.$$

**Figure 2-3.** *The Ecliptic and Zodiac.*
The ecliptic is the apparent path of the sun against the background stars. It is tilted by $23\frac{1}{2}°$ from the celestial equator. The constellations of the zodiac are centered approximately on the ecliptic.

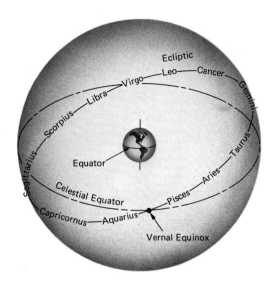

The sun, moon, and planets are constantly in motion and slowly change their locations with respect to the background stars. For example, the sun spends six months (March 22 to September 22) north of the celestial equator and six months (September 24 to March 20) south of the celestial equator. The apparent path of the sun against the background stars is called the *ecliptic*. As shown in Figure 2-3, the ecliptic is inclined to the celestial equator by $23\frac{1}{2}°$. The sun takes one year to go all the way around the ecliptic. Since there are $360°$ in a circle and 365 days in a year, the sun appears to move eastward along the ecliptic at a rate of about $1°$ per day.

The moon and planets are never very far from the ecliptic. Although their individual motions over a period of time appear complicated, the planets never stray more than a few degrees north or south of the sun's apparent path. For convenience, astronomers have divided the entire sky into 88 groupings of stars called *constellations*. There are twelve constellations centered approximately on the ecliptic. They are called the constellations of the *zodiac* and were believed to have some special, mystical significance in ancient times. The sun, moon, and planets can usually be found in these constellations.

The ecliptic is inclined to the celestial equator because the earth's axis of rotation is not perpendicular to its orbit. As shown in Figure 2-4, the earth's axis of rotation is tilted by $23\frac{1}{2}°$ from the perpendicular. This inclination of the earth's axis has a profound effect on the earth; it is responsible for the seasons.

To see why the inclination of the earth's axis causes the seasons, consider someone living in the continental United States. During the summer months, the north pole of the earth is tilted toward the sun. This means that the sun is north of the celestial equator and therefore the sun appears high in the sky for most of the midday. But during the winter months, the north pole of the earth is tilted away from the sun. As a result, the sun is south of the celestial equator and even at noontime the sun will appear low in the sky, near the southern horizon.

As a result of the varying positions of the sun in the sky over the course of a

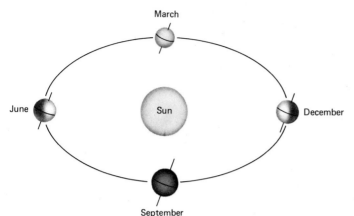

March

June

Sun

December

September

**Figure 2-4.** *The Inclination of the Earth's Axis.* The earth's axis of rotation is *not* perpendicular to the earth's orbit. The axis is tilted at an angle of $23\frac{1}{2}°$ and therefore the sun appears south of the celestial equator for six months of the year and north of the celestial equator for the remaining six months.

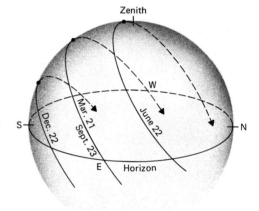

Zenith

Dec. 22

Sept. 23

Mar. 21

June 22

W

S

N

E    Horizon

**Figure 2-5.** *The Path of the Sun Across the Sky.* As seen by a typical observer in the United States, the path of the sun across the sky varies during the year. In the summer months, the sun passes high overhead, while during the winter it is low in the southern sky.

year, the number of daylight hours also changes from month to month. During the summertime, the sun rises in the northeast, passes high overhead, and sets in the northwest. But during the wintertime, the sun rises in the southeast, passes low in the southern sky, and sets in the southwest. As shown in Figure 2–5, the path of the sun in June across the sky is much longer than the path of the sun across the sky in December. Therefore, in order to make its long journey across the sky in the summer, the sun must spend more hours above the horizon than in the winter. Thus, a typical person in the United States might have 16 hours of daylight in June as opposed to only 8 hours of daylight in December. On days when the sun is in the sky for a long period of time, the air and the ground will be heated to a higher temperature than on days when the duration of daylight is short. This is one of the primary reasons why the summers are hot and the winters are cold.

   While the duration of daylight plays an important role in the seasons, there is a second important effect. When the sun is high in the sky, a patch of the earth's surface receives more concentrated sunlight than when the sun is low, nearer the horizon. As shown in Figure 2–6, at high noon in summer, one square foot of

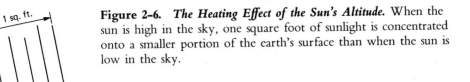

**Figure 2-6. *The Heating Effect of the Sun's Altitude.*** When the
sun is high in the sky, one square foot of sunlight is concentrated
onto a smaller portion of the earth's surface than when the sun is
low in the sky.

sunlight illuminates only $1\frac{1}{4}$ square feet of the earth's surface. But at high noon in
the winter, since the sun is low in the sky, this same square foot of sunlight is spread
over nearly $2\frac{1}{2}$ square feet of ground. Thus, the sun heats up the ground much
more efficiently in July than in January. The seasons are caused by the combined
effect of the duration of daylight and the elevation of the sun above the horizon.

As the sun moves along the ecliptic, it reaches its most northerly point on June
22. On that date, the sun is $23\frac{1}{2}°$ north of the celestial equator and this point on
the ecliptic is called the *summer solstice.* Therefore, June 22 is called the *date of the
summer solstice;* it is the first day of summer. Since the sun is at its most northern
declination at the time of the summer solstice, the sun is above the horizon longer
than at any other time during the year. Roughly speaking, June 22 is the "longest
day of the year."

Similarly, we notice that the sun is at its southernmost declination, $23\frac{1}{2}°$ south
of the celestial equator, on December 22. The location of the sun on the ecliptic on
this day is called the *winter solstice.* On this *date of the winter solstice,* the sun spends the
least amount of time above the horizon. December 22 is the "shortest day of the
year"; it is the first day of winter.

In between the summer and winter solstices there are two times when the sun is
exactly on the celestial equator. One of these points is called the *vernal equinox.* The
sun is at the vernal equinox on March 21, the first day of spring. Similarly, on the
first day of autumn, September 23, the sun is again on the celestial equator at a
point called the *autumnal equinox.* On the first day of spring and the first day of fall,
the sun rises exactly in the east and sets exactly in the west. Daytime and nighttime
each last for 12 hours.

For the purposes of illustration, it was useful to restrict our discussion of the
seasons to the northern hemisphere. In Australia, however, it is hot in December
and cool in June. The first day of spring for someone in Canada is the first day of
autumn for someone in New Zealand. The seasons in the southern hemisphere are

simply reversed. When the north pole is tilted away from the sun, giving rise to winter at the northern latitudes, the south pole is tilted toward the sun, thus producing summer for someone living "down under."

In thinking about the seasons and the inclination of the ecliptic to the celestial equator, there are some places on the earth where interesting things happen. At the North Pole and at the South Pole, daylight lasts for 6 months and nighttime lasts for 6 months. At the North Pole, the sun rises on the first day of spring and stays above the horizon until the first day of autumn, half a year later. The maximum altitude of the sun above the horizon is $23\frac{1}{2}°$, an altitude which occurs on the date of the summer solstice. The situation is reversed for the South Pole.

Actually, there is a substantial region of the earth surrounding the North Pole where the sun stays above the horizon for a full 24 hours on at least one day during the year. On such days, the "midnight sun" can be seen. The southernmost limit of this region is called the *arctic circle,* $23\frac{1}{2}°$ south of the pole (at a latitude of $66\frac{1}{2}°$ north). On the arctic circle, the midnight sun can be seen only on one day, the date of the summer solstice.

A similar situation exists for a region surrounding the South Pole. The northernmost limit of the midnight sun is called the *antarctic circle,* $23\frac{1}{2}°$ north of the pole (at a latitude of $66\frac{1}{2}°$ south). In between the arctic and antarctic circles, the midnight sun can never be seen.

On the equator, the sun is directly overhead at high noon on the first day of spring and the first day of autumn. During the summer months, the sun always passes north of the zenith, while during the winter months, the sun is always south of the zenith at high noon. There must be a region of the earth surrounding the equator where the sun appears directly overhead on at least one day during the year. The northernmost limit of this region is called the *Tropic of Cancer* at a latitude of $23\frac{1}{2}°$ north. On the Tropic of Cancer, the sun is at the zenith at high noon only

**Figure 2–7.  *The Earth at the Solstices.*** At the time of a solstice, the sun is at its greatest declination, either north or south of the celestial equator. At this time of the year, one of the poles is experiencing 24 hours of continuous daylight, while the other is in constant darkness.

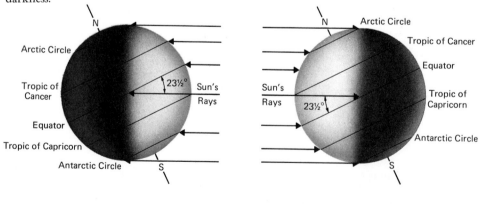

Winter Solstice                    Summer Solstice

on the date of the summer solstice. Similarly, the southernmost limit of this region is called the *Tropic of Capricorn* at 23½° south. At this tropic, the sun is directly overhead only at noon on the date of the winter solstice. These various regions of the earth are illustrated in Figure 2–7, for the dates of December 22 and June 22.

From this brief discussion of the seasons note that some of the most fundamental properties of our environment are directly related to astronomical considerations. Whether a certain region on the earth will support a tropical jungle or will be eternally covered with ice and snow is entirely a result of the relationship between the sun and our planet.

## *The Telling of Time* **2.2**

EVER since men first looked up into the sky, one of the primary tasks of astronomers has been establishing ways of telling time. The need for accurate telling of time is universal. The Pharaoh wanted to know when the Nile would flood. The timing of religious events was extremely important to the Hebrews and Moslems. Even today, if we were to lose track of April 15 we would find ourselves in deep trouble with the Internal Revenue Service.

For ancient man there were three natural clocks in the sky. The rising and setting, or *diurnal motion,* of the sun and stars provided a way of measuring day and night. The phases of the moon defined the length of a month. And the apparent motion of the sun along the ecliptic gave rise to the concept of a year.

Unfortunately, however, nature has not been very kind to us. For example, there is not an integer number of days in the year; the year lasts for approximately 365¼ days. There is not an integer number of days in the lunar month; it takes the moon about 29½ days to go through all its phases. As a result, there is not an integer number of lunar months in a year. Since 365¼ divided by 29½ equals 12⅖,

**Figure 2–8.** *The Meridian.* The meridian is an imaginary line running from north to south and passing through the zenith. The transits of celestial objects can be used to measure time.

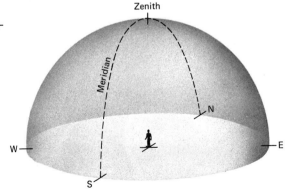

the moon goes through slightly more than twelve cycles of its phases as the earth goes once around the sun.

These and similar complications meant that better ways of measuring time had to be devised. Since the earth rotates, it is quite natural that we would want to measure time in such a way that it is related to the earth's rotation. This was accomplished by defining the *meridian,* an imaginary line in the sky passing through the north and south points of the horizon and the zenith overhead. With the aid of the meridian, a means of measuring the diurnal motion of objects such as the sun and stars is established. Indeed we can now invent different kinds of "days."

The simplest and most obvious kind of day is the *apparent solar day.* A solar day is the length of time it takes for the sun to go from one passage or *transit* across the meridian to the next. It is therefore the time from one high noon to the next. A clock based on the apparent solar day would measure *apparent solar time.* Sundials measure apparent solar time.

Astronomers find it useful to define a different kind of "day" relative to successive transits of some object on the celestial sphere other than the sun. In other words, for the purpose of his observations, the astronomer prefers to measure time according to the stars rather than by the sun. To accomplish this, recall that right ascension is measured in units of time ($0^h\ 00^m\ 00^s$ to $24^h\ 00^m\ 00^s$) eastward from the vernal equinox. We can therefore define the *sidereal day* as the time between successive transits of the vernal equinox. *Sidereal time,* which is based on the sidereal day, is very useful to the astronomer. There is a simple relationship between the right ascension of an object and the sidereal time at which the astronomer wishes to observe that object. This allows the astronomer to point his telescope accurately at whatever he wishes to study.

To a high degree of accuracy, we may think of the solar day as measured relative to the sun and the sidereal day as measured relative to the stars. They are not equal. The solar day is about 4 minutes longer than the sidereal. The reason for this is shown in Figure 2–9. As the earth rotates about its axis, it also revolves about the sun. During the course of one sidereal day, the earth rotates once with respect to the stars. But during a solar day, the sun appears to move slightly (approximately 1° per day, eastward) along the ecliptic. Therefore, to complete one solar day, the earth has to rotate a little bit more to catch up with the sun. During a sidereal day, the earth rotates through 360° with respect to the stars. But during a solar day the earth rotates through 361° with respect to these same stars. This extra 1° of rotation translates into 4 minutes of time.

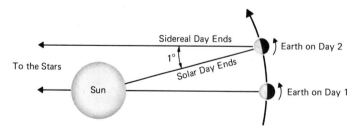

**Figure 2–9. The Sidereal and Solar Days.** The sidereal day is measured in reference to the stars, while the solar day is measured in reference to the sun. Because the earth is moving along in its orbit, the solar day is a little longer than the sidereal day.

At this point you might think that we have come to the end of the story. You might be inclined to say "Let's leave sidereal time to the astronomers. I shall measure my day by the sun, since the sun and not the stars determine when I am awake or asleep. I shall define the apparent solar day to be exactly 24 hours long and be done with it." No such luck. Nature does not oblige us by taking such a simple approach. The earth's orbit around the sun is not a perfect circle; rather, the orbit is an ellipse with the sun at one focus. Kepler's second law says that the earth is moving faster at *perihelion,* when it is nearest the sun, than at *aphelion,* when it is farthest from the sun. This variable speed of the earth along its orbit means that the speed of the sun along the ecliptic will not be constant throughout the year. In other words, the exact length of a solar day changes slightly over the course of a year. If you had said that there will be 24 hours in an apparent solar day, then you would find that the duration of an "hour" or a "minute" would vary slightly from one day to the next in order to agree with the sun. This would be terribly inconvenient.

So the sun is really not a good timekeeper. Timekeeping would be a lot simpler if the sun moved at a constant speed across the sky. To circumvent this difficulty, an imaginary "average" sun was devised. This imaginary sun, which is called the *mean sun* is sometimes a little ahead or behind the real sun in the sky. The real sun and the mean sun are displaced by just enough so that the mean sun does, in fact, move along the celestial equator at a constant speed. The *mean solar day* is, therefore, the time between successive transits of the mean sun. By doing things in this fashion, astronomers have guaranteed that the mean solar day is equal to the average length of an apparent solar day. On any given date, the apparent solar day will be a little longer or little shorter than this average value, but the mean solar day is constant throughout the entire year. Twenty-four hours on your wristwatch or clock is exactly one mean solar day. *Mean solar time* is based on the mean solar day and passes at a uniform rate throughout the year.

From what has been said, it might seem that your wristwatch measures mean solar time. This is not true, and the reason does not have anything to do with astronomy. Recall that time is measured by transits of the meridian. For example, 12 o'clock noon in mean solar time is defined as the instant that the center of the imaginary mean sun crosses the meridian. But there are different meridians for different locations on the earth. Someone in San Francisco has a different zenith and meridian than someone in Los Angeles. Therefore, strictly speaking, there is a small difference in mean solar time at these cities. If you travelled from San Francisco to Los Angeles and if you wanted your wristwatch to read mean solar time, you would constantly be having to reset your clock during your journey. This would be extremely inconvenient.

To cope with this difficulty, it would be advantageous if everyone in nearby cities would agree on the same time. For this purpose, *time zones* were invented. By mutual agreement, all the clocks of everyone living in a particular time zone are synchronized. Thus, all the clocks from Seattle to San Diego read the same time, Pacific Standard Time.

By dividing the world into time zones we say that we have "standardized"

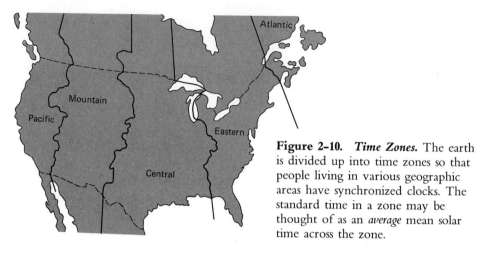

**Figure 2–10. *Time Zones.*** The earth is divided up into time zones so that people living in various geographic areas have synchronized clocks. The standard time in a zone may be thought of as an *average* mean solar time across the zone.

mean solar time for particular regions. There are four such time zones in the continental United States and clocks across the country read one of four types of time: Eastern Standard Time (EST), Central Standard Time (CST), Mountain Standard Time (MST), and Pacific Standard Time (PST). Each time zone differs by one hour from each adjacent time zone so that, for example, 10:00 A.M. EST in New York is the same as 7:00 A.M. PST in Los Angeles. These time zones are selected so that the standard time in a zone is approximately equal to the average mean solar time for all cities in that time zone. Thus, Central Standard Time on a clock in Chicago is near but not exactly equal to mean solar time for that precise location on earth.

During the summer months, the sun is in the sky longer than during the winter months. The sun rises earlier and sets later in June than it does in December. To take advantage of these extra daylight hours, *daylight savings time* was invented. From April through October clocks are set ahead by one hour. The reason for doing this is that it seems to be more convenient for the average person. For example, on a particular day during the summer the sun would rise at 4:00 A.M. and set at 8:00 P.M. if clocks measured standard time. But with daylight savings time, the sun on that day would rise at 5:00 A.M. and set at 9:00 P.M., which more closely reflects when the typical person is awake or asleep.

At this point, our understanding of time may be summarized as follows:

**1.** Apparent solar time is based on the sun. The length of an apparent solar day varies throughout the year from about 23 hours, $59\frac{1}{2}$ minutes to 24 hours, $\frac{1}{2}$ minute.

**2.** Sidereal time is based on the stars. The length of a sidereal day is 23 hours, 56 minutes, 4 seconds.

**3.** Mean solar time is based on an imaginary mean sun. The length of a mean solar day is exactly 24 hours and is constant throughout the year.

4. Apparent solar time, sidereal time, and mean solar time are measured with reference to the local meridian of an observer. They all depend on the location of the observer on the earth.

5. Standard time was invented so that all the clocks in a time zone would be synchronized. In a time zone, standard time may be thought of as the *average* mean solar time.

6. Daylight savings time was invented to take advantage of the extra daylight hours during the summer months. To get daylight savings time, add one hour to standard time.

Therefore, the time on a wristwatch is based partly on astronomy and what is seen in the skies, and partly on what is convenient and useful. The same is true with the calendar and the year. The problem with the year is that the earth rotates $365\frac{1}{4}$ times on its axis in one complete revolution about the sun. More precisely, the *tropical year* is defined as the time it takes the sun to go once around the ecliptic with respect to the vernal equinox. The tropical year turns out to be equal to 365.242199 mean solar days. It is this fraction of a day that causes some complications.

First of all, suppose that calendars were always printed with exactly 365 days. This means that we would be in error by about 1 day every 4 years. As a result, the date of the first day of spring would gradually change. After a few decades, the first day of spring would actually occur in April, rather than around March 21. And after a few centuries, summer would occur during December and January. It is generally agreed that this would be an undesirable situation and so a method was invented of setting up the calendar to assure that this does not happen.

About two thousand years ago, it was apparent that the telling of time was a complete mess. Every town and city in the Roman Empire had its own calendar, few of which were the same. Thus, a traveller would find himself going from year to year as he went from one town to the next. To straighten things out, Julius Caesar instituted calendar reform in 46 B.C. According to the best information available, the length of the year was $365\frac{1}{4}$ days. Therefore, in order to account for this fraction of a day, Caesar decreed that every four years an extra day should be added. These years with 366 days are called *leap years,* and this method is still used today. Once every four years we add February 29. Every year which is evenly divisible by 4 is a leap year. Thus, for example, 1976, 1980, and 1984 are leap years.

This system would work just fine if the length of a year was 365.250000 mean solar days instead of 365.242199. But this difference amounts to the loss of one day every 128 years. Thus, by the time of the Renaissance, the calendar was again in need of revision. Indeed, during the reign of Pope Gregory XIII, the first day of spring was occurring on March 11.

In 1582, Pope Gregory instituted a further calendar reform. In order to get closer to the true value of the tropical year, it was decreed that in addition to Caesar's method for establishing leap years, centuries not divisible by 400 will not be leap years. Thus 1700, 1800, and 1900 are evenly divisible by 4 but not by 400.

Therefore, while they would have been leap years in Caesar's system, they are common years in Pope Gregory's system.

In the Gregorian calendar, the length of the year comes out to be 365.2425 mean solar days. The difference between this value and the true value amounts to an error of only 1 day in 3300 years.

The calendar we use today is the Gregorian calendar with one small modification. The years 4000, 8000, 12000, and so on, which would have been leap years according to Pope Gregory, will now be common years. The calendar is therefore accurate to 1 day in 20,000 years. As a result of all these rules, reforms, and regulations, we may now all rest comfortably, secure in the knowledge that for many years to come, it will snow on Christmas and be hot and sticky on the Fourth of July, not vice versa.

# 2.3 *Tides and Precession*

IN DISCUSSING the seasons, we found that the relationship between the earth and the sun has a profound effect on the environment. Yet this discussion dealt with only the relative orientations of the earth, its axis of rotation, and the sun. To understand the cause of the seasons, it was not necessary to talk about the gravitational forces between the earth and the sun. Of course, all the bodies in the solar system exert gravitational forces on the earth. Since the sun is tens of thousands of times more massive than anything else in the solar system, the orbits of the planets are completely dominated by the enormous gravitational field of the sun. But to a much lesser extent, all the other bodies in the solar system are pulling on each other. Thus, as we learned at the end of Chapter 1, the orbits of the planets are not exactly perfect ellipses. The earth's orbit, for example, deviates slightly from an ideal ellipse as a result of the gravitational perturbations of the moon and other planets. But there are other ways in which the gravitational fields of the sun and the moon profoundly affect the earth.

Consider the earth and the moon, as shown in Figure 2–11. There are some parts of the earth that are nearer to the moon than other parts. Since the strength of gravity varies with distance, those parts of the earth nearest the moon experience a

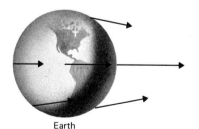

Earth

Moon

**Figure 2–11. *The Moon's Pull on the Earth.*** The gravitational force of the moon varies from one side of the earth to the other. The force is strongest nearest the moon and weakest on the opposite side of the earth.

To Moon

**Figure 2-12.** *The Shape of the Oceans.* The tidal force of the moon deforms the shape of the oceans. High tides occur at locations nearest and farthest from the moon.

greater attraction toward the moon than those parts that are farther away. In other words, there will be a difference in the gravitational pull of the moon from one side of the earth to the other. This variation across the earth is called a *differential gravitational force,* or *tidal force;* it tries to deform the earth by stretching it slightly in the direction of the moon. Of course the earth does not stretch too easily, but the earth's surface is covered with water. Since water flows more easily than rocks, the oceans of the earth are deformed by this differential gravitational force. The final result is that we observe *tides*.

Due to the fact that the moon is much closer to the earth than any other body in the solar system, the differential gravitational force of the moon dominates and controls the tides. Obviously a high tide occurs on that part of the earth closest to the moon since at that region, the moon's pull is greatest and the water in the oceans is drawn to that area. As a result, the time of high tide at a certain location will be near the time at which the moon is on the meridian. But in addition, a second high tide will occur 12 hours later. This is because the oceans are deformed into the shape of an "oblate spheroid," which looks like a flattened ball, as shown in Figure 2-12. While the pull of the moon is strongest on that part of the earth closest to the moon, the pull is weakest on the opposite side of the earth, weaker than at the earth's center. Since the earth receives an intermediate pull from the moon, the earth takes an intermediate position between the oceans on either side. Therefore, the net effect produces a high tide nearest the moon and a high tide farthest from the moon. Thus, if one high tide occurs at 10:00 A.M., a second high tide will occur around 10:00 P.M. with low tides in between at approximately 4:00 A.M. and 4:00 P.M.

Although the tides are dominated by the moon, our sun does have a noticeable, but smaller effect. As you might expect, when the sun and moon are lined up, the combined tidal forces of both bodies tend to exaggerate the tides. Thus at new moon and at full moon, the high tides are very high and the low tides are very low. This situation is referred to as *spring tides,* shown in Figure 2-13. But when the sun and moon are at right angles, at first and last quarter, the tidal forces of these two bodies act against each other. The variation in the heights of the tides is least pronounced at these times, resulting in the so-called *neap tides*.

Ancient man was well aware of the relationship between the tides and the moon, thousands of years before Newton and gravity. Any sailor knew, for example, that high tide occurs when the moon is high in the sky. Low tides were known to occur at approximately the time of moonrise and moonset. Yet these

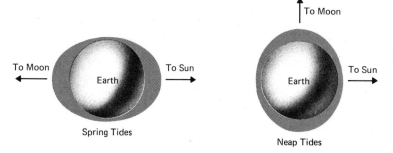

**Figure 2–13.** *Spring and Neap Tides.* Spring tides occur when the sun, moon, and Earth are lined up. Neap tides occur when the moon and sun are at right angles, at first and last quarter.

ancestors were also aware of a far more obscure effect of the moon and the sun on the earth. Over centuries of observations it had become apparent that the vernal equinox is moving slowly among the fixed stars.

While the tides are an obvious, easily observable phenomena related to the gravitational pull of the moon and sun, the gravity of these two bodies has another important effect on the earth. It is well known from geography that the earth is not a perfect sphere but rather has an *equatorial bulge.* Our earth is fatter at the equator than at the poles; the diameter of the earth at the equator is about 27 miles longer than the distance from pole to pole. But it was also known that the earth's axis of rotation is tilted by $23\frac{1}{2}°$ away from the perpendicular direction to the earth's orbit. Therefore, this equatorial bulge is tilted by $23\frac{1}{2}°$ out of the plane of the ecliptic. Since the moon is always very near the ecliptic, both the sun and the moon are usually either above or below this equatorial bulge, as shown in Figure 2–14. Because of the fact that the equatorial bulge of the earth is inclined to the ecliptic, the gravitational forces of the sun and moon try to "straighten up" the earth; they try to tip the earth up so that the equatorial bulge lies in the plane of the ecliptic. If the earth did not rotate, billions of years ago the sun and moon would have succeeded in doing this, and today the earth's axis would be exactly perpendicular to its orbit. But since our planet still has seasons, it is obvious that this has not happened. To see why this is so, we shall examine the behavior of a child's spinning top.

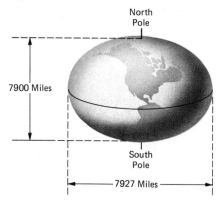

**Figure 2–14.** *The Equatorial Bulge.* The earth is fatter at the equator than at the poles (exaggerated in this drawing). This bulge is tilted at an angle of $23\frac{1}{2}°$ with respect to the ecliptic.

**46**

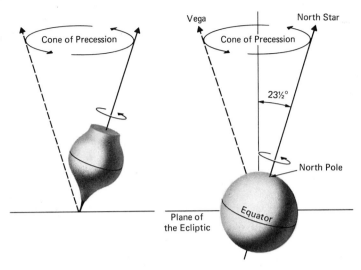

**Figure 2-15.** *Precession.* As a result of the action of gravity on a spinning top, the top precesses. Similarly, because of the action of the gravitational forces of the moon and sun on the earth's equatorial bulge, the earth's axis of rotation also precesses.

Imagine a child playing with a toy top on the sidewalk. With the aid of a string, he starts the top spinning and sets it down on the ground. Of course, gravity is acting on the top and we might expect to see the top promptly fall over. This does not happen. Rather, the spinning top begins to wobble around in a circle even though the top is inclined at a precarious angle. The axis of rotation of the top slowly traces out a circle, as shown in Figure 2–15. This motion is called *precession*. The reason for this unique motion is, in essence, that the axis of rotation of the top tries to remain fixed in space. But simultaneously, the earth's gravity tries to pull the top over. The combination of these two actions produces precession. Precession may be thought of as a compromise between the top's desire to keep its axis of rotation pointed in a constant direction and gravity's desire to pull the top over.

This situation is analogous to the earth. The child's top is not spherical; it is fatter in one direction than the other. The earth is not spherical; it has an equatorial bulge. Just as gravity does not succeed in immediately knocking over the rotating top, the sun and moon do not succeed in straightening up the rotating earth. And just as the top slowly precesses, our earth also slowly precesses.

The rate of precession in the case of the earth is extremely slow. Gradually from one year to the next, the north pole of the earth traces out a circle among the stars. Right now, the north celestial pole is near the star Polaris at the end of the handle of the Little Dipper. When the pyramids were built in ancient Egypt, the star Thuban in the constellation of Draco was the north star. In 12,000 years, the bright star Vega will be very near the celestial pole. It takes 26,000 years for the earth's pole to complete one circle among the stars, as shown in Figure 2–16.

There is one very important side effect that results from the precession of the earth. Obviously, if the north celestial pole moves among the stars, the south celestial pole must also be moving among the stars, tracing out a similar circle in the southern skies. But if both poles are moving, then the celestial equator must be gradually shifting its orientation in space. As the celestial equator changes its

**47**

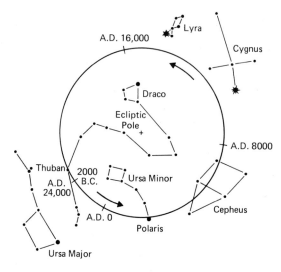

**Figure 2-16.** *The North Celestial Pole.* Due to precession, the north celestial pole is gradually moving among the stars. At this time, Polaris is the "pole star."

position, the vernal equinox, which is the point where the ecliptic and the celestial equator cross, must also be gradually moving. In other words, over the centuries, the location of the sun on the first day of spring moves slowly among the constellations of the zodiac. The vernal equinox is now in the constellation of Pisces and will soon enter the constellation of Aquarius. This is what astrologers and other misguided people mean by the "dawning of the Age of Aquarius."

More importantly, recall that we tied our coordinate system of right ascension and declination to the location of the vernal equinox. Right ascension is measured eastward from the vernal equinox, which we now realize is moving. As a result of this gradual motion of the vernal equinox, the coordinates of the stars listed in a star catalogue will change by small amounts from one year to the next. For example, the right ascension and declination of the bright star Regulus in Leo, the lion, are listed as

$$a = 10^h \ 3^m \ 3^s$$
$$\delta = +12°27'21''$$

in a star catalogue prepared for 1900. But in 1950, the position of this same star was given as

$$a = 10^h \ 5^m \ 43^s$$
$$\delta = +12°12'44''$$

The star has not moved; the coordinate system moved. This does *not* mean that we have done things wrong. We are located here on the earth and it is very convenient and natural to use a coordinate system tied to the earth in order to express the locations of the stars. Unfortunately there is a minor complication. The earth precesses. This is dealt with by simply noting in all our star catalogues the year or so-called *epoch* for which the catalogue was prepared. Most star charts and catalogues in use today are for epoch 1950, which means that the positions given in the books are precise for the first day of the year 1950. Of course, such positions are a little wrong today, but the error is very small and a professional astronomer who needed precise positions would know how to make the necessary tiny corrections.

This phenomenon of precession was first discovered over two thousand years ago by Hipparchus who correctly deduced that the north celestial pole is gradually changing its position in the sky. Both Ptolemy and Hipparchus measured the rate at which the earth precesses and obtained values that are fairly accurate. This rate of precession may be expressed by saying that the vernal equinox moves westward along the ecliptic at a speed of approximately $1\frac{1}{2}°$ per century. It therefore takes about 26,000 years for the vernal equinox to go once all the way around the zodiac.

## *Eclipses* **2.4**

FOR ancient man, eclipses were certainly among the most terrifying of all natural phenomena. Without warning, the sun would begin to disappear from the sky, as though it were being swallowed by some vast demonic creature and "day turned into night." As darkness descends, the stars come out one by one, and all of nature responds as though it were late in the evening. Birds return to their nests, crickets begin to sing, chickens go to roost, and the cows start their slow journey back to the barn. Only man seems to realize that something very strange is happening. In only a few minutes, the sun which supplies us with light and warmth has mysteriously vanished. Many primitive cultures believe that these events must be remedied by making a lot of noise to scare off the demon who is swallowing the sun. Of course, they always succeed.

The ancient Greeks seem to have been first in really understanding what is going on during an eclipse. During a *solar eclipse,* the moon passes directly between the earth and the sun, and people standing in the path of the moon's shadow see an eclipse of the sun. During a lunar eclipse, the moon moves through the earth's shadow, and people on the nighttime side of the earth see an eclipse of the moon. Obviously, solar eclipses can occur only at the time of new moon while lunar eclipses can only occur at full moon.

It is well known that eclipses are rare events. Most people go through their entire lives without ever seeing a solar eclipse, and few have ever experienced an eclipse of the moon. Yet there is a full moon or a new moon every two weeks. At first glance, therefore, we might wonder why we do not see an eclipse every 14 days. The reason has to do with the orientations of the orbits of the moon and earth.

The moon, of course, is in orbit about the earth. However, the moon's orbit is tilted slightly with respect to the earth's orbit, as shown in Figure 2–17. The angle between the plane of the earth's orbit (i.e., the plane of the ecliptic) and the plane of the moon's orbit is about 5°. These two planes intersect along a very important line called the *line of nodes*. As a result of this inclination of the moon's orbit, the moon is usually above or below the ecliptic. Most of the time, new moon and full moon phases occur under conditions when the sun, moon, and earth are *not* in perfect alignment. A perfect alignment between the sun, moon, and earth is *only* possible when new moon and full moon occur along the line of nodes, as shown in Figure 2–18. The times when perfect alignments are possible are called *eclipse seasons,* and only at such times can eclipses occur. An eclipse of the sun is seen only when the new moon is very near the line of nodes, because only then will the moon's shadow fall directly on the earth. Similarly, lunar eclipses are seen only when the full moon is very near the line of nodes because only then will the moon pass directly through the earth's shadow.

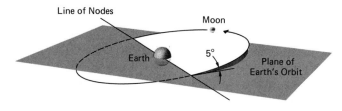

**Figure 2–17.  *The Line of Nodes.*** The moon's orbit is tilted slightly with respect to the earth's orbit. The plane of the moon's orbit intersects the plane of the earth's orbit along the line of nodes.

**Figure 2–18.  *When Eclipses Occur.*** Eclipses can only happen when the full moon or new moon occurs on or very near the line of nodes. At all other times, a perfect alignment between the sun, moon, and Earth is not possible. (*Adapted from Dr. Abell*)

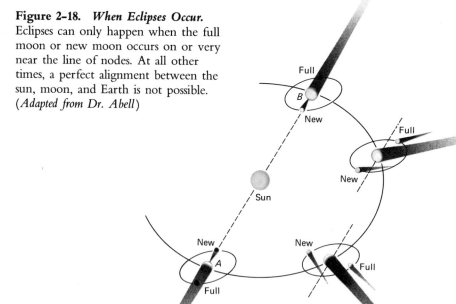

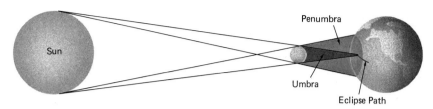

**Figure 2-19.** *The Geometry of a Total Solar Eclipse.* During a total solar eclipse, the moon's umbra traces an eclipse path across the earth. Totality is seen only inside the eclipse path; a partial eclipse is seen in those regions covered by the moon's penumbra.

During a solar eclipse, only the very tip of the darkest part of the moon's shadow touches the earth, as shown in Figure 2–19. As the earth rotates and as the moon moves along in its orbit, the moon's shadow moves across the earth's surface at a speed slightly greater than 1000 miles per hour. The path of the moon's shadow on the earth is called the *eclipse path,* and only those people located in the eclipse path will be treated to a total eclipse of the sun. The eclipse path is usually only a few miles wide, and due to the speed of the moon's shadow across the earth, the *duration of totality* is only a few minutes. Theoretically, the eclipse path is never more than 167 miles wide and the maximum possible duration of totality is $7\frac{1}{2}$ minutes.

Because the sun is a sphere and not a "point source" of light, there are two parts to the moon's shadow. The darkest part of the moon's shadow is called the *umbra.* No sunlight at all falls inside the umbra; the sun is completely covered by the moon for anyone located in the umbra. By contrast, for a person standing in the *penumbra,* only part of the sun is covered. Such a person sees only part of the sun obscured by the moon; he is observing a *partial solar eclipse.* As shown in Figure 2–19, the penumbra covers a large region of the earth's surface during a solar eclipse. People living in a zone 2000 miles on either side of the eclipse path will see a partial eclipse of the sun.

Since the orbit of the moon about the earth is known, astronomers can calculate when and where eclipses occur. The results of these calculations are published in books such as the *Ephemeris.* Figure 2–20 shows a typical eclipse map taken from the *Ephemeris.* From a map like this, you can immediately learn where you must go on the earth's surface to observe a total eclipse of the sun.

An interesting complication with regard to solar eclipses becomes apparent with a little more thinking about the details of the moon's orbit. Neglecting perturbations, the moon's orbit is an ellipse with the earth at one focus. This means that sometimes the moon is nearer the earth than at other times. Indeed, the earth–moon distance over the course of a lunar month varies by about 10 per cent from the average value of 238,900 miles. The point in the orbit nearest the earth is called *perigee,* while the point farthest from the earth is called *apogee.* This means that the apparent size of the moon, as seen from the earth, varies over the course of a lunar month. At perigee, the moon actually looks a little bigger in the sky than at apogee.

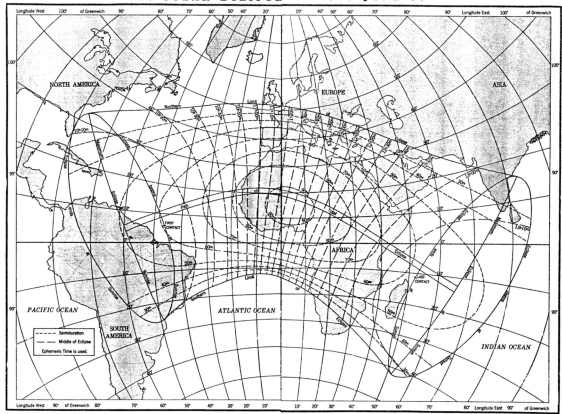

**Figure 2–20.** *A Typical Eclipse Map.* Eclipse predictions are conveniently expressed in the form of a map that shows the eclipse path and the region of the earth covered by the penumbra along with related information.

The average apparent angular diameter of *both* the sun and moon as seen from the earth is very nearly $\frac{1}{2}°$. This coincidence is responsible for the fact that during a total solar eclipse, the moon appears to "fit" exactly over the sun. However, if a solar eclipse happens to occur when the moon is near apogee, then the moon has an apparent angular diameter slightly less than $\frac{1}{2}°$. This means that the moon, as seen in the sky, is not big enough to cover the entire disc of the sun completely. As shown in Figure 2–21, under these circumstances, the umbra does *not* reach all the way down to the earth's surface. Therefore, no one sees a completely total eclipse of the sun. Rather, at mid-eclipse we can still see the outer edge of the sun around the moon. Such an eclipse is called an *annular eclipse* because a ring or annulus of sunlight is seen around the moon at mid-eclipse.

While very few people have ever seen a solar eclipse, the average person might observe several lunar eclipses during his life. This does not mean that lunar eclipses

**52**

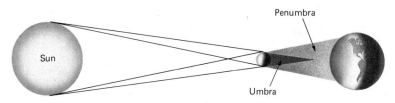

**Figure 2–21.** *The Geometry of an Annular Eclipse.* Annular eclipses are seen when the new moon is on the line of nodes *and* at apogee. Under these conditions, the moon's umbra does not reach all the way down to the earth.

occur more frequently than solar eclipses. Quite to the contrary, there are at least two (but never more than five) solar eclipses somewhere on the earth each year. The number of lunar eclipses during a twelve-month period ranges from zero to three. However, in order to see a total eclipse of the sun, you must be standing in the path of totality. But when a total eclipse of the moon occurs, *anyone* living on the nighttime side of the earth can see the eclipse. This is apparent from the geometry of a lunar eclipse shown in Figure 2–22.

A total lunar eclipse begins when the moon moves into the earth's penumbra. If you were standing on the moon, you would see the earth begin to cover up the sun, yet come sunlight would still be falling on you. Since the moon at this point is still in partial sunlight, the full moon appears just a little dimmer than usual. It is very difficult to notice this slight dimming when the moon is in the penumbra. However, when the moon moves into the umbra and totality begins, all direct sunlight is cut off. The moon seems to disappear almost completely from the sky. Totality ends when the moon leaves the umbra and the entire eclipse is over when the moon exits the penumbra. To the unaided eye, only the umbral phase of the eclipse is easily seen.

A *partial lunar eclipse* occurs when the moon does not move completely into the earth's umbra. Some part of the lunar surface always stays in the penumbra and some piece of the moon always remains easily seen. A *penumbral eclipse* occurs when

**Figure 2–22.** *The Moon and the Earth's Shadow.* If the full moon passes completely through the umbra, we see a total eclipse. If only part of the moon passes through the umbra, there is a partial eclipse. Penumbral eclipses occur when the moon passes only through the penumbra.

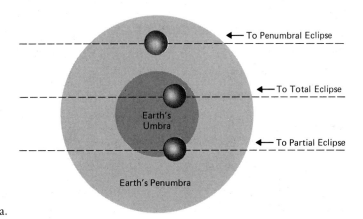

53

**Figure 2-23.** *Partial Lunar Eclipse.* The photograph was taken during a partial eclipse of the moon in March 1961. The illuminated portion of the moon is inside the earth's penumbra. The obscured portion is in the earth's umbra. (*Paul Roques, Griffith Observatory*)

the moon moves only through the penumbra and not at all through the umbra. Penumbral eclipses go virtually unnoticed. Even at mid-eclipse, a trained observer would see only a slight overall dimming in the appearance of the full moon.

Lunar eclipses are of virtually no interest to the modern astronomer. Of course, it is always nice to see an eclipse, but there is nothing new that we could learn from observing an eclipse of the moon. On the other hand, solar eclipses are extremely important. The sun is the only star in the sky that does not look like a pin-point of light. It is the only star that we can study "up close." Most of the time, the solar astronomer is concerned with examining the complex phenomena on the surface of the sun, and many of these phenomena will be discussed in Chapter 6. During a total solar eclipse, the moon blocks out the blinding solar surface. Only under such conditions can astronomers see and study the tenuous structure of the sun's upper atmosphere. During the precious few minutes of totality, the sun's *corona* flashes into view. The corona consists of extremely thin, hot gases extending hundreds of thousands of miles out from the sun's surface. From examining the nature of the corona, a great deal can be learned about the nature of our star. For example, the shape of the corona, seen in Figure 2-24, varies remarkably from one eclipse to the next. Such changes reflect corresponding changes in the structure of the sun's magnetic field.

**Figure 2-24.** *The Solar Corona.* The shape of the solar corona varies from one eclipse to the next. Astronomers have learned that these variations are caused by changes in the structure of the sun's magnetic field. (*Hale Observatories*)

**Figure 2-25.** *Prominences.* During a solar eclipse, it is possible to see huge jets of gas at the sun's edge surging up many thousands of miles into space. These jets of gas erupting from the sun are called "prominences." (*Lick Observatory photograph*)

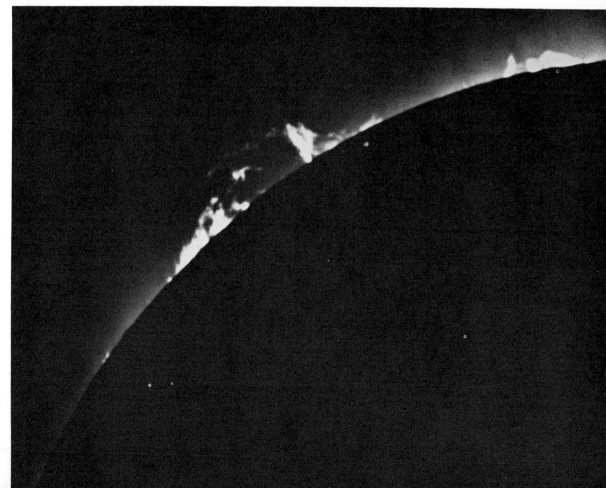

Since studying the entire corona is only possible during a total solar eclipse, astronomers go to great lengths to insure good observations. Jet airplanes are loaded with telescopes and related equipment and the solar astronomer flies along the path of totality during mid-eclipse. By chasing the moon's shadow across the earth, he can make totality last a little longer than it does on the ground. And, of course, by being above the clouds he is guaranteed that he will not be "rained out."

Man has been observing eclipses for thousands of generations. For centuries, these dramatic natural phenomena have been a source of awe and wonder. Yet only recently have we developed the insight and technology to use eclipses for probing the secrets of the universe.

## Review Questions and Exercises

1. What is the celestial sphere?
2. Where are the north and south celestial poles?
3. How are right ascension and declination measured?
4. What is the difference between the vernal equinox and the autumnal equinox?
5. What is the ecliptic?
6. Present an argument to show that the ecliptic is inclined to the celestial equator for the same reason that the earth's axis of rotation is not perpendicular to the plane of the earth's orbit about the sun.
7. What is the meridian?
8. What is the difference between a solar day and a sidereal day?
9. Why is the solar day not constant throughout the year?
10. What is mean solar time and why is the mean sun more useful for measuring time than the real sun?
11. Why do we find it useful to have time zones?
12. Why do we have leap years?
13. Describe the two basic causes for the seasons.
14. What are the solstices and the equinoxes?
15. What is meant by the arctic and antarctic circles?
16. What is the cause of the tides in the oceans?
17. What is the difference between spring tides and neap tides?
18. Describe the phenomenon of precession.
19. What are some of the effects of precession?
20. How long does it take for the vernal equinox to travel once around the ecliptic due to precession?
21. Who first discovered the phenomenon of precession?
22. What is the line of nodes?
23. What role does the line of nodes play with regard to eclipses?
24. What is meant by the eclipse season?

25. What is the maximum possible duration of totality in a solar eclipse?
26. What is the difference between the umbra and penumbra?
27. Under what conditions do annular eclipses occur?
28. Why can't an eclipse occur at first quarter phase of the moon?
29. What is a penumbral eclipse of the moon?
30. Give two reasons why it is advantageous for modern astronomers to observe solar eclipses from airplanes.
31. What is the maximum number of solar eclipses that could possibly occur in one year?

# 3 Surveying the Solar System

*The Sun and Its Companions*

FOR thousands of years, our ancestors regarded the vast and immobile earth to be at rest at the center of the universe. The sun, moon, and planets performed an intricate celestial ballet among the constellations of the zodiac as the heavens themselves revolved about our earth. Only four hundred years ago did mankind gradually begin to accept the idea that a fanciful geocentric cosmology could not possibly be a valid description of reality. Beginning with the work of Copernicus, Kepler, Galileo, and Newton, it became obvious that both reason and observations clearly pointed to a heliocentric picture of the solar system in which all the planets orbit the sun. It is the sun and not the earth that dominates the solar system.

The sun dominates the solar system simply because it is so very massive. Almost 99.9 per cent of the mass of the entire solar system is found in the sun, which means that only $\frac{1}{10}$ per cent is left over for all the planets combined. The planets therefore could be regarded as microscopic impurities in the vast vacuum of space surrounding our star.

The mass of the sun is 330,000 times the mass of the earth. Everything else in the solar system adds up to a total mass of about 440 earths. The sun is also the largest object in the solar system. The sun's diameter is 846,000 miles, or about 109 times the diameter of the earth. This means that $1\frac{1}{3}$ million earths could fit inside the sun.

From observing stars in the sky, astronomers have come to realize that the sun is a very typical star. Like all stars, the sun is a huge sphere of hot luminous gases. Temperatures on the sun range from 6000°K (11,000°F) at its surface to over 12 million degrees Kelvin at the sun's center. The sun radiates light because *thermonuclear reactions,* the same reactions that occur in a hydrogen bomb, are taking place at the sun's core. This produces 500 sextillion horsepower, which provides all the heat and light for the rest of the solar system. In virtually every sense, the sun completely dominates the solar system.

There are nine planets orbiting the sun. Six of these planets were known to the ancients (Mercury, Venus, Earth, Mars, Jupiter, Saturn) while the remaining three

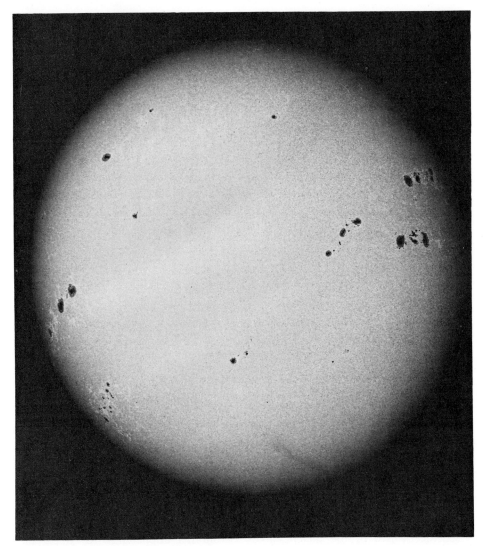

**Figure 3-1.** *The Sun.* The sun contains almost 99.9 per cent of the mass of the solar system. Compared to the sun, the combined mass of all the planets is therefore very small. (*Hale Observatories*)

(Uranus, Neptune, Pluto) were discovered with telescopes. How well you can see any of these planets on a particular night depends on where the planet is. Obviously, the planet must be located in the nighttime sky. If the planet appears too near the sun, it will be impossible to discern the planet in the blinding glare of the sun. This is often the case with Mercury, the innermost member of the solar system. The best time to observe Mercury and Venus is when they are farthest from the sun: at greatest eastern or western elongation. With the outer planets, the best time to

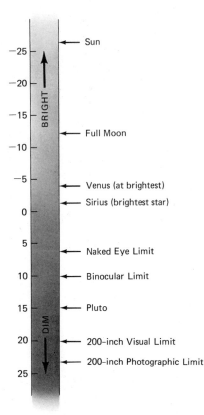

**Figure 3-2.** *The Apparent Magnitude Scale.* Astronomers denote the apparent brightness of an object in the sky by giving its "apparent magnitude." Most stars visible to the naked eye have apparent magnitudes between 1 and 6. About a dozen stars are slightly brighter than first magnitude.

observe is when they are at opposition. At such times, the planets are well placed in the sky and at or near *greatest brilliancy.*

The locations of planets in their orbits have a direct effect on how bright the planets appear in the sky. In general, if a planet such as Mars is near the earth, it appears as a bright reddish "star" in the nighttime sky. If it is far from the earth, the planet can be quite dim. Astronomers have invented a numerical scale for denoting the brightness or *apparent magnitude* of objects in the sky. This scale is adopted from a scheme devised by the ancient Greek astronomer Hipparchus. In Hipparchus' scheme, the brightest stars in the sky were called "first-magnitude stars," while the dimmest visible to the naked eye are called "sixth-magnitude stars." Stars with intermediate brightnesses received intermediate designations.

The apparent magnitude scale has been revised and expanded since the time of Hipparchus. As before, typical bright stars, such as Aldebaran (in Taurus, the bull) and Altair (in Aquila, the eagle), have apparent magnitudes of 1. About a dozen stars are slightly brighter than first magnitude. The dimmest stars visible to the naked eye still have apparent magnitudes of 6. But to include very bright objects, the scale was expanded to negative numbers. Thus, the full moon has a magnitude of $-12\frac{1}{2}$ while the magnitude of the sun is $-26\frac{1}{2}$. To include very dim objects, such as those seen only through a telescope, the scale was expanded in the opposite

**Table 3-1**

**61**

*Surveying the
Solar System*

| Planet | Apparent Magnitude at Greatest Brilliancy |
|---|---|
| Mercury | −1.9 |
| Venus | −4.4 |
| Mars | −2.8 |
| Jupiter | −2.5 |
| Saturn | −0.4 |
| Uranus | +5.7 |
| Neptune | +7.6 |
| Pluto | +14.9 |

direction. Thus, with a good pair of binoculars it is possible to see tenth-magnitude stars while photography with the world's largest telescope reveals stars whose magnitude is 24. Table 3–1 gives the magnitudes of the planets (except that of the earth, of course) at the time of greatest brilliancy.

Clearly, Venus can be one of the brightest objects in the sky, outshone by only the sun and moon. In addition, since the *limiting magnitude* for the naked eye is 6.0, it should be possible to see Uranus on a clear, dark night. Provided you know exactly where to look, Uranus should appear as a very dim star just barely discernable to the unaided eye.

Obviously, astronomers would like to know a lot more about the planets. Their masses, sizes, distances from the sun, number of satellites, and so forth, all constitute important data about the members of the solar system.

By applying Newtonian mechanics to observed orbits of objects in the solar system it is possible to deduce the masses of the planets. Table 3–2 summarizes the results of such calculations expressed in terms of earth masses.

Clearly, Jupiter is by far the most massive planet, followed by Saturn, Neptune,

**Table 3-2**

| Planet | The Masses of the Planets (Earth = 1) |
|---|---|
| Mercury | 0.06 |
| Venus | 0.82 |
| Earth | 1.00 |
| Mars | 0.11 |
| Jupiter | 317.9 |
| Saturn | 95.2 |
| Uranus | 14.6 |
| Neptune | 17.2 |
| Pluto | 0.1(?) |

**Figure 3–3.** *Pluto.* Even with the most powerful telescopes, Pluto always looks like a dim star. It never is brighter than fifteenth magnitude. (*Lick Observatory photograph*)

and Uranus. The remaining five planets (Mercury, Venus, Earth, Mars, Pluto) by comparison have very low masses.

In addition to having a wide range of masses, the planets are scattered over a wide range of distances from the sun. Since the orbits of the planets are ellipses rather than circles, astronomers choose to express the size of a planet's orbit by

**Figure 3–4.** *The Planetary Orbits.* The scale drawing shows the relative sizes of the orbits of the planets. Mercury, Venus, Earth, and Mars are crowded close to the sun. By contrast, the orbits of the outer planets are spread out.

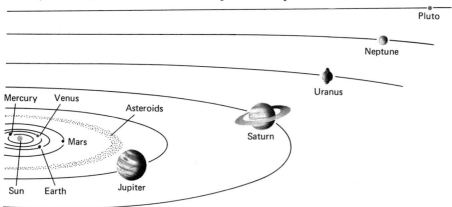

**Table 3–3**

**63**

*Surveying the
Solar System*

| Planet | Semimajor Axis (i.e., average distance to the sun) | | |
|--------|------|------------------------|-------------------|
|        | AUs  | Millions of Kilometers | Millions of Miles |
| Mercury | 0.39 | 58 | 36 |
| Venus | 0.72 | 108 | 67 |
| Earth | 1.00 | 149 | 93 |
| Mars | 1.52 | 228 | 142 |
| Jupiter | 5.20 | 778 | 483 |
| Saturn | 9.54 | 1426 | 886 |
| Uranus | 19.18 | 2868 | 1782 |
| Neptune | 30.06 | 4494 | 2793 |
| Pluto | 39.44 | 5896 | 3664 |

giving the length of its semimajor axis. This length may be thought of as the "average" distance between the sun and a planet. Figure 3–4 shows a scale drawing of the solar system while Table 3–3 gives the numerical values of the sizes of planetary orbits. These distances can be expressed in either AUs (recall that 1 AU = 93 million miles = semimajor axis of earth's orbit) or kilometers or miles.

Using Kepler's third law, it is possible to calculate the sidereal periods of the planets from the sizes of their semimajor axes. Recall that the sidereal period is the "true" orbital period of the planet. It is how long the planet takes to complete one full orbit around the sun. Table 3–4 gives the sidereal periods of the nine planets.

Knowing the distance to a planet, it is fairly easy to calculate the planet's size or diameter. The astronomer simply measures the *angular size* of the planet as seen through his telescope. From knowing how far away the planet is, he can translate the angular size into miles or kilometers. Figure 3–5 shows a scale drawing of the planets against the sun while Table 3–5 gives the sizes of the planets. These

**Table 3–4**

| Planet | Sidereal Period |
|--------|-----------------|
| Mercury | 87.97 days |
| Venus | 224.70 days |
| Earth | 365.26 days |
| Mars | 1.88 years |
| Jupiter | 11.86 years |
| Saturn | 29.46 years |
| Uranus | 84.01 years |
| Neptune | 164.79 years |
| Pluto | 247.69 years |

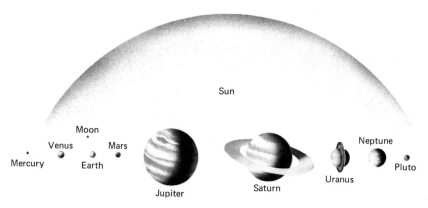

**Figure 3–5.** *The Sun and the Planets.* This scale drawing shows the relative sizes of the nine planets and the sun.

dimensions are expressed in earth diameters (i.e., earth = 1.00), kilometers, and miles.

From the information about the sizes and masses of the various planets, we notice a striking fact about the solar system. There are really *two* kinds of planets. Mercury, Venus, Earth, Mars, and Pluto all have small diameters and low masses. By contrast, Jupiter, Saturn, Uranus, and Neptune all have large diameters and much higher masses. For this reason, the five smaller planets are called *terrestrial planets;* they all superficially resemble the earth and have hard rocky surfaces. On the other hand, the four larger planets are called *Jovian planets;* they all superficially resemble Jupiter and do not have hard surfaces. In looking at the Jovian planets, the astronomer is seeing the surface of a huge ball of gas typically consisting of hydrogen, helium, methane, and/or ammonia. If these planets have solid surfaces at all, they must be buried far below all these gases.

The fact that terrestrial and Jovian planets have very different chemical compositions (i.e., rocks vs. gases) is readily reflected in the *average density* of the planets.

**Table 3–5**

| Planet | Planet's Diameter | | |
|--------|-----------|------------|-------|
| | Earth = 1 | Kilometers | Miles |
| Mercury | 0.38 | 4,880 | 3,032 |
| Venus | 0.95 | 12,112 | 7,526 |
| Earth | 1.00 | 12,742 | 7,918 |
| Mars | 0.53 | 6,800 | 4,230 |
| Jupiter | 11.19 | 143,000 | 88,900 |
| Saturn | 9.47 | 121,000 | 75,200 |
| Uranus | 3.69 | 47,000 | 29,000 |
| Neptune | 3.50 | 45,000 | 28,000 |
| Pluto | 0.47(?) | 6,000(?) | 4,000(?) |

**Figure 3-6. *Jupiter.*** Jupiter is the largest planet. Its mass is greater than all the other planets combined. (*Lick Observatory photograph*)

Density is simply a measure of how much matter is contained in a particular volume. For example, the density of water under normal conditions is 1 gram per cubic centimeter, which means that one cubic centimeter of water has a mass of one gram. The average density of a planet is therefore obtained by dividing its mass by its volume. The results (expressed in $gm/cm^3$) are given in Table 3–6. Notice that the average density of Saturn is less than the density of water ($= 1.0 \ gm/cm^3$). This means that Saturn would float if it were placed in a huge bathtub.

Finally, an important piece of information about a planet is its *surface gravity*. Surface gravity tells how much things would weigh on a planet or other object. For example, the surface gravity of the sun is 28, which simply means that something on the sun would weigh 28 times more than it does on the earth; if you weigh 150 pounds on the earth, you would weigh $150 \times 28 = 4200$ pounds on the sun. Similarly, the surface gravity on the moon is 0.16, which means that an astronaut walking on the lunar surface weighs only 16 per cent of what he did before leaving the earth. Like density, surface gravity depends on both the mass and size of the planet. Table 3–7 gives the surface gravities of the nine planets.

**Table 3–6**

| Planet | Average Density (gm/cm³) |
|---|---|
| Mercury | 5.1 |
| Venus | 5.3 |
| Earth | 5.5 |
| Mars | 3.9 |
| Jupiter | 1.3 |
| Saturn | 0.7 |
| Uranus | 1.6 |
| Neptune | 2.3 |
| Pluto | ? |

**Table 3–7**

| Planet | Surface Gravity (Earth = 1) |
|---|---|
| Mercury | 0.4 |
| Venus | 0.9 |
| Earth | 1.0 |
| Mars | 0.4 |
| Jupiter | 2.6 |
| Saturn | 1.1 |
| Uranus | 1.1 |
| Neptune | 1.4 |
| Pluto | ? |

Notice that the surface gravities of the most massive planets are not especially high. Although these planets are many times more massive than the earth, their diameters are also huge. Since, according to Newton, gravity depends on both mass and distance, the large sizes of the Jovian planets results in moderate surface gravities. Thus, a person who weighs 150 pounds on Earth would tip the scales at 390 pounds on Jupiter, even though Jupiter is hundreds of times more massive than Earth.

To an imaginary astronomer from some advanced race of creatures in some distant part of the universe, it would be practically impossible to know about the existence of the planets. Only if he were to travel toward our star in his spacecraft would his careful scrutiny reveal some tiny pieces of matter in orbit about the sun. As he came within a few trillion miles of the solar system, he would first see Jupiter, the largest of all the planets. Then he would notice Saturn. Only with very great difficulty might he discover Earth.

# 3.2 *The Inner Planets*

FOUR of the smallest planets in the solar system are crowded fairly close to the sun. The innermost planet, Mercury, is only slightly larger than our moon. At an average distance of 58 million kilometers (36 million miles) from the sun, daytime temperatures are very high. Lead and tin would melt and flow like water under the noontime Mercurian sun.

Since Mercury's orbit is so close to the sun, it is often very difficult to observe this planet from earth. At best, Mercury is seen in the east just before sunrise or in the west just after sunset. As a result, astronomers have to cope with the twilight

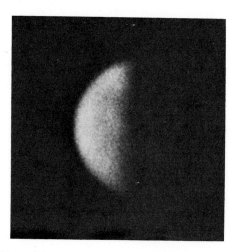

**Figure 3–7. *Mercury Seen from Earth.*** Because it is so close to the sun, Mercury is a very difficult planet to observe. This is a good example of one of the best Earth-based photographs. No major surface features can be seen. (*New Mexico State University Observatory*)

**Figure 3-8.** *Mercury Seen from Mariner 10.* In 1974, a spacecraft flew within 700 kilometers (450 miles) of Mercury's surface, sending back numerous photographs. Astronomers were amazed to discover that Mercury looks very much like our moon. (*JPL; NASA*)

glare of the sun slightly below the horizon in order to study Mercury. Figure 3–7 shows an excellent earth–based photograph of Mercury. No major surface features can be distinguished. Indeed, very little was known about this tiny planet until the mid–1970s when Mariner 10 came within a few hundred kilometers of Mercury's surface and sent back numerous photographs. These photographs revealed a heavily cratered terrain that looks very much like the surface of our moon. Figure 3–8 shows a typical view from Mariner 10. More was learned about Mercury during a few hours on March 29, 1974, than the total amount of knowledge accumulated over the entire course of recorded history. For this reason, a full chapter in this book is devoted to this important spaceflight to the innermost planet.

While Mercury is often very difficult to observe from earth, the next planet from the sun frequently dominates the nighttime sky. Venus is also an inferior planet; its orbit is smaller than the Earth's orbit. But since the average distance between the sun and Venus is 108 million kilometers (67 million miles), this planet can appear much farther from the sun than Mercury. At greatest elongation, Venus is 47° from the sun compared to 28° for Mercury. Consequently, Venus can set in the west several hours after sunset or rise in the east several hours before sunrise.

**67**

**Figure 3-9.** *Venus.* Aside from the sun and moon, Venus is often the brightest object seen in the sky. This planet is always covered with thick clouds. (*Hale Observatories*)

Astronomers therefore are often not hampered by the glare of twilight in observing this planet.

An excellent earth-based photograph of Venus is shown in Figure 3-9. The entire planet is perpetually covered with thick clouds, which makes it impossible to see the planet's surface. This thick cloud-cover is also responsible for Venus' brilliancy. The clouds efficiently reflect lots of sunlight.

Astronomers speak of the *albedo* of a planet as a measure of how efficiently the planet reflects sunlight. Albedo is simply the amount of sunlight reflected by the planet. An albedo of 1.00 means that an object reflects 100 per cent of the light falling on it, like a shiny mirror. An albedo of zero means that an object is completely black and reflects no light at all. Table 3-8 gives the albedos of the planets. Venus has the highest albedo of all the planets and reflects 76 per cent of the light from the sun. This is an important reason why Venus appears so bright in the sky.

The third planet from the sun is the largest of the inner planets. In many respects, the earth is very different from any other planet in the solar system. First of all, almost 71 per cent of the earth's surface is covered with water. If alien creatures were to send probes to the earth (such as the Surveyors which went to the moon or the Viking spacecrafts to Mars), chances are that their vehicles would land in the water and sink to the bottom of the ocean. By contrast, Mercury and Venus have virtually no water and Mars has only a very small amount of water.

The earth is also the first planet from the sun that has a satellite. This does not seem unusual until we compare our moon with all the other satellites in the solar system. There are a total of 34 moons distributed among six planets. Mercury, Venus, and Pluto do not have any known satellites.

Most of the satellites in the solar system are quite small; their diameters are less

**Figure 3-10.** *A Typical Close-up of the Earth's Surface.* Almost 71 per cent of the earth's surface is covered with water. By contrast, there is little or no water on Mercury, Venus, or Mars.

than about 1000 kilometers (600 miles). There are, however, seven satellites that are comparatively large and have diameters of about 3000 kilometers (2000 miles) or more. These seven objects could be called "giant satellites." They are listed in Table 3-9 along with the planet that they orbit. It is interesting to note that six of the giant satellites belong to the largest planets in the solar system. The seventh orbits the earth. In the case of Jupiter, Saturn, and Neptune, the planets are typically a thousand times more massive than even their largest satellites. These satellites appear very tiny when compared to the huge Jovian planets that they orbit. This is not the case with the earth and its moon. The earth is only 81 times more massive than the moon. Compared to all the other planets and their satellites, the earth and

**Table 3-8**

| Planet | Albedo |
|---|---|
| Mercury | 0.06 |
| Venus | 0.76 |
| Earth | 0.39 |
| Mars | 0.15 |
| Jupiter | 0.51 |
| Saturn | 0.50 |
| Uranus | 0.66 |
| Neptune | 0.62 |
| Pluto | ? |

**Table 3-9**

| Planet | Giant Satellite | Diameter of Satellite Kilometers | Diameter of Satellite Miles |
|---|---|---|---|
| Earth | Moon | 3476 | 2160 |
| Jupiter | Io | 3640 | 2260 |
| Jupiter | Europa | 3050 | 1890 |
| Jupiter | Ganymede | 5270 | 3270 |
| Jupiter | Callisto | 4900 | 3040 |
| Saturn | Titan | 5800 | 3600 |
| Neptune | Triton | 6000 | 3700 |

**Figure 3–11.** *The Double Planet.* Most planets are much more massive than any of their satellites. The earth–moon system is the only exception; the earth and the moon are not vastly different in size. (*NASA*)

the moon are not extremely different in size. The earth–moon system is therefore sometimes called a *double planet.*

The fourth planet from the sun, Mars, has two moons. Both satellites are very tiny. The largest is called Phobos and is about 25 kilometers (16 miles) in diameter. The smaller is Deimos and measures about 13 kilometers (9 miles) from one side to the other. The names Phobos and Deimos come from the Greek words meaning "fear" and "panic"—appropriate companions of the god of war.

Of all the terrestrial planets, Mars is the only one whose surface features can be seen from earth. Near the time of oppositions, when Mars is relatively near the earth, even amateur telescopes reveal a polar cap and dark greenish markings against the rust-colored planet. A good earth-based photograph is shown in Figure 3–12.

Seasons occur on Mars just as seasons occur here on earth. During the Martian spring and summer, the polar cap shrinks and the greenish areas become more pronounced. During the fall and winter, the dark areas fade while the polar cap increases in size. These climatic changes suggested the possibility of some sort of plant-like life to astronomers in the nineteenth century. Scientists are still fascinated by the prospect of finding life on Mars. This was the primary motivation for sending the Viking spacecrafts to this planet in the mid-1970s.

**Figure 3–12.** *Mars.* The size of the polar caps and the dark greenish-appearing areas change with the Martian seasons. This led some scientists to believe that there is life on Mars. (*Lick Observatory photograph*)

Although the Martian surface can be seen from earth, it is still impossible to distinguish any details. Clear views can only be obtained from spacecrafts that either pass near or orbit the red planet. Astronomers were therefore amazed to discover that Mars has craters when Mariner 4 swung within 9850 kilometers (6110 miles) of the planet's surface in 1965 sending back 22 pictures. During later missions, thousands of photographs of the Martian surface were sent back to earth. These views reveal huge volcanoes and immense canyons. This planet is so varied and fascinating that two full chapters in this text are devoted to Mars.

**Figure 3–13.** *The North Polar Cap.* In August of 1972, Mariner 9 took this remarkable picture of Mars' northern polar cap. At this time it was late spring in the northern hemisphere and much of the polar cap had evaporated. (*JPL; NASA*)

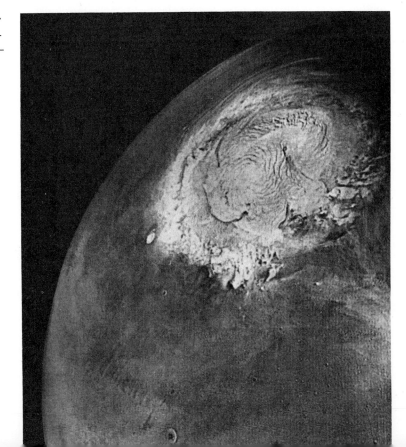

One way in which the four inner planets differ greatly from each other is with regard to their atmospheres. Mercury has no atmosphere. Venus has a massive atmosphere primarily composed of carbon dioxide ($CO_2$), which weighs down on the planet with immense pressure. Earth has a moderate atmosphere consisting of about 80 per cent nitrogen ($N_2$) and 20 per cent oxygen ($O_2$). The Martian atmosphere is very thin and consists mostly of carbon dioxide.

Whether or not a planet can have an atmosphere depends on two important factors: temperature and escape velocity. To the physicist, temperature is a measure of the average speed of atoms or molecules in a substance. If something is cold, the average speed of its atoms or molecules is low. If something is hot, its atoms or molecules are moving very fast. Indeed, there is a direct mathematical relationship between the temperature of a substance and the average speed of its atoms or molecules.

Scientists prefer to use a temperature scale that is directly associated with the average speeds of atoms and molecules. The familiar Fahrenheit and Centigrade temperature scales are based on phenomena such as the temperature of the human body (approximately 100°F) or the melting and boiling points of water (0°C and 100°C, respectively). Temperatures measured on the Kelvin scale reflect the average speeds of atoms or molecules. For example, at 0°K all motion stops. This is the lowest possible temperature; it is called *absolute zero*. Figure 3–14 and Table 3–10 relate some typical temperatures from the Kelvin, Centigrade, and Fahrenheit

**Figure 3–14.** *Different Temperature Scales.* This drawing illustrates the relationship between the three most commonly used temperature scales. Scientists prefer the Kelvin scale.

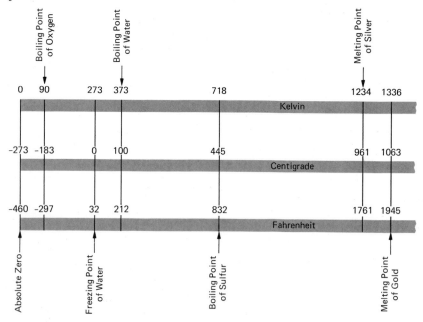

**Table 3-10**

|  | Kelvin | Centigrade | Fahrenheit |
|---|---|---|---|
| Absolute Zero | 0°K | −273°C | −460°F |
| Water Freezes | 273°K | 0°C | 32°F |
| "Room Temperature" | 293°K | 20°C | 68°F |
| Water Boils | 373°K | 100°C | 212°F |

scales. Throughout the rest of this book frequent use of the Kelvin scale will be made in discussing temperature.

Consider a bottle of gas, such as oxygen ($O_2$). Some of the oxygen molecules are moving slowly, and some are moving rapidly. Nevertheless, most of the molecules have an average speed that directly depends on the temperature of the gas. In other words, there is a *distribution* of molecular velocities that is peaked near the average speed. Figure 3–15 shows how the distribution of molecular speeds in a bottle of gas varies with temperature. At room temperature, the average speed of the oxygen molecules in the air is about $\frac{1}{2}$ kilometer per second.

The average speed of the molecules of a particular gas also depends on the mass of the molecules. For example, hydrogen is the lightest gas. Hydrogen molecules ($H_2$) have very small masses. Consequently, at a particular temperature, the average velocity of hydrogen molecules will be much higher than the average velocity of more massive molecules. Low-mass molecules move around more easily than high-mass molecules. At room temperature, the average speed of hydrogen mole-

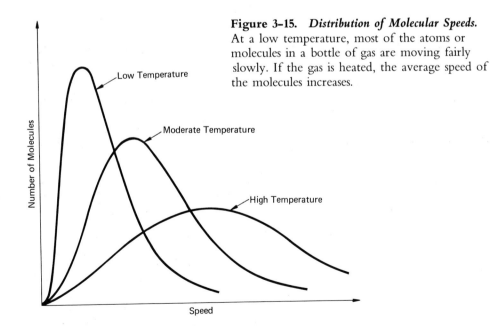

**Figure 3–15.** *Distribution of Molecular Speeds.* At a low temperature, most of the atoms or molecules in a bottle of gas are moving fairly slowly. If the gas is heated, the average speed of the molecules increases.

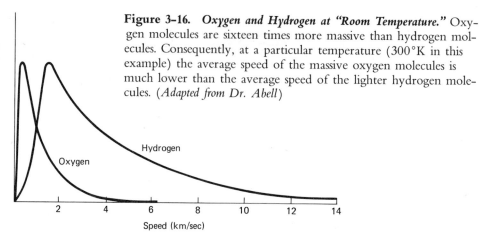

**Figure 3–16.** *Oxygen and Hydrogen at "Room Temperature."* Oxygen molecules are sixteen times more massive than hydrogen molecules. Consequently, at a particular temperature (300°K in this example) the average speed of the massive oxygen molecules is much lower than the average speed of the lighter hydrogen molecules. (*Adapted from Dr. Abell*)

cules is about 2 kilometers per second. By way of illustration, Figure 3–16 shows the distribution of hydrogen molecules and oxygen molecules at room temperature.

Now let's think about a planet. As a result of radiation from the sun, the surface of the planet will have a particular temperature. If this temperature is low, the molecules in the planet's atmosphere will be moving slowly. If the temperature is high, the molecular speeds will be high. Whether or not the planet can hold on to its atmosphere depends on the strength of the planet's gravitational field. If the planet's surface gravity is low, and if the average speed of the molecules is high, the atmosphere will escape. This is why Mercury has no atmosphere. Mercury is so hot and its surface gravity is so low that this planet is incapable of retaining an atmosphere.

In the case of planets with moderate surface gravities and moderate temperatures, it is possible to retain some gases while other gases escape. For example, the earth's gravitational field is strong enough to keep an atmosphere of oxygen and

**Table 3–11**

| Planet | Escape Velocity | |
| --- | --- | --- |
| | Kilometers/Second | Miles/Second |
| Mercury | 4.3 | 2.7 |
| Venus | 10.3 | 6.4 |
| Earth | 11.2 | 7.0 |
| Mars | 5.1 | 3.2 |
| Jupiter | 60 | 37 |
| Saturn | 35 | 22 |
| Uranus | 22 | 14 |
| Neptune | 25 | 16 |
| Pluto | ? | ? |

nitrogen. But at the earth's temperature, light-weight hydrogen molecules are moving so rapidly that they can escape to outer space.

The escape velocity of a planet can be used to decide what gases that planet can hold on to. Recall that escape velocity is the minimum speed needed to overcome the pull of a planet's gravitational field. For example, in order for Apollo astronauts to journey to the moon, their spacecraft must achieve a speed greater than the escape velocity of the earth (11 kilometers per second). Table 3–11 lists the escape velocities of the various planets.

A good rule of thumb is: A planet can retain a particular gas in its atmosphere only if the escape velocity is at least six times larger than the average molecular velocity of that gas. Only then is the escape velocity so large that a gas cannot appreciably leak off into outer space. The best way to display this rule is with a diagram such as Figure 3–17. The surface temperatures of the planets are plotted against their escape velocities. On top of this plot, lines are drawn for different gases and represent six times the average molecular speeds of these gases. The planet can retain a particular gas only if its point on this graph lies above the line for that gas. For example, the points for the outer planets lie above all of the lines. These planets are so cool and so massive that they can hold onto all gases in their atmospheres. The point representing our moon lies below all of the lines. Our moon cannot retain any gases and therefore has no atmosphere. The point for Mars lies below the lines for hydrogen and helium; these gases easily escape from the planet. But Mars lies above the lines for heavier gases such as water vapor ($H_2O$) and carbon dioxide ($CO_2$); these gases are retained.

The four inner planets are small, have rocky surfaces, and are crowded close to

**Figure 3–17. *Conditions for Retaining an Atmosphere.*** A planet can retain a particular gas in its atmosphere only if the escape velocity from the planet is at least six times greater than the average molecular speed of that gas. See text for discussion. (*Adapted from Dr. Abell*)

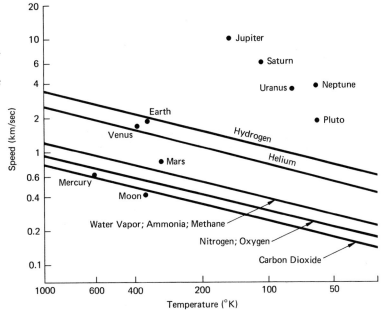

the sun. The next four planets are huge, have gaseous surfaces, and are spread out at great distances from the sun. The inner terrestrial planets have a total of only three satellites. The remaining 31 satellites in the solar system are distributed among the four Jovian planets. We shall now turn our attention to the largest members of the solar system.

# 3.3  The Outer Planets

AT A distance of 778 kilometers (almost half a billion miles) from the sun is Jupiter, by far the largest planet in the solar system. Jupiter is so massive and so far from the sun that it easily retains all of the gases in its atmosphere, including hydrogen—the lightest of all elements. The chemical composition of Jupiter's atmosphere is therefore probably representative of the primordial materials out of which the solar system was created some $4\frac{1}{2}$ billion years ago.

Unlike the inner terrestrial planets, the "surface" of Jupiter, as seen through a telescope, consists of thick turbulent gases. Indeed, most of the planet is made of hydrogen ($H_2$), helium (He), methane ($CH_4$), and ammonia ($NH_3$). The appearance of Jupiter's atmosphere is dramatically affected by the planet's rapid rate of rotation. As shown in Table 3–12, Jupiter rotates faster than any other planet. The circulation patterns in Jupiter's atmosphere are therefore drawn out into bands parallel to the planet's equator. This gives Jupiter its characteristic striped appearance, as shown in Figure 3–18. In addition to light and dark bands, the Great Red Spot is a prominent feature of Jupiter's surface.

Jupiter does not rotate like a rigid body. The rate of rotation at the equator (9 hours, 50 minutes) is faster than near the poles (9 hours, 55 minutes). Additionally, this high rotation speed is responsible for Jupiter being somewhat flattened. The

**Table 3–12**

| Planet | Period of Rotation |
|---|---|
| Mercury | 58.6 days |
| Venus | 242.6 days |
| Earth | 23.9 hours |
| Mars | 24.6 hours |
| Jupiter | 9.8 to 9.9 hours |
| Saturn | 10.2 to 10.6 hours |
| Uranus | 10.8 hours |
| Neptune | 16  hours(?) |
| Pluto | 6.4 days |

**Figure 3–18.** *Jupiter.* This photograph is an excellent Earth-based view of the largest planet. Ganymede, one of Jupiter's largest satellites, is seen along with its shadow on the planet's surface. (*Hale Observatories*)

**Table 3–13**

| Planet | Number of Satellites |
|---|:---:|
| Mercury | 0 |
| Venus | 0 |
| Earth | 1 |
| Mars | 2 |
| Jupiter | 14 |
| Saturn | 10 |
| Uranus | 5 |
| Neptune | 2 |
| Pluto | 0 |

equatorial diameter of Jupiter is about 9500 kilometers (6000 miles) larger than the distance from pole to pole.

Jupiter has more satellites than any other planet in the solar system. Four of Jupiter's moons are very large while the remaining ten satellites are quite small. In the mid-1970s, two satellites were discovered orbiting Jupiter that probably have diameters less than 20 kilometers. Table 3–13 lists the planets along with the total number of known satellites each possesses.

At almost twice the distance of Jupiter from the sun is Saturn, a ringed planet. Like Jupiter, Saturn rotates rapidly, is slightly flattened, and has many satellites. But in addition, Saturn is surrounded by a beautiful system of rings, as shown in Figure 3–19. Because stars can be seen shining through Saturn's rings, astronomers have realized that the rings are not solid. It is believed that the rings consist of billions of tiny particles (probably grains of rock and ice) orbiting the planet like billions of tiny moons.

**Figure 3–19. *Saturn.*** This Earth-based photograph of Saturn shows the magnificent ring system along with faint cloud belts on the planet's surface. (*Lick Observatory photograph*)

Saturn has ten satellites one of which, Titan, is very large. Astronomers have good reason to believe that Titan has an atmosphere containing methane. The nature of Saturn's rings as well as Titan's atmosphere will be important subjects of investigation when Pioneer 11, an interplanetary spacecraft, arrives at this mysterious planet in September of 1979.

The next two planets from the sun, Uranus and Neptune, are very similar. Both planets are approximately the same size and have roughly the same mass, density, surface gravity, and albedo. In many respects they are twins. The only obvious difference is that Uranus, like Saturn, is surrounded by a system of rings.

While Jupiter and Saturn appear very bright and are easily identified in the nighttime sky, Uranus and Neptune are telescopic objects. Yet even with the most powerful telescopes, these two remote planets exhibit small greenish-blue discs devoid of any surface features.

The sizes of objects in the sky (such as the apparent diameters of planets) are expressed in units of angular measure: degrees, minutes, and seconds of arc. A full circle contains 360°. Each degree is divided into 60 minutes of arc and each minute

**Figure 3–20. *Uranus.*** The seventh planet from the sun, Uranus, is so far away that no surface details can be seen from Earth. (*New Mexico State University Observatory*)

**Figure 3–21. *Neptune.*** Nepture is very similar to Uranus in size, mass, and chemical composition. As with Uranus, no surface details can be seen from Earth. (*New Mexico State University Observatory*)

**Table 3-14**

| Jovian Planet | Apparent Diameter (in seconds of arc) |
|---|---|
| Jupiter | 47″ |
| Saturn* | 20″ |
| Uranus* | 4″ |
| Neptune | 2″ |

* Excluding rings

of arc is further subdivided into 60 seconds of arc. The apparent diameter of the moon is $\frac{1}{2}$°. The apparent diameters of the four Jovian planets at opposition (when they appear largest) are given in Table 3–14. For comparison, a silver dollar viewed from a distance of one mile has an angular diameter of about 5 seconds of arc. A dime viewed from a distance of one mile has nearly the same angular size as the planet Neptune seen from the earth.

Most of the planets in the solar system are oriented so that their axes of rotation are approximately perpendicular to the planes of their orbits. This is not the case with Uranus. Uranus is tilted more than any other planet. To express the extent of this tilting, astronomers speak of the angle between the plane of the planet's equator and the plane of its orbit, as shown in Figure 3–22. This angle is called the *obliquity*. If this angle is zero degrees, then the planet's axis of rotation is exactly perpendicular

**Figure 3-22.** *Obliquity.* The angle between the orbital plane and equatorial plane of a planet is called the obliquity. The obliquity is a measure of how much a planet's axis of rotation is tilted with respect to its orbit.

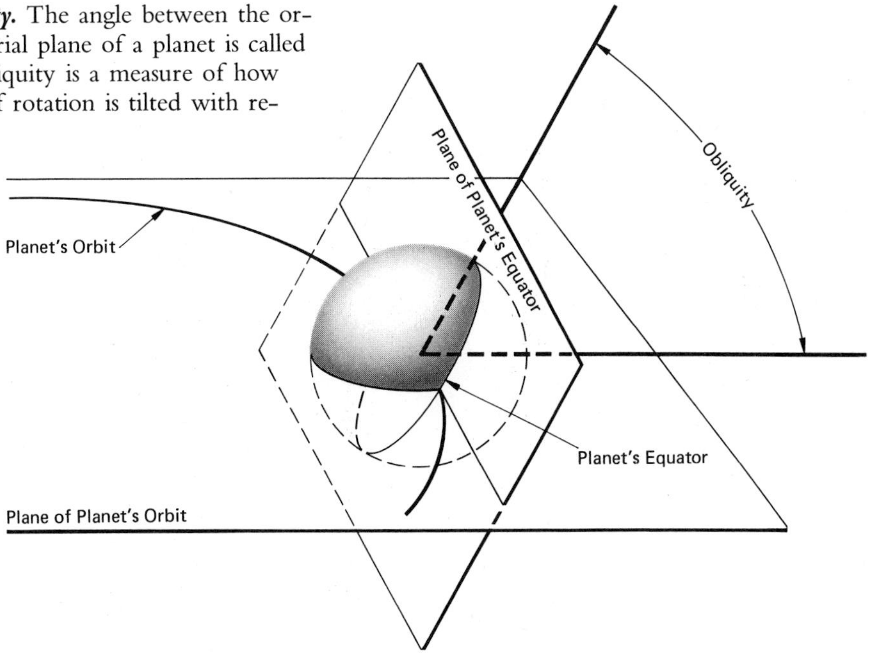

Plane of Planet's Equator

Obliquity

Planet's Orbit

Planet's Equator

Plane of Planet's Orbit

**Table 3–15**

| Planet | Obliquity |
|---|---|
| Mercury | 7° |
| Venus | 23° |
| Earth | 23½° |
| Mars | 24° |
| Jupiter | 3° |
| Saturn | 27° |
| Uranus | 98° |
| Neptune | 29° |
| Pluto | ? |

**Table 3–16**

| Planet | Orbital Eccentricity |
|---|---|
| Mercury | 0.206 |
| Venus | 0.007 |
| Earth | 0.017 |
| Mars | 0.093 |
| Jupiter | 0.048 |
| Saturn | 0.056 |
| Uranus | 0.047 |
| Neptune | 0.009 |
| Pluto | 0.250 |

to the plane of its orbit. In the case of the earth, the obliquity is 23½°, which causes the seasons, as discussed in Chapter 2. Table 3–15 gives the obliquities of the nine planets. Thus, unlike any other planet, Uranus' axis of rotation is almost parallel to its orbital plane. As we shall see later in the text, this inclination results in very strange seasons on Uranus.

Pluto, the farthest planet from the sun, is so small and so remote that even with the most powerful telescope it looks like a tiny star. Figure 3–23 shows a typical view of Pluto. For this reason, very little is known about Pluto.

There is some speculation that Pluto may not be a true planet. Instead, it may be a former satellite of Neptune that managed to escape and go into orbit about the sun. One of the primary reasons for this speculation is that Pluto's orbit is more

**Figure 3–23.** *Pluto.* Pluto is so small and so far away that it looks like a dim star. Very little is known about this remote planet. (*Hale Observatories*)

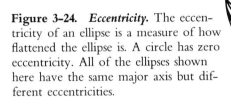

**Figure 3-24.** *Eccentricity.* The eccentricity of an ellipse is a measure of how flattened the ellipse is. A circle has zero eccentricity. All of the ellipses shown here have the same major axis but different eccentricities.

elliptical and more tilted than the orbits of any of the other planets. Astronomers speak of the *orbital eccentricity* as a measure of how much a planet's orbit departs from circularity. A circle has zero eccentricity. The larger the eccentricity, the more elliptical is the planet's orbit. Figure 3–24 shows various ellipses with different eccentricities. Table 3–16 gives the orbital eccentricities of the nine planets. Clearly, Pluto has the most highly elliptical orbit of any planet. Indeed, its orbit is so elliptical that sometimes Pluto is closer to the sun than Neptune.

In addition to being somewhat flattened, Pluto's orbit is more steeply inclined than the orbit of any other planet. Recall that the ecliptic defines the plane of the earth's orbit about the sun. The earth's orbital plane and the plane of the ecliptic are identical. The orbital planes of all the other planets are tilted slightly with respect to the plane of the ecliptic. As shown in Figure 3–25, the *inclination* of a planet's orbit to the ecliptic is a direct measure of how much the planet's orbit is tilted out of the

**Figure 3-25.** *Inclination of an Orbit to the Ecliptic.*
The angle between the plane of the ecliptic and the plane of a planet's orbit is called the orbital inclination. Most of the planets lie very nearly in the same plane as the earth's orbit.

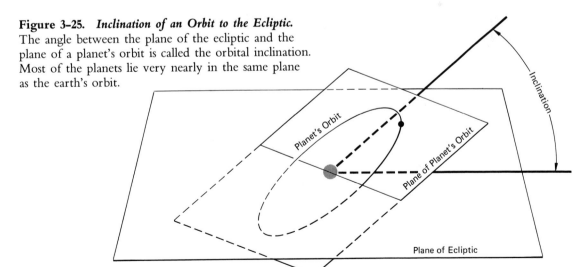

**Table 3–17**

| Planet | Orbital Inclination |
|--------|---------------------|
| Mercury | 7.0° |
| Venus | 3.4° |
| Earth | 0.0° |
| Mars | 1.8° |
| Jupiter | 1.3° |
| Saturn | 2.5° |
| Uranus | 0.8° |
| Neptune | 1.8° |
| Pluto | 17.2° |

plane of the ecliptic. Table 3–17 gives the orbital inclination of the nine planets. Most of the planets lie very near the plane of the ecliptic. For this reason, they are always seen in the zodiac. Pluto's orbit, however, is steeply inclined to the ecliptic.

Quantities such as the length of the semimajor axis, the orbital inclination, and the eccentricity are some of the numbers that make up the *orbital elements* of a planet's orbit about the sun. The unusual values of some of Pluto's orbital elements lead astronomers to suspect that Pluto is an escaped satellite.

Besides the sun, the planets are the most massive objects in the solar system. Nevertheless, there are numerous very tiny objects orbiting the sun such as comets, meteoroids, and asteroids. We shall now turn our attention to these smaller objects.

# 3.4 *Vagabonds of the Solar System*

AS WE have seen, the sun is by far the most massive object in the solar system. Almost 99.9 per cent of the mass of the solar system is contained in the sun. The

**Table 3–18**

| Object | Percentage of Mass |
|--------|--------------------|
| Sun | 99.86 |
| Planets | 0.135 |
| Satellites | 0.00004 |
| Comets | 0.00003 |
| Asteroids | 0.0000003 |
| Meteoroids | 0.0000003 |

second most massive objects are the planets. Altogether the planets account for slightly more than 0.1 percent of the solar system's mass. This leaves very little matter for everything else. Table 3–18 lists the major constituents of the solar system along with their mass percentages. Aside from the sun, the planets and their satellites, there is very little matter contained in the remaining objects that orbit the solar system.

In between the orbits of Mars and Jupiter there is a comparatively large region of space. This gap in the solar system marks the boundary between the small terrestrial planets orbiting near the sun and the huge Jovian planets whose orbits are spread out at great distances. In the early 1800s, astronomers began discovering numerous small objects in this region between Mars and Jupiter. These objects are sometimes called *minor planets* or *planetoids*. We shall usually refer to them as *asteroids*.

The biggest asteroids were discovered first. They were simply the easiest to see. The largest asteroid is Ceres, which has a diameter of 785 kilometers (488 miles). Ceres orbits the sun once every 4.6 years at an average distance of about 2.8 AU. Most asteroids are considerably smaller than Ceres. Less than a dozen asteroids have diameters greater than 100 miles. Only a few hundred are larger than 25 miles across. There are thousands of asteroids which are approximately 1 mile in size.

As a result of their small size, asteroids look like tiny stars when seen through a telescope. Most asteroids are discovered entirely by accident. For example, an astronomer might be photographing a certain region of the sky. Upon developing his photographic plate, he finds that one of the "stars" has moved slightly during the time he was exposing the film. Figure 3–26 shows a good example. The astronomer immediately realizes that his field of view contained an asteroid whose motion about the sun resulted in the blurred image.

Most asteroids orbit the sun at distances from 2.3 to 3.3 AU and have sidereal

**Figure 3–26.** *Asteroids.* An asteroid usually appears as a blurr on an astronomical photograph. While the astronomer takes a time exposure of the stars, the motion of the asteroid in orbit about the sun results in a blurred image. The "trails" of two asteroids are shown here. (*Yerkes Observatory photograph*)

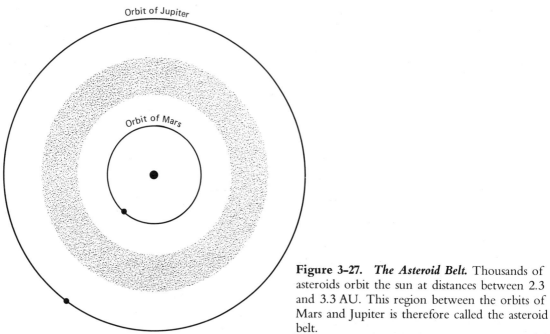

**Figure 3–27.** *The Asteroid Belt.* Thousands of asteroids orbit the sun at distances between 2.3 and 3.3 AU. This region between the orbits of Mars and Jupiter is therefore called the asteroid belt.

**Figure 3–28.** *A Comet.* This comet, named Comet Mrkos after its discoverer, produced a dramatic sight in the sky during August 1957. For more than a full week it could be easily seen with the naked eye. (*Hale Observatories*)

84

periods ranging from 3.5 to 6 years. Since there are literally thousands of asteroids between the orbits of Mars and Jupiter, this region is called the *asteroid belt,* as shown in Figure 3–27. Scientists were concerned about a possible mishap when Pioneer 10 and Pioneer 11 journeyed to Jupiter in the mid-1970s. Fortunately both spacecrafts passed through the asteroid belt without suffering any collisions. Although there may be 50,000 to 100,000 asteroids, these space flights proved that the typical distances between individual asteroids must be comparatively large.

While asteroids revolve about the sun in nearly circular orbits approximately in the plane of the ecliptic, *comets* move through the solar system along highly eccentric orbits inclined at all angles. Asteroids are made out of rock. Comets, on the other hand, contain ice in addition to some rocky material. Because of their ice content, comets can create a dramatic sight in the nighttime sky. As a comet approaches the sun, the solar heat vaporizes some of the ice in the comet's *nucleus.* The resulting gases streaming off of the cometary nucleus produce a long flowing *tail,* as shown in Figure 3–28.

Comets are usually discovered when they are comparatively far from the sun. At a distance of several AUs, most of the comet's nucleus is still frozen and therefore the comet looks like a dim, fuzzy spot through a telescope. A dramatic tail develops only after the comet has passed near the sun.

Astronomers typically discover a dozen comets each year. Yet, most of these comets are so far from the earth and the sun that they never produce a dramatic sight in the nighttime sky. In order to give an easily visible *apparition,* the comet must pass near the sun (to vaporize a large fraction of the comet's ices) and near the earth (to give a big, bright image). This usually occurs once every few years. During a favorable apparition, a comet can be seen for several consecutive nights drifting slowly against the background stars. After several weeks, the comet fades from sight as it journeys back out into the depths of space.

For some unfortunate reason, comets are sometimes confused with *meteors.* They are entirely different. A comet consists of a comparatively large chunk of rock and ice, typically ten to twenty miles across. The vaporization of its ices produces a tail that can be millions of miles long. A meteor, on the other hand, is a brief flash of light caused by a tiny piece of rock striking the earth's upper atmosphere. Meteors are sometimes called "shooting stars" or "falling stars."

There are many tiny pieces of rock orbiting the sun. Quite often, one of these tiny rocks collides with the earth. Upon striking the earth's upper atmosphere with a high speed, the tiny rock is vaporized. The resulting flash of light lasts for only a fraction of a second. Figure 3–29 shows a good example of a meteor. Most meteors are produced by a tiny rock approximately the same size as a grain of sand. The rocks that produce meteors are called *meteoroids.* They are essentially very small asteroids.

While most meteors are produced by tiny meteoroids, occasionally a much larger rock strikes the earth. Under such conditions, the larger meteoroid can survive the impact with the earth's atmosphere without being completely vaporized. The remaining material simply falls to the ground. The resulting fragments are called *meteorites.*

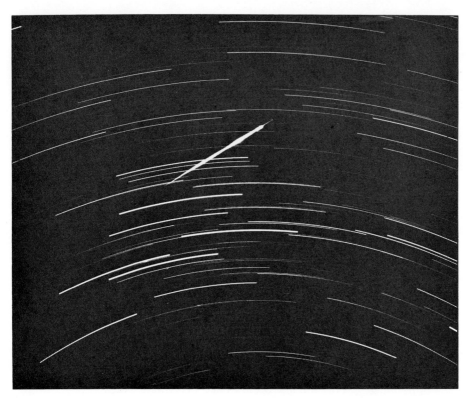

**Figure 3–29.** *A Meteor.* Meteors are produced by tiny rocks striking the earth's atmosphere at very high speeds. The rock is vaporized in a fraction of a second producing a brief flash of light. The diurnal motion of the background stars near the north celestial pole appear as blurred arcs of circles in this time exposure. (*Ronald A. Oriti, Griffith Observatory*)

**Figure 3–30.** *A Meteorite.* Occasionally a large meteoroid can survive its flight through the earth's atmosphere without being completely vaporized. This particular meteorite was found in Arizona and consists of nickel and iron. (*Collection of Ronald A. Oriti, photograph by Paul Roques, Griffith Observatory*)

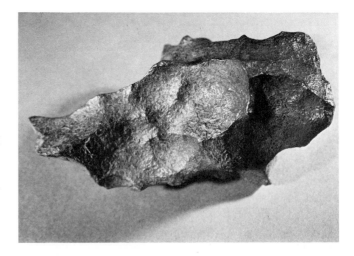

If you know what to look for and where to search, it is possible to find meteorites. Figure 3–30 shows a typical example. Most meteorites are made of stone. These are the most difficult to find because they are easily mistaken for ordinary rocks. Other meteorites contain large quantities of nickel and iron. These are much easier to find because they can be picked up with magnets. The meteorite shown in Figure 3–30 is composed almost entirely of iron and nickel.

Meteorites are important to scientists because they consist of matter from outer space. Aside from the moon rocks brought back by the Apollo astronauts, meteorites are the only samples of extraterrestrial material available for study. Some meteorites are probably as old as the solar system itself. Earth rocks, on the other hand, are considerably younger and are constantly being altered by wind, rain, and erosion. Analysis of meteoritic material can therefore play an important role in obtaining information about the very earliest history of the solar system.

In these few pages we have surveyed the major constituents of the solar system. Comparatively little was known about the planets until a very few years ago. Then, with the advent of interplanetary space flight, vast quantities of information were suddenly available. Mankind was able to see and examine the planets with a clarity and detail never before possible. Most of the rest of this text is devoted to these new discoveries.

# Review Questions and Exercises

1. Briefly discuss the distribution of mass in the solar system.
2. How many planets are there? List them in order from the sun.
3. What is meant by apparent magnitude?
4. List the planets in order of decreasing size, with the largest first and the smallest last.
5. List the planets in order of decreasing mass.
6. What is meant by density?
7. What is meant by surface gravity?
8. List the planets in order of decreasing surface gravity. On which planets would you weigh most nearly the same as you do on earth?
9. What is meant by a terrestrial planet? What is meant by a Jovian planet?
10. Contrast and compare the Jovian and terrestrial planets.
11. Why is Mercury a difficult planet to observe from earth?
12. What is meant by albedo?
13. Which planet has the highest albedo? Which has the lowest?
14. What surface details can earth-based astronomers see on Mercury with their telescopes?
15. What surface details can earth-based astronomers see on Venus with their telescopes?

16. Name one way in which the earth's surface is very different from the surface of any other planet.
17. Why is the earth sometimes called a double planet?
18. What surface details can earth-based astronomers see on Mars with their telescopes?
19. Which planet has seasons like the earth?
20. Why do scientists prefer to use the Kelvin temperature scale?
21. What is meant by absolute zero?
22. Discuss how the average molecular speed in a gas depends on temperature.
23. Discuss how the average molecular speed in a gas depends on the masses of the molecules of the gas.
24. Discuss how the escape velocity of a planet can be used to determine whether or not that planet can retain an atmosphere.
25. In what way does Jupiter's rapid rate of rotation affect the appearance of the planet?
26. Which planet has the largest number of satellites? Which planets have none?
27. Which planets have rings and what are the rings probably made of?
28. How are Uranus and Neptune similar? How are they different?
29. What is meant by obliquity?
30. Why is it that very little is known about Pluto?
31. Contrast and compare the eccentricity and inclination of Pluto's orbit with the orbits of the other planets.
32. Briefly discuss the typical sizes of asteroids.
33. What is the asteroid belt?
34. What is a comet?
35. What is the difference between a meteor, a meteoroid, and a meteorite?

# Light and Matter

**4**

## *The Nature of Light*  4.1

ASIDE from meteorites and a few hundred pounds of moon rocks, almost everything we know about the universe beyond the earth comes to us in the form of light. Understanding the nature and properties of light is therefore critical if we are to appreciate the wealth of information contained in the radiation from the stars and planets.

Modern understanding of light began with the pioneering work of Sir Isaac Newton. In the mid–1600s, Newton discovered that by passing a beam of white light through a glass prism, the white light was broken up into the colors of the rainbow, as shown in Figure 4–1. This demonstrates that white light is actually composed of all the colors of the rainbow combined. In addition, from the numerous experiments performed during the eighteenth century, it was realized that light exhibits many phenomena normally associated with *waves*. For example, if you shine a beam of light on the edge of a razor blade, careful examination of what happens as the light passes the sharp edge of the razor shows dark and light patterns similar to the ripples set up behind the edge of a pier in the ocean being pounded by water waves. The analogy between water waves and the behavior of light was so striking that physicists concluded that light must be a *wave phenomenon*.

**Figure 4–1.  *A Prism.***
When a beam of white light passes through a glass prism, the light is broken up into the colors of the rainbow. The resulting rainbow is called a *spectrum.*

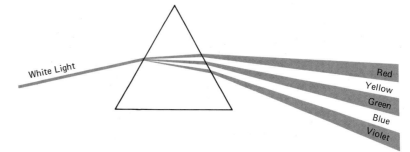

White Light

Red
Yellow
Green
Blue
Violet

**Figure 4–2.** *Wavelength.* Light is a wave phenomenon. The wavelength of a particular color of light is the distance between successive crests in the wave. This distance is usually symbolized by the Greek letter $\lambda$.

As scientists began to understand what was going on in experiments with light shining on razor blades, past thin wires, through pin holes and slits, they realized that they could talk about the *wavelength* of various colors of light. In ocean waves, the wavelength is simply the distance between successive wave crests. Similarly, the distance between successive crests in light waves is the wavelength of the light, as shown in Figure 4–2. Wavelength is usually symbolized by the Greek letter $\lambda$ (lambda). And, just as the size of wavelengths in the ocean is determined by the properties of water, so the wavelength of a particular beam of light is determined by the color of the light. While the distance between wave crests in the ocean is very large, the distance between crests in a light wave is extremely small, so small that a new "unit of measure" was invented to express these tiny distances. To see why this is necessary, we realize that it would be absurd for someone to say that he is 0.00104 miles tall; the mile is not a convenient unit of measure to express one's height. It would be far more reasonable to say that your height is $5\frac{1}{2}$ feet; the foot is a convenient unit of measure. It turns out that the wavelengths of visible light cover the range from 0.000016 inches (violet) to 0.000028 inches (red). To express the wavelength of light in inches is just as inconvenient as expressing your height in miles. To solve this problem, physicists invented a new unit of measure called the *angstrom* (abbreviated Å). One angstrom is equal to one ten millionth of a millimeter or about four billionths of an inch. In terms of the angstrom, visible light has wavelengths ranging from about 4000 Å to 7000 Å. Light of a particular wavelength reacts with the cells in the retina of the eye to give the sensation of a particular color. In general terms, colors and their corresponding wavelength ranges are given in Table 4–1.

One question which comes to mind is: What is at wavelengths shorter than

**Table 4–1**

| Color | Wavelength Range (Å) |
| --- | --- |
| Violet | 3,900–4,400 |
| Blue | 4,400–5,000 |
| Green | 5,000–5,600 |
| Yellow | 5,600–5,900 |
| Orange | 5,900–6,400 |
| Red | 6,400–7,400 |

3900 Å, and what is at wavelengths longer than 7400 Å? Another major break-through in understanding the physics of light was necessary to answer this obvious yet profound question.

During the early 1800s, many scientists focused their attention on the phenomena of electricity and magnetism. Prior to the work of physicists such as Michael Faraday and Hans Christian Oersted, it was believed that these two phenomena in nature were entirely separate and unrelated. In a number of important experiments, however, Faraday and Oersted proved that electricity and magnetism are intimately associated with each other. For example, an electric current running through a wire creates a magnetic field around the wire. Conversely, if a wire loop is moved through a magnetic field, an electric current begins flowing through the wire.

These discoveries prompted many physicists to begin experimenting and theorizing about the nature of electricity and magnetism. Their investigations culminated with the brilliant work of James Clerk Maxwell in the mid-1800s. Maxwell, a Scottish physicist, discovered that all of the properties of electricity and magnetism could be stated in four simple equations called *Maxwell's electromagnetic field equations*. These are among the most important equations in physics today. Everything known about electricity and magnetism can be understood from the viewpoint of Maxwell's pioneering work.

Shortly after Maxwell discovered his four equations, it was realized that these equations could be combined to give a new set of equations that described waves of energy traveling at 300,000 kilometers per second (186,000 miles per second). These new equations actually describe all of the wave properties of light. Indeed, Maxwell's brilliant analysis of electricity and magnetism gave mankind a profound insight into the true nature of light.

One interesting feature of the Maxwellian description of light is that it involves *no* restrictions on wavelength. It was therefore apparent that electromagnetic radiation should be possible at *all* wavelengths, from thousands of miles down to fractions of an angstrom, not just in the range from 4000 Å to 7000 Å. The reason why such radiation cannot be seen is that the human eye does not respond to anything but visible light. Physicists therefore set about the business of trying to detect and discover electromagnetic radiation at all wavelengths, radiation that might be called "invisible" light.

In a series of experiments during the last half of the nineteenth century, many of the exotic types of radiation were discovered. For example, at wavelengths longer than red light, just beyond 7000 Å, is a form of invisible light called *infrared radiation*. Hot objects such as a kitchen stove emit large amounts of infrared radiation. While the human eye cannot see this form of light, human skin does respond to infrared radiation, giving rise to the sensation of warmth. Infrared radiation covers the range from about 7000 Å to 100,000 Å. Astronomers who observe infrared light from stars and planets usually prefer to express wavelengths of infrared radiation in *microns* (abbreviated $\mu$) rather than angstroms. One micron equals ten thousand angstroms. Thus, the range of infrared radiation extends from about 0.7 $\mu$ ( $= 7000$ Å) to 10 $\mu$ ( $= 100,000$ Å).

At wavelengths shorter than those of violet light, just beyond 4000 Å, another form of invisible light called *ultraviolet radiation* exists. While the human eye does not respond to this type of light, ultraviolet radiation can be very destructive. If you look at a source of ultraviolet radiation, although nothing can be seen or felt you will rapidly go blind. This radiation destroys the cells in the retinae of your eyes. Ultraviolet radiation is used to sterilize instruments in hospitals. Lying on the beach during the summer, the ultraviolet radiation from the sun causes chemical reactions to occur in exposed skin, producing a sunburn. Ultraviolet radiation covers the range from about 4000 Å down to 100 Å.

In the 1880s, Hertz discovered very long wavelength radiation called radio waves. In essence, there is no difference between a common flashlight and a radio transmitter. The former is a source of visible light waves and the latter is a source of invisible radio waves. Just as the human eye detects visible light, radios and television sets detect radio light (of course, additional machinery in radios and TVs convert the invisible radio waves into audible sounds and pictures).

Finally, around the turn of the century, the work of Roentgen and Curie led to the discovery of X rays (100 Å down to 1 Å) and γ rays (wavelengths shorter than 1 Å). These forms of invisible light are unique in that they easily pass through matter. When a doctor "takes an X ray," he is literally shining a beam of light through the body. Denser parts of the body, such as bones and teeth, show up as shadows in photographs.

All these forms of light, ranging from the very short wavelength γ rays to the longest wavelength radio waves, together make up the *electromagnetic spectrum*. The only real difference between all these types of radiation is their wavelength. They all behave according to the same physical laws; they all are described by Maxwell's equations.

A drawing of the electromagnetic spectrum, as shown in Figure 4–3, shows that visible light constitutes only a very small fraction of the entire spectrum. An overwhelming percentage of all types of electromagnetic radiation is totally invisible to the human eye.

We see objects in the sky because they emit light. Stars and planets are visible to the human eye because much of their radiation lies between 4000 Å and 7000 Å. Yet, surely there could be unusual stars or galaxies which emit primarily radio waves or X rays. Such objects might be totally invisible to the human eye. We would be in a good position to know what to look for and expect if we understood exactly how and why matter radiates light. Such an understanding begins with a simple laboratory experiment.

Imagine that you have an object, such as a bar of iron, in your laboratory. Suppose this bar of iron is supplied with energy, perhaps with the aid of a blowtorch. As the bar of iron is heated, the temperature of the iron begins to rise; the iron atoms begin to vibrate faster and faster. Soon the bar of iron begins to glow with a dull red color. As more energy is applied in the form of heat, the color and brightness of the light from the iron begin to change. As the temperature rises, the dull red becomes a bright red, then the bright red turns into a blinding yellowish-white, sometimes called "white hot." If the bar of iron could be prevented from

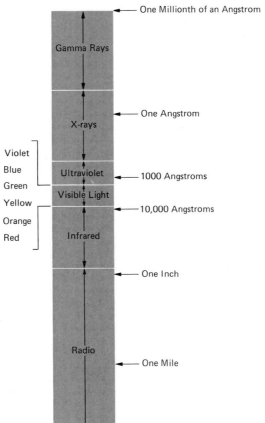

One Millionth of an Angstrom

Gamma Rays

One Angstrom

X-rays

Violet
Blue
Green

Ultraviolet — 1000 Angstroms

Visible Light

Yellow
Orange
Red

— 10,000 Angstroms

Infrared

One Inch

Radio

One Mile

**Figure 4–3.** *The Electromagnetic Spectrum.* The electromagnetic spectrum is the complete array of all types of electromagnetic radiation. Note that visible light constitutes only a very small fraction of the entire range.

melting and vaorizing as it approached temperatures of tens of thousands of degrees, the bar would actually glow with an incredibly brilliant bluish color.

This simple experiment, which is almost part of everyone's common experience, embodies some of the most fundamental radiation laws in physics. Such laws reveal how the intensity and color of light emitted by a hot object vary with temperature. As might be expected from this kind of experiment, the chemical composition of the heated object has some effect on the precise nature of the light emitted. A brass bar or an aluminum bar would give slightly different results than an iron bar. Nevertheless, the changes in color and brightness would be nearly the same in all cases.

In order to describe these changes in color and brightness, physicists prefer to imagine that they are dealing with an *ideal* object rather than bars made of iron or brass. With an ideal object, the details of the structure of the atoms out of which the object is composed would not play a role in the results we obtain. The imaginary ideal object invented for this purpose is called a *blackbody*. A blackbody is an object which absorbs *all* the electromagnetic radiation which falls on it. This is why it is called "black"; all light is absorbed and none is reflected. A blackbody is also a "perfect radiator" in that it emits radiation that depends *only* on its temperature; there is no dependence on the chemical properties of this ideal radiator. The radiation that is given off is called *blackbody radiation*. Radiation emitted by objects in the real world (bars of iron or the atmospheres of stars) differs from blackbody radiation only because of the particular chemical and physical properties.

During the second half of the nineteenth century, physicists spent a considerable amount of time trying to study blackbodies experimentally. In their experiments they made blackbodies out of carbon or covered objects with soot and found the same kind of relationship between temperature and radiation as was described in the observations of a hot bar of iron. For example, the German physicist Wilhelm Wien discovered a way of expressing the manner in which the colors of radiation emitted by hot objects depend on temperature.

Every object that is above absolute zero ($0°K$) emits some type of radiation. Cool objects emit very long wavelength radiation. For example, the human body at a temperature of about $300°K$ is a source of infrared radiation. Very hot objects with temperatures in millions of degrees emit very short wavelength radiation, such as X rays. Only if the temperature of an object ranges from about $3000°K$ to $10,000°K$ will the object emit primarily visible light.

The radiation emitted by objects above absolute zero covers a range of wavelengths. However, there is a particular wavelength at which *most* of the radiation is emitted. This particular wavelength, called $\lambda_{max}$, depends only on the temperature of the blackbody. As suggested by the experiment with an iron bar, the wavelength at which most of the radiation is emitted depends inversely on the temperature. The higher the temperature, the shorter the wavelength. The lower the temperature, the longer the wavelength. This relationship between the temperature (T) of an object and the wavelength ($\lambda_{max}$) at which most of the radiation is emitted is

**Table 4–2**

| Temperature (°K) of Blackbody | Wavelength ($\lambda_{max}$) at Which Most Radiation Is Emitted | Type of Radiation |
|---|---|---|
| 3° | 1,000 $\mu$ | Radio Waves |
| 300° | 10 $\mu$ | "Far" Infrared |
| 3,000° | 1 $\mu$ | "Near" Infrared |
| 4,000° | 7,500 Å | Red Light |
| 6,000° | 5,000 Å | Yellow Light |
| 8,000° | 3,750 Å | Violet Light |
| 10,000° | 3,000 Å | "Near" Ultraviolet |
| 30,000° | 1,000 Å | "Far" Ultraviolet |
| 300,000° | 100 Å | "Soft" X Rays |
| 1½ Million Degrees | 20 Å | "Hard" X Rays |
| 3 Billion Degrees | $\frac{1}{100}$ Å | Gamma Rays |

called *Wien's law*. Table 4–2 gives $\lambda_{max}$ for blackbodies at various temperatures. It should be reemphasized that a hot object emits a *range* of wavelengths; Wien's law tells us where *most* of the radiation is emitted. Thus, a bar of iron at 1000°K gives off a dull red glow; some visible red light is emitted. However, according to Wien's law, most of the radiation from this object is in the invisible infrared range at about 10,000 Å.

Physicists thus understood why a bar of iron appears to change color (red ⟶ orange ⟶ yellow ⟶ blue) as it is heated to higher and higher temperatures. There is, however, a second equally important effect. As the temperature of the bar of iron increased, the brightness of light given off also increased. Objects at low temperatures emit only small amounts of radiation, while objects at high temperatures emit huge amounts of radiation. This relationship between the temperature and brightness of radiation from an ideal blackbody was first formulated by Josef Stefan in 1879. The resulting *Stefan's law* simply states that the amount of energy emitted by a blackbody increases as the "fourth power" of the temperature. Thus, according to Stefan's law, if you double the temperature of an object (for example, go from 2000° to 4000°K) the amount of energy given off in the form of radiation will go up by a factor of 16, because $2 \times 2 \times 2 \times 2 = 16$. If you triple the temperature of a blackbody (for example, go from 3000° to 9000°K), the blackbody will give off 81 times more energy because $3 \times 3 \times 3 \times 3 = 81$.

From both Wien's law and Stefan's law we now can appreciate why cool objects emit small amounts of reddish light while hot objects emit large amounts of bluish light. Stefan's law tells us about the total amount of energy emitted, while Wien's law tells us at what wavelength most of this energy is concentrated. However, neither of these laws tells us precisely how the radiation from a blackbody is distributed over *all* wavelengths. For example, according to Wien's law, a blackbody at 6000°K emits most of its radiation in the form of yellow light at 5000 Å. From Stefan's law, the total amount of energy at *all* wavelengths can be calculated; physicists can calculate how many "watts" are given off by the blackbody. But how much energy is given off by this blackbody in blue light? What is the intensity of the radiation at 4000 Å? Or at 8000 Å? Questions of this type are unanswered by either law. Indeed, by the end of the nineteenth century, it was apparent that from the "classical" understanding of light based on Maxwell's equations, it was *impossible* to answer such questions. A crisis had developed in physics. Classical theory gave absurd answers. The nature of light was once again a mystery.

In 1900, the brilliant German physicist Max Planck resolved this dilemma by proposing that light is *quantized*. According to Planck, light can exist only in discrete packets called *photons*, whose energy depends on the wavelength of the light. The shorter the wavelength of a photon, the higher the energy contained in the photon. An X-ray photon carries a lot more energy than a radio-wave photon. Up until the turn of the century, physicists were entirely content to think of light as waves. No one had ever performed any experiments to look for the discrete or quantized nature of light. Yet, the idea that a beam of light actually consists of a stream of very tiny photons, each carrying a specific quantity of energy, proved to be very

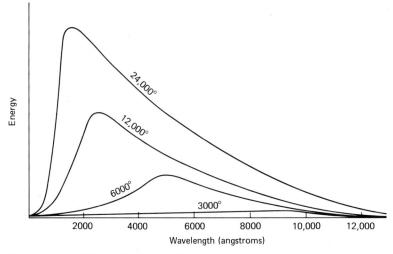

**Figure 4-4.** *Blackbody Curves.* The curves on this graph are drawn according to Planck's radiation law. The precise amount of energy emitted by an ideal blackbody at a specific temperature and at a specific wavelength is given by these curves.

powerful. Planck was then able to prove both Stefan's and Wien's law as well as to calculate precisely how much energy is given off at each and every wavelength by a blackbody at a specific temperature. The final result of Planck's work is the so-called *Planck blackbody radiation law.*

Perhaps the best way to illustrate the Planck blackbody radiation law is with the aid of a graph, as shown in Figure 4–4. This graph simply shows how much energy is emitted at various wavelengths by blackbodies at various temperatures. Energy is plotted vertically and wavelength is plotted horizontally, and the effects of both Wien's and Stefan's laws are immediately apparent. At low temperatures, the curves given by Planck's law "peak" at long wavelengths. At higher temperatures, the peak in the curve (at $\lambda_{max}$) moves to shorter wavelengths. In addition, the curves for high temperatures are much higher than the curves for lower temperatures. This is in accordance with Stefan's law, which says that the hotter an object is, the more energy it emits. But, more importantly, Planck's law tells precisely how much energy is emitted at every wavelength by a blackbody at a particular temperature.

Of course, stars and planets are not ideal blackbodies. Nevertheless, radiation emitted by real objects in the sky is usually not far from the Planck curves in Figure 4–4. Planck's law is therefore often a good approximation of what goes on in the real universe.

About a hundred years ago, scientists realized that visible light was only a very small portion of the entire electromagnetic spectrum. Simultaneously, astronomers realized that they were missing a lot of information. When we look through a telescope, we are seeing only the visible light from planets and stars. Man's understanding of the universe would be dramatically affected if somehow we could "see" what the heavens look like in X rays, radio waves, and other wavelengths far removed from those of visible light.

A hundred years ago, scientists simply did not have the equipment and machines to detect nonvisible light. In addition, after some considerable experimentation, physicists found that there was another complication that would further hamper the efforts of astronomers. The earth's atmosphere is *opaque* to most of the electromagnetic spectrum. Visible light easily gets through the air we breath. Air is also transparent to certain radio waves. But virtually no other types of radiation can get through the 10-mile-high layer of air above our heads. For example, if the human eye could see only X rays, the sky would look completely black. The X rays from stars and galaxies cannot get through all that air. Of course, this is a very fortunate situation. By being opaque to most of the electromagnetic spectrum, our atmosphere protects us from many of the deadly radiations. For example, the sun emits great quantities of ultraviolet radiation. But a layer of ozone ($O_3$) gas high in the earth's atmosphere absorbs almost all of this type of light. If this ozone layer suddenly disappeared, life on the earth would be destroyed in less than 24 hours. The earth's surface would literally be sterilized by the sun's ultraviolet rays. Figure 4–5 shows what portions of the electromagnetic spectrum can penetrate the earth's atmosphere.

The earth's atmosphere is a protective blanket shielding all life from deadly radiations. Simultaneously, however, this protective blanket has kept astronomers in ignorance. From the earth's surface astronomers cannot see what the universe looks like in X rays, γ rays, infrared and ultraviolet radiation. But within the past few years the space program has given mankind the opportunity to place scientific equipment in orbit, many miles above the earth's atmosphere. Telescopes have been built to detect and record exotic radiations from planets, stars, and galaxies. Mankind has literally been given a new set of eyes with which to see his universe. It is reasonable to suppose that almost every idea in astronomy will be profoundly affected by this new wealth of information. For this reason, modern astronomy is one of the most exciting and rapidly changing fields of human knowledge.

**Figure 4–5.** *Transparency of the Earth's Atmosphere.* Only a small portion of the electromagnetic spectrum can penetrate the earth's atmosphere. Visible light and radio waves can get through (shown as light regions). The air is opaque to most other wavelengths (shown as dark regions).

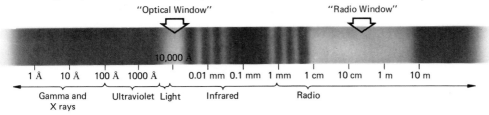

# 4.2 Telescopes

ALMOST everything which interests the astronomer is very far away. Huge planets, massive stars, and immense galaxies are so distant that they appear as faint pinpoints of light in the nighttime sky. It is therefore not surprising that the *telescope* is the astronomer's most important tool.

The central idea behind a telescope is to take incoming starlight and focus it into a big, bright, sharp image. This can be done in one of two ways. An ordinary convex lens (like a "magnifying glass") can be used to focus light rays, as shown in Figure 4–6. The light rays converge at the *focal point* and the distance between the lens and the focal point is called the *focal length*.

Alternatively, incoming starlight can be focused using a mirror rather than a lens. As shown in Figure 4–7, a concave mirror focuses incoming light at a *focal point* and the distance between the mirror and the focal point is called the *focal length*. Either of these two methods of focusing light can be used as the fundamental principle in the design of a telescope.

There are essentially two types of telescopes: those which are based entirely on the use of lenses and those which are based on the use of mirrors. Galileo's telescope, as well as many of the major telescopes built before the twentieth century, were of the first type. Such telescopes are called *refractors* because, by means of lenses, light is bent or "refracted." The refractor consists of a large *objective lens* mounted, for

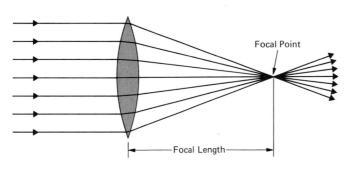

**Figure 4–6.** *A Convex Lens.* A convex lens focuses incoming starlight to a *focal point*. The distance between the lens and the focal point is called the *focal length*.

**Figure 4–7.** *A Concave Mirror.* A concave mirror focuses incoming starlight to a *focal point*. The distance between the mirror and the focal point is called the *focal length*.

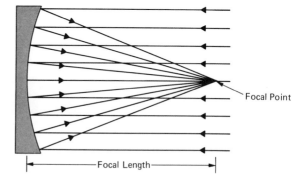

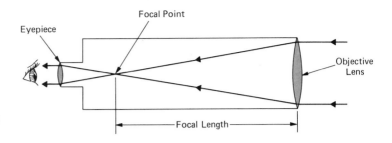

**Figure 4-8. A Refracting Telescope.** The image formed at the focal point of the objective lens can be examined and magnified with the aid of an "eyepiece." The resulting optical arrangement is called a refracting telescope.

convenience, at one end of a telescope tube. This convex objective acts like a giant magnifying lens and focuses the incoming starlight to the focal point. At the other end of the telescope, the astronomer places an eyepiece with which to examine the image formed around the focal point. The complete optical design of a refracting telescope is shown in Figure 4–8.

When the astronomer refers to the "size" of a refracting telescope, he is usually speaking of the diameter of the objective lens. The largest refractor in the world is the 40-inch refractor at Yerkes Observatory in Wisconsin, while the second largest is the 36-inch refractor at the Lick Observatory on Mount Hamilton in California.

As with so many topics and inventions in astronomy, the design of the *reflecting telescope* originated with the work of Sir Isaac Newton. It was Newton who discovered that a concave mirror could be used to focus light rays, as shown in Figure 4–7. A reflecting telescope is built around a *primary mirror* rather than an objective lens. Such a mirror is usually made out of a slab of glass, which is ground until it is concave and then coated with a thin layer of aluminum or silver. Obviously, however, if the astronomer tried to observe the image formed at the focal point by placing his eye or an eyepiece at the focal point, his head would block out the incoming starlight and he would just see the reflection of his face. To circumvent this difficulty, a second mirror called a *diagonal mirror* is placed in front

**Figure 4-9. The World's Largest Refractor.** The largest refractor in the world is located at Yerkes Observatory. The objective lens is 40 inches in diameter and has a focal length of 63 feet. (*Yerkes Observatory photograph*)

**99**

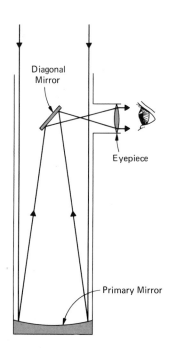

Diagonal
Mirror

Eyepiece

Primary Mirror

**Figure 4–10.** *A Newtonian Reflector.* This type of a reflecting telescope makes use of a *diagonal mirror,* which reflects the converging light rays to one side so that the astronomer can better view the image with the aid of an eyepiece.

of the focal point and reflects the converging light rays to one side, as shown in Figure 4–10. This allows the eyepiece to be mounted on the side of the telescope for convenient viewing. This type of a telescope is called a *Newtonian reflector.*

When the astronomer refers to the "size" of a reflecting telescope, he is usually speaking of the diameter of the primary mirror. One of the largest reflecting telescopes in the world is located on Palomar Mountain in California. The mirror in the Palomar telescope contains $14\frac{1}{2}$ tons of glass and has a diameter of 200 inches.

The 200-inch Palomar telescope has been in operation since the late 1940s. Until that time, the 100-inch reflector on Mt. Wilson near Los Angeles had been the world's largest telescope. Other major reflectors include the 120-inch at Lick Observatory near San Francisco and the 107-inch at McDonald Observatory in Texas. Construction has been completed on two 150-inch reflectors, one of which is at the Kitt Peak National Observatory in Arizona, while the other is located at the Cerro Tololo Inter-American Observatory in Chile.

There are several major advantages of reflectors over refractors. For example, the glass used in the refractor's objective lens must be of the highest quality, but this is not required of glass used in a reflector. Light does not pass through the glass used in a reflector's mirror and, therefore, bubbles and other defects inside the glass cannot affect the quality of the image. In addition, a mirror reflects *all* colors of light in exactly the same way. Lenses made of glass absorb "near ultraviolet" light in the wavelength range from 3000 Å to 4000 Å. A refractor's objective lens is therefore opaque to this radiation. The astronomer must use a reflector if he wishes to make observations at wavelengths shorter than 4000 Å.

**Figure 4-11. *Palomar Observatory.*** This dome on Palomar Mountain houses one of the world's largest reflecting telescopes. The diameter of the primary mirror is 200 inches. (*Hale Observatories*)

**Figure 4-12. *Different Types of Reflectors.*** Four popular types of reflectors are shown. The four optical arrangements are (a) prime focus, (b) Newtonian, (c) Cassegrain, and (d) Coudé focus.

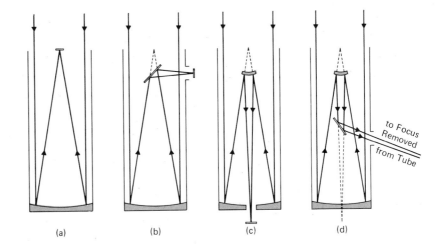

(a)  (b)  (c)  (d)

to Focus
Removed
from Tube

One difficulty commonly encountered with Newtonian reflectors is that in practice, the focal point is often high off the ground. This is especially true if the mirror has a very long focal length. Observing with a Newtonian reflector is frequently cumbersome and inconvenient. The astronomer copes with this difficulty by using mirrors to reflect the light rays to a more accessible location, thereby modifying the design of the telescope.

One of the most popular modifications results in the so-called Cassegrain reflector. In the Cassegrain system, the diagonal mirror is replaced with a convex mirror, which reflects the converging light rays back toward the primary mirror through which a small hole has been drilled. The light rays pass through this hole and come to focus just behind the primary mirror. This is often a far more convenient location from which to do observations.

Another popular modification frequently employed with very large telescopes involves the use of the so-called Coudé focus. Such a system starts off like a Cassegrain telescope. Light goes from the concave primary to a convex secondary mirror. As the light rays head back toward the primary mirror, the beam is intercepted by a flat, diagonal mirror which sends the light to some distant location in the observatory usually called the *Coudé room*. In the Coudé room, the astronomer can analyse the starlight with complex equipment that is too large or heavy to attach to the side or back of the telescope.

A final alternative arrangement of the reflecting telescope that will be discussed here is perhaps the simplest of all. In the case of very large telescopes such as the Palomar telescope, the mirror is so big that it is possible to place observing apparatus directly at the primary focus without blocking out very much of the incoming light. Such an arrangement is called the *prime focus*. The 200-inch telescope has an *observing cage* at the prime focus in which the astronomer rides while making his observations. Some of the most important work in astronomy today is done by astronomers sitting all night long in the observer's cage of the 200-inch telescope.

Telescopes do three things: they magnify, they resolve, and they gather light. Magnification is the property most commonly attributed to the telescope. It is well known that when looking through a telescope, distant objects appear larger. This, of course, is the primary advantage of the telescope. The *magnifying power* of a telescope tells how many times larger a telescopic image is compared to the naked-eye view. For example, to the naked eye, the diameter of the moon is $\frac{1}{2}°$. If a telescope gives an image of the moon $20°$ in diameter, we say that the magnifying power of the telescope is 40, written as "$40\times$," because the moon appears 40 times larger through the telescope. Thus, if a telescope has a power of $300\times$, images seen through this telescope will appear 300 times larger than to the naked eye.

Sometimes commercially available telescopes sold in department stores are advertized according to their "power." Often this constitutes a clear case of misleading advertising. The magnifying power of a telescope can be changed simply by using different eyepieces. The more powerful the eyepiece, the higher the magnification of the telescope. It is possible to take a very poorly constructed telescope, equip it with a powerful eyepiece, and arrive at a very high magnifying

**Figure 4–13.** *The 200-inch Telescope.* This view looks "down" the 200-inch telescope toward the primary mirror. An astronomer rides in the observer's cage at the prime focus. (*Hale Observatories*)

power. Unfortunately, the customer who buys such a telescope is often severely disappointed. If he looks at Jupiter through his "450× telescope for under $75," he will see a large blob instead of a clear image of the planet. While Jupiter may have been magnified, so have all the defects and shortcomings of the telescope. While a high magnification is frequently desirable, there are numerous other qualities which must be considered in a well-designed telescope.

In addition to magnifying, telescopes also have the ability to *resolve*. The *resolving power* of a telescope tells about the sharpness and clarity of the telescopic image. A telescope with a low resolving power gives fuzzy and indistinct images, while one with a high resolving power gives sharp, clear images. For example, through a small inexpensive telescope with a low resolving power, Castor, the second brightest star

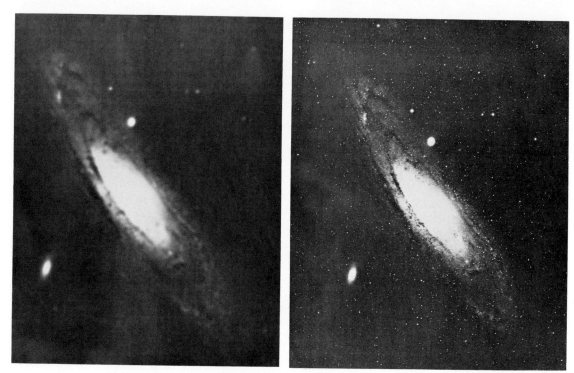

**Figure 4-14.** *A Demonstration of Resolving Power.* These two views of the same galaxy illustrate the effects of resolving power. The view on the left corresponds to a low resolving power. With a high resolving power, as shown on the right, sharp images are obtained. (*Courtesy of Dr. O'Dell*)

in Gemini, looks like any ordinary star, much the same as it looks to the naked eye. However, a larger telescope with a higher resolving power reveals Castor to be a *double star,* two stars very close together.

The resolving power of a telescope is expressed as the smallest angle between two stars for which separate, recognizable images are produced. The smaller the angle, the better the resolution. For example, a typical good amateur telescope has a resolving power of one second of arc (1″). Through this telescope, stars separated by more than 1″ produce clear, distinct images, while the images of stars separated by less than 1″ are blurred together.

The resolving power of an individual telescope depends to a large degree on the quality and precision of the telescope's optics. However, because of the wave nature of light, there is a *limiting resolving power* for all telescopes. No telescope can ever give an infinitely sharp image. Instead, there is a limiting resolving power and no telescope can do better than this limit, no matter how good the optics. This limiting resolving power increases with the diameter of the primary mirror or lens of the telescope. The 200-inch telescope on Palomar Mountain has twice the resolving power of the 100-inch telescope at Mt. Wilson. Theoretically, the Palomar telescope should be able to resolve stars separated by only $\frac{1}{50}$ seconds of arc. In practice, such things as weather conditions and the stability of the atmosphere prevent the astronomer from ever achieving the theoretical limit.

Finally, telescopes gather light. The primary lens or mirror of a telescope takes all the light covering an area of several square inches or square feet and focuses this light into a small, bright image. It is this focusing (or gathering of light) that results in an increased brightness of the resulting image. Stars that are too faint to be seen with the naked eye easily show up through a telescope. As shown schematically in Figure 4–15, the bigger the primary lens or mirror of a telescope, the more light it gathers. The area of the 200-inch mirror at Mt. Palomar is four times the area of the 100-inch mirror at Mt. Wilson. The Palomar telescope, therefore, gathers four times more light and its resulting images are four times brighter than the Mt. Wilson telescope. Improved resolution and increased light gathering are the two primary reasons why astronomers build larger and larger telescopes.

Up until only a few decades ago, everything mankind knew about the cosmos was based entirely on observations with visible light. The visible light to which the human eye responds was the only source of information. Yet, during the nineteenth century it became clear that visible light was only a very small fraction of the entire electromagnetic spectrum. There must be many forms of "invisible" light with wavelengths both shorter and longer than those of the visible light. Since the astronomer's telescopes were designed for use with visible light, it became pain-

**Figure 4–15.** *Light-Gathering Power.* The light-gathering power of a telescope depends on the size of the telescope's primary lens or mirror. Bigger telescopes intercept more starlight and therefore produce brighter images.

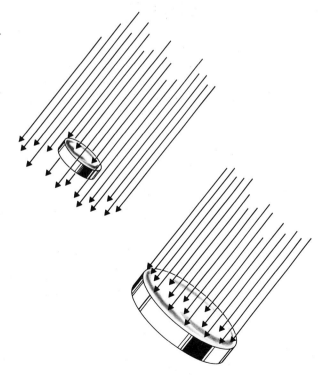

fully obvious that he was observing only a small fraction of all possible types of radiation from celestial objects.

In the early 1930s, Karl Jansky, a young engineer at Bell Telephone Laboratories was experimenting with very long radio antennas when he noticed that he was receiving static from astronomical sources beyond the earth. The importance of this discovery cannot be overemphasized. Up until 1931, astronomers only knew about visible light coming from celestial objects. Yet, now radio light had also been detected from cosmic sources. Astronomers then realized that if they could figure out how to build special antennas, which might be called *radio telescopes,* they could pin point these cosmic sources of static. Building such telescopes would be like giving new eyes to a blind person. After all, mankind had literally been blind to radio waves from outer space. With Jansky's discovery, astronomers had the hope of being able to "see" the radio universe just as clearly as they could see the visible universe.

The first antenna designed to pick up cosmic radio noise was built by Grote Reber in 1936. However, because of the Second World War, real progress was delayed until the late 1940s. At the end of the war, astronomers realized that many of the recent advances in electronics and electrical engineering could be directly applied to the problem of building radio telescopes. Radio telescopes began to spring up in England, the Netherlands, Australia, and later in the United States.

The basic idea behind the design of a radio telescope is very simple. Radio telescopes are reflectors. A radio telescope consists of a large concave "dish" that concentrates radio waves at a focus. This dish, which reflects the radio waves to the focal point, can be made out of metal or a fine wire mesh. At the focal point there is a radio receiver that converts the radio waves into an electric current. This current is carried through wires to amplifiers and electronic recording devices. If the radio telescope is pointed toward a source of radio noise, a strong "signal" is detected informing the *radio astronomer* that he is "looking" at a bright spot in the *radio sky*. If the radio telescope is not aimed at a source of radio waves, no static comes through the receiver.

Radio waves comprise a very large part of the electromagnetic spectrum. Radio waves have wavelengths extending from about $\frac{1}{10}$ inch to thousands of miles. The earth's atmosphere is, however, transparent only to wavelengths between $\frac{1}{10}$ inch to about 100 feet. It is in this range that the radio astronomer makes his observations. Just as you can "tune" your car radio to specific frequencies or wavelengths, the radio astronomer can select the precise wavelength at which he wishes to make observations. The resulting views of the radio sky often look very different at various wavelengths.

Radio telescopes are a lot bigger than their optical counterparts. While the largest optical reflector has a mirror 200 inches in diameter, the world's largest radio telescope has a dish 1000 feet in diameter. There are several reasons for this. First of all, it is much easier to build a large radio telescope than a large optical telescope. Due to the long wavelength of radio radiation, minor defects in the precise shape of the concave dish are unimportant. Secondly, radio astronomers *must* build large radio telescopes. The energy carried by a photon of radio light is much less than the energy carried by a photon of visible light. The dish must, therefore, be large so that enough radio energy is collected to produce a detectable signal. In addition, in order to obtain good resolution, the radio astronomer has no choice but to make his observations with a large dish. If a radio astronomer tried to make observations with a small telescope, he would find it impossible to pinpoint the locations of radio sources in the sky. Because the wavelength of radio waves is so long, the diameters of radio telescopes must be large to give a clear view of the radio sky. All the major radio telescopes in the world have diameters of over 100 feet. Some of the most well-known radio telescopes are at the Arecibo Observatory in Puerto Rico, the National Radio Astronomy Observatory in West Virginia, the Jodrell Bank Station and Mullard Radio Observatory in England, and the Radiophysics Laboratory in Australia.

**Figure 4–16.** *A Typical Radio Telescope.* The 140-foot radio telescope at the National Radio Astronomy Observatory in Greenbank, West Virginia, is an example of one of the finest radio telescopes in operation today. (*NRAO*)

Radio astronomers do not work under some of the restrictions that hamper the observations of ordinary optical astronomers. For example, the sun emits a lot of light but very little radio radiation. Therefore, the radio astronomer can make observations at any time, day or night. In addition, he is not hampered by weather conditions. After all, a radio or television set works just fine even when it is raining or snowing.

In observing the universe, the ground-based astronomer has two options. He can use optical telescopes and make observations of visible light, or he can use radio telescopes and make observations of radio light. These options are possible because the earth's atmosphere is transparent to visible and radio light. The astronomer therefore speaks of the "optical window" and the "radio window" in referring to the transparency of the air to these types of radiation (see Figure 4–5). Unfortunately for the astronomer, these are the only two "windows" that exist. The air we breathe is opaque to all other wavelengths in the electromagnetic spectrum. For example, water vapor in the air efficiently absorbs almost all incoming infrared radiation and a layer of ozone high in the atmosphere equally efficiently absorbs ultraviolet radiation from outer space. Trying to observe the universe at infrared, ultraviolet, and X-ray wavelengths from the earth's surface is like trying to look through a brick wall. None of these radiations can pass through our atmosphere. The only salvation is to go *above* the earth's atmosphere with airplanes, rockets, balloons, or satellites. Only at high altitudes can the astronomer make observations unhampered by the opaque air.

In between visible light and radio light in the electromagnetic spectrum is a type of light called *infrared radiation*. Infrared light has wavelengths from about 8000 Å to 1 mm. Some of the so-called "near" infrared radiation (around 10,000 Å) can pass through the earth's atmosphere. Special photographic plates sensitive to infrared light around 10,000 Å can be used to produce pictures at these wavelengths that lie just beyond the range to which the human eye responds (the human eye does not see light with wavelengths longer than about 7,500 Å). Best results with near infrared photography are obtained from observatories at high altitudes such as the new 84-inch reflector on Mauna Kea in Hawaii.

At wavelengths longer than 10,000 Å, the earth's atmosphere is opaque to virtually all infrared light and the astronomer must turn to new techniques. To make observations, the infrared astronomer places all his equipment aboard an airplane or attaches it to a balloon, which enables him to get above the water vapor in the earth's atmosphere (see Figure 4–17). Very successful observations, especially by Dr. Frank Low at the University of Arizona, are carried out in airplanes that fly at altitudes up to 50,000 feet. Such observations have discovered that many stars emit unusually large amounts of infrared light. This seems to suggest that these stars are surrounded by large quantities of dust, which might be in the process of condensing into planets. In addition, several galaxies have been discovered to be emitting exceptionally huge amounts of infrared light. The source of this light is not understood; perhaps some violent processes or explosions are occurring at the centers of these galaxies. But, as of the late 1970s, the source of infrared light from galaxies is one of the baffling mysteries in modern astronomy.

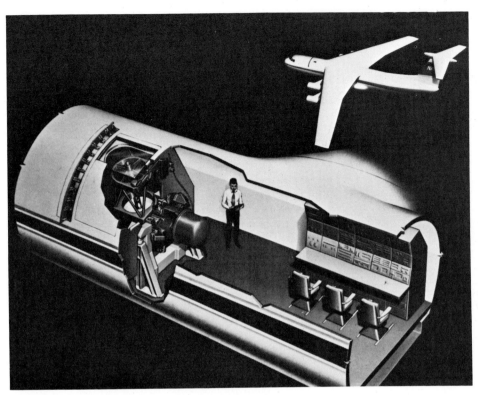

**Figure 4-17.** *The Gerard P. Kuiper Air-borne Observatory.* In order to observe the infrared sky, astronomers fly their equipment at altitudes of 50,000 feet. Only at such altitudes is the astronomer above most of the water vapor in the earth's atmosphere. (*NASA*)

While infrared radiation lies just to the long wavelength side of visible light, ultraviolet radiation is to the short wavelength side of visible light. Ultraviolet light covers the wavelength range from 4000 Å down to about 100 Å. Just as the near infrared light manages to squeeze through the long wavelength side of the optical window, the near ultraviolet (3000 Å to 4000 Å) also gets through the earth's atmosphere. Astronomers are accustomed to making observations in the near ultraviolet from ordinary optical observatories. However, to observe at wavelengths shorter than 3000 Å, it is absolutely necessary to get far above the earth's atmosphere.

Some of the most successful far ultraviolet observations of celestial objects have been performed by Dr. George Carruthers and his associates at the Naval Research Laboratory. These observations have been made from rocket flights, from Apollo missions to the moon, and from Skylab.

The kinds of photographs obtained by Dr. Carruthers' ultraviolet telescopes are incredible. For example, Figure 4–19 shows a photograph of the nearest galaxy, the

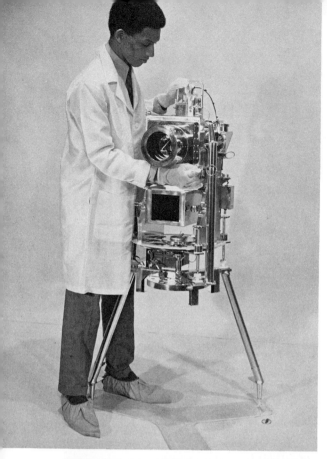

**Figure 4–18.** *An Ultraviolet Telescope.* Dr. Carruthers is shown here with the ultraviolet telescope that he designed. This telescope was taken to the moon by the Apollo 16 astronauts. (*Naval Research Laboratory*)

**Figure 4–19.** *The Large Magellanic Cloud.* The photograph on the left shows the appearance of the Large Magellanic Cloud in the ultraviolet light. An ordinary optical photograph, shown on the right, demonstrates that there are many stars in the sky that are invisible to the human eye. (*Naval Research Laboratory and Lick Observatory photograph*)

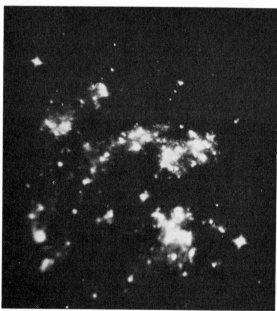

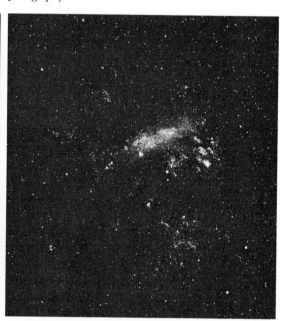

so-called Large Magellanic Cloud. This photograph was taken by Apollo 16 astronauts from the moon in April of 1972. For comparison, an excellent earth-based photograph of exactly the same region of the sky is also shown. In the earth-based photograph we see this galaxy in visible light (4000 Å to 8000 Å) the way it looks to our eyes. In the lunar-based photograph we see this same galaxy in invisible ultraviolet light (1250 Å to 1600 Å). Incredibly hot stars, which emit most of their radiation in the ultraviolet, blaze forth in this remarkable photograph brought back from the moon. Conversely, the cooler stars that prominently appear in the visible image are not seen at all in the ultraviolet view.

Another dramatic example of ultraviolet astronomy is shown in Figure 4–20. Two views of Comet Kohoutek are shown here. Both photographs were taken from rocket flights early in January of 1974 as Comet Kohoutek passed near the sun. Both photographs are shown to exactly the same scale. The view on the left is in visible light; it looks like any ordinary comet. However, the ultraviolet view reveals that the head of Comet Kohoutek is surrounded by a huge halo. The halo shown in this photograph is 4 million miles in diameter. In ultraviolet light, Comet Kohoutek is four times larger than our sun, which has a diameter of about 1 million miles.

When we look at these photographs, we are seeing what no man has ever seen before. We are seeing that which the human eye cannot see. We are seeing the invisible universe.

**Figure 4–20.** *Comet Kohoutek.* The photograph on the left shows the visual appearance of Comet Kohoutek. The photograph on the right is the ultraviolet view. (*Johns Hopkins University; Naval Research Laboratory*)

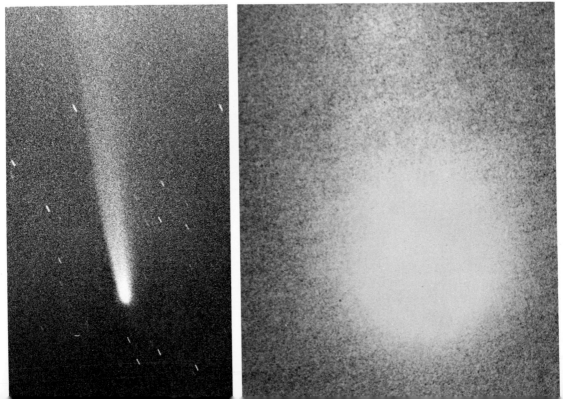

At wavelengths shorter than those of ultraviolet light lies the domain of X rays. No X rays at all get through the earth's atmosphere; all X-ray observations must be done from outer space. Fortunately for the astronomer, there are devices that easily detect X rays. Such devices are called *proportional counters,* as shown in Figure 4–21, and are very similar to ordinary Geiger counters. In essence, a proportional counter consists of a bottle of gas. When an X ray passes through the gas, electrons are knocked off the atoms in the gas and electronic equipment sends signals down to astronomers on the earth.

Unquestionably, one of the most important advances in the history of astronomy centers about the flight of *Explorer 42.* Explorer 42 is an X-ray detecting satellite. Its entire purpose is to map the X-ray sky, discover new sources, and examine their characteristics. The design and construction of the satellite was a team effort of many scientists. From Explorer 42, mankind obtained its first organized and complete view of the X-ray sky. By 1974, no less than 161 X-ray sources have been discovered. Some sources correspond to nearby stars while others are associated with distant galaxies and quasars.

Beginning with the 1970s, astronomers have for the first time been able to

**Figure 4–21.** *An X-Ray Telescope.* An X-ray telescope consists of a series of proportional counters that produce signals when exposed to X rays. This particular X-ray telescope will be used in the High Energy Astronomical Observatory (called "HEAO") scheduled to be launched in the late 1970s. (*Naval Research Laboratory*)

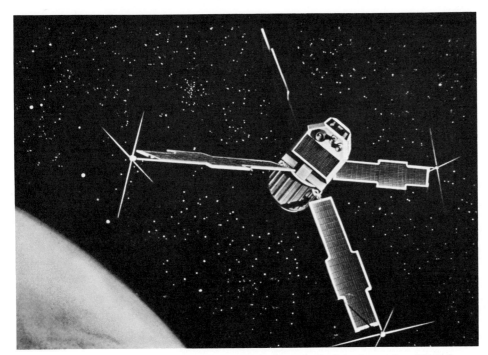

**Figure 4-22.** *Explorer 42.* This X-ray satellite was launched in December 1970 from Kenya. Almost 200 X-ray sources in the sky had been discovered during the first four years of operation. (*NASA*)

observe the heavens at virtually all wavelengths. No longer will the astronomer's work be based only on what his eyes can see. The last quarter of the twentieth century will be devoted to bringing together all the observations at various wavelengths, synthesizing and unifying the data to give mankind its first truly complete picture of the universe. Virtually every idea, every theory, and every concept of the cosmos will be profoundly affected by this new knowledge.

## Atoms and Spectroscopy   4.3

M A N's modern understanding of light began when Newton passed a beam of white light through a glass prism and found that the light was broken up into the colors of the rainbow. This rainbow of colors, ranging from violet at 4000 Å to red at 7000 Å, is called a *spectrum.* Studying the spectra of stars and galaxies is one of the most important tools the astronomer has for unlocking the secrets of the universe.

**Figure 4–23. *A Spectrograph.***
By means of lenses, incoming
light is passed through a prism
and the resulting spectrum is
focused onto a photographic
plate.

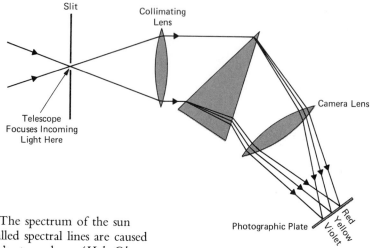

**Figure 4–24. *The Solar Spectrum.*** The spectrum of the sun
shows many dark lines. These so-called spectral lines are caused
by the chemical elements in the sun's atmosphere. (*Hale Observatories*)

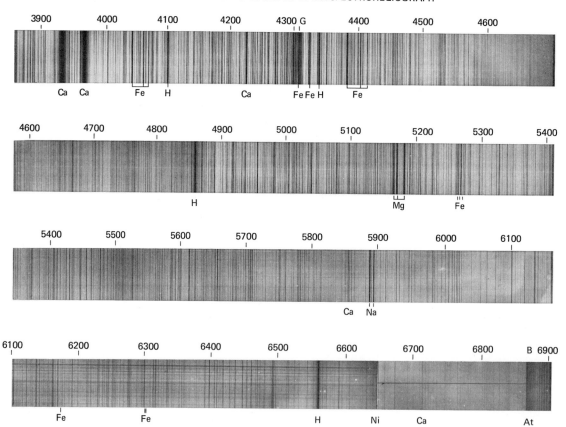

SOLAR SPECTRUM made with the 13-foot SPECTROHELIOGRAPH

In view of the importance of spectra, scientists have designed and constructed instruments to aid in the examination of spectra. A device used to observe spectra visually is called a *spectroscope,* while a device that produces a photograph of a spectrum is called a *spectrograph.*

The design of a typical spectrograph is shown in Figure 4–23. Incoming light is directed toward a prism by means of a *collimating lens.* The rainbow of colors that leave the prism are focused onto a photographic plate by a *camera lens.*

In the early 1800s, William Wollaston was observing the spectrum of the sun when he noticed that there were some thin dark lines among the colors. Several years later, this discovery was confirmed by Joseph Fraunhofer who found a total of 600 such lines. A high-quality photograph of the solar spectrum taken with a modern spectrograph is shown in Figure 4–24.

At first these so-called *spectral lines* were a complete mystery. What is the meaning of these lines? What could cause these thin dark lines? Wollaston's idea that the lines might designate the boundaries between various colors did not appear to be a satisfactory explanation.

A major breakthrough in understanding spectra occurred when physicists realized that spectral lines could be produced artificially in the laboratory. If a beam of white light is passed through a bottle of gas, the spectrum of the light passing through the gas will contain spectral lines, as shown in Figure 4–25. Further, each gas produces its own unique pattern of spectral lines. The spectrum of light passing through hydrogen gas, for example, shows a pattern of spectral lines that is unlike the patterns produced by any other gas. The hydrogen gas somehow manages to absorb light from the spectrum at specific wavelengths, producing a distinctive pattern of dark lines among the colors. Furthermore, it logically follows that if this same pattern of lines exists in the solar spectrum, there must be hydrogen in the gases that make up the atmosphere of the sun. Clearly mankind now had the ability to discover what the stars are made of; the science of *spectral analysis* was born.

The fundamental idea behind spectral analysis is very simple. Pass white light through jars containing various familiar gases, examine the resulting spectra, and

**Figure 4–25. *The Formation of Spectral Lines.*** The spectrum of white light seen through a cloud of gas contains dark spectral lines. The pattern of spectral lines depends on the chemical composition of the gas.

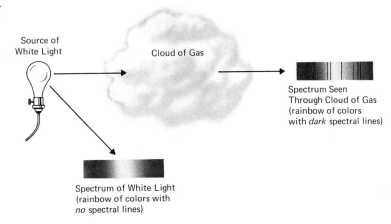

Source of White Light

Cloud of Gas

Spectrum Seen Through Cloud of Gas (rainbow of colors with *dark* spectral lines)

Spectrum of White Light (rainbow of colors with *no* spectral lines)

record all the distinctive patterns. Obviously there are some chemicals that are not normally gases, such as calcium and potassium. In such cases simply heat a small amount of the solid until it vaporizes and then use the vaporized chemical in experiments. The end results of these experiments are contained in a lot of books that describe the patterns of spectral lines that various chemicals produce. When an astronomer takes a spectrum of the sun or a star, he can use these reference books to *identify* spectral lines with known chemicals. He therefore concludes that these chemicals are present in the sun or star. In the solar spectrum shown in Figure 4–24, a person familiar with the patterns of spectral lines produced by certain elements recognizes these lines and therefore realizes that hydrogen (H), calcium (Ca), iron (Fe), nickel (Ni), chromium (Cr), strontium (Sr), astatine (At), silicon (Si), and so forth, must be present in the solar atmosphere. Some of the more prominent spectral lines in Figure 4–24 are identified with their chemical symbols.

During the second half of the nineteenth century, astronomers began applying spectral analysis to the objects in the sky. Each and every chemical leaves its own unique and indelible fingerprints in the form of spectral lines on the light. Therefore, this new knowledge allowed astronomers to discover what the planets and stars are made of. But this new and powerful tool did not tell them *how* and *why* the spectral lines were formed in the first place.

Around the turn of the twentieth century, physicists began to talk seriously about the concept of *atoms*. It seemed quite reasonable to suppose that everything is made up of very small objects called atoms, which probably have diameters of about 1 Å.

In 1911, Sir Ernest Rutherford proposed a "nuclear" or solar-system model of the atom that forms the basis of our current understanding of *atomic structure*. The central idea behind Rutherford's concept of the atom is that, crudely speaking, atoms can be thought of as miniature solar systems. At the center of each atom is the *nucleus,* composed entirely of massive particles called *protons* and *neutrons.* Protons and neutrons have very nearly the same mass. Protons, however, carry a positive electric charge while neutrons are electrically neutral. Orbiting the nucleus are very tiny low-mass particles called *electrons.* Electrons carry a negative electric charge and revolve about the massive nucleus like planets around the sun. Although gravity provides the force which holds the solar system together, the electric forces between the negatively charged electrons and the positively charged protons hold the atom together.

Rutherford's solar-system model of a typical atom is shown schematically in Figure 4–26. Under normal conditions, the number of negatively charged electrons equals the number of positively charged protons in the nucleus, so that the atom is electrically neutral. It is the number of electrons in the atom of a particular element that determines the chemical properties of that element. Hydrogen atoms have one electron, helium atoms have two electrons, lithium atoms have three electrons, and so forth.

After Rutherford proposed his model of the atom, it was realized that there were some severe problems. If electrons are in circular or elliptical orbits about the nucleus, they must be constantly undergoing accelerations, just as planets going

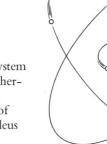

**Figure 4–26.** *An Atom.* A nuclear or solar–system model of the atom was first proposed by Rutherford in 1911. Most of the mass of an atom is contained in the nucleus, which is composed of protons and neutrons. Electrons orbit the nucleus in circular or elliptical orbits.

around the sun are constantly accelerated by the sun's gravitational field. According to Maxwell's equations, charged particles that are accelerated must lose energy by emitting electromagnetic radiation. Therefore, electrons in orbit should rapidly spiral into the nucleus as they emit a flash of light and Rutherford's atoms, if they exist, should collapse almost at once! This was not good news for Rutherford who thought he had a pretty good idea of what atoms were like. Fortunately, however, in 1913, Niels Bohr saved the day by proposing a simple idea that offered a way out of this dilemma and led to the modern theory of *quantum mechanics.*

Think for a moment about the solar system. In principle, planets, asteroids, or meteoroids can orbit the sun at *any* distance whatsoever. A spacecraft, such as Mariner 10, can be placed in any orbit just by aiming a rocket in the appropriate direction. While there are no fundamental restrictions on orbits about the sun, Niels Bohr came up with the idea that this might *not* be the case with orbits about the nuclei of atoms. Suppose that there are strict conditions on the orbits of electrons in atoms, that only certain orbits are "allowed" and all others are prohibited. In this way, the orbits of electrons about the nucleus are *quantized* because the speed and distance of electrons from the nucleus can have only certain allowed quantities.

Niels Bohr then applied this revolutionary concept of quantization to Rutherford's model of the atom. Bohr assumed that only certain orbits were permitted and that when an electron is in a permitted orbit it does not radiate energy. An electron radiates or absorbs energy only when it "jumps" from one allowed orbit to another allowed orbit. Hydrogen best illustrates Bohr's model of the atom because, with only one electron, it is the simplest of all atoms. Nevertheless, the fundamental ideas can be applied to any atom containing any number of electrons.

According to Bohr, there are certain permitted orbits for the electron in the hydrogen atom, as shown in Figure 4–27. We shall call them by names such as "orbit #1," "orbit #2," "orbit #3," and so forth, in increasing order from the nucleus. Bohr found that the orbit nearest the nucleus (orbit #1) has the lowest energy. As a result, this orbit is called the *ground state* of the atom. If an electron is to go into a higher orbit, it must be supplied with energy to "boost" it up to that higher orbit. But the electron *cannot* be given any random amount of

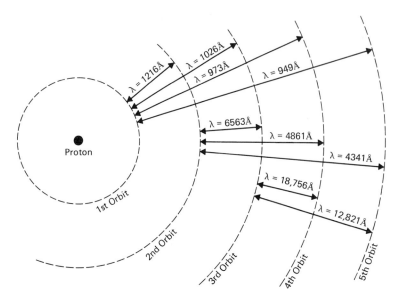

**Figure 4-27.** *The Hydrogen Atom.* The hydrogen atom is the simplest of atoms. Its nucleus contains only one proton, which is orbited by only one electron. In going from one allowed orbit to another, the electron absorbs or emits light at very specific wavelengths.

energy. If the electron is to go from one specific low orbit to another specific higher orbit, it must be supplied with a very precise amount of energy. This amount of energy corresponds exactly to the difference in the energy of the electron in the lower and higher orbits.

When Bohr calculated the amounts of energy required to boost the electron from the second orbit (orbit #2) to higher orbits (orbit #3, orbit #4, and so forth), he found that the needed amounts of energy corresponded to the energy in light of very specific wavelengths. For example, to go from orbit #2 to orbit #3 requires the energy contained in red light at a wavelength of exactly 6563 Å. To go from orbit #2 to orbit #4, we need the energy of blue light having a wavelength of exactly 4861 Å. To go from orbit #2 to orbit #5 requires the energy of light at a wavelength of 4341 Å, and so forth. These are exactly the same wavelengths at which spectral lines are observed in the spectrum of hydrogen gas! Finally, scientists had an explanation of how and why spectral lines are formed.

Understanding of the Bohr model of the atom and the formation of spectral lines up to this point may be summarized as follows:

1. An atom contains a nucleus made up of protons and neutrons.
2. Electrons orbit the nucleus of an atom in very specific allowed orbits.
3. When an electron jumps from one allowed lower orbit to another allowed higher orbit, the atom must absorb a very precise amount of energy.
4. If energy is available to the atom from white light, the atom extracts energy from the light as a specific wavelength.
5. If a source of white light illuminates a gas, the atoms in the gas extract energy from the white light at a series of wavelengths, thereby giving rise to the spectral lines we observe.

Atoms of different chemical elements have different numbers of electrons and therefore different allowed orbits. For example, helium with two electrons in orbit about its nucleus has a very different structure from iron atoms, which have 26 electrons. The allowed orbits in the helium atom therefore occur at different locations than the allowed orbits in the iron atom. Consequently, it should be no surprise that the spectrum of helium looks nothing like the spectrum of iron. This is why each chemical produces its own unique pattern of spectral lines. The subject of quantum mechanics deals with calculating the structure of atoms and discovering the allowed orbits. Every spectral line can now be understood in terms of electrons jumping from one orbit to another.

The work of Niels Bohr enabled scientists to understand spectral lines. In particular, all the experimental work of Gustav Kirchoff, who formulated some of the most fundamental laws of spectral analysis, can be completely explained. Perhaps the best way to examine and understand these laws is with the aid of the experiment shown in Figure 4–28. In this experiment there is a source of white light, such as a light bulb. Looking at the spectrum of this white light through a spectroscope, we find that the spectrum contains *no* spectral lines at all. We therefore say that the hot filament in the light bulb emits a *continuous spectrum*. If this light is shined on a cloud of gas, the spectrum of the light passing through the gas still shows the original, continuous rainbow of colors. But now a series of dark, thin spectral lines is superimposed on these colors. The atoms in the gas *absorb* light at certain specific wavelengths as the electrons in these atoms jump from low orbits to higher orbits. The resulting spectrum is called an *absorption line spectrum*. As the atoms in the gas absorb light from the continuous spectrum, a lot of atoms find themselves with electrons in very high orbits, far above their ground states. Such

**Figure 4–28.** *Kirchoff's Laws.* A continuous source of white light shows no spectral lines. When white light is passed through a cloud of gas, the atoms absorb light at specific wavelengths resulting in an absorption line spectrum. When this cloud of gas is viewed with a spectroscope at an angle away from the beam of white light, an emission line spectrum is seen.

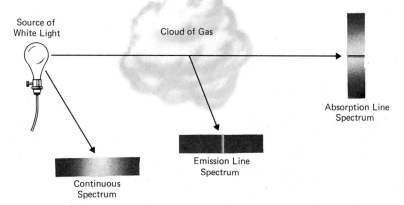

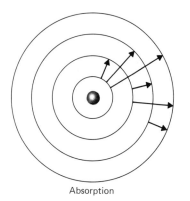
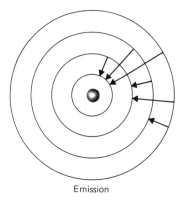

Absorption

Emission

**Figure 4-29.** *Absorption and Emission.* When electrons jump from a low orbit to a high orbit, they must absorb energy. When electrons jump from a high orbit back down to a low orbit, they emit energy. This energy is either absorbed or emitted at *very* specific wavelengths, as dictated by quantum mechanics.

atoms are said to be *excited*. It is logical that the atoms prefer to return to their ground states; the ground state is the normal, stable condition for an atom. An atom becomes *de-excited* and returns to its ground state when electrons jump from high orbits back down to lower orbits. In undergoing such transitions, the atoms must lose energy and they *emit* light at certain specific wavelengths. The wavelengths of the light emitted as electrons cascade back down to their ground states are exactly the same as the wavelengths of the light originally absorbed by the atoms in the first place. In going back to the lower orbit, an electron surrenders exactly the same light that it absorbed. But the atoms have no way of remembering which way the original beam of light from the light bulb was going. Therefore this light is emitted in all directions. In other words, if the cloud of gas is observed in our experiment at some angle away from the beam of white light, an *emission line spectrum* is seen. Such a spectrum is mostly black, but superimposed on this blackness is a series of bright *emission lines*. The light in these lines comes from the atoms emitting radiation as they become de-excited. The pattern of bright lines in the emission spectrum is exactly the same as the pattern of dark lines in the absorption spectrum.

The wealth of information contained in the spectra of light received from the stars cannot be overemphasized. Although the objects which the astronomer studies are incredibly remote, virtually everything he would want to know has left an indelible mark on the light which reaches his telescope. In a very real sense, the astronomer is limited only by the reliability of his instruments and the depth of his knowledge of the properties of light.

Very frequently, it is noticed that a familiar pattern of spectral lines from some well-known chemical look slightly peculiar. The spectral lines that the astronomer observes might be shifted from their usual positions, or they might appear very fuzzy and broadened, or a familiar line might be split up into a number of closely spaced lines. In such situations the astronomer is quick to realize that something very unusual is going on with the star or galaxy he is observing. The pathology of the light he receives bears witness to these unusual conditions. At this point the astronomer turns to the physicist to tell him about the various factors (motion, turbulence, magnetic fields, and so forth) that can affect the appearance of spectral lines.

When an astronomer observes the spectrum of some celestial object, he frequently notices that familiar patterns of spectral lines are *not* where they should be. For example, the skilled astronomer is always quick to recognize the prominent two lines of calcium (the so-called H and K lines) in the spectrum of a star or a galaxy. In the laboratory, these two lines *always* show up at wavelengths of 3968 Å and 3934 Å. Yet, in a stellar spectrum, these lines frequently appear in a very different region of the spectrum. Careful scrutiny of other recognizable spectral lines reveals that they *all* have been shifted by corresponding amounts, either toward the red or the blue end of the spectrum. In order to understand what could cause a shift in all the lines in a spectrum, simply appeal to an everyday common experience:

Imagine that you are standing on the sidewalk of a busy street when suddenly you hear the siren of an approaching ambulance or police car. As the vehicle comes toward you, the wail from the siren seems to have a very high pitch. But, after the ambulance or police car passes you, the pitch of the siren appears to drop significantly. This change in pitch can also be noticed by someone standing near a railroad track listening to the bell or whistle of a passing train. As the source of noise approaches, the pitch seems high; as the source of noise recedes, the pitch seems low.

This phenomenon is known as the *Doppler effect.* The sound waves are "bunched up" in front of an approaching source of noise and thus the listener hears a high frequency pitch. On the other hand, the sound waves are "spread out" behind a receding source of noise and thus the listener hears a low frequency pitch.

This same sort of phenomenon also occurs with light. If a source of light is coming toward an observer, the light waves will be slightly compressed, as shown schematically in Figure 4–31. If the light from the source contains spectral lines, these lines will appear at wavelengths *shorter* than usual; they will be shifted toward the blue end of the spectrum. Conversely, if a source of light is receding from an observer, the light waves will be slightly expanded, again as shown in Figure 4–31. Spectral lines from this receding source will therefore appear at wavelengths *longer*

**Figure 4–30.** *The Doppler Effect.* The pitch of a source of sound is affected by the source's motion. An approaching ambulance siren seems to have an unusually high pitch while a receding siren appears to have a much lower pitch.

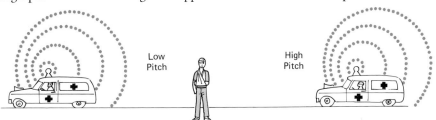

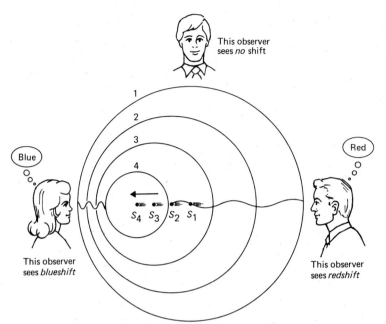

**Figure 4–31. *Redshifts and Blueshifts.*** The precise wavelength of a particular spectral line is affected by the motion of the source. An approaching source of light has its spectral lines shifted to shorter wavelengths (toward the blue end of the spectrum). A receding source of light has its spectral lines shifted to longer wavelengths (toward the red end of the spectrum).

than usual; they will be shifted toward the red end of the spectrum. In other words, if an astronomer finds that all the lines in the spectrum of a star are shifted from their usual "laboratory" positions, he immediately concludes that the star is moving. If the lines are shifted to shorter wavelengths, resulting in a so-called *blueshift,* he knows that the star is coming toward him. If the lines are shifted to longer wavelengths, resulting in a so-called *redshift,* he knows that the star is moving away from him. The bigger the shift, the higher the speed.

While shifts in spectral lines are usually caused by the motion of the source, many physical processes can occur in the source that change the appearance of the spectral lines. For example, around the turn of the century, Pieter Zeeman discovered that the presence of a magnetic field can cause a single spectral line to split up into several spectral lines. If a source of light in the laboratory is placed between the poles of a magnet, the spectral lines from that light will be split into two or more "components." The exact number of component lines depends on the details of the structure of the atoms in the source of light. This phenomenon is appropriately called the *Zeeman effect.*

All atoms in magnetic fields exhibit Zeeman splitting of their spectral lines. For example, in the absence of a magnetic field, the spectrum of sodium is dominated by two very strong lines (the so-called sodium D-lines) at 5890 Å and 5896 Å. In the presence of a magnetic field, however, the 5896 line splits into four lines while the 5890 line splits into six lines. As with all atoms, the separation of the components depends on the strength of the magnetic field. The stronger the field, the bigger the separation. By measuring the separation between components of a line

**Figure 4–32.** *A Spectroscopic Plate.* The spectrum of a star or galaxy is photographed on a spectroscopic plate. An astronomer is shown holding a typical spectroscopic plate on which the spectrum of a star has been exposed.

split caused by the Zeeman effect, a physicist can calculate the strength of the magnetic field acting on the atoms.

The Zeeman effect has important applications to astronomical observations. If an astronomer discovers several closely spaced lines in the spectrum of a star near the same wavelength at which he normally would expect to see only one spectral line, he immediately concludes that a magnetic field must be present. In this way, astronomers have discovered that sunsports possess strong magnetic fields, as do so-called "magnetic stars" and "white dwarf" stars.

The discussion to this point has focused on two effects that can dramatically change the appearance of a spectrum. According to the Doppler effect, motion of a source of light causes displacement of spectral lines from their usual (laboratory) wavelengths. The higher the speed, the greater the shift. Magnetic fields at a source of light causes Zeeman splitting of spectral lines into two or more component lines. The separation between the components increases with increasing field strength. There are, however, numerous more subtle effects that can alter spectral lines. These effects do not split or shift the lines. Instead, these effects cause the spectral lines to *broaden*. This *spectral line broadening* causes the lines to appear wide and "washed out." Different physical processes result in different types of line broadening. For example, if the gas from which the spectral lines originate is very turbulent or under extreme pressure, the spectral lines will be broadened. By studying the precise shape or *profile* of spectral lines, the astronomer can deduce many of the details concerning conditions in a star's atmosphere.

While atoms give rise to patterns of spectral lines, molecules give rise to bands of

Water Molecule
(2 hydrogen atoms
and 1 oxygen atom)

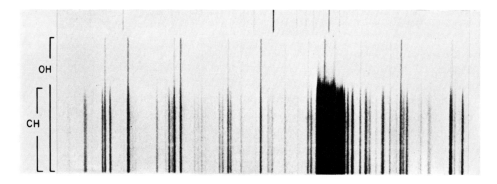

Carbon Dioxide Molecule
(1 carbon atom and 2
oxygen atoms)

Ammonia Molecule
(1 nitrogen atom and
3 hydrogen atoms)

**Figure 4–33.** *Some Molecules.* Molecules consist of atoms. Molecules of water ($H_2O$), carbon dioxide ($CO_2$), and ammonia ($NH_3$) are shown schematically.

**Figure 4–34.** *Molecular Spectrum.* Vibrational and rotational bands caused by CH and OH observed in an oxyacetylene blowtorch flame are shown here. (*Lick Observatory photograph*)

OH

CH

lines. Molecules are combinations of atoms such as water ($H_2O$), carbon dioxide ($CO_2$), and ammonia ($NH_3$), which was mentioned in connection with the atmospheres of planets. When dealing with individual atoms, transitions of electrons between various allowed orbits give rise to *line spectra*. However, vibrations and rotations of molecules give rise to *band spectra*. Just as the orbits of electrons are quantized, the manner in which molecules can vibrate and rotate is also quantized. Molecules can vibrate and rotate only in certain specific allowed fashions, and each mode of vibration or rotation gives rise to a huge number of spectral lines having very nearly the same wavelength. These vibrational and rotational spectral lines are bunched together giving the appearance of bands in the spectrum. An example is shown in Figure 4–34.

The central idea behind the quantum theory of molecules is that molecules, which are simply combinations or chains of atoms, cannot vibrate and rotate at just any speed or frequency. Vibrations and rotations can occur only at specific allowed rates. Energy must be added to a molecule in order to speed up the rate at which it is vibrating and rotating. But the molecule will accept energy only in certain specific amounts, amounts precisely equal to the quantity of energy needed for the molecule to go from a slower vibrational or rotational state to a faster vibrational or rotational state. If this energy is available from white light, a spectral line is formed. Detailed calculations of molecules show that there are literally thousands of allowed vibrational and rotational states for each molecule, hence there are great numbers of lines that together give the appearance of bands.

If a band of lines is seen in the spectrum of a star or planet, then there must be molecules present in the atmosphere of that planet or star. It is the task of the physicist in the laboratory to discover which molecules produce which bands. In this fashion, astronomers discovered that the atmosphere of Venus contains carbon dioxide ($CO_2$). A spectrum of the reflected light from Venus showing the carbon dioxide bands appears in Figure 4–35.

Over the past several years astronomers have discovered that there are many molecules in space. In the late 1960s, astronomers began discovering many different kinds of organic molecules in large clouds of gas floating between the stars. These molecules, such as formaldehyde ($H_2CO$), cyanogen ($CN$), hydrogen cyanide ($HCN$), and cyanoacetylene ($HC_3N$) are usually associated with living matter. While no one believes that there might be animals or planets floating around in space, these observations do suggest that the basic building blocks for life may be present in the interstellar gas. Some of the basic molecules out of which we ourselves are made may have existed even before the formation of the earth.

Starlight received here on earth contains an incredible wealth of information. With patience and care, this light can be analysed and deciphered to reveal some of the most intimate details of even the most remote objects in the universe. Astronomers are limited only by the quality of their instruments and their own skill and resourcefulness.

**Figure 4-35.** *Carbon Dioxide.* This spectrum of Venus shows molecular bands caused by carbon dioxide, indicating the presence of this gas in the planet's atmosphere. (*Lick Observatory photograph*)

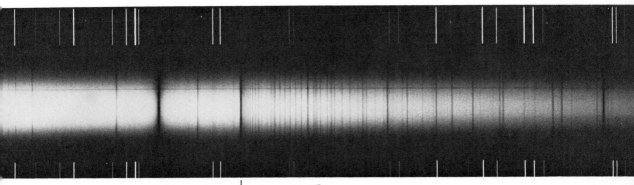

8689 A

# 4.4 Atoms and Nuclei

ONE OF the primary goals of astronomy is simply to find out what is going on elsewhere in the universe. More specifically, the astronomer wants to know all about conditions and properties of the moon, the planets, stars, and galaxies. As we learned earlier in this chapter, the astronomer obtains the desired information primarily through studying light from celestial objects. But in order to understand data from starlight, the scientist must be intimately aware of the chemistry and physics of matter. Only then will he be capable of interpreting and appreciating the full meaning of his observations.

Man's modern understanding of matter began in the nineteenth century when scientists started talking seriously about *atoms* and *elements*. An element is a substance that cannot be decomposed by chemical reactions into simpler substances. Elements are therefore the simplest and most fundamental substances in nature; they are the chemical building blocks of matter. There are 92 elements that occur naturally in the universe ranging from hydrogen and helium to uranium. These elements are sometimes found by themselves in a pure state (e.g., gold in a gold nugget) but most usually exist in chemicals that are combinations of elements (e.g., sodium and chlorine are found together as common salt).

As we learned earlier, matter consists of atoms. An atom is the smallest particle of an element that still retains the properties that characterize that element. Since there are 92 naturally occurring elements, there must be a total of 92 different kinds of naturally occuring atoms. Atoms of different elements combine to form more complicated chemicals. Such combinations of atoms are called *molecules*. Molecules therefore consist of two or more atoms that are chemically bound to each other.

A major breakthrough in man's understanding of the elements occurred in the late 1800s with the work of the Russian scientist D. I. Mendeleev. By Mendeleev's time, a large number of elements had been discovered. He began by numbering them in order of increasing weight. Hydrogen, the lightest element, is "number 1." Helium, the second lightest element, is "number 2." Lithium is "number 3," beryllium is "number 4," and so forth. These important numbers are today called *atomic numbers*. The atomic number of the heaviest naturally occurring element, uranium, is 92.

Mendeleev then noticed that certain elements have very similar chemical properties. For example, hydrogen, lithium, sodium, and potassium all behave in a very similar fashion in chemical reactions. Helium, neon, argon, and krypton all have similar chemical properties. Mendeleev discovered that the elements could be arranged to display their similarities and differences. Such an arrangement is called the *periodic chart of the elements,* as shown in Figure 4–36. Each box contains the *chemical symbol* for a particular element (e.g., H for hydrogen, He for helium) along with the element's atomic number. Reading the chart horizontally, from left to right, we see that the elements are simply arranged in order of increasing atomic number. Reading the chart vertically, however, we notice that elements with

126

| 1<br>H | | | | | | | | | | | | | | | | | 2<br>He |
|---|---|---|---|---|---|---|---|---|---|---|---|---|---|---|---|---|---|
| 3<br>Li | 4<br>Be | | | The Periodic Chart | | | | | | | | 5<br>B | 6<br>C | 7<br>N | 8<br>O | 9<br>F | 10<br>Ne |
| 11<br>Na | 12<br>Mg | | | | | | | | | | | 13<br>Al | 14<br>Si | 15<br>P | 16<br>S | 17<br>Cl | 18<br>A |

| 19<br>K | 20<br>Ca | 21<br>Sc | 22<br>Ti | 23<br>V | 24<br>Cr | 25<br>Mn | 26<br>Fe | 27<br>Co | 28<br>Ni | 29<br>Cu | 30<br>Zn | 31<br>Ga | 32<br>Ge | 33<br>As | 34<br>Se | 35<br>Br | 36<br>Kr |
|---|---|---|---|---|---|---|---|---|---|---|---|---|---|---|---|---|---|
| 37<br>Rb | 38<br>Sr | 39<br>Y | 40<br>Zr | 41<br>Nb | 42<br>Mo | 43<br>Tc | 44<br>Ru | 45<br>Rh | 46<br>Pd | 47<br>Ag | 48<br>Cd | 49<br>In | 50<br>Sn | 51<br>Sb | 52<br>Te | 53<br>I | 54<br>Xe |
| 55<br>Cs | 56<br>Ba | 57<br>La | 72<br>Hf | 73<br>Ta | 74<br>W | 75<br>Re | 76<br>Os | 77<br>Ir | 78<br>Pt | 79<br>Au | 80<br>Hg | 81<br>Tl | 82<br>Pb | 83<br>Bi | 84<br>Po | 85<br>At | 86<br>Rn |
| 87<br>Fr | 88<br>Ra | 89<br>Ac | 104 | 105 | 106 | | | | | | | | | | | | |

| 58<br>Ce | 59<br>Pr | 60<br>Nd | 61<br>Pm | 62<br>Sm | 63<br>Eu | 64<br>Gd | 65<br>Tb | 66<br>Dy | 67<br>Ho | 68<br>Er | 69<br>Tm | 70<br>Yb | 71<br>Lu |
|---|---|---|---|---|---|---|---|---|---|---|---|---|---|
| 90<br>Th | 91<br>Pa | 92<br>U | 93<br>Np | 94<br>Pu | 95<br>Am | 96<br>Cm | 97<br>Bk | 98<br>Cf | 99<br>Es | 100<br>Fm | 101<br>Md | 102<br>No | 103<br>Lr |

**Figure 4-36.** *The Periodic Chart of the Elements.* There are 103 elements. The first 92 elements are found in nature while the remaining 11 are made by nuclear physicists in the laboratory. By arranging the elements in the fashion shown in this chart, elements with similar chemical properties are grouped together in vertical rows.

similar chemical properties are arranged in vertical columns. Thus, for example, fluorine, chlorine, bromine, and iodine behave in a very similar fashion in chemistry experiments. If one element in a particular vertical column is a caustic, corrosive gas, then all of the elements in that column are usually caustic corrosive gases. If one element in a particular column is a shiny metal with good electrical conductivity, then all of the elements in that column are usually shiny metals that would make good wires for carrying electric current. For example, notice that copper (Cu), silver (Ag), and gold (Au) are together in a vertical column.

In many respects, this simple arrangement of the elements in the periodic chart literally opened the door for discovering the detailed atomic structure of matter. In the late 1800s, it was discovered that there are very tiny particles—much smaller than atoms—that help make up the composition of atoms. These particles are called *electrons*. An electron has a very small mass and carries a negative electric charge. As we learned earlier, Ernst Rutherford proposed a "solar system" model of the atom in which electrons orbit a massive positively charged nucleus. Niels Bohr further discovered that electrons could revolve about the nucleus only in certain "allowed" orbits. From this atomic model and the periodic chart of the elements, scientists were in a position to understand exactly how atoms are constructed and thereby to obtain a deep insight into the true meaning of the periodic chart of the elements.

Under "normal" conditions (i.e., under conditions where an atom is not missing any electrons as a result of extreme temperature or pressure) the number of electrons is given by the element's atomic number. Thus, the hydrogen atom has one electron, the helium atom has two, and so forth. A uranium atom normally has 92 electrons orbiting its massive nucleus. The allowed orbits permitted to electrons in an atom are grouped together in *shells.* The innermost shell can contain a maximum of two electrons. The second shell can contain up to eight electrons. The third shell can also have eight electrons while the fourth shell can contain eighteen electrons. The reasons for these shells and for the maximum number of electrons they can hold comes from the details of *quantum mechanics,* which originated with the work of Niels Bohr.

Moving across the periodic chart from left to right is a progression from lighter elements to heavier elements. It is a progression of increasing atomic number and therefore increasing numbers of electrons. The top row on the periodic chart contains only two elements: hydrogen and helium. Hydrogen atoms have one electron and helium atoms have two. In its ground state, the single electron in a hydrogen atom is orbiting the nucleus in the first shell. In the ground state, the two electrons in a helium atom are orbiting the nucleus in the first shell. At most, this shell can contain only two electrons. With helium, therefore, the first shell is said to be *filled.* To get to the next element, lithium, the third electron must go into the next highest shell. The lithium atom has one electron in the second shell, while the beryllium atom has two electrons in the second shell, and so forth, up to neon. Neon (atomic number = 10) has two electrons in the first shell and eight electrons in the second shell. The second shell can contain a maximum of eight electrons and thus with neon this shell is filled. To get to the next atom, sodium, it is necessary to start filling the third shell. Sodium is the first element on the third horizontal row of the periodic chart. In moving from left to right along the third row, one new electron is added to the third shell for each new element. Argon, the last element on the third row, has its shells filled with 18 ( = 2 + 8 + 8) electrons.

From this shell model of the atom it is possible to understand many of the chemical properties of the elements. For example, consider the right-hand column on the periodic chart (helium, neon, argon, krypton, and so forth). The atoms of each of these elements have completely filled shells. A filled shell is an extremely stable configuration. Atoms with completely filled shells do not have any extra electrons that they could "lend" to other atoms. Neither do these atoms have any "gaps" in their shells and therefore cannot "borrow" any electrons from other atoms. Consequently, helium, neon, argon, krypton, and xenon are *inert gases.* They are never found in chemical compounds with other atoms. They exist in nature only in pure form.

Now consider the second column from the right in the periodic chart (fluorine, chlorine, bromine, iodine, and so forth). The atoms of these elements are lacking only one electron for closed shells. Each of these atoms may be thought of as having a "gap." By contrast, all the atoms of the elements in the left-hand column (hydrogen, lithium, sodium, potassium, and so forth) have one spare electron orbiting above the inner closed shells. Each of these atoms can be thought of as having one electron it can "lend." Atoms of elements in these two columns can therefore easily combine to form a multitude of chemicals. For example, common salt is sodium chloride (NaCl). In each molecule of salt, the spare electron from the sodium atom is shared by the chlorine atom.

In general, the atoms of elements on the left side of the periodic chart are "lenders" of electrons, while the atoms on the right side are "borrowers." For example, hydrogen is a good "lender." Oxygen atoms need two electrons to complete their shells. Consequently, two hydrogen atoms can each lend one electron to an oxygen atom. The resulting molecule has the formula $H_2O$. The resulting substance is called water.

Atoms with one or two spare electrons (e.g., sodium, magnesium, potassium, calcium, and so forth) are good "lenders" while atoms requiring one or two electrons to fill a shell (e.g., oxygen, fluorine, sulfer, chlorine, and so forth) are good "borrowers." For elements near the middle of the periodic table, this distinction is not so clear cut. For example, carbon can either borrow or lend four electrons. Carbon can act like a "lender" of its four spare electrons and form, for example, carbon tetrachloride ($CCl_4$) with chlorine. Conversely, carbon can act like a "borrower" taking the spare electrons of four hydrogen atoms to form methane ($CH_4$). This dual ability of carbon to operate in both ways makes this element very versatile. This is one of the reasons why carbon is found in the complex molecules of biological material.

The atomic number of an element tells how many electrons (under normal conditions) are contained in atoms of that element. The atomic number therefore is an important piece of information about atoms. It allows scientists to gain a perspective on the elements and how they are related to each other. There is, however, additional important information scientists can obtain about atoms and electrons. For example, by listing the elements in order of increasing atomic number, we obtain a progression from the lighter to the heavier elements. But in this progression the actual weights of elements do not increase as simply as the atomic numbers.

Hydrogen is the lightest element. Suppose we take the weight of a hydrogen atom to be 1. An atom of the next element, helium, is then found to have a weight of 4. A lithium atom has a weight of 7, a beryllium atom is 9, a boron atom is 11, a carbon atom is 12, and so forth.

Actually, scientists today prefer to express the weights of atoms in *atomic mass units,* rather than pounds or grams. In this system, hydrogen is 1.008, helium is 4.004, lithium is 6.940, beryllium is 9.013, and so forth. These numbers are called the *atomic weights* of the elements. The atomic weight of uranium, the heaviest naturally occurring element, is 238.07.

Atoms are held together by electric forces just as the solar system is held together by gravitational forces. The negatively charged electrons orbiting the nucleus of an atom do not spontaneously fly off into space because the nucleus contains a positive electric charge. Under normal conditions, when an atom has its proper number of electrons, the total negative charge of all the electrons is exactly balanced by equal and opposite positive charge on the nucleus. The atomic number therefore tells the total positive electric charge carried by the nucleus of an atom. A hydrogen nucleus has 1 unit of positive charge, helium nuclei each have 2 units, lithium nuclei each have 3 units, and so forth. In chemical reactions and at high temperatures or pressures, an atom can lose one or more of its electrons. Such atoms are then said to be *ionized*. The processes which strip atoms of their electrons cannot, however, alter the charge on the nucleus. Thus, even if an atom is ionized, it does not lose its identity as being a piece of a particular element.

In the 1920s, scientists discovered particles in the nuclei of atoms which are responsible for the nucleus' positive charge. These particles are called *protons*. A proton is much more massive than an electron. Protons weigh 1840 times more than electrons. Nevertheless, the electric charge carried by a proton is exactly equal and opposite to that carried by an electron. We, therefore, arrive at the important realization that the *atomic number of an element tells exactly the total number of protons in the nuclei of atoms of that element.*

With this much knowledge about nuclei, scientists began arriving at some important conclusions. Since the atomic number gives the number of protons in a nucleus, hydrogen has one proton, helium has two protons, lithium has three protons, and so forth. But recall that most of the mass of an atom is located in its nucleus, just as most of the mass of the solar system is in the sun. The electrons orbiting an atom contribute only a very tiny amount to the atom's total mass. Now also recall the atomic weights of the elements. A helium atom weighs four times more than a hydrogen atom. A lithium atom weighs seven times more than a hydrogen atom.

How could it be that a helium atom, whose nucleus contains two protons, can weigh four times more than a hydrogen atom with only one proton? How could it be that a lithium atom, whose nucleus contains three protons, can weigh seven times more than a hydrogen atom with one proton? To solve this dilemma, nuclear physicists theorized that there must be another type of particle in the nucleus. This new particle must have almost exactly the same mass as a proton. But unlike a proton, this particle cannot have any electric charge. This massive, electrically neutral particle is called a *neutron*.

In 1932, James Chadwick reported the results of his experiments with nuclei. He had succeeded in discovering the neutron. Man's understanding of matter had reached the point where the existence of unknown subatomic particles could be successfully predicted.

With the discovery of the neutron, scientists were in a position to obtain a much deeper understanding into the nature of the elements. The nucleus of a hydrogen atom contains only one proton. The nucleus of a helium atom contains two protons, but because its atomic weight is 4, it must also contain two neutrons. A lithium nucleus must have three protons and four neutrons to give it an atomic weight of 7. In other words, it should be possible to understand the atomic weights of the elements in terms of the total number of protons and neutrons in the nuclei. The total number of protons and neutrons in the nucleus is called the *mass number*.

At this point it seems that we are faced with a puzzling dilemma in examining the periodic chart. Consider copper, the twenty-ninth element. Since the atomic number is 29, there must be 29 protons in each copper nucleus. But the atomic weight of copper is 63.54. Does this mean that each copper nucleus contains $34\frac{1}{2}$ neutrons? No!

The number of protons in a nucleus of an atom determines what element the atom is. The exact number of neutrons in a nucleus does not particularly affect which element the nucleus belongs to. For example, almost all of the hydrogen in nature contains only one proton in its nucleus. There is, however, a rare second type of hydrogen that has a neutron in addition to the proton in its nucleus. This "heavy" form of hydrogen, called deuterium, has an atomic weight of 2. Both forms of hydrogen behave in exactly the same way in chemical reactions. As far as the chemical properties are concerned, it is almost impossible to tell them apart. Atoms of a particular element that have the same number of protons but different numbers of neutrons are called *isotopes*. There are two isotopes of hydrogen found in nature, the common type with one proton in the nucleus, and the rare type with a proton and a neutron in the nucleus.

Returning to the example with copper, experiments reveal that two isotopes of copper are found in nature. The lighter isotope has a mass number of 63, which means that its nuclei contain 29 protons and 34 neutrons. The heavier isotope has a mass number of 65 and thus its nuclei have 29 protons and 36 neutrons. These same experiments also reveal that 69 per cent of the copper found in nature is of the lighter isotope while the remaining 31 per cent is of the heavier type. Consequently, the atomic weight of copper found in nature and containing both isotopes must lie between 63 and 65. Since the lighter isotope is more common, the atomic weight of copper (63.54) lies closer to 63.

Since a particular element often can have several isotopes, numbers are usually written along with the element's symbol to indicate a specific isotope. A subscript to the lower left of the symbol gives the element's atomic number ( = number of protons) while a superscript to the upper left of the symbol gives the isotope's mass number ( = number of protons + number of neutrons). It is then always easy to figure out how many neutrons are in the isotope's nuclei: simply subtract the atomic number from the mass number. Thus, for example, the two naturally occurring isotopes of copper are $^{63}_{29}$Cu and $^{65}_{29}$Cu. Similarly, there are three naturally-occurring isotopes of neon: $^{20}_{10}$Ne, $^{21}_{10}$Ne, $^{22}_{10}$Ne.

For the light-weight elements most isotopes have very nearly the same number of neutrons and protons in their nuclei. But beyond $^{40}_{20}$Ca, naturally occurring isotopes have more neutrons than protons. The reason for this is that neutrons do not have any electric charge. Unlike protons, neutral particles do not repel each other. It is therefore possible to pack lots of neutrons into the nuclei of the heavier elements. For example, there are two common isotopes of silver: $^{107}_{47}$Ag and $^{109}_{47}$Ag whose nuclei contain 47 protons along with either 60 or 62 neutrons, respectively. Figure 4–37 displays this tendency for heavy elements to have more neutrons than protons in their nuclei. The dotted region indicates the locations of all the naturally occurring isotopes. Isotopes that lie outside of this dotted region have either too many neutrons or too many protons. As a result, these nuclei are *unstable* and therefore subject to *radioactive decay*.

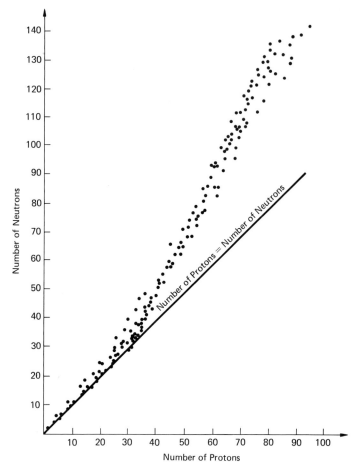

**Figure 4-37.** *Stable Nuclei.* The nuclei of light-weight isotopes
have nearly the same number of neutrons and protons. More
massive nuclei have more neutrons than protons. The dotted
region indicates the location of stable nuclei. Isotopes outside of
this region have either too many protons or neutrons and are
therefore radioactive.

   A nucleus is unstable because it contains either too many protons or neutrons.
Such nuclei can decay in a variety of ways. For example, they can simply cast off the
extra particles. Alternatively, the excess protons can be converted into neutrons or
excess neutrons can be converted into protons. Such conversions occur by means of
*beta decay*. If a nucleus has too many neutrons, excess neutrons can change into
protons by emitting electrons. When a negatively charged electron is cast off from
a neutron, its mass does not change appreciably but it now has a net positive charge.
The final result is a proton. The emission of electrons from neutron-rich isotopes

was discovered around 1900 as a result of the work of Henri Becquerel and Mme. Curie.

After World War I, scientists began wondering if unstable proton-rich nuclei could not convert some of their excess protons into neutrons by a process similar to the conversion of neutrons into protons. In order for this conversion to occur, the proton would have to emit a *positively* charged electron. Such a particle would carry away the proton's positive charge, leaving behind a neutron.

In 1933, Carl D. Anderson discovered a new particle whose mass is the same as an electron but has a positive charge. These positively charged electrons are called *positrons* or *anti-electrons*. With the discovery of this new subatomic particle, the suspicions of scientists were confirmed. Excess protons can be converted into neutrons by the emission of positrons. This process is sometimes called *inverse beta decay*.

When a nucleus radioactively decays by emitting electrons or positrons, the electric charge on the nucleus changes. But recall that the total positive charge on a nucleus ( = atomic number) determines which chemical element that nucleus is. Consequently, radioactive decay changes one element into another. For example, the common isotope of uranium, $^{238}_{92}$ U, radioactively decays into thorium ($^{234}_{90}$ Th) by casting off a helium nucleus, $^4_2$He. Thorium is also radioactive and beta decays into protactinium ($^{234}_{91}$ Pa), which further beta decays into a new isotope of uranium, $^{234}_{92}$ U.

The rate of radioactive decay may be slow or fast depending on how unstable the nucleus is. This rate is expressed in terms of a *half-life*. The half-life of an isotope is simply how long it takes for half of the nuclei in a sample of the isotope to decay into a new element. The half-life of $^{238}_{92}$ U is $4\frac{1}{2}$ billion years, which means that if you started with a pound of $^{238}_{92}$ U, after $4\frac{1}{2}$ billion years only half a pound of $^{238}_{92}$ U would be left. The remaining half pound of material would be other elements or isotopes. By contrast, the half-life of $^{234}_{90}$ Th is fairly short, only 24.1 days. The resulting isotope of protactinium, $^{234}_{91}$ Pa, has a half-life of 1.14 minutes while the half-life of $^{234}_{92}$ U is almost a quarter of a million years.

Nuclear physicists can create new isotopes and even new elements in the laboratory using huge machines such as *cyclotrons* or *betatrons*. These machines accelerate particles like protons to very high speeds, near the speed of light. By directing the beam of high-speed protons at a substance, some of the protons penetrate the nuclei of the atoms in the "target" thereby creating new isotopes. Thus, either by natural radioactive decay or by artificial means, one element can be turned into another element.

Astronomers have very good reason to believe that when the universe was created some 15 billion years ago, all that existed was hydrogen and some helium. Of course, the earth and everything around us consist of many heavier elements. Where, then, did these heavier elements come from if the universe began with just hydrogen and helium? The answer lies in the fact that at the enormous temperatures and pressures at the centers of stars conditions are such that one element can be turned into another. As a star goes through its life cycle, lighter elements are converted into heavier elements. Finally, at the end of its life, a dying star frequently explodes, casting off the processed material into space. This material enriches the interstellar medium with many heavy elements. It is from this enriched interstellar medium that planets such as our earth later condense. It is therefore clear that every heavy atom in our bodies, every heavy atom of food we eat, of air we breath and of water we drink was created billions of years ago in some long-since-dead star.

## Review Questions and Exercises

1. What happens when a beam of white light passes through a glass prism. What does this prove?
2. What is an angstrom? What is a micron?
3. Approximately what range of wavelengths correspond to visible light?
4. What is meant by the electromagnetic spectrum?
5. Who was James Clark Maxwell?
6. List the various types of radiation that make up the electromagnetic spectrum.
7. What is a blackbody?
8. What does Wien's law tell us?
9. What does Stefan's law tell us?
10. Who was Max Planck?
11. What is a photon?
12. How does Planck's blackbody radiation law tell us *more* then Wien's law and Stefan's law combined?
13. Briefly discuss the effects of the earth's atmosphere on the astronomer's efforts to observe stars and planets at various wavelengths.
14. Describe the optical design of a refracting telescope.
15. Describe the optical design of a Newtonian reflector.
16. Contrast and compare the various possible optical arrangements of a reflecting telescope.
17. Contrast and compare magnifying power, resolving power, and light gathering power.
18. What is a radio telescope?
19. What is meant by optical and radio windows?
20. Compare and contrast a radio telescope and an ordinary reflecting telescope.
21. Briefly discuss how astronomers make astronomical observations at ultraviolet, infrared, and X-ray wavelengths.
22. What is a spectrograph and what is it used for?
23. What is an atom?
24. What are electrons, protons, and neutrons? Where are they located in an atom?
25. How does Bohr's model of the atom account for spectral lines?
26. Discuss the difference between absorption line spectra and emission line spectra.

**27.** What is the Doppler effect?

**28.** What is the Zeeman effect?

**29.** How does the appearance of the spectrum of molecules differ from that of individual atoms?

**30.** What is an element? What is an isotope?

**31.** What is the significance of atomic numbers?

**32.** Describe the periodic chart of the elements.

**33.** Explain why helium, neon, argon, and krypton are inert gases from the viewpoint of their atomic structure.

**34.** What is meant by atomic weight?

**35.** What is meant by the mass number of an isotope?

**36.** Why can one element change into another element as a result of radioactive decay?

**37.** What is meant by the half-life of a radioactive isotope?

# 5 Our Star–The Sun

## 5.1 The Sun and Its Cycles

As far as our lives are concerned, the sun is the single most important object in the sky. The sun provides the heat and light that make life possible on our planet. Energy from the sun is responsible for the currents in the oceans and the weather in the atmosphere. Without the sun, our planet would be a barren and frozen wasteland drifting aimlessly through space.

The sun is a huge, massive sphere of gas some 1,400,000 kilometers (about a million miles) in diameter. The sun's gases are extremely hot, ranging from about 5,800°K at the solar surface to roughly 15 million degrees Kelvin at the sun's center. Most sunlight we receive on earth comes from the sun's surface. According to Wien's law (see Section 4.1) for a blackbody at about 5,800°K, sunlight is centered in the middle of the range of wavelengths for visible light. This is not an accident. Millions of years of evolution have given creatures on earth eyes that respond to the visible light from the sun.

The first person to make detailed observations of the sun's surface was Galileo in the early 1600s. Using the recently invented telescope, Galileo discovered *sunspots* on the sun. As shown in Figure 5–1, sunspots are small, dark regions seen against the bright solar surface. Contrary to first impressions, sunspots are not really black. They appear dark only in contrast to their very bright surroundings. Astronomers today realize that a sunspot is actually a cool region on the sun's surface. The temperature in a sunspot is about 4,000°K, almost 2000° cooler than the neighboring solar surface. As a result of this lower temperature (recall Stefan's law in Section 4.1), a square foot of sunspot emits much less light than a corresponding square foot of unblemished solar surface. If a sunspot could be seen without the surrounding area for contrast, the sunspot would actually appear fairly bright.

Only the outer layers of the sun can be seen with the naked eye. These outer parts of the sun are called the *solar atmosphere*. Astronomers find it convenient to divide the atmosphere into three distinct regions that have very different properties. These three layers in the solar atmosphere are called the *photosphere, chromosphere,* and *corona.*

Virtually all the light received from the sun comes from the photosphere.

136

When one speaks of the sun's "surface," he is usually referring to this layer in the solar atmosphere. Indeed, the word "photosphere" means "sphere of light." An ordinary photograph of the sun, such as Figure 5-1, is a photograph of the photosphere.

Sunspots are the most easily recognizable feature to be seen in the photosphere. By observing sunspots from day to day, astronomers are able to observe how the sun rotates. Surprisingly, the sun does *not* rotate like a "rigid body" such as the earth. Sunspots near the solar equator take 25 days to go once around the sun. By contrast, sunspots at latitudes near 30° north or south of the equator take $27\frac{1}{2}$ days to go once around the sun. And near the polar regions, at latitudes of 75°, sunspots take 33 days to return to their starting place. At the north and south poles, the rotation period is perhaps as long as 35 days. This phenomenon is called *differential rotation;* the sun rotates faster at the equator than the poles.

Sunspots are not permanent features on the sun. Rather, they start off as a small blemish on the photosphere called a *pore,* grow to a maximum size which can be tens of thousands of miles across, and then fade away. This entire process takes several months. In addition, the number of sunspots seen on the photosphere varies. Sometimes the sun is covered with literally hundreds of spots while at other times no sunspots can be seen.

Since the time of Galileo, astronomers have been counting the number of

**Figure 5-1.** *The Sun.* Most of the light from the sun comes from the solar photosphere. An ordinary photograph of the sun is a photograph of the photosphere. Sunspots are often seen in such photographs. (*Hale Observatories*)

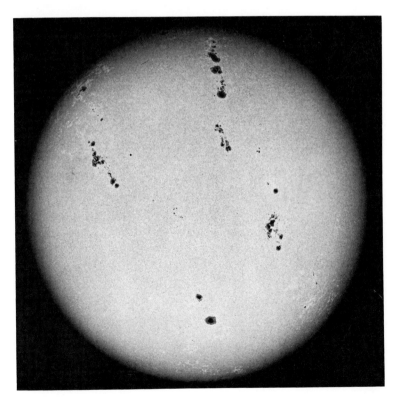

**Figure 5-2.** *The Sunspot Cycle.* The number of sunspots seen on the sun varies with a period of 11 years. Sunspot maxima occurred in 1948, 1959, and 1970. The sun is almost devoid of sunspots in 1954, 1965, and 1976.

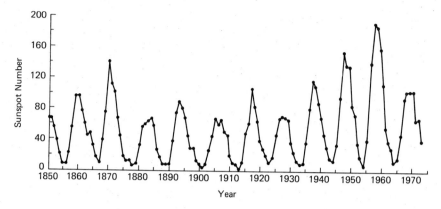

**Figure 5-3.** *A Sunspot Group.* This photograph of a group of sunspots was taken from a balloon carrying a telescope. The darkest areas of a sunspot are called the umbra, while the surrounding greyish regions are the penumbra. Notice that the photosphere is covered with granules. (*Project Stratoscope, Princeton University supported by NSF, ONR, and NASA*)

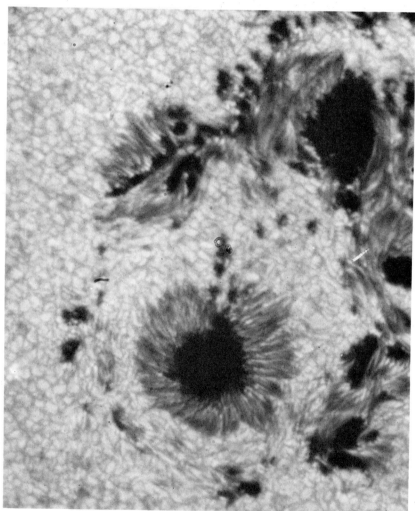

sunspots observed on the solar surface. By the mid–1800s it had become clear that the number of sunspots varies periodically. This phenomenon is called the *sunspot cycle*. Over the course of one sunspot cycle, which lasts about 11 years, the number of sunspots goes from a maximum (hundreds of spots) to a minimum (virtually no spots) back to a maximum again. The time of the greatest number of sunspots is called *sunspot maximum* (in 1948, 1959, 1970, and so forth) while the time of the least number of sunspots is called *sunspot minimum* (in 1954, 1965, 1976, and so forth). Figure 5–2 shows a graph of the number of sunspots observed over the past century.

Figure 5–3 shows an excellent photograph of a group of sunspots. This photograph was taken from a balloon flown at a high altitude to overcome the limitations imposed by the earth's atmosphere. A sunspot consists of two parts: a dark, inner region called the *umbra* surrounded by a lighter region called the *penumbra*.

Pictures taken from balloons at altitudes of 80,000 feet above the earth provide some of the best photographic records of the solar atmosphere. Notice in Figure 5–3 that the photosphere is not uniformly smooth. Instead, the photosphere is covered with numerous *granules* ranging in size from 300 kilometers to 1000 kilometers across. The average diameter of a granule is about 700 kilometers (450 miles).

Granulation occurs uniformly everywhere on the sun. It is also independent of the 11-year solar cycle. Granulation is therefore a permanent feature of the photosphere.

The motion of the gases in the granules can be studied by detailed examination of their spectra. Such spectra exhibit Doppler shifts (see Section 4.3) indicating vertical speeds of about 2 kilometers per second. At the bright center of a granule, hot gases are rising upward while near the darker edge of a granule, cooler gas is sinking. Granules are therefore *convective cells* in which hot gases rise up in the center of a column, flow over the top, dissipate heat, and then sink back down along the sides of the column. Indeed, the central bright region of a granule is about 100°K hotter than the cooler sides. Crudely speaking, the solar photosphere can be likened to a boiling pot of soup on a kitchen stove.

Studying the spectra of light from sunspots has revealed another important fact. In 1908, George E. Hale noticed Zeeman splitting (see Section 4.3) of the spectral lines in sunspot spectra. This means that there must be very intense magnetic fields associated with sunspots. The average strength of the background magnetic field of the sun is roughly the same as the strength of the earth's geomagnetic field. However, inside a sunspot, the magnetic field can be hundreds or even thousands of times more intense than the average background strength. This discovery provided the first important clue in explaining the cause of sunspots.

Many sunspots occur in *bipolar groups,* that is the entire spot consists of two separate and pronounced groups of smaller spots. Figure 5–4 shows a large bipolar group over a series of several weeks revealing solar rotation. Notice how the sunspot actually seems to have two major points of concentration. Following the direction of solar rotation, these two regions of concentration are called the *preceding (p) spot* and the *following (f) spot*. From detailed examination of the Zeeman splitting across the entire group, it is found that these two spots are of opposite polarity. If the

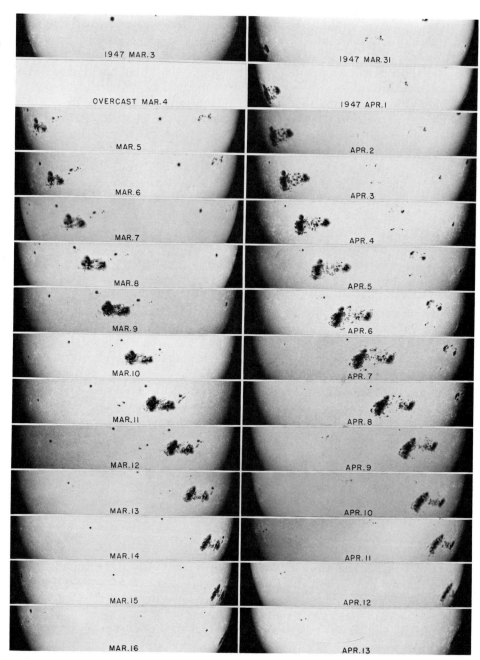

**Figure 5-4.** *A Large Bipolar Group.* This series of photographs taken over several weeks in 1947, shows a large bipolar group during two complete solar rotations. The spot to the right is the preceding (p) spot while the larger one to the left is the following (f) spot. (*Hale Observatories*)

preceding spot has a north magnetic pole, then the following spot has a south magnetic pole, or vice versa. In addition, bipolar groups in the northern and southern hemispheres of the sun have their magnetic polarities ordered in the opposite sense. If the preceding spot north of the solar equator has a north magnetic pole, then the preceding spot south of the equator has a south magnetic pole.

Finally, after observing the Zeeman splitting in sunspots over many years, it was realized that the ordering of the polarity in bipolar groups changes over the solar cycle. If at one time, the preceding spots in the northern hemisphere have north magnetic poles, then eleven years later during the next solar cycle, preceding spots in the northern hemisphere have south magnetic poles.

All these important magnetic properties of bipolar groups comprise the *law of polarity*. Stated succinctly: (1) p spots and f spots in a bipolar group have opposite polarity, (2) p spots in the northern and southern hemispheres have opposite polarity, and (3) the polarity of p spots in each hemisphere reverses sign with each new solar cycle. In view of the last fact, we see that it actually takes 22 years to complete a full solar cycle. Only after 22 years are the polarities of bipolar groups back to the way they were before the cycle began.

In the 1960s, H. A. Babcock proposed a theory to account for the complicated behavior of the magnetic properties of sunspots. Noting that the sun exhibits differential rotation, Babcock realized that this would cause the magnetic field in the sun to become very twisted up. To see why this is so, consider a single *magnetic field line* in the photosphere running from the north to south poles of the sun, as shown in Figure 5–5. Initially, the magnetic field runs directly north to south. But after one solar rotation, since the equatorial regions rotate faster than the polar regions, the field line near the equator becomes stretched out slightly in the direction of the sun's rotation. After two or three more rotations, this stretching becomes even more pronounced. Indeed, after many rotations, the magnetic field in the photosphere becomes completely wrapped around the sun many times. It is important to realize that in the process of wrapping, the magnetic field lines start off crowded together at high latitudes, far from the sun's equator. As the wrapping continues, the crowded portions of the stretched magnetic field lines in both hemispheres migrate toward the equator.

**Figure 5–5.** *Differential Rotation and the Wrapping of Magnetic Field Lines.* Because of the differential rotation of the sun, magnetic field lines become stretched and wrapped around the sun.

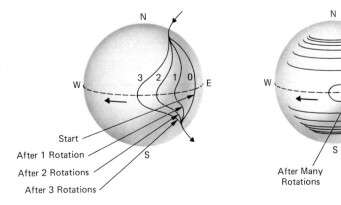

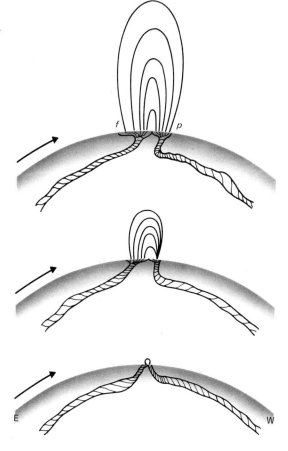

**Figure 5-6.** *The Formation of a Bipolar Sunspot Group.* Three views of the cross-section of the photosphere are shown as a flux rope of tangled magnetic field erupts through the solar surface. The arrows indicate the direction of solar rotation. (*Adapted from Dr. Babcock*)

Now think for a moment about these crowded regions. As the magnetic field lines become increasingly stretched and crowded together, the local intensity of the magnetic field becomes higher and higher. Finally, after becoming extremely twisted, the intensity of the magnetic field is several hundred times stronger than the sun's overall magnetic field. At this stage, the twisted magnetic field becomes buoyant and erupts outward through the photosphere. As illustrated in Figure 5-6, this happens when the tangled magnetic field lines—sometimes called a "flux rope"—develop a kink that punctures the photosphere.

Magnetic fields inhibit and restrict the motions of atoms, ions, and electrons. Thus, where the kink in the flux rope rises through the sun's surface, the gases normally boiling at a temperature of 5,800°K find that their random thermal motions are inhibited. Since the atoms in the gases are confined to move at lower speeds, the temperature of the gas drops. Consequently, according to Stefan's law, this location of concentrated magnetic field emits less light and appears darker than the surrounding solar surface. A sunspot is born!

There are numerous features of Babcock's theory that agree remarkably well

**Figure 5-7. *Bipolar Groups in the Northern and Southern Hemisphere.*** By simply following the direction of the magnetic field lines (upward = north pole; downward = south pole) it is possible to understand why bipolar groups in the northern and southern hemispheres of the sun have their polarities oriented in the opposite sense. (*Adapted from Dr. Babcock*)

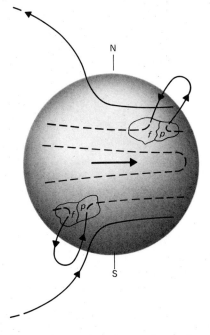

with observation. For example, where the kink punctures the photosphere the magnetic field lines rise up, form a loop, and descend back down into the sun's surface, as shown in Figure 5-6. There are, therefore, two locations where the magnetic field penetrates the photosphere: one where the magnetic field lines come up and one where they go down. Consequently, the resulting sunspot group exhibits two major centers of concentration. One center is the preceding spot while the other is the following spot, according to the direction of solar rotation. In addition, by following the direction of the magnetic field lines, as shown in Figure 5-7, it is immediately clear why bipolar groups in the north and south hemispheres have their polarities ordered in the opposite sense. As shown in Figure 5-7, if the magnetic field is directed upward in the preceding spot in the northern hemisphere then the field lines are pointed downward in the preceding spot in the southern hemisphere.

During the course of a solar cycle, sunspots do not appear randomly all over the sun. Instead, sunspots are confined to a specific range of latitudes north and south of the solar equator. In addition, the exact locations of the sunspots at any one time depend critically on the stage of the solar cycle. Just after sunspot minimum, the first spots appear at relatively large latitudes, 30° north and south of the equator. A few years later, around the time of sunspot maximum, most sunspots occur at moderate latitudes 15° from the equator. And near the end of the solar cycle, the last few sunspots appear comparatively near the equator. This migration of sunspots over the solar cycle is best displayed in a *Maunder butterfly diagram,* as shown in Figure 5-8.

Now recall that as the differential rotation begins wrapping magnetic field lines

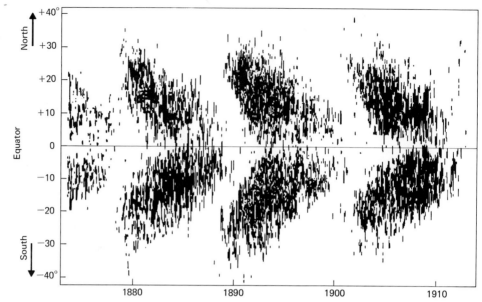

**Figure 5-8. *The Maunder Butterfly Diagram.*** This graph shows the latitudes of sunspots over several solar cycles. At the onset of a solar cycle, just after sunspot minimum, spots appear far north or south of the solar equator. Over the course of the cycle, spots appear nearer and nearer to the equator. (*Adapted from Dr. Maunder*)

around the sun (Figure 5–5), the magnetic field starts piling up at comparatively high latitudes. Consequently, the first sunspots at the beginning of the solar cycle appear comparatively far from the solar equator. As the wrapping continues over the years, the region of concentrated magnetic field migrates towards the equator. And so do the sunspots. The Maunder butterfly diagram therefore has a straightforward explanation in terms of Babcock's theory.

In order to appreciate the full extent of the processes and phenomena on the sun, it is necessary to observe parts of the solar atmosphere other than just the photosphere. An important instrument that makes such observations possible is called a *monochrometer.*

A monochrometer is simply a device, primarily consisting of a filter, that allows the astronomer to look at the sun at *one* specific wavelength. The most fruitful observations are made at wavelengths of 6563 Å (in the middle of an absorption line of hydrogen) and 3934 Å (in the middle of an absorption line of singly ionized calcium). When an astronomer uses a monochrometer (from the Latin meaning "one color") *all* the light from the sun is blocked out *except* light at the desired wavelength. It turns out that the light emitted by the sun at 6563 Å and 3934 Å does *not* come from the photosphere. Rather, the light at these wavelengths comes from higher altitudes in the solar atmosphere, from the so-called chromosphere. In other words, when the astronomer takes a photograph of the sun through a monochrometer, he gets a picture of the chromosphere, *not* the photosphere.

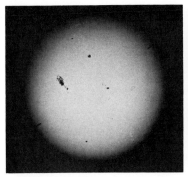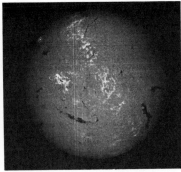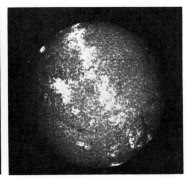

**Figure 5-9.** *Three Views of the Sun.* All three photographs shown here were taken on the same day. At the left is a white-light view. The middle photograph shows the sun as seen in the light of the hydrogen atom while the appearance of the sun in the light of the calcium atom is seen at the right. (*Hale Observatories*)

Photographs taken in the red light of the hydrogen atom at 6563 Å are called *Hα photographs* or *Hα filtergrams,* while those taken in the violet light of the calcium atom are said to be photographed in the *calcium K-line.* Figure 5–9 shows three views of the sun taken on the same day. The first view is just a white light photograph showing the photosphere. The second and third photographs were taken in the light of hydrogen and calcium, respectively, and show the structure of the sun's upper atmosphere. So-called *plages* (bright areas) and *filaments* (dark areas) are seen.

Photographs of the edge of the sun or *solar limb* in the light of calcium or hydrogen atoms are particularly interesting. Such filtergrams show dramatic *prominences* that consist of huge jets of gas gushing up hundreds of thousands of miles out of the solar surface. Prominences can take many different and beautiful shapes that can change very rapidly as the gas rises and falls. Figures 5–10, 5–11, and 5–12 show some different views of typical prominences.

Careful study of the sun at various wavelengths, even in the ultraviolet and X-ray region of the electromagnetic spectrum (see Figure 5–12), reveals that many of the phenomena astronomers observe are closely related. For example, prominences appear bright against the dark sky. But, if an astronomer "looks down" on a prominence and sees it against the sun's disk, by contrast the prominence appears dark and is called a filament. Prominences and filaments are the same thing. In addition, plages and prominences (or filaments) often occur over sunspots. The twisted magnetic field erupting through the photosphere producing a sunspot also pushes gas thousands of miles above the sun's surface giving rise to prominences. This indicates that sunspots are only *one* phenomenon associated with the life cycle of a so-called *activity center* on the sun. As a piece of the twisted magnetic field of the sun bursts through the solar surface, a whole range of phenomena is produced: plages, sunspots, filaments, prominences, and even X-ray bursts. These phenomena associated with an activity center typically last as long as 270 days or ten solar rotations.

**145**

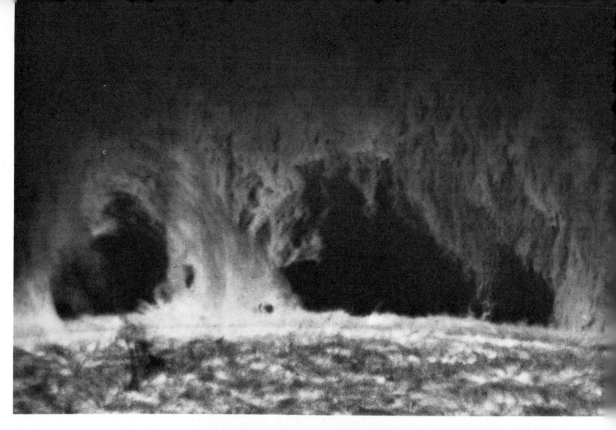

**Figure 5–10. *A Prominence.***
This magnificent view of a solar prominence was photographed on March 31, 1971, in the red light of the hydrogen atom. The prominence rises to an altitude of 40,000 miles above the solar surface. (*Hale Observatories*)

**Figure 5–11. *A Prominence.***
This prominence was photographed in 1958 in the violet light of the calcium atom. The prominence extends over half a million miles above the sun's surface. (*Hale Observatories*)

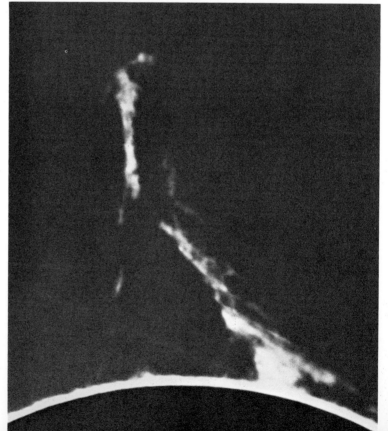

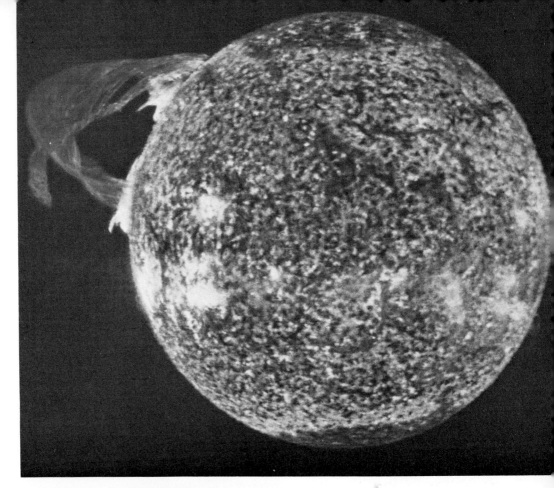

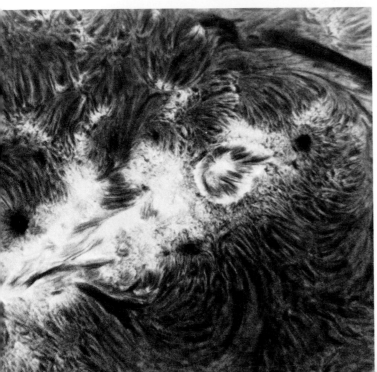

**Figure 5-12.** *The Invisible Sun.* This remarkable photograph was taken by astronauts onboard Skylab in 1973. This view shows the appearance of the sun in the light of singly ionized helium in the far ultraviolet ($\lambda = 304$ Å). A beautiful prominence arching hundreds of thousands of miles into space is seen. (*Naval Research Laboratory/NASA*)

**Figure 5-13.** *An Activity Center.* Sunspots are only one of many phenomena associated with activity centers. Flares, plages, prominences, and filaments occur around activity centers like the one shown in this photograph. (*Hale Observatories*)

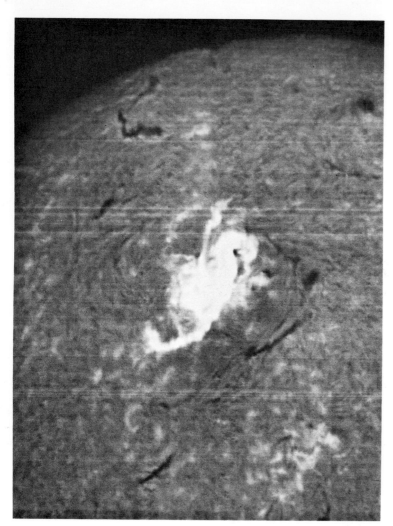

**Figure 5-14.** *A Solar Flare.* A solar flare is a sudden, temporary outburst of energy on the sun's surface. Flares are most easily recognized in photographs taken in the red light of hydrogen, such as the one shown here. This outburst occurred on July 16, 1959. (*Hale Observatories*)

Some of the most violent phenomena associated with activity centers are solar *flares*. Flares are sudden outbursts of energy that are often seen in the vicinity of huge, complex sunspots. Small flares last for only 20 minutes while the most violent flares can last for up to 3 hours. These explosions occur in the chromosphere and can most easily be seen in H$\alpha$ photographs. Figure 5–14 is a good example of one such photograph. A single flare can emit ten trillion megawatts ($10^{27}$ ergs/sec) of energy over the course of its brief life. It is generally believed that flares occur when the magnetic field lines above a huge sunspot group pinch off, break, and reconnect.

Finally, the sun's corona constitutes the outermost regions of the solar atmosphere. The corona is best seen at the time of a total solar eclipse when the blinding disc of the sun is blocked out by the moon. Spectroscopic studies of the light of the

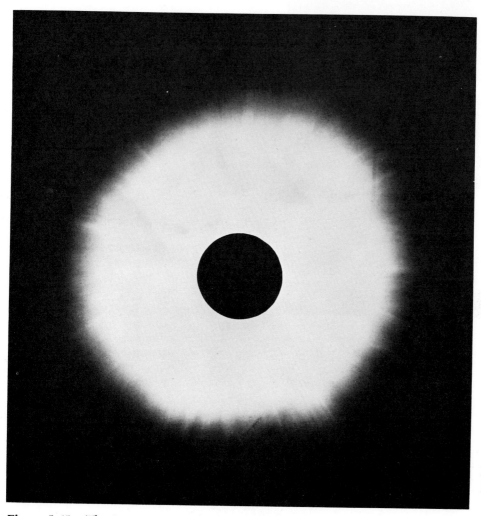

**Figure 5–15.** *The Corona at Sunspot Maximum.* During sunspot maximum, the solar corona has a very circular, undistorted appearance. (*Yerkes Observatory photograph*)

corona reveal that the temperature of this outermost part of the sun's atmosphere is between one and two million degrees! Solar astronomers believe that noise (so-called "acoustic energy") from the violent bubbling and boiling of the gases in the sun's surface heats up the thin gases in the corona to incredibly high temperatures. In other words, the sounds produced by the boiling sun give rise to the beautiful delicate corona we see during a solar eclipse.

As mentioned earlier (see Section 2.4), the shape of the corona changes dramatically over the years. At the time of sunspot maximum, the corona has a very circular and uniform appearance, as shown in Figure 5–15. By contrast, at or near

**Figure 5-16.** *The Corona at Sunspot Minimum.* At sunspot minimum, the solar corona has a very distorted shape. The shape of the sun's corona is primarily affected by the magnetic field of the sun. (*Yerkes Observatory photograph*)

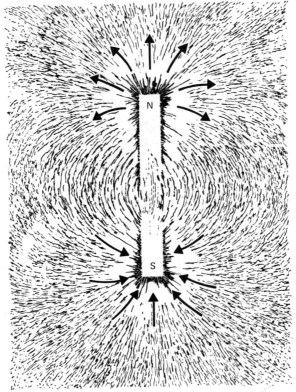

**Figure 5-17.** *Iron Filings Around a Magnet.* When iron filings are scattered around a magnet, they line up with the magnetic field. In this way the iron filings reveal the paths followed by the magnetic field lines.

sunspot minimum, the corona is extremely distorted. As shown in Figures 5–16 and 5–17, the shape of the corona is quite similar to the patterns formed by scattering iron filings around a magnet in the laboratory.

**151**

*Our Star—*
*The Sun*

These changes in the shape of the corona clearly indicate the influence of a magnetic field. Specifically, near the time of sunspot maximum, most of the sun's magnetic field is being twisted up in the photosphere to form sunspots. Therefore, in the absence of a magnetic field far above the sun, the corona has a uniform circular shape. But during the course of the solar cycle, the breaking and reconnecting of magnetic field lines push more and more of the sun's magnetic field up into the outermost layers of the solar atmosphere. Finally, at the time of sunspot minimum, almost all of the sun's magnetic field is in the corona giving it the characteristic appearance shown in Figure 5–16. A new solar cycle starts once again as the magnetic field descends back into the photosphere, this time with its polarity reversed from the previous cycle.

The interrelations between the various levels of the solar atmosphere are truly remarkable. Of course, the sun is the only star in the universe which astronomers can examine in minute detail. It is the only star whose surface features and phenomena can be studied at close range. Man will never know any other star as intimately as the one that is responsible for his very existence.

## *Thermonuclear Energy* 5.2

IN THE early 1900s, scientists realized that they were faced with a puzzling dilemma. They simply could not explain why the sun shines. The source of this dilemma was not astronomy but rather geology. Beginning in the late nineteenth century, geologists were uncovering evidence that rocks were very old, perhaps hundreds of millions of years old. Indeed, the oldest rocks found on the earth today are three billion years old.

The reason for the dilemma was that scientists could not think of any process by which the sun could shine for billions of years. A lump of coal the size of the sun would burn for only a few thousand years. H. von Helmholtz and Lord Kelvin showed that a gradual contraction of the sun could convert gravitational energy into thermal energy thereby producing the sun's light. Unfortunately this process cannot go on for more than a hundred million years and Lord Kelvin concluded that the sun's age was about 20 million years. Kelvin was so puzzled by the discrepancy between his calculations and geological evidence that he suggested the geologists might simply be in error. Undaunted, geologists continued to discover impressive evidence that the earth must be extremely ancient. The great age of the earth—and therefore, logically, the sun—plunged physics into a major crisis around the turn of the century.

One of the major hang-ups in science during the time of Helmholtz and Kelvin was the fact that everyone believed that atoms were indestructable. It was inconceivable that one element could be converted into another. In 1898, however, Henri Becquerel unexpectedly discovered natural radioactivity. Ensuing work by Mme. Curie, Ernst Rutherford, and Frederick Soddy proved that uranium radioactively decays into a host of other elements by emitting particles. This breakthrough in physics provided the first clue to the source of the sun's energy.

In 1905, in an attempt to resolve certain logical inconsistencies in Maxwell's electromagnetic theory, Albert Einstein proposed the *special theory of relativity*. One of the important consequences of this theory is the *equivalence of mass and energy*. Contrary to beliefs of the nineteenth century, Einstein showed that (in theory, at least) matter could be converted into energy. The amount of energy released by such a conversion is staggering. An amount of mass, *m*, is converted into an amount of energy, *E*, according to the famous equation $E = mc^2$, where $c$ is the speed of light. Since the speed of light is so huge (300,000 kilometers per second = 186,000 miles per second), the resulting energy is enormous.

The amount of uranium in the sun is very tiny. Natural radioactive decay could therefore not be the source of the sun's light. But about 73 per cent of the sun's mass is hydrogen. Hydrogen is by far the most plentiful element in the sun.

About 1920, A. S. Eddington and R. d'E. Atkinson independently suggested that perhaps hydrogen is the source of the sun's energy. At the sun's center, temperatures and pressures are so high that the nuclei of hydrogen atoms (i.e., protons) must be constantly colliding with tremendous violence. Normally, protons repel each other because of their positive charge. But in the sun's center, protons are travelling at such high speeds that in spite of electrostatic repulsion, two protons can get very close to each other. At this point *strong nuclear forces* take over. These are the same forces that hold the nuclei of heavy atoms together. Thus, perhaps, protons could combine to form heavier nuclei. Specifically, Eddington proposed the reaction

$$4H \longrightarrow He + energy.$$

To see why this reaction releases energy, recall that the mass of a hydrogen atom is 1.008 atomic mass units while the mass of a helium atom is 4.004 atomic mass units (see Section 4.4). Now compute the difference between the initial and final mass

$$
\begin{aligned}
4 \times 1.008 = \quad & 4.032 \text{ amu (mass of initial hydrogen)} \\
- \; & 4.004 \text{ amu (mass of final helium)} \\
\hline
& 0.028 \text{ amu (mass lost in this transaction).}
\end{aligned}
$$

According to Einstein, the mass lost is converted into energy. For every four hydrogen atoms converted into one helium atom, 0.028 atomic mass units are liberated in energy. The lost mass, 0.028 amu, is 0.7 per cent of the mass of the initial hydrogen. Thus, if 1 gram of hydrogen is converted into helium, then 0.07

grams of matter is turned into energy. The amount of energy released by this conversion of 1 gram of hydrogen is 20,000 kilowatt hours.

In order to produce the sun's energy, 600 million tons of hydrogen must be converted into helium every second! The final result of this reaction is 596 tons of helium with the remaining 4 tons of matter transformed into energy. There is enough hydrogen in the sun to keep this reaction going for a total of about ten billion years. By the 1920s, the resolution of the dilemma of the sun's energy was at hand. But notice that the resolution required the discovery of new laws of physics and a deep understanding of the true nature of physical reality.

It is important to realize that in the thermonuclear reaction 4H $\longrightarrow$ He, four protons combine to give one helium nucleus. Since a helium nucleus consists of two protons and two neutrons, two of the initial hydrogen nuclei must have been converted into neutrons. As mentioned in the previous chapter, this process is called *beta decay*. Each time a proton is converted into a neutron, a positron ($e^+$) is liberated. This positron, or antielectron, carries away the proton's positive electric charge. As a result, electric charge is said to be *conserved* in the reaction. For each unit of electric charge that goes into the reaction, one unit of electric charge comes out.

Around 1930, the famous nuclear physicist Wolfgang Pauli was studying the phenomenon of beta decay. There are certain quantities in physics that traditionally are conserved in all processes. Electric charge is one. Energy is another. The total amount of energy that goes in (including "mass energy" according to $E = mc^2$) must equal the total amount of energy that comes out.

In examining the beta decay of neutrons according to the reaction n $\longrightarrow$ p + e$^-$, Pauli realized that energy was not conserved. The energy possessed by the neutron was not equal to the sum of the energies possessed by the resulting proton and electron. Pauli was therefore faced with two choices. Either energy is not conserved in nuclear physics, or a third unknown particle is liberated in addition to the proton and electron. This unknown particle must be very illusive; at that time there was not one shred of experimental evidence that a third particle was produced by beta decay. Nevertheless, Pauli's deep conviction in the basic conservation laws of physics lead him to choose the second alternative. He postulated the existence of *neutrinos* and *antineutrinos*. One of the primary purposes of these massless, uncharged particles is to carry away some of the energy of the reaction so that there is an overall conservation of energy.

In 1953, two decades after Pauli's enlightened guess, Drs. Cowan and Reines succeeded in detecting antineutrinos. And a few years later, neutrinos were detected. Consequently, physicists now realize that the reaction by which protons are converted into neutrons in thermonuclear processes is properly written as

$$p \longrightarrow n + e^+ + \nu$$

where the Greek letter nu, $\nu$, stands for a neutrino. Every time four hydrogen nuclei are converted into one helium nucleus, two neutrinos must be liberated because two of the protons turn into two neutrons. It took over twenty years to discover neutrinos simply because they are so illusive. Neutrinos easily pass through great quantities of matter. Matter is almost completely transparent to neutrinos.

Unlike the situation in 1920, physicists are now aware of several ways of converting four protons into one helium nucleus. Eddington's proposal that $4H \longrightarrow He$ is just too simple. Furthermore, the probability that four hydrogen nuclei would collide simultaneously is almost zero. Instead there are a few intermediate steps. In 1938, H. A. Bethe and C. L. Critchfield proposed a three-step process called the *proton–proton chain.*

The proton–proton chain begins with the collision of two hydrogen nuclei. One of the protons beta decays into a neutron. The final result is a nucleus of the heavy isotope of hydrogen, $^2H$. This isotope is called *deuterium* and the particle consisting of one proton and one neutron is called a *deuteron.*

The next step in the proton–proton chain is that the deuteron collides with another proton simply resulting in the nucleus of a light isotope of helium, $^3He$. The light helium nucleus contains two protons and one neutron.

The final step involves the collision of two $^3He$ nuclei. When two $^3He$ nuclei collide, four of the six particles stick together as a $^4He$ nucleus and the two remaining protons simply fly off. Notice that the first two steps must occur twice in order for the third step to occur once. The entire proton–proton chain can be written as the following series of reactions

$$^1H + {}^1H \longrightarrow {}^2H + e^+ + \nu$$

$$^2H + {}^1H \longrightarrow {}^3He$$

$$^3He + {}^3He \longrightarrow {}^4He + 2{}^1H.$$

Six protons go in and one helium nucleus and two protons come out. Consequently, the final result of the reaction is still

$$4{}^1H \longrightarrow {}^4He.$$

In 1938, H. A. Bethe proposed a second series of reactions whereby four protons are converted into one helium nucleus. Unlike the proton–proton chain, however, this second scheme involves the use of isotopes of carbon, nitrogen, and oxygen as *catalysts.* The reactions begin when a proton collides with the nucleus of a carbon atom ($^{12}C$). In the following steps, three more protons are swallowed up by the resulting isotopes of carbon, nitrogen, and oxygen. The final result produces a helium nucleus ($^4He$) and gives back the original carbon nucleus. These nuclear reactions are therefore called the *CNO cycle* and may be written as follows

$$^{12}C + {}^1H \longrightarrow {}^{13}N$$

$$^{13}N \longrightarrow {}^{13}C + e^+ + \nu$$

$$^{13}C + {}^1H \longrightarrow {}^{14}N$$

$$^{14}N + {}^1H \longrightarrow {}^{15}O$$

$$^{15}O \longrightarrow {}^{15}N + e^+ + \nu$$
$$^{15}N + {}^1H \longrightarrow {}^{12}C + {}^4He.$$

The proton–proton chain is the dominant nuclear reaction in low-mass stars like the sun. More massive stars, whose central temperatures are higher than 15 million degrees K, are powered by the CNO cycle.

Once the source of energy inside the sun was understood and formulated, scientists could begin making a theoretical model of the sun's interior structure. A central feature of such a model explains how energy is transported from the nuclear reactions at the sun's *core* to the photosphere. Actually, it takes almost a million years for energy from the sun's core to make its way to the solar surface. Photons suffer numerous collisions inside the sun in the arduous journey to the photosphere.

Two different processes are responsible for carrying energy from the core to the surface of the sun. Throughout most of the sun's interior, energy flows outward by *radiative transport*. From the core out to a distance of 0.85 solar radii, photons migrate outward by scattering, absorption, and re-emission. At 0.85 solar radii from the sun's center (that is, at a depth of about 66,000 miles below the photosphere) a second type of energy transfer becomes important. In this outer 15 per cent of the sun, energy is transported by *convection*, a bubbling and boiling process whereby hot gases rise toward the solar surface while cooler gases sink downward.

The sun's interior can therefore be divided into three basic regions, as shown in Figure 5–18. Nuclear reactions (primarily the proton–proton chain) produce the sun's energy at the core. Energy from the core is transported outward through the

**Figure 5-18. *The Structure of the Sun.*** Energy is produced at the sun's core by thermo-nuclear reactions (mostly the proton–proton chain). For 85 per cent of the outward jour-ney, energy is transported by radiative transfer. During the remaining 15 per cent, energy is transported by convection.

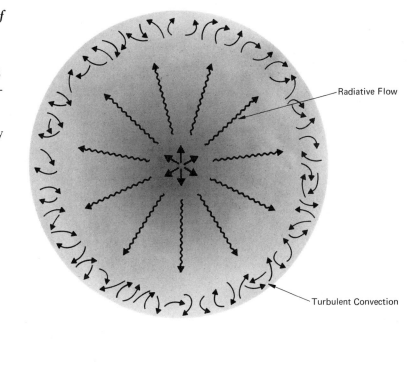

Radiative Flow

Turbulent Convection

*radiative zone* by radiative transport. Finally, on the last leg of the outward journey, the sun's matter becomes quite opaque to photons and radiative transfer is no longer an efficient process. Convection is the dominant means of energy transport for the outer 15 per cent of the sun's interior. This outermost region of the solar interior is therefore called the *convective zone.* The boiling of the sun's gases in this zone is related to many of the phenomena associated with the solar cycle.

As mentioned earlier, it takes a very long time for photons to migrate from the sun's core to the photosphere. As a result of the numerous collisions suffered by the light in its outward journey, photons that finally leave the sun's surface no longer contain any information about their origin. By observing light from the sun, astronomers therefore cannot hope to obtain any direct information about the nuclear reactions at the sun's center. Everything scientists believe about the sun's interior is *inferred* from the laws of physics and complicated mathematical calculations. Surely it would be advantageous to develop some means of "observing" the sun's center to verify some of the ideas about the sun's internal structure.

Photons suffer numerous collisions in their arduous outward migration, but most matter is totally transparent to neutrinos. Every time four hydrogen nuclei are converted into a helium nucleus, at least two neutrinos are released. These neutrinos easily penetrate the sun and escape into space with virtually no collisions at all. With almost no collisions, the neutrinos do not have the opportunity to lose any information about the sun's center where they were created. If, somehow, scientists could build "neutrino telescopes," they would have the opportunity to "see" what is going on at the sun's center.

The proton–proton chain mentioned earlier in this section is believed to be the dominant nuclear reaction in the sun. In this reaction, two neutrinos are liberated every time four hydrogen nuclei are converted into one helium nucleus. Nuclear physicists were quick to realize that detecting these neutrinos would be extremely difficult. Only very energetic neutrinos would possibly be detected here on earth, and the neutrinos produced by the fusion of two protons into a deuteron have comparatively little energy.

The final reaction in the proton–proton chain discussed earlier ($^3$He + $^3$He $\longrightarrow$ $^4$He + $2^1$H) occurs 91 per cent of the time. During the remaining 9 per cent of the reactions, one of two different processes occur. One of these two reactions turns out to be especially interesting to scientists who want to detect neutrinos from the sun.

Once in a thousand times, the proton–proton chain terminates in a reaction different from the way described earlier. The process starts off just as before, namely,

$$^1H + {}^1H \longrightarrow {}^2H + e^+ + \nu$$
$$^2H + {}^1H \longrightarrow {}^3He$$

but now the $^3$He nucleus combines with a $^4$He nucleus to produce an isotope of beryllium, $^7$Be, according to

$$^3\text{He} + {}^4\text{He} \longrightarrow {}^7\text{Be}.$$

At this point, three protons have been used up. The fourth proton necessary for the conversion of hydrogen into helium now combines with the $^7\text{Be}$ nucleus according to

$$^7\text{Be} + {}^1\text{H} \longrightarrow {}^8\text{B}$$

and results in an isotope of boron that beta decays into another isotope of beryllium, $^8\text{Be}$, according to

$$^8\text{B} \longrightarrow {}^8\text{Be} + e^+ + \nu.$$

The $^8\text{Be}$ nucleus is very unstable and promptly breaks into two helium nuclei according to

$$^8\text{Be} \longrightarrow 2{}^4\text{He}.$$

Thus, the process gives back the original $^4\text{He}$ nucleus plus a second $^4\text{He}$ nucleus resulting from the combination of four protons. The final result is therefore the same as before, namely $4{}^1\text{H} \longrightarrow {}^4\text{He}$, except that in this rare reaction, a helium nucleus is used as a catalyst and involves the momentary creation and destruction of isotopes of boron and beryllium.

The important point to realize is that in the middle of this reaction (specifically, $^8\text{B} \longrightarrow {}^8\text{Be} + e^+ + \nu$) highly energetic neutrinos are produced. In the mid-1960s, it was realized that it should be possible to detect the neutrinos from this beta decay of boron.

In 1967, Dr. Raymond Davis and his colleagues at Brookhaven National Laboratories took up the challenge of detecting energetic neutrinos from this rare branch of the proton–proton chain. Davis realized that these neutrinos should convert a chlorine isotope ($^{37}\text{Cl}$) into agron ($^{37}\text{A}$) by the reaction

$$^{37}\text{Cl} + \nu \longrightarrow {}^{37}\text{A} + e^-.$$

Therefore, if Davis could get enough chlorine together in one place, the production of argon would confirm the occurrence of the expected nuclear reactions at the sun's center. By detecting the creation of argon in a large quantity of chlorine, Davis would be indirectly observing the nuclear "fires" at the solar core.

The experiments began in 1968. Dr. Davis constructed a huge tank containing 100,000 gallons of common cleaning fluid, tetrachloroethylene ($C_2Cl_4$). To shield it from extraneous radioactivity, the tank (shown in Figure 5–19) is buried 4,850 feet in the ground in a mine in South Dakota.

The central idea behind Davis' experiment is very simple. Every once and a while, an energetic solar neutrino should hit a chlorine nucleus in the cleaning fluid, thereby converting it into argon. At regular intervals, Davis would flush the tank

**Figure 5–19.** *A Neutrino Trap.* This tank contains 100,000 gallons of common cleaning fluid. This fluid is composed mostly of chlorine and Dr. Davis expected to confirm
the existence of neutrinos from the sun by detecting the conversion of chlorine to
argon inside the tank. (*Courtesy of Dr. Davis, Brookhaven National Laboratory*)

and pass the cleaning fluid through sensitive chemical devices that measure the amount of argon produced in the liquid. From detailed theoretical solar models, Davis knew exactly how much argon to anticipate.

This solar neutrino experiment has been operating for almost a decade. In spite of precautions and meticulous attention to every conceivable detail, it is now entirely clear that Davis and his colleagues are *not* detecting the anticipated amount of argon. Indeed, many times Davis finds little or no argon in the cleaning fluid. Since an unusually small amount of argon has been detected in the experiment, it seems logical to conclude that neutrinos are simply *not* coming from the sun. Something is very wrong!

Up until the work of Dr. Davis, astronomers felt that they had a good understanding of the sun. Thermonuclear reactions (that is, the reactions in the proton–proton chain) convert hydrogen into helium. In these reactions, neutrinos are produced. Neutrinos from one "branch" of the proton–proton chain should convert chlorine to argon here on earth. The fact that careful scientists have failed to detect the anticipated amount of argon is a source of great concern. This entire matter is now referred to as "the mystery of the missing solar neutrinos." Scientists are beginning to suspect that either they do not understand nuclear physics, or they really do not understand the sun . . . or both!

What could be wrong? One possibility is that in traveling the 93 million miles from the sun, neutrinos radioactively decay into other particles. In other words, neutrinos *are* being created at the sun's center, but by the time they get to the earth they have turned into some other kind of particle. This speculation, while it would explain why Davis is not detecting neutrinos, is not based on one shred of experimental evidence.

Another possibility is that no neutrinos are being created in the sun at all. But if nuclear physicists understand anything, this means that *the sun has shut off*. In desperation, several astrophysicists explored this alternative during the mid-1970s. If true, it means that the sun will grow dimmer during the next several thousand years and plunge the earth into an incredible ice age.

Another approach involves introducing complications into theoretical solar models that would inhibit the beryllium–boron branch of the proton–proton chain. For example, if the sun had a rapidly spinning core or a large interior magnetic field, the reactions in the beryllium–boron branch would very rarely occur, thereby producing very few neutrinos. One team of scientists have even proposed that a small *black hole* might exist at the sun's center and is swallowing up all of the neutrinos.* In any case, it is clear that astrophysicists are becoming quite desperate in reconciling their understanding of the sun with the negative results of Davis' experiment.

As implausible as it may sound, this bizarre experiment with cleaning fluid buried in South Dakota could have some far-reaching implications for the future of

---

* For a discussion of black holes, the reader is referred to a recent book by the author of this text: *The Cosmic Frontiers of General Relativity* (Little, Brown & Co., 1977).

mankind. It is conceivable that, at the current rate of usage, all the reserves of fossil fuels (that is, coal, oil, and natural gas) will be depleted before the end of this century. Scientists have therefore begun to search for alternative sources of energy. Nuclear reactions using uranium or plutonium have the severe disadvantage that they produce large amounts of radioactive waste.

Thermonuclear fusion looks like an attractive alternative. Back in the 1950s, scientists succeeded in building devices that reproduce the reactions in the sun in an uncontrolled fashion. These devices that convert hydrogen into helium with the accompanying release of vast quantities of energy are called *hydrogen bombs*. More recently, physicists have turned their attention to producing *controlled* thermonuclear fusion. It is their hope that sometime before the end of the century, they will be able to build machines that convert hydrogen into helium at a controllable and manageable rate. The primary fuel, hydrogen, for such machines will come directly from the ocean.

All efforts at producing controlled thermonuclear fusion have met with little success. Scientists still have a long way to go. And now they are beginning to get very worried. All their efforts are based on modern nuclear physics—the same nuclear physics that predicts neutrinos from the sun. There seems to be a major flaw in our understanding of why the sun shines. Could this same flaw be related to the lack of success in producing controlled thermonuclear fusion?

Resolving the mystery of the missing solar neutrinos may be one of the most important developments in modern science. The full explanation will give a deep insight into the nature and properties of the sun's interior. And it may provide mankind with virtually limitless energy for many centuries into the future.

# 5.3  *A Comparison with Other Stars*

FOR thousands of years men have looked up into the sky and wondered at what they saw. Man's preoccupation with the stars, the planets, and the sky dates back to the earliest times and almost certainly was a part of prehistoric cultures. Yet in spite of this universal preoccupation, it is truly remarkable to realize how incredibly recent man's current understanding of the universe really is. When you pick up a modern astronomy book, to a large degree you are faced with facts and theories only a few decades old. Indeed large portions of this text were unthinkable in the 1960s.

A good example of man's changing concepts of the universe deals with the stars. Up until the mid-1800s, astronomers really did not know where the stars are. Of course, there were many catalogues specifying right ascensions and declinations, but the exact positions in space were unknown. Are the stars fairly nearby, just beyond the edge of the solar system? Or are they incredibly remote, trillions upon trillions of miles from the earth?

In 1600, Giordano Bruno, one of Galileo's contemporaries, was burned at the stake for his heretical belief that the stars were like suns (perhaps with planets orbiting them) at very great distances. He was at least two centuries ahead of his time; it was not until 1838 that astronomers first succeeded in measuring vast stellar distances.

In the early 1800s, astronomers finally had the necessary optical and mathematical tools necessary to measure the distances to the nearest stars. The method used in these observations is called *parallax* and the basic idea is very simple. Imagine looking at a nearby object, such as a telephone pole, against a distant background of mountains, as shown in Figure 5–20. As you move from one location to another, the nearby object will appear to shift in relation to the distant background. This technique of parallax can be applied directly to the measurement of stellar distances. The earth orbits the sun and therefore, as seen from the earth, nearby stars should appear to move slightly with respect to distant background stars, as shown in Figure 5–21. In other words, if the astronomer observes a star field containing a nearby star on two separate occasions a few months apart, he will notice a small shift in the location of that nearby star. The bigger the shift, the nearer the star. By measuring

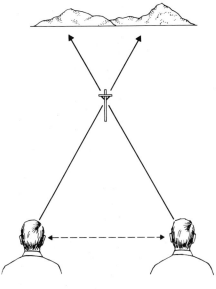

**Figure 5–20. *Parallax.*** Parallax is simply the effect whereby near objects appear superimposed against different parts of a distant background, depending on the location of the observer.

**Figure 5–21. *Stellar Parallax.*** As the earth orbits the sun, the location of nearby stars against the background of distant stars will appear to change slightly. From careful observations of these slight changes, astronomers can calculate the distances to nearby stars.

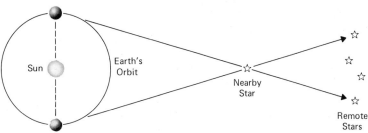

the size of this shift over a known period of time, the astronomer can easily calculate the distance to the star.

The problem with parallactic measurements is that all stars, even the nearest ones, are very far away. Therefore, the *parallactic angle* that the astronomer tries to measure is always very small. Today, astronomers have measured the parallaxes of thousands of stars. However, only about 700 stars are near enough so that their distances have been determined with reliable accuracy. These distances are often expressed in *light years,* the distance light travels in one year, equal to 6 trillion miles. For example, the nearby star, α Centauri, is about 5 light years away.

The measurement of stellar distances was profoundly important to the course of astronomy. To see why this is so, imagine looking up at the star-filled nighttime sky. You see bright stars and you see dim stars. But after a moment's thought, you realize that the apparent magnitudes of the stars do not tell you fundamental facts about the stars themselves. A truly bright star may appear dim in the sky simply because it is extremely far away. On the other hand, a bright-appearing star may actually be a dim star that just happens to be relatively nearby. In order to discover the true nature of stars, the astronomer would prefer to know the star's "real brightness" or *absolute magnitude.* This would tell him how bright the star really is. It would tell him the star's *luminosity,* that is, how many times brighter or dimmer a star is compared to the sun.

When astronomers were finally able to measure the distances to the stars, they realized that they could calculate the absolute magnitude of stars. If an astronomer knows both the apparent magnitude (from measuring how bright it appears in the sky) and the distance (from measuring the parallax), he can easily calculate the star's real brightness. For example, Polaris in the Little Dipper, has an apparent magnitude of $2\frac{1}{2}$. To the naked eye, the star has a moderate brightness. But the distance to Polaris is about 900 light years; it is quite far away. From knowing Polaris' apparent magnitude and distance, calculations reveal that the star is 10,000 times brighter than the sun. Polaris must be shining with a luminosity of 10,000 suns in order for it to appear as a $2\frac{1}{2}$ magnitude star in the sky at a distance of 900 light years.

Today, astronomers realize that the absolute magnitudes of stars cover a wide range. The brightest known stars cover a wide range. The brightest known stars have a luminosity of one million suns. And the dimmest stars shine with a luminosity of only a ten thousandth that of the sun. It would take 10,000 of the dimmest stars to give off as much light as our sun.

In addition to knowing the luminosities, astronomers can also discover the surface temperatures of stars. From the discussion of the nature of light (see Section 4.1), recall that the temperature of an object has a direct effect on the wavelength of light emitted by the object. According to Wien's law, a blackbody with a temperature of 4,000°K emits primarily red light, while a blackbody at a temperature of 10,000°K emits lots of blue light. Realizing this, astronomers conclude that reddish-appearing stars in the sky (for example, Betelgeuse and Antares) must be cool with surface temperatures around 3,000°K. On the other hand, bluish-appearing stars (for example, Rigel and Spica) must be hot; their surface temperatures must be about 15,000°K. Indeed, there is a direct correlation between the color of a star and its surface temperature. Just by measuring the colors of stars, the astronomer can deduce their surface temperatures.

Another approach to determining the temperatures of stars involves spectra. By passing the light from a star through the prism of a spectroscope, the light is broken up into the colors of the rainbow. The resulting spectrum shows absorption lines caused by the chemicals in the star's atmosphere. Naturally, a particular chemical must be present in the star in order for its spectral lines to show up. But additionally, the temperature of the star's atmosphere has a profound effect on which chemicals produce strong, dark lines and which produce faint, weak lines, or no lines at all. To see why this is so, we must think for a moment about the atoms in a stellar atmosphere.

Consider a cool star with a surface temperature around 3,000°K. Stars are composed of mostly hydrogen and helium while other elements are present only in "trace" amounts. But the hydrogen and helium in the atmosphere of a cool star cannot produce any spectral lines. Even though they are the most abundant chemicals, the temperature of the star is so low that the electrons in hydrogen and helium atoms are almost never excited up out of their ground states. Since spectral lines are caused by electrons jumping from one allowed orbit to another in an atom (see Section 4.3), hydrogen and helium do not form any spectral lines.

In cool stars, the temperature is so low that molecules can form. In particular, titanium combines with oxygen to form titanium oxide (TiO). Titanium oxide produces numerous spectral lines at visible wavelengths when heated to a temperature of around 3,000°K. Consequently, even though titanium oxide is a relatively rare chemical, the spectra of cool stars show broad bands of spectral lines caused by this molecule. Any time an astronomer sees titanium oxide bands in a stellar spectrum, he immediately concludes that he is observing a cool star.

In very hot stars, with temperatures of 20,000°K or higher, most atoms are highly ionized. In the hottest stars, even hydrogen has lost its electrons. Thus, again hydrogen cannot produce spectral lines. Only helium manages to retain one or both of its electrons and thus the spectra of very hot stars show helium lines.

Only if the surface temperature of a star is around 10,000°K are conditions just right for the formation of hydrogen lines. The temperature is high enough to excite electrons out of their ground states but not so high that the hydrogen atoms become ionized.

Even though all stars have roughly the same chemical composition, their spectra differ widely. Figure 5–22 shows some typical stellar spectra. Which spectral lines appear depends on how hot the star is. By identifying the lines in a star's spectrum, astronomers can immediately deduce the star's surface temperature. The coolest stars in the sky have surface temperatures around 3,000°K while the hottest stars have surface temperatures of roughly 25,000°K.

In 1911, the Danish astronomer Ejnar Hertzsprung came up with the idea of drawing a graph to compare the luminosities and surface temperatures of stars. This work was repeated independently by the American astronomer Henry Norris Russell two years later. In many respects, the graph first devised by these two

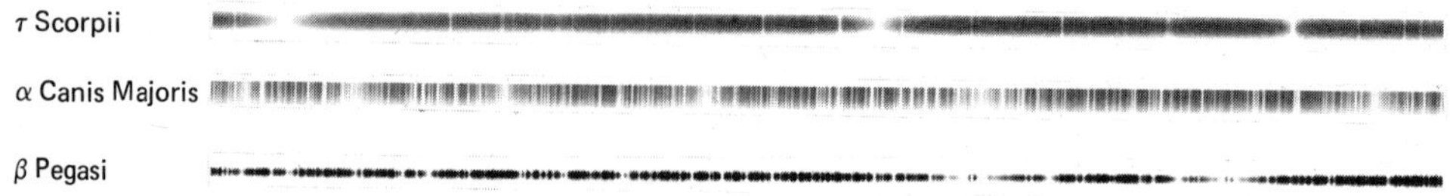

τ Scorpii

α Canis Majoris

β Pegasi

**Figure 5-22.** *Typical Stellar Spectra.* The appearance of stellar spectra can vary widely from one star to the next. The upper spectrum is of a very hot star (τ Scorpii), whose surface temperature is about 25,000°K. The middle spectrum is of Procyon (α Canis Minoris), whose surface temperature is about 7,000°K. The lower spectrum is of a cool star (β Pegasi), whose surface temperature is about 3,000°K. (*Hale Observatories*)

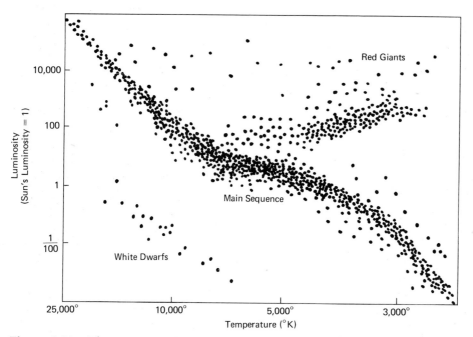

**Figure 5-23.** *The Hertzsprung-Russell Diagram.* In plotting the luminosities of stars against their surface temperatures, the dots on the resulting graph cluster into three main groups: main-sequence stars, red giant stars, and white dwarf stars.

scientists is perhaps the most important diagram in all of astronomy and astrophysics! It is appropriately called the *Hertzsprung–Russell diagram* or *H–R diagram* for short.

One way of drawing an H–R diagram is simply to plot the luminosities of stars against their surface temperatures, as shown in Figure 5–23. Every dot represents a star whose luminosity and surface temperature are known.

The remarkable feature of the H–R diagram is that the dots are not scattered randomly all over the graph. Instead, three prominent groupings stand out. The dots representing most of the stars visible to the naked eye fall along a band stretching from the upper left (hot, bright, bluish stars) to the lower right (cool, dim, reddish stars) on the H–R diagram. This band is called the *main sequence.* Stars whose dots lie in this band are called *main sequence stars.* The sun's luminosity (1 sun) and surface temperature (6,000°K) place it on the main sequence. Our sun is therefore a typical main sequence star.

A second major grouping on the H–R diagram occurs near the upper right-hand corner. These dots represent cool stars (3,000 to 4,000°K) with very high luminosities. These stars are very bright because they are extremely large. They are therefore called *red giant stars* (*red* for their color and *giants* for their size). Almost every bright, reddish-appearing star seen in the nighttime sky is a red giant. These stars are so big that if one were placed at the center of the solar system, its surface would extend almost out to the orbit of Mars!

Finally, there is a small cluster of dots in the lower left-hand corner of the H–R diagram. These dots represent dim, hot stars. They are called *white dwarf stars* due to their color and size. A typical white dwarf is about the same size as the earth and has a surface temperature of around 25,000°K so that it shines with a bluish-white light.

Stars shine by emitting light. In order to give off light, they use up energy. As they consume their nuclear fuels, their structure changes. When the luminosity and temperature of a star changes, its dot on the H–R diagram must move around. In practice, no such changes have ever been observed. The reason is that the life cycles of stars extend over billions of years. Major milestones representing significant changes in the structure of a star are separated by hundreds of millions of years, far longer than the span of recorded history or the memory of men.

The H–R diagram may be compared to one frame of a motion picture film. Suppose someone were to give you one frame from a reel of movie film and ask you: "From what you see in this one picture, tell me the plot of the movie." To the astronomers, the H–R diagram is that one picture. We simply do not live long enough to see the whole movie or even a small segment. Yet, changes in the structure of a star must obey the laws of physics. By applying the laws of physics to theoretical models of stars, astrophysicists are able to calculate how stars are born, how they mature, and what happens to them when they grow old and die. In doing so, astrophysicists learn how the dot representing a star moves around on the H–R diagram. Understanding the meaning of the H–R diagram is therefore the same as understanding the process of *stellar evolution.*

In order to calculate theoretical models of stars, the astrophysicist requires an additional important piece of information. He needs to know how much matter is contained in a star; he needs to know the *masses* of stars. Fortunately, about half of the stars in the sky are not single stars like the sun. Instead, they are *double stars,* two stars orbiting their common center like the earth and the moon. In some cases, astronomers can actually see the two stars revolving about each other. A series of

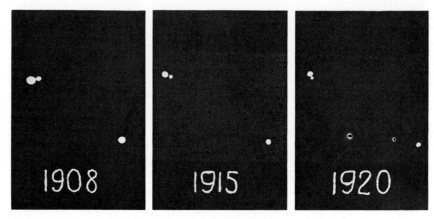

**Figure 5–24.** *Kruger 60.* These photographs spanning many years reveal two stars orbiting each other. This stellar system is called Kruger 60 and is a good example of a binary star. (*Yerkes Observatory photograph*)

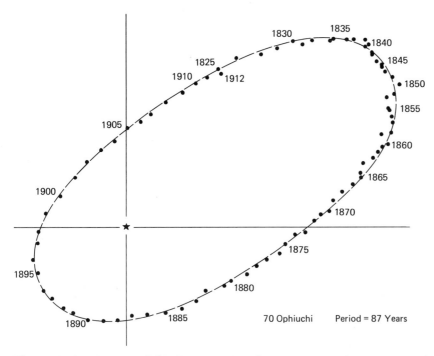

**Figure 5–25.** *A Binary Orbit.* By measuring the separation and orientation of the two stars in a binary, the astronomer can draw their orbit. The orbit of 70 Ophiuchi, whose period is 87 years, is shown here.

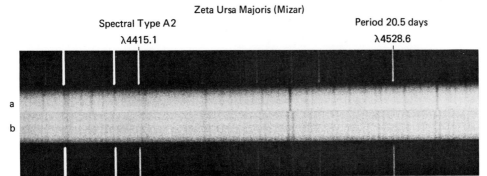

Zeta Ursa Majoris (Mizar)

Spectral Type A2       Period 20.5 days

$\lambda 4415.1$       $\lambda 4528.6$

(a) June 11, 1927. Lines of the two components superimposed. (One star is moving to the left while the other star is going to the right)

(b) June 13, 1927. Lines of the two components separated. (One star is coming toward the earth while the other star is receeding)

**Figure 5-26.** *A Spectroscopic Binary.* The star Mizar in the Big Dipper is a spectroscopic binary. Two spectra are shown. In the lower view all the spectral lines are doubled. One set of lines is from the approaching star (slightly blueshifted) while the other set is from the receding star (slightly redshifted). In the upper view, one star is moving to the left while the other is moving to the right. Hence there is no net shift for either set of lines and they appear superimposed on each other. (*Hale Observatories*)

photographs in Figure 5–24 shows a typical double star. Observations spanning many years result in drawings of the orbits, as shown in Figure 5–25. Double stars whose motions can be observed are called *visual binaries.*

Most double stars are so far away or so close together that it is impossible to see two separate stars. Nevertheless, the astronomer can discover unresolvable double stars by taking a spectrum. As the two stars in a binary system revolve about their common center, they are alternately coming toward or moving away from the earth. Consequently, their spectral lines are alternately redshifted and blueshifted as a result of the Doppler effect (see Section 4.3). Such double stars are called *spectroscopic binaries* and from studying the shifting of the spectral lines, the astronomer can obtain information about the orbital speeds and periods of the stars.

A third important type of double star is the *eclipsing binary.* Sometimes the orbit of a double star is oriented in space so that, as seen from earth, one star alternately eclipses the other. Although the astronomer often cannot resolve two separate stars, he is easily convinced that he is observing an eclipsing binary by the characteristic way in which the total magnitude of the system varies in a periodic fashion, as shown schematically in Figure 5–27. By measuring the total magnitude of the system for many nights, the astronomer can draw a *light curve* that is simply a plot of the brightness of the binary over a period of time. By noting the times, shapes, and

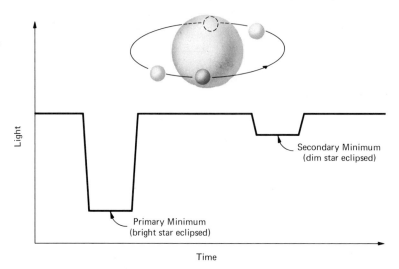

Figure 5–27. *An Eclipsing Binary.* An eclipsing binary is a double star in which the two stars alternately pass in front of each other. By measuring the total magnitude of an eclipsing binary for many nights, the astronomer can draw a light curve. From the shape of the light curve, many of the properties of the binary can be deduced.

locations of the *minima* (that is, the "dips" in the light curve that occur during eclipses) many of the properties of the binary can be deduced.

The whole point behind studying binary stars is that from careful observation, many of the features of the stars' orbits about each other can be discovered. If the orbits are known, the astronomer can use Newtonian mechanics to calculate the masses of the individual stars. It is like the "inverse" problem of calculating the trajectories of Apollo astronauts to the moon. In that case, the masses of the earth and moon are known, and the orbit must be calculated. With a binary, in principle the orbit is known and the masses must be calculated. In practice, information about the orbit of the two stars in a binary is usually incomplete. As a consequence, the astronomer often cannot unambiguously determine the individual masses of each of the two stars. Best results are obtained when a double star is, for example, both an eclipsing and spectroscopic binary. This provides two separate sources of data from which the calculations can be attached.

The final result of many years of observing binaries is that the masses of many stars have been determined. It is convenient to express stellar mass in terms of *solar masses,* that is, how many times larger or smaller the mass of a star is compared to the sun. The least massive stars typically contain only $\frac{1}{10}$ solar mass while the most massive stars have 60 solar masses.

In discovering the stellar masses, it was noticed that the masses of main sequence stars are correlated with their luminosities. Specifically, low-mass stars are dim while high-mass stars are very bright. Figure 5–28 shows this correlation in the form of a graph called the *mass–luminosity relation*. This relation is an important tool in astronomy. When the astronomer discovers a main-sequence star in the sky, he can turn to this graph to obtain a reliable estimate of the star's mass. For example, from Figure 5–28, a main sequence star that is 100 times brighter than the sun has a mass about 3 times greater than the sun.

We now have a good idea about the ranges of properties of stars. Stellar luminosities extend from $\frac{1}{10,000}$ suns to 1 million suns. Surface temperatures extend from 3,000°K to about 25,000°K. And stellar masses lie between $\frac{1}{10}$ suns to 60 suns. But in examining random stars in the sky, astronomers find that their properties are not uniformly distributed over these ranges. Stars of moderately low mass, moderately low temperature and moderately low luminosity are much more common than stars with extreme temperatures, masses, and luminosities. Thus, for example, there are a lot more 1 and 2 solar mass stars than there are with masses of 40 or 50 suns. Our sun (luminosity = 1 sun, mass = 1 sun, temperature = 6,000°K) is therefore a very ordinary, typical, "garden variety" star.

Knowing the masses, temperatures, and luminosities of stars, the astrophysicist finds that he is in a position to calculate the life cycles of stars. By applying the laws of physics to a hypothetical ball of gases, he can discover what must be going on in order for that ball of gas to look like a star.

Newly created stars contain great quantities of hydrogen. When the temperatures at the centers of these stars reach about ten million degrees, thermonuclear

**Figure 5-28.** *The Mass–Luminosity Relation.* The masses and luminosities of main-sequence stars are correlated, as shown in this graph. The brighter the star, the higher its mass.

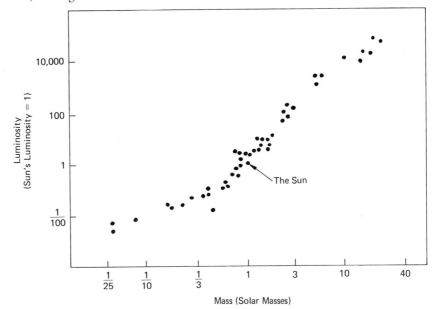

reactions are ignited at their cores. Hydrogen nuclei are fused into helium nuclei by either the proton–proton chain or the CNO cycle. This is a very stable situation for a star; the sun has enough hydrogen at its core to support this nuclear reaction for a total of ten billion years. Calculations concerning theoretical stars with *hydrogen burning* at their centers reveal that their resulting luminosities and surface temperatures place them on the main sequence in the H–R diagram. This is the true meaning of the main sequence. All main-sequence stars are comparatively young; they all have hydrogen burning occurring at their centers.

Eventually, all the hydrogen at the core of a main sequence star is used up. When all the hydrogen in a star's core is gone, hydrogen burning (that is, either the proton-proton chain or the CNO cycle) shuts off. With no more outpouring of energy, the star's core begins to collapse inward as a result of the enormous weight of the star's outer layers. The temperature deep inside the star therefore gets higher and higher. Soon it becomes so hot above the star's core that hydrogen burning can be ignited in a thin shell surrounding the core. After all, there is still plenty of fresh hydrogen between the star's core and its surface. As *hydrogen shell burning* is ignited, the new outpouring of energy begins pushing the star's atmosphere outward and the star starts to expand dramatically.

When the temperature at the center of the star's contracting core reaches 100 million degrees, nuclei of helium atoms are fused together to form carbon according to the reaction

$$3\,^4He \longrightarrow \,^{12}C.$$

This is called *helium burning* or the *triple alpha process*. By the time that helium burning is ignited at a star's core, the outer atmosphere of the star has expanded to an enormous size. This in turn caused the star's atmosphere to become quite cool. Stars powered by helium burning therefore have hot cores but huge cool atmospheres. Since they are gigantic, they are bright. Since they are cool (3,000 to 4,000°K) they emit lots of red light. They are red giants!

This is the true meaning of the *red giant branch* on the H–R diagram. These stars have exhausted all the hydrogen at their cores and are shining as a result of helium burning. They are "middle-aged" stars. In about five billion years, our sun will begin to turn into a red giant.

Eventually all the helium at the core of a red giant is used up and helium burning shuts off. If the star is very massive, temperatures can get high enough to ignite even more complex nuclear reactions. *Carbon burning, oxygen burning,* and *silicon burning* occur at the cores of the most massive stars near the end of their life cycles. In these reactions, many of the heavy elements are created.

Near the end of a star's life cycle, complicated nuclear reactions inside an old star can result in *instabilities*. The structure of the star becomes unstable and the star blows up, ejecting large quantities of matter into space. The burned-out core of the dying star no longer contains any nuclear fuels. It simply contracts under the inward force of gravity. This contraction stops when the star has shrunk down to the size of the earth. Such an object is hot (surface temperature around 25,000°K) and, because of its small size, is quite dim. It is a white dwarf! White dwarfs are dead stars containing about one solar mass of matter.

Astronomers have good reason to believe that hydrogen and helium were the only elements present in the universe shortly after the "creation event" some 20 billion years ago. Yet trees, rocks, people, and planets are made out of many heavy elements. These chemicals must have been created by the nuclear processes deep inside ancient stars. As these stars ended their lives, they cast out many of these heavy elements into space thereby *enriching* the *interstellar medium.* Billions of years later, new stars condensed out of the enriched gas floating in space. Our sun is one of these "later-generation" stars. For this reason, the sun's spectrum contains many lines from heavy elements. More importantly, however, the existence of heavy elements permitted the formation of the other objects in the solar system. The earth, the moon—indeed virtually every atom in your body—were created long ago deep inside an ancient, dead star.

## *Our Corner of the Universe* 5.4

LOOKING up at the night sky, we see a wide range of phenomena. The moon, planets, stars—perhaps an occasional comet or meteor—and the Milky Way are among the objects that inspired our ancestors to take up the study of astronomy. Of all these objects, the Milky Way was destined to remain the most mysterious for the longest period of time. This hazy band of light stretches all the way around the sky and was a source of fanciful myths and legends in ancient cultures.

In the early 1600s, Galileo turned his telescope toward the Milky Way and

**Figure 5–29.** *The Milky Way.* This mosaic of several wide-angle photographs along the Milky Way extends from Sagittarius to Cassiopeia. (*Hale Observatories*)

discovered that it consisted of millions and millions of very dim stars. Over the next two centuries, astronomers became convinced that our star, the sun, was just one of billions of stars in a large disk-shaped object called the *Milky Way Galaxy*. Indeed, in 1785, William Herschel attempted to discover the sun's true location in the Galaxy. Herschel reasoned that if we are near the edge, a "thinning" in the numbers of stars should be seen along one portion of the Milky Way. Alternatively, if we are near the center, the same number of stars should be seen in any given portion along the Milky Way. After examining 683 selected regions over the sky, Herschel (erroneously!) concluded that we are at the center of the Galaxy. He did not realize that his view through the Milky Way was severely restricted by interstellar gas and dust. Nevertheless, this was man's first attempt to discover his location in relation to the large-scale structure of the universe.

An accurate determination of our position in the Milky Way Galaxy was delayed until the early part of the twentieth century. Around the time of World War I, the famous American astronomer, Harlow Shapley, was observing clusters of stars called *globular clusters*. From observing certain types of stars in these clusters, Shapley was able to deduce how far away the clusters are. Virtually all of these clusters lie far above or below the plane of the Milky Way and therefore Shapley's view of these globulars was not hampered by the interstellar matter in the Milky Way itself. When Shapley mapped out the positions of these 93 globular clusters, he found that they formed a spherical distribution that was *not* centered on the sun. Instead, the system of clusters was centered on a location toward the direction of Sagittarius some 30,000 light years away. Shapley boldly and correctly concluded that this location is the true center of our Galaxy.

The primary difficulty with visual observations of the structure of the Galaxy is the gas and dust between the stars in the plane of the Milky Way. It is simply impossible to see very far through all that interstellar material. But think for a moment about a very foggy, rainy day. Although you may not be able to see very far with your eyes, your radio and television set work just fine. Visible light does not get very far, but radio waves easily penetrate the obscuring material. The next major advance in analyzing the Milky Way therefore was destined to come from radio astronomy.

In the early 1950s, radio astronomers had finally perfected their telescopes to such a degree that they could detect radio waves from hydrogen gas in the Galaxy. These radio waves easily penetrate the interstellar "smog" and by noting the locations of concentrations of hydrogen gas, radio astronomers were able to map the Galaxy. The final picture to emerge is that we live in a huge spiral system of billions of stars approximately 100,000 light years in diameter. As shown schematically in Figure 5–30, the sun is about two thirds of the way from the *galactic nucleus*. If we could view the Galaxy face on, we would see *spiral arms* containing dust, gas, and stars. The edge on view would look something like two fried eggs back to back. Just as the planets orbit the sun, the sun itself is in orbit about the galactic nucleus. It takes roughly 200 million years to make one full revolution about the center of our Galaxy.

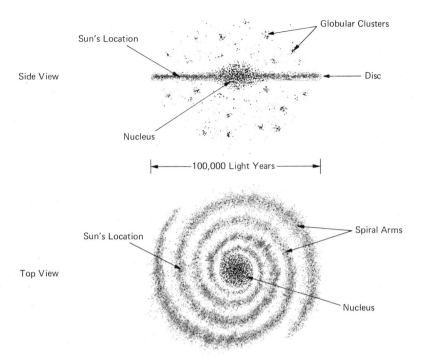

**Figure 5–30.** *The Milky Way Galaxy.* The sun is one of billions of stars in the Galaxy. Seen face on, the Galaxy has a spiral appearance. Seen edge on, it would resemble two fried eggs back to back.

One of the primary purposes of astronomy is to give mankind a perspective on his location in the universe. The course of modern astronomy over the past few centuries has been a series of widening horizons. It began with the realization that Earth is the third planet from the sun. In the eighteenth century, the distances to the nearest stars were determined. Then a few decades ago, the sun's true position in the Milky Way was discovered. But what lies beyond the Galaxy? What, if anything, lies beyond the most remote stars in the Milky Way Galaxy?

Back in the early 1920s, a heated debate was raging among professional astronomers. The controversy centered about fuzzy-looking objects in the sky called *nebulae* (from the Latin word meaning "clouds"). By that time, catalogues of thousands of nebulae had accumulated. Many nebulae were large clouds of glowing gases inside our own Milky Way. No argument there. But many nebulae exhibited a pinwheel or spiral appearance. These were the controversial objects. Some astronomers (for example, H. Shapley) strongly believed that the spiral nebulae were nearby, scattered around the Galaxy with the globular clusters. Others (for example, H. D. Curtis) felt that these nebulae were separate huge *galaxies*, like our own, but at very great distances.

**Figure 5-31.** *An Elliptical Galaxy.* Elliptical galaxies, such as this one in Virgo, have a featureless circular or elliptical shape. Elliptical galaxies do not have any dust lanes or spiral arms. (*Hale Observatories*)

**Figure 5-32.** *A Spiral Galaxy.* Spiral galaxies, such as this one in Ursa Major, have beautiful spiral arms. Our Milky Way Galaxy would probably look very much like this if it could be viewed from a very great distance. (*Hale Observatories*)

**Figure 5-33.** *A Barred-Spiral Galaxy.* In this type of galaxy, the spiral arms originate at the ends of a bar that passes through the nucleus of the galaxy. This particular barred spiral is in Eridanus. (*Hale Observatories*)

**Figure 5-34.** *An Irregular Galaxy.* The so-called Large Magellanic Cloud, a nearby galaxy visible from southern latitudes is an irregular. Irregular galaxies have distorted, peculiar shapes. (*Lick Observatory photograph*)

**174**

This heated debate was settled once and for all with the work of a gifted young astronomer, Edwin Hubble, at Mount Wilson Observatory. Hubble succeeded in finding certain types of stars in several spiral nebulae.

These stars, called *Cepheid variables,* change their magnitudes in a regular fashion with a period that is correlated with their real brightness. In reality, these variable stars are extremely bright, typically thousands of times brighter than the sun. However, they appear so incredibly dim in spiral nebulae that they must be extremely far away. Indeed, one of the nearest spiral nebulae, the Andromeda nebula, is two million light years away. They aren't "nebulae" at all; they are separate huge stellar systems, *galaxies,* like our own.

During the late 1920s, Hubble realized that all the galaxies in the sky could be divided into four distinct classes or categories. Perhaps the most common type are *elliptical galaxies.* Elliptical galaxies show no spiral structure at all. They look like blobs. They can have a very circular appearance (see Figure 5–31) or they can be quite flattened.

Another common type are *spiral galaxies.* Spiral galaxies exhibit beautiful spiral arms winding outward from a central nucleus. Figure 5–32 shows a typical spiral. We live in a spiral galaxy.

The third class is called *barred-spiral galaxies.* In these galaxies, the spiral arms originate at the ends of a "bar" running through the center of the galaxy. A good example of a barred-spiral is shown in Figure 5–33.

And finally, any galaxy that does not conveniently fit into one of the above three categories usually has a very peculiar, distorted appearance. They are called *irregular galaxies.* One of the nearest galaxies to our Milky Way Galaxy, the so-called Large Magellanic Cloud, is an irregular.

Whenever an astronomer discovers something in the sky, one of his first impulses is to attach a spectrograph to his telescope and take a spectrum. Even before Hubble discovered the distances to galaxies, it was known that many of these "spiral nebula" exhibited redshifts of their spectral lines. In the late 1920s, Hubble began investigating the spectra of galaxies in a systematic fashion and arrived at one of the most important astronomical discoveries of the twentieth century.

Perhaps the best way to appreciate Hubble's discovery is to look at some data, as shown in Figure 5–35. Photographs of five elliptical galaxies are shown on the left. Since these photographs are all to the same scale, the size of the image is a measure of the distance to the galaxy. The largest-appearing galaxy (at the top) is the nearest and the smallest-appearing galaxy (at the bottom) is the most distant. The spectra of the galaxies are shown to the right of Figure 5–35. In this illustration, the actual spectra of the galaxies are the fuzzy bands of light in the middle of each spectrum. The bright spectral lines above and below are reference markers (that is, the so-called *comparison spectrum*) artificially placed on the photographic film by the astronomer at the telescope so that he can line up the spectra properly. Notice that in the spectra of the galaxies (that is, in the fuzzy bands in between the comparison spectra) only two spectral lines are easily seen. They are the so-called H and K lines of singly ionized calcium. These two spectral lines normally appear in the blue colors of the spectrum in laboratory experiments.

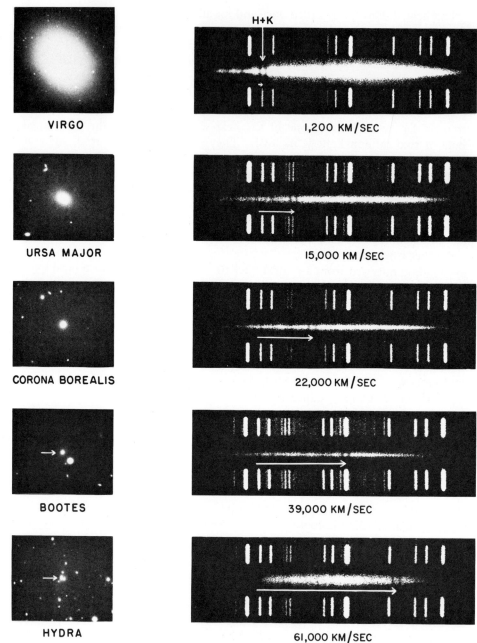

**VIRGO**

**H+K**

**1,200 KM/SEC**

**URSA MAJOR**

**15,000 KM/SEC**

**CORONA BOREALIS**

**22,000 KM/SEC**

**BOOTES**

**39,000 KM/SEC**

**HYDRA**

**61,000 KM/SEC**

**Figure 5-35.** *The Redshifts of Galaxies.* Photographs of five elliptical galaxies are shown on the left. Their corresponding spectra are shown on the right. Nearby galaxies (which look big) have low redshifts. Distant galaxies (which look small) have higher redshifts. (*Hale Observatories*)

In the spectrum of the nearest galaxy (in Virgo), the two spectral lines are shifted only slightly toward the red end of the spectrum. But in analyzing the light from more and more distant galaxies, notice that the spectral lines are shifted further and further toward the red end of the spectra (that is, toward the right side of the spectra in Figure 5–35). Indeed, in the spectrum of the most distant galaxy (in Hydra) in Figure 5–35, the H and K lines are shifted all the way across the spectrum and lie among the red colors.

The obvious interpretation of the redshifting of spectral lines comes from the Doppler effect. Recall that the Doppler effect (Section 4.3) attributes a redshift to the speed of a source of light away from an observer. The greater the redshift, the higher the speed. Thus, since all the galaxies in Figure 5–35 exhibit redshifts, they are all moving away from us. Furthermore, the nearest galaxy has the smallest redshift while more distant galaxies have progressively larger redshifts. From the Doppler effect, Hubble concluded that nearby galaxies are moving away from us slowly while more distant galaxies are rushing away from us much more rapidly.

In the early 1930s, Dr. Hubble and his colleague, M. L. Humason, had accumulated enough data to demonstrate that there is a simple and direct relationship between the distances and redshifts of galaxies. By interpreting the redshifts as recessional velocities, this relationship is easily displayed in the form of a graph as shown in Figure 5–36. This simple, but all-important relationship between the distances and speeds of galaxies is called the *Hubble law*.

To understand the true significance of the Hubble law, imagine someone blowing up a balloon, as shown in Figure 5–37. Furthermore, suppose there are a number of spots on the balloon. As the balloon expands, each spot recedes from every other spot. All the spots are moving away from each other. In addition, as

**Figure 5–36.** *The Hubble Law.* The distances and recessional velocities of galaxies are related in a simple, linear fashion. Nearby galaxies are moving away from us slowly while more distant galaxies are moving away from us much more rapidly.

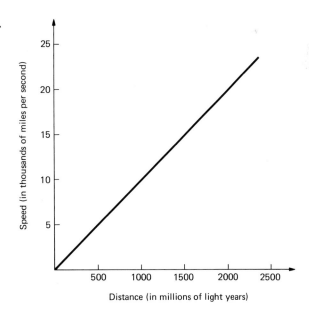

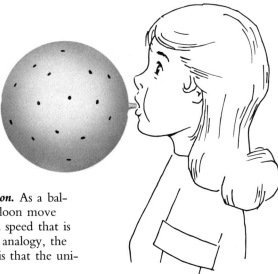

**Figure 5-37.** *An Expanding Balloon.* As a balloon expands, all points on the balloon move away from any given point with a speed that is proportional to their distances. By analogy, the true meaning of the Hubble Law is that the universe is expanding.

seen from *any* spot, all the other spots are moving away with a speed that is directly proportional to distance. Nearby spots are moving away slowly, while more distant spots are moving away more rapidly. The reason, of course, is that the balloon is expanding. The details of this little story about an expanding balloon are strongly reminiscent of the Hubble law. Indeed, by analogy, the true meaning of the Hubble law is that *the universe is expanding.*

In view of the fact that the universe is expanding, it is possible to think backward in time to when the galaxies were all piled on top of each other. From the observed rate of expansion, it logically follows that approximately 20 billion years ago, all the matter in the universe must have been concentrated into a state of infinite density. Then, for some (poorly understood) reason, a stupendous explosion must have occurred throughout all space which started the universe expanding. This was the "creation event." This was the *big bang.*

Most astronomers believe that the *big-bang cosmology* is correct . . . at least, it's the best theory anyone has ever proposed. The universe originated in a primordial explosion that started the expansion we see today. The big-bang cosmology is the simplest cosmological model that can be inferred from the Hubble law and is entirely consistent with all observational evidence. For example, in the late 1960s, astronomers discovered that there is a weak radiation field spread out through space. This radiation field has a blackbody distribution (Section 4.1) corresponding to a temperature of 3°K. It is therefore called the *3° blackbody background radiation.* This universal radiation field can be explained as the cooled-off remains of the *primordial fireball* out of which the universe was born. Figuratively speaking, astronomers have detected the "echo" of the big bang.

We are now in a position to draw an overall picture of our place in space and time. The current party line is that we live in an expanding universe that began with a primordial explosion 20 billion years ago. As the primordial fireball cooled, stars and galaxies formed. Galaxies are not scattered uniformly through space but rather are grouped together in *clusters*. Figure 5–38 shows a typical cluster of galaxies. Our Milky Way Galaxy is located in a cluster of about 20 galaxies affectionately called the *Local Group*. Our Galaxy is one of the largest members of the Local Group and contains billions upon billions of stars distributed around spiral arms rotating about the galactic nucleus. Our star, the sun, is about two thirds of the way from the nucleus and is orbited by nine planets.

Although the universe is 20 billion years old, the age of the solar system is only about 5 billion years. As we shall see in the next two chapters, this age is based primarily on geological evidence from the earth and the moon. It was necessary to wait for billions of years after the big bang to form planets. Hydrogen and helium are the only two elements that could have survived the primordial fireball. Billions

**Figure 5–38.** *A Cluster of Galaxies.* Galaxies are grouped together in clusters, such as this one in Hercules. Some clusters are "poor" in that they contain only a dozen or so galaxies. Others are "rich" and contain thousands of members. (*Hale Observatories*)

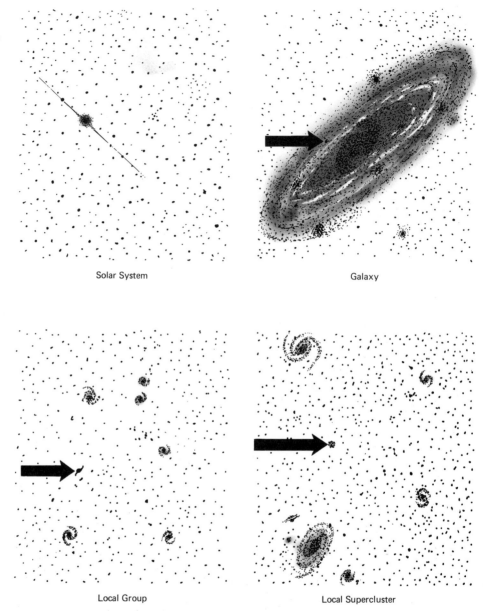

Solar System

Galaxy

Local Group

Local Supercluster

**Figure 5-39.** *Our Place in the Universe.* A view 10 billion miles across would encompass the entire solar system. Expanding the view to 100,000 light years, we would see our entire Galaxy. A volume of space several million light years across would contain the Local Group, while a view several hundreds of millions of light years across would include several clusters of galaxies that comprise the Local Supercluster.

of years elapsed while heavier elements were created inside ancient, dead stars. As these stars ended their life cycles, they cast heavy elements out into space, thereby enriching the interstellar medium. After about 15 billion years, our part of the Milky Way Galaxy contained enough of these elements so that our sun and its entourage of planets, satellites, asteroids, meteoroids, and comets could form.

## Review Questions and Exercises

1. List the three major layers in the solar atmosphere.
2. Explain why sunspots appear darker than the surrounding photosphere.
3. Describe the rotation of the sun.
4. What is meant by the solar cycle? When will the next sunspot maximum occur? When will the next sunspot minimum occur?
5. What is granulation?
6. Describe the appearance and properties of a bipolar sunspot group.
7. What is the law of polarity?
8. How does Babcock's theory account for the solar cycle and the law of polarity?
9. What is the Maunder butterfly diagram?
10. What is a monochromator?
11. How are filaments and prominences related?
12. What is a solar flare?
13. Describe the appearance of the corona over the course of a solar cycle.
14. Explain why the solar cycle really lasts for 22 years rather than 11 years.
15. Explain why the thermonuclear fusion of four hydrogen nuclei into one helium nucleus produces energy.
16. How much hydrogen must be converted into helium each second in order to account for the sun's luminosity?
17. What is a neutrino?
18. What is the proton–proton chain?
19. What is the CNO cycle?
20. In what parts of the sun are radiative transport and convection important in transferring energy outward from the center of the sun?
21. Describe Dr. Davis' experiment. What method does he use to detect neutrinos from the sun?
22. Discuss some of the implications of the "mystery of the missing solar neutrinos."
23. How do astronomers use parallax to determine the distances to nearby stars?
24. Give an argument to show that if the distance to a star and its apparent magnitude are known, the star's luminosity can be inferred.

25. Why are the colors of stars related to their surface temperatures?
26. Discuss the reasons why the temperature of a star's atmosphere influences the strengths of spectral lines.
27. What is the H–R diagram?
28. Why are double stars important? What important data can astronomers obtain from studying binaries?
29. What is the mass-luminosity relation?
30. Briefly discuss the evolution of a typical star like the sun. At what stage in its life cycle is a star on the main sequence? When does a star become a red giant? When does it become a white dwarf?
31. Describe the Milky Way Galaxy.
32. Where is the sun located in the Galaxy?
33. What did Edwin Hubble do to resolve the controversy about the distances to the spiral nebulae?
34. Briefly describe the four basic types of galaxies seen in the sky. What kind of galaxy is the Milky Way?
35. What is the Hubble law and what does it tell us about the structure of the universe?
36. What is meant by the big-bang cosmology?

# Our Planet Earth 6

## *Minerals and Rocks* 6.1

OF ALL the objects in space, man is most familiar with his own planet, Earth. We live on its surface, breathe its atmosphere, drink its water, and can easily pick up and examine its rocks. Since our planet is so readily accessible for examination, large quantities of information and fields of study have developed over the past few centuries. In this brief chapter we shall discuss some of the highlights of what is known about Earth.

When we examine the solid ground beneath our feet, most of the material we find is *rock*. Whether we study a rock on the earth's surface or analyze samples from the deepest mines or wells, which sometimes descend to depths of five miles, we find material very similar to specimens from the surface. Often we find geological formations involving rocks that were once buried many miles below the earth's surface. Volcanic rock is probably from depths of down to 60 miles. Yet in all cases, these samples are very similar to the material directly below our feet. Rocks that we can pick up off the ground are therefore quite representative of the entire earth's *crust*.

Rocks are made up of *minerals*. Minerals, in turn, consist of elements or chemical combinations of elements in the form of crystals. Very few elements occur in a pure or *native* state in nature. Platinum, gold, and sulfur are sometimes found by themselves (for example, gold "nuggets," or sulfur crystals shown in Figure 6–1), but most other elements exist in the form of chemical compounds. Geologists

**Figure 6–1.** *Native Sulfur.* Sulfur is one of the few elements which occurs in nature in a pure form. Sulfur crystals are bright yellow. (*Ward's Natural Science Establishment, Inc.*)

**183**

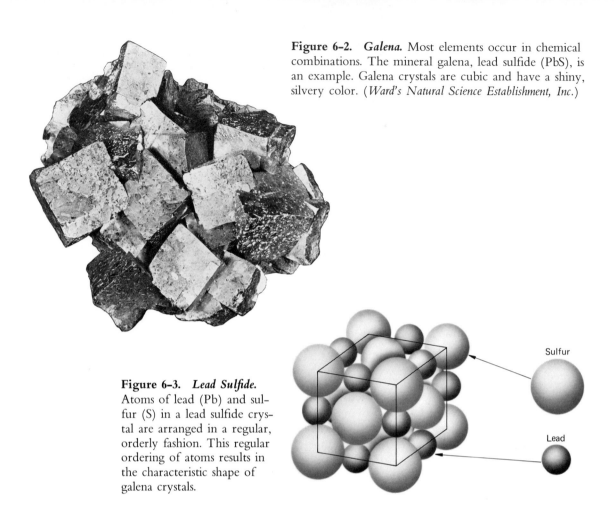

**Figure 6-2. *Galena.*** Most elements occur in chemical combinations. The mineral galena, lead sulfide (PbS), is an example. Galena crystals are cubic and have a shiny, silvery color. (*Ward's Natural Science Establishment, Inc.*)

Sulfur

Lead

**Figure 6-3. *Lead Sulfide.*** Atoms of lead (Pb) and sulfur (S) in a lead sulfide crystal are arranged in a regular, orderly fashion. This regular ordering of atoms results in the characteristic shape of galena crystals.

usually do not use the chemical name of the compound but instead have their own special vocabulary. For example, the mineral consisting of lead sulfide (PbS) is called galena. Galena crystals are shown in Figure 6–2.

Crystals of a mineral can form by the cooling of molten rock or by the evaporation of water that contains a particular chemical compound solution. For example, when salt water evaporates, the sodium and chlorine in the solution form halite crystals. The mineral halite is common table salt (NaCl). As the crystals form, the atoms in the chemical compound arrange themselves in a regular, orderly fashion. The precise details of this regular ordering depend on the particular properties of the chemical compound. Lead sulfide is a very simple compound; the arrangement of lead and sulfur atoms is shown in Figure 6–3. This particular ordering of lead and sulfur atoms dictates that galena crystals must be cubic. The shape of a crystal is a direct result of how the atoms in the mineral are arranged.

In addition to having a definite chemical composition and a characteristic crystalline form, minerals also possess specific physical properties that aid in their identification. These properties include cleavage, hardness, specific gravity, color, luster, and streak.

The forces between the atoms in a crystal are not necessarily equal in all directions. When subjected to a sharp blow, a crystal will fracture along directions where the forces between the atoms are weakest. If planes of weakness exist, a mineral will repeatedly break or *cleave* along these *cleavage planes*. Crystals may have one, two, three, four, or six cleavage planes. For example, halite has three cleavage planes at right angles to each other. Examining the directions of *cleavage* is one of the techniques geologists use in identifying and classifying minerals.

It should be emphasized that cleavage planes and crystal faces are *not* the same thing. Both are distinctive properties of crystals. But cleavage results from planes of weakness while crystal structure reflects the geometry of how the mineral's atoms are arranged.

Another test commonly used in the identification of minerals is *hardness*. Some minerals are very soft (for example, talc and gypsum) while others are very hard (for example, diamond). When two minerals are rubbed together, the harder one scratches the softer. By simply rubbing one specimen against a variety of standard minerals or substances, the geologist can determine the hardness of the specimen.

During the nineteenth century, the German mineralogist Friedrich Mohs devised a hardness scale that is still used today. As shown in Figure 6–4, the softest minerals are assigned a value of 1 while the hardest has a value of 10. This scheme of designating hardness is called the *Mohs scale*. By way of example, both galena and halite have hardnesses of $2\frac{1}{2}$, which means that they will scratch gypsum but not calcite.

**Figure 6–4.** *The Mohs Hardness Scale.* The hardness of a mineral can range from 1 (the softest) to 10 (the hardest). Ten commonly recognized minerals used to distinguish various degrees of hardness are listed on the left. For comparison, several nonmineral substances are given on the right.

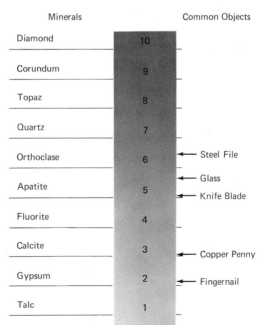

Another important characteristic of a mineral is its *specific gravity*. Specific gravity is simply the ratio of the mass of a mineral to the mass of an equal volume of water. For example, a cubic inch of halite weighs twice as much as a cubic inch of water. We therefore say that the specific gravity of halite is 2. Most rock-forming minerals have specific gravities between 2.3 and 2.6. Galena has a very high specific gravity of 7.6. One cubic inch of galena weighs 7.6 times more than a cubic inch of water.

Additional techniques used by geologists in identifying minerals involve *color, streak,* and *luster.* Color is one of the most obvious properties of a mineral. Galena is grey, azurite is blue, and olivine is green—to name a few. When a mineral is powdered, however, it often exhibits a more diagnostic color than when it is in large pieces. By vigorously rubbing a mineral on an unglazed porcelain plate, a streak is obtained. The color of the streak can further aid the geologist in identifying a specimen.

In addition to color and streak, the luster of a mineral is used in identification. Galena and pyrite (commonly called "fool's gold") have a metallic luster. Non-metallic lusters are sometimes termed pearly, silky, resinous, or greasy.

Minerals are the building blocks of rocks. Rocks consist of one or more minerals in various combinations. Although there are more than 2,000 different kinds of minerals, most of these are very rare. Consequently, in order to understand common rocks, it is necessary to concentrate on only a few very abundant minerals.

*Silicates* are by far the most common type of rock-forming minerals. The term "silicates" applies to a large number of different kinds of minerals, all of which are based on the element silicon. *Carbonates* (that is, substances involving $CO_3$) form another important class of minerals. Additional classes include *oxides* (for example, iron oxide = hemitite; aluminum oxide = bauxite) and *sulfides* (for example, lead sulfide = galena; zinc sulfide = sphalerite).

Quartz is one of the most common silicate minerals on earth. It is a major constituent of many different kinds of rocks. Quartz is silicon dioxide ($SiO_2$) and forms characteristic six-sided crystals, as shown in Figure 6–5. They exhibit no cleavage and shatter along random directions. Depending on impurities and crystal size, quartz may take on a wide range of colors and appearances. Clear quartz, smoky quartz, milky quartz, rose quartz, amethyst, flint, jasper, opal, and petrified wood are all forms of the same basic silicate, quartz.

The so-called feldspar group constitutes the most abundant minerals on our planet. Like quartz, feldspar is also a silicate. Almost 60 per cent of the total weight of the earth's crust is made of feldspar.

There are two basic kinds of feldspar. The class of silicates that contains potassium and aluminum is called orthoclase while the series of silicates of sodium, calcium, and aluminum is termed plagioclase. Feldspar is slightly harder than glass but not so hard as quartz. It is the single most abundant mineral found in rocks.

Another common rock-forming mineral is mica. Mica is easily recognized by its pearly luster and conspicuous cleavage along one plane. It is a very soft mineral and can be cleaved into sheets with the flick of a fingernail. Two types of mica are recognized: white mica (muscovite), which is a silicate of hydrogen, potassium, and aluminum, and black mica (biolite), which in addition contains iron and magnesium. Figure 6-6 shows a rock containing both feldspar and mica.

Calcite ($CaCO_3$) is a common carbonate mineral. It occurs in a wide variety of forms. Marble, chalk, and limestone are all composed of calcite. It has a hardness of 3 on the Mohs scale and commercially serves as a source of lime for glass, mortar, and cement.

Numerous other rock-forming minerals are recognized by the geologist. They include ferromagnesian minerals (for example, olivine) and clay minerals (for example, kaolin). Different abundances of various minerals in rocks as well as different conditions under which the rocks themselves were formed result in the varied appearance of the material in the earth's crust.

**Figure 6-5 (*left*). *Quartz.*** Quartz ($SiO_2$) is a very common silicate mineral. Its crystals have a characteristic hexagonal shape. (*W. Cunningham, San Bernardino Valley College*)

**Figure 6-6 (*below*). *Feldspar and Mica.*** Feldspar is the most abundant rock-forming mineral. It is shown here (the light mineral) along with mica (the dark, pearly mineral) in pegmatite. (*W. Cunningham, San Bernardino Valley College*)

**Figure 6-7 (*right*). *Olivine.*** Olivine is one of the less common rock-forming minerals. It is usually dark green and contains iron and/or magnesium along with silicon and oxygen. Olivine is found in lunar rocks as well as in certain types of meteorites. (*Ward's Natural Science Establishment, Inc.*)

**Figure 6–8.  Granite.** Granite is a common coarse-grained igneous rock. It is composed of quartz, feldspar, and ferromagnesian minerals (hornblend and mica). Granite blocks are often used as a building material. (*Ward's Natural Science Establishment, Inc.*)

**Figure 6–9 (*left*).  Basalt.** Basalt is a common fine-grained igneous rock. A microscope must be used to see the crystalline structure. Basalt is composed of plagioclase feldspar and relatively large amounts of ferromagnesian minerals that give the rock its dark color. Unlike granite, there is no quartz. (*Ward's Natural Science Establishment, Inc.*)

**Figure 6–10 (*right*).  Obsidian.** Obsidian is a jet-black natural glass that forms when molten rock of the proper composition cools very rapidly. Obsidian has no crystalline structure and its dark color is due to ferromagnesian minerals such as magnetite. (*W. Cunningham, San Bernardino Valley College*)

While minerals are classified according to their characteristic properties (for example, crystal form, cleavage, color, hardness, and so forth), it is not obvious how rocks can be categorized. The properties of rocks vary so widely that it is not easy to classify them in the same way that minerals are distinguished. Geologists instead choose to divide all rocks into three main groups according to their origin. The three groups are igneous, sedimentary, and metamorphic.

*Igneous rocks* are those that have cooled from a molten state. Igneous rocks can be observed forming from cooling volcanic lava. Almost two thirds of the rocks in the earth's crust are igneous. Bedrock under the oceans and beneath the continents falls into this category.

As molten rock cools, crystal growth of the various minerals interferes with other. Consequently, the structure of igneous rocks is characterized by ragged crystals and a random arrangement of grains. If the cooling process is slow, then large crystals have the chance to form. This results in *coarse-grained* igneous rocks such as granite. If the cooling is more rapid, fine-grained igneous rocks, such as basalt, are formed. The mineral grains may be so small that the geologist needs a microscope to see them. Sometimes, the molten rock has the proper composition and cools rapidly enough to form a black natural glass called obsidian. Obsidian has no grain structure at all. Grain size therefore tells us about how igneous rocks were formed and can be used as a basis for classification.

In addition to grain size, mineral composition can be used to classify igneous rocks. For example, granite contains quartz and basalt does not. Specifically, all igneous rocks contain feldspar and ferromagnesian minerals. In one class of igneous rock, feldspar is very abundant compared to the ferromagnesian minerals while in another class, the ferromagnesian minerals predominate over feldspar. Table 6–1 lists six common igneous rocks classified according to grain size and mineral content.

**Table 6–1.** *Some Igneous Rocks*

| Mineral Composition | Rock Name | |
|---|---|---|
| | (Fine-Grained) | (Coarse-Grained) |
| Quartz<br>Feldspar<br>Ferromagnesian Minerals | Rhyolite | Granite |
| Feldspar*<br>Ferromagnesian Minerals | Andesite | Diorite |
| Ferromagnesian Minerals*<br>Feldspar | Basalt | Gabbro |

\* Asterisk indicates predominant minerals.

While igneous rocks form from a molten state, *sedimentary rocks* are laid down by wind, water, and ice. Overlying deposits compress and cement tiny particles and grains together. Sedimentary rock therefore often has a banded appearance such as sandstone has (Figure 6–11).

There are two basic sources of sedimentary rock. First of all, sedimentary rocks can form simply by the cementing of smaller rock fragments. If the rock fragments are comparatively large—sometimes a foot in diameter—the resulting rock is called conglomerate. Conglomerate is literally cemented gravel. If the fragments are smaller, the resulting rock is called sandstone. If the fragments are microscopic, shale is formed.

A second kind of sedimentary rock can be formed by precipitation. Limestone, a fine-grained rock composed of calcite, is made in this fashion. Limestone often

**Figure 6–11.** *Sandstone.* Sandstone is a sedimentary rock formed by the cementing of tiny rock fragments (mostly quartz). The cementing of larger rock fragments results in conglomerate, while shale is made from microscopic fragments. (*Ward's Natural Science Establishment, Inc.*)

**Figure 6–12.** *Limestone.* Limestone is formed by precipitation and consists almost entirely of calcite. This particular sample must have formed in a warm, shallow sea where marine life was abundant. (*Ward's Natural Science Establishment, Inc.*)

**Table 6-2.** *Some Sedimentary Rocks*

| Origin | Primary Characteristic | Rock Name |
|---|---|---|
| Cemented Fragments | Coarse-Grained | Conglomerate |
| | Medium-Grained | Sandstone |
| | Very Fine Grained | Shale |
| Precipitate | Calcite ($CaCO_3$) | Limestone, Chalk |
| | Chalcedony ($SiO_2$) | Chert, Flint |
| | Salt ($NaCl$) | Halite |

contains fragments of shells that were the source of the calcium carbonate. Figure 6-12 shows a good example. Chalk, another well-known sedimentary rock, is a variety of limestone usually made from tiny one-celled animals. Precipitation of nonbiological material can also form sedimentary rocks. Chert, flint, and jasper are examples of sedimentary rock composed of microcrystalline quartz. Evaporation of salt water (that is, water containing $NaCl$) results in halite. Table 6-2 lists some common sedimentary rocks according to their properties and origin.

Finally, if *either* igneous or sedimentary rocks become buried far below the earth's surface, enormous pressures and temperatures can cause profound changes. Rocks which have undergone such changes are called *metamorphic rocks*. For example, heat and pressure transform limestone into marble. Similarly sandstone is turned into quartzite and shale becomes slate. Marble and slate are common building materials. At first glance, quartzite sometimes looks like sandstone, but close inspection reveals that the quartz grains are firmly interwoven. Table 6-3 lists some common metamorphic rocks along with their possible origins.

When examining a rock, one of the primary questions in the geologist's mind often concerns the *age* of the specimen. Is the rock very ancient, perhaps two or three billion years old? Or was it recently formed, only a few million years ago? Often the layering of different kinds of rock provides a clue. In an exposed

**Table 6-3.** *Some Metamorphic Rocks*

| Origin | Rock Name |
|---|---|
| Shale | → Slate |
| Limestone | → Marble |
| Sandstone | → Quartzite |
| Shale *or* fine-grained igneous rocks. | → Schist |
| Various (e.g., almost anything *except* pure limestone or pure quartz sandstone). | → Gneiss |

mountainside or a canyon, it is obvious that one layer of rock must have been laid down before another. This, however, tells only about the *relative ages* of rocks: which is younger and which is older. To obtain the *absolute age* measured in years, geologists have two methods.

First of all, the fossil remains (that is, bones and shells) of extinct creatures can provide important clues in dating rocks. For example, trilobites, extinct ancestors of shrimp and lobsters, flourished 600 million years ago during the so-called Cambrian period of the Paleozoic era. The discovery of trilobite fossils in a sedimentary rock clearly dates the rock to within a few million years. Obviously, this method of dating can be used only on our planet. Lunar rocks and specimens that might someday be returned from other planets must be dated by some other technique.

Certain properties of minerals, especially color, are caused by *trace* amounts of various elements. For example, even though all quartz is silicon dioxide ($SiO_2$), rose quartz gets its distinctive color from very small amounts of titanium. Rubies are red because of small amounts of chromium. Similarly, a rock can contain trace amounts of radioactive isotopes. These isotopes provide an important method of accurately dating rocks.

An isotope of an element is termed *radioactive* because it spontaneously *decays* into another isotope, usually by the emission of particles such as electrons (see the discussion of beta decay in Section 4.4). In this way, one element is converted into another by a natural process.

The rate of radioactive decay varies from one isotope to another. Scientists denote this rate by the *half-life;* that is, how long it takes for exactly one-half of the atoms of a particular isotope to turn into another isotope. For example, the half-life of a particular isotope of uranium ($^{235}U$) is 700 million years. After 700 million years, exactly half of the $^{235}U$ atoms in a particular substance have turned into other elements. The *decay series* for $^{235}U$, a naturally occurring isotope of uranium, is shown in Table 6–4. The terms *parent isotope* and *daughter isotope* are used to denote the initial and final isotopes of a particular radioactive decay. The final result of this

**Table 6–4.** *The Decay Series of* $^{235}U$

| Parent Isotope | Half-Life | Daughter Isotope |
|:---:|:---:|:---:|
| $^{235}U$ | 700 million years | $^{231}Th$ |
| $^{231}Th$ | 25 hours | $^{231}Pa$ |
| $^{231}Pa$ | 32 thousand years | $^{227}Ac$ |
| $^{227}Ac$ | 22 years | $^{227}Th$ |
| $^{227}Th$ | 18 days | $^{223}Ra$ |
| $^{223}Ra$ | 11 days | $^{219}Rn$ |
| $^{219}Rn$ | 4 seconds | $^{215}Po$ |
| $^{215}Po$ | $\frac{1}{500}$ second | $^{211}Pb$ |
| $^{211}Pb$ | 36 minutes | $^{211}Bi$ |
| $^{211}Bi$ | 2 minutes | $^{207}Tl$ |
| $^{207}Tl$ | 5 minutes | $^{207}Pb$ |

Table 6–5. *Radioisotopes Used in Rock-dating*

| Radioactive Parent Isotope | Half-life (billions of years) | Stable Daughter Isotope |
|---|---|---|
| Potassium ($^{40}K$) | 1.3 | Argon ($^{40}A$) |
| Rubidium ($^{87}Rb$) | 47.0 | Strontium ($^{87}Sr$) |
| Uranium ($^{235}U$) | 0.7 | Lead ($^{207}Pb$) |
| Uranium ($^{238}U$) | 4.5 | Lead ($^{206}Pb$) |

sequence of radioactive decays is an isotope of lead ($^{207}Pb$) that is *stable;* it is not radioactive at all. Also notice that the half-life of the initial uranium isotope is extremely long—much longer than any of the other isotopes in the chain. Since the *geological time-scale* (that is, the ages of rocks) is typically measured in hundreds of millions of years, as far as the geologist is concerned, this sequence looks like $^{235}U \longrightarrow {}^{207}Pb$. As far as rocks are concerned, the intermediate steps are so brief that they can virtually be ignored.

In view of the manner in which $^{235}U$ decays into $^{207}Pb$, geologists realize that they can discover the ages of rocks by measuring the relative abundances of these two iotopes. Young rocks contain relatively large amounts of $^{235}U$ compared to $^{207}Pb$. Ancient rocks contain very little $^{235}U$ compared to $^{207}Pb$.

In addition to measuring the $^{235}U/^{207}Pb$ relative abundances, geologists often use three other pairs of isotopes. These are listed in Table 6–5 along with their half-lives.

From measurements of the relative abundances of these isotopes, geologists believe that the solar system is 4.6 billion years old. As we shall see in the next chapter, this technique of age-dating has been used in the analysis of lunar rocks. It will most certainly be used on samples from Mercury and Mars at sometime in the distant future.

## *Probing the Earth's Interior* 6.2

EVERYTHING discussed up to this point has dealt only with the material in the earth's crust. The deepest mines and wells reach only a couple of miles below the earth's surface. Volcanoes spew forth molten rock from depths of a few dozen miles. This is, however, only a tiny fraction of the earth's diameter. The earth's interior is hidden and unaccessible. Learning about the earth's interior is just about as difficult as discovering the properties of the most distant galaxies in the universe.

Although the earth's interior is not accessible for direct examination, much can be learned by studying *earthquakes*. As we shall see in the next section, earthquakes occur primarily in regions where large land masses are moving past each other in

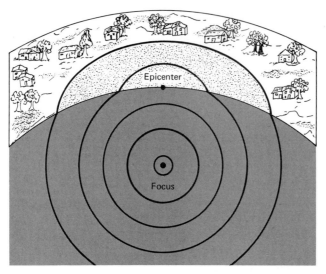

**Figure 6–13.** *The Focus and Epicenter of an Earthquake.*
Earthquakes originate at a focus, usually far below the
earth's surface. The point on the ground directly above
the focus is called the epicenter.

a process called *plate tectonics*. Stresses build up over years or centuries until the
earth's surface can no longer withstand the enormous pressures. Suddenly, and
almost without warning, violent slippage occurs. In a few seconds, the stresses are
released with a rapid vibratory motion in the earth's crust. The resulting phenomenon is called an earthquake.

Earthquakes occur well below the earth's surface. The place where an earthquake originates is called the *focus*. As shown in Figure 6–13, the point on the
earth's surface directly above the focus is called the *epicenter*.

The severity of an earthquake can be studied and recorded using instruments
called *seismographs*. In essence, a seismograph consists of a mass suspended from a
spring. As shown in Figure 6–14, when the earth shakes, the mass remains very
nearly stationary while its housing (and scale used for recording) oscillates up and
down. This particular arrangement responds to vertical motion of the earth's
surface. Horizontal movements can be measured using a horizontal pendulum
seismograph, as shown in Figure 6–15.

The magnitude of an earthquake is usually measured on the *Richter scale,*
invented by Charles F. Richter at the California Institute of Technology. The
Richter scale is based on the maximum amplitude of an earthquake's vibrations.
Each step on the Richter scale corresponds to a change of 10 in amplitude and a
factor of 30 in energy released. In an inhabited area, an earthquake of 4.5 will cause
some damage, while one of magnitude 6 or more will result in widespread
destruction. One of the largest earthquakes ever recorded occurred in Alaska in
1964. It measured 8.6 on the Richter scale and the total amount of energy released

**Figure 6-14.** *The Seismograph.* In essence, a seismograph consists of a weight suspended from a spring. During an earthquake, the building which houses the weight shakes up and down, while the weight remains relatively stationary.

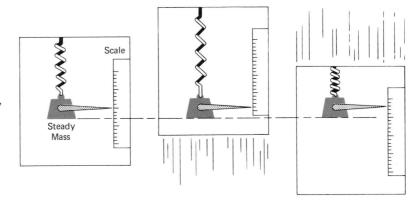

Scale

Steady Mass

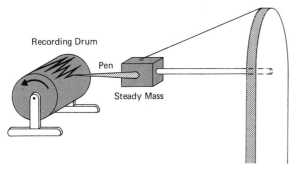

Recording Drum

Pen

Steady Mass

**Figure 6-15.** *A Pendulum Seismograph.* Horizontal motions of the earth's crust can be detected and measured with a horizontal pendulum seismograph.

**Figure 6-16.** *An Earthquake Recording.* The author is examining the recording of an earthquake on the drum of a seismograph.

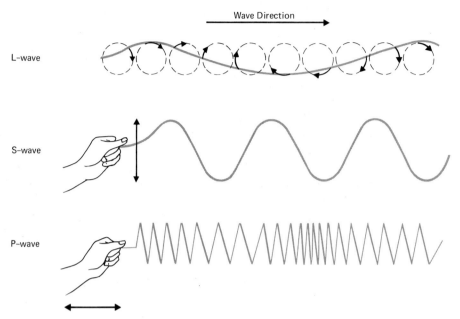

**Figure 6–17. *Three Types of Waves.*** Earthquakes produce three types of waves. L-waves travel on the earth's surface. S-waves and P-waves travel through the earth.

was about double the energy content in all the coal and oil produced each year throughout the entire world. Roughly a million earthquakes occur each year, but most have magnitudes less than 2.5. About 15 violent earthquakes with magnitudes greater than 7 occur each year. Only 9 earthquakes with magnitudes 8.4 to 8.6 have occurred since 1899.

Earthquakes produce three kinds of waves. First of all, during an earthquake, the ground oscillates up and down in little circles in the same way that water waves cause the surface of the ocean to go up and down. These are called *L-waves* and travel *around* the earth. Unlike L-waves, which are confined to the earth's surface, the remaining two types of vibration travel *through* the earth. These two types of vibration are called *S-waves* and *P-waves*.

S-waves are termed *transverse* because the oscillations are perpendicular to the direction in which the wave is traveling. S-waves are analogous to the waves traveling down a rope which is being shaken up and down, as shown in Figure 6–17.

By contrast, P-waves are termed *longitudinal* because the oscillations are parallel to the direction in which the wave is traveling. P-waves consist of successive regions of compression and rarefaction analogous to waves traveling down a spring which is being pushed and pulled, as shown in Figure 6–17.

**196** The three types of waves produced by earthquakes travel at different speeds.

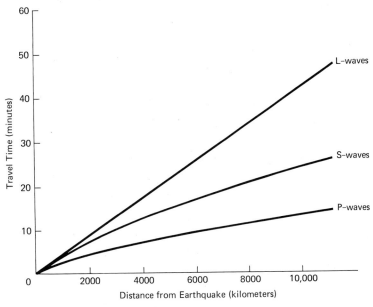

**Figure 6–18.** *Travel Times of Seismic Waves.* The three types of waves produced by earthquakes travel at different speeds. P-waves move most rapidly through the earth while L-waves travel much more slowly. S-waves travel at an intermediate speed, which ranges between 3 and 7 kilometers per second.

P-waves are the fastest and can move with a velocity up to 14 kilometers per second through the earth. L-waves are the slowest and travel over the earth's surface at about 4 kilometers per second. Figure 6–18 shows the travel times for the three types of waves plotted against the distance from the earthquake. At a seismographic recording station a few thousand kilometers from the epicenter of an earthquake, the P-waves always arrive first, then the S-waves, and finally the L-waves. By measuring the time-delay between the arrival of the various types of waves at three or more recording stations, seismologists can calculate the location of the epicenter. From knowing the distance to the epicenter, and from studying the amplitude of the oscillations on their seismographs, seismologists can deduce the Richter scale magnitude of the earthquake.

S-waves ad P-waves do not travel along straight lines through the earth. Instead, the varying density and composition throughout the earth's interior cause the waves to be bent or *refracted*. By studying exactly how S-waves and P-waves are bent as they travel through the earth, geologists can discover the structure of the earth's interior.

When an earthquake occurs somewhere on the earth, all seismological stations within a few thousand kilometers of the epicenter record both S-waves and

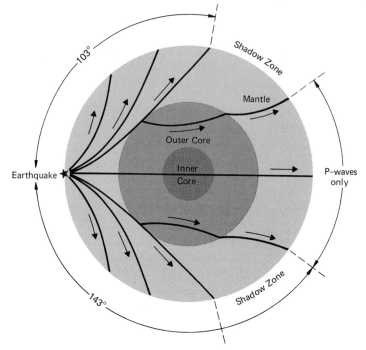

**Figure 6-19.** *Seismic Waves Traveling Through the Earth.*
S-waves are detected at angles up to 103° from the epicenter of
an earthquake. Only P-waves can penetrate the earth's liquid
core. Because of the refraction of P-waves by the core, there is
a region (103° to 143°) where no waves are detected.

P-waves. But far from the epicenter, no S-waves at all can be detected. Specifically,
as shown in Figure 6–19, all S-waves are absent at an angular distance greater than
103° from the epicenter. By the late nineteenth century, it was clear that the
structure of the earth's interior must be very different from the material immediately below the crust in order to explain the absorption of S-waves that try to travel
through or near the earth's center.

In addition to the complete absorption of S-waves deep inside the earth,
P-waves are severely refracted as they pass within a few thousand kilometers of the
earth's center. Specifically, as shown in Figure 6–19, at an angular distance between
103° and 143° from an epicenter, no P-waves are detected from an earthquake.
This region, from 103° to 143°, where neither S-waves nor P-waves are felt, is
called the *shadow zone*. In 1906, the seismologist R. D. Oldham proposed that the
earth was divided into a solid *mantle* surrounding a *liquid core*. It was realized that
transverse waves (that is, S-waves) cannot travel very far through liquids, and a
molten core 7,000 kilometers (4,300 miles) in diameter could explain the existence
of shadow zones.

As the quality and sensitivity of seismographs improved, geologists realized that they could detect faint traces of P-waves in the shadow zones. This led Inge Lehmann in 1936 to propose the existence of a small solid *inner core* deep inside the liquid core which refracts some of the P-waves into the shadow zones.

Based on numerous seismological records of earthquakes, a fairly reliable picture of the earth's interior has emerged, as shown in Figure 6–20. The earth has a diameter of 12,800 kilometers (8,000 miles). The outermost layer is called the crust and varies in thickness from 35 kilometers under the continents to about 5 kilometers under the oceans. Beneath the crust there is a change in composition, called the Mohorovicic discontinuity, which marks the surface of the mantle. The mantle probably consists of igneous rock rich in ferromagnesian minerals and is about 2,900 kilometers (1,800 miles) thick. Inside the mantle is a liquid core primarily composed of iron. The diameter of the liquid core is 7,940 kilometers (4,320 miles). The inner, solid core has a diameter of 2,500 kilometers (1,560 miles) and is probably composed of crystalline iron or iron–nickel.

The existence of both solid and liquid cores can be understood by thinking about the changes in pressure, temperature, and the resulting melting point of matter inside the earth. In going from the earth's surface to its center, both the temperature and pressure rise steadily, as shown in Figure 6–21. On this same graph, the melting point of the material inside the earth is plotted. In going deeper and deeper into the earth, the pressure rises. As the pressure rises, so does the melting point of the rock. Notice that in the mantle, the melting point of the igneous rock is far above the temperature of the rock. Hence the rock is solid. But at the

**Figure 6–20.** *The Structure of the Earth.* From studying the propagation of S-waves and P-waves, it is believed that the earth consists of a thin crust, a solid mantle, a liquid outer core, and a solid inner core.

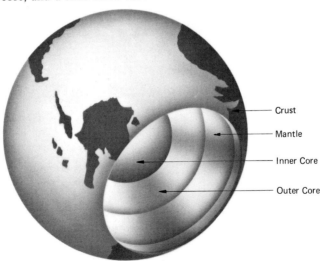

Crust

Mantle

Inner Core

Outer Core

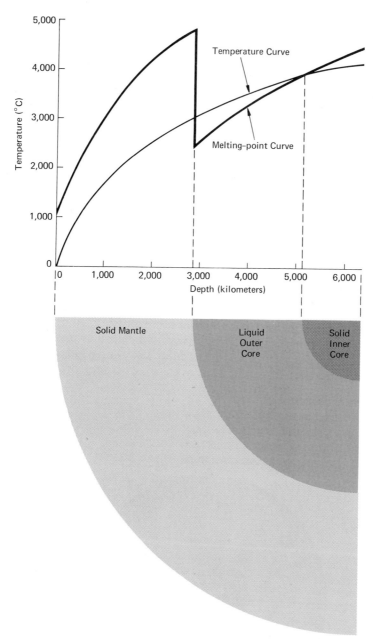

**Figure 6-21.** *Temperature and Melting-Point Curves.* The temperature rises steadily from the surface to the center of the earth. By plotting the melting-point curve on this graph it is possible to deduce over what regions the earth's matter is solid or liquid.

boundary between the mantle and the core, there is a dramatic change in the chemical composition of the earth, from igneous rock to iron. The melting point curve suddenly drops in response to this change in composition. Notice that the melting point curve now lies below the temperature curve. Hence the iron is liquid. Going to still greater depths, the melting point curve continues to rise as the pressure inside the earth increases. At a depth of about 5,200 kilometers below the earth's surface, the two curves cross. In the innermost portions of the earth, the pressure is so great that the melting point curve again lies above the temperature curve. The inner core is therefore solid. These effects account for the curious situation of a liquid layer sandwiched between a solid inner core and a solid mantle.

There are several reasons for believing that the earth's core consists primarily of iron. First of all, iron is an abundant element in the universe. But perhaps more importantly, the overall density of the earth (that is, the mass of the earth divided by its volume) is much higher than the density of rocks in the crust or mantle. The average density of the earth is 5.5 grams per cubic centimeter while the density of typical rocks in the crust and mantle is only about 2.7 grams per cubic centimeter. In order to come out with an overall density of 5.5 grams per cubic centimeter, the material in the core must be much denser than the material in the crust and mantle. Iron fulfills this requirement. Indeed, as we shall see in later chapters, it is precisely this argument involving density that leads scientists to believe that the planet Mercury possesses an iron or nickel–iron core while our moon does not.

## Our Changing Planet 6.3

ALMOST as soon as reliable maps of the earth were first available, geographers began noticing that certain continents seemed to fit together like huge pieces of a giant jig-saw puzzle. Most notably, South America would appear to fit very snugly against Africa. This "fitting" is especially apparent if we turn our attention to the *continental shelves* surrounding the continents rather than the present-day water levels of the oceans. The dry land gradually slopes into the ocean resulting in a "shelf" surrounding each of the continents before the final precipitous plunge to the ocean floor. By simply matching up the outlines of the continental shelves, it is clear that North America, Greenland, Europe, Africa, and South America fit together remarkably well, as shown in Figure 6–22.

In the early 1900s, the German meteorologist Alfred Wegener proposed the concept of *continental drift*. Wegener was motivated to propose this theory on both biological and geological evidence. Specifically, until about 200 million years ago, new species of plants and animals seemed to develop almost simultaneously on now-distant places around the world. For example, fossils of the extinct lizard Lystrosaurus are found in Africa, Antarctica, and India. In addition, similar geo-

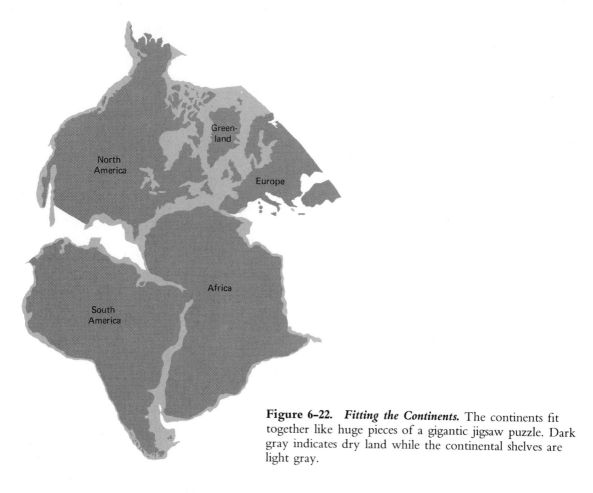

**Figure 6-22.** *Fitting the Continents.* The continents fit together like huge pieces of a gigantic jigsaw puzzle. Dark gray indicates dry land while the continental shelves are light gray.

logical specimens are found on the west coast of Africa and the east coast of South America.

To account for these remarkable phenomena, Wegener hypothesized that all the land masses originally formed one gigantic super-continent called *Pangaea,* which began to break up into smaller continents some 200 million years ago. He felt that the continents, composed of light rock, floated on the heavier basalt. As early as 1915, Wegener wrote: "The continents must have shifted. South America must have lain alongside Africa . . . the two parts must have then become increasingly separated over millions of years."

Wegener's ideas were met with vicious scorn and ridicule. For half a century after the publication of his book, *The Origin of the Continents and Oceans,* geologists believed that Wegener's theory of continents wandering around on the earth was complete nonsense. Indeed, in the 1920s, the president of the prestigious American Philosophical Society labeled Wegener's hypothesis as "utter, damned rot!"

Beginning in the 1960s, however, evidence started accumulating that directly supported Wegener's notion of drifting continents. In the mid-1950s, Dr. Bruce C. Heezen from Columbia University and his colleagues discovered a huge mountain range 40,000 miles long circling the globe under the oceans. By studying samples from the ocean floors, it became increasingly clear that the ocean floor was actually spreading. In the mid-1960s, Drs. Harry H. Hess of Princeton University and Robert S. Dietz at the U.S. Navy Electronics Laboratory proposed that matter from the earth's interior is surging upward in oceanic rifts pushing the continents apart, as shown in Figure 6–23. During the late 1960s and early 1970s, geological evidence supporting this contention continued to mount and today Wegener stands totally vindicated.

The general picture to emerge from Wegener's early work and recent sup-

**Figure 6–23.** *The Spreading Sea Floor.* In the 1960s, evidence accumulated that indicated that molten rock surging up from the earth's interior is causing the sea floor to spread. As the sea floor spreads, the continents (South America and Africa, shown here) are pushed apart.

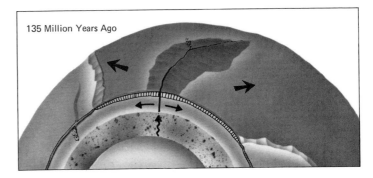

135 Million Years Ago

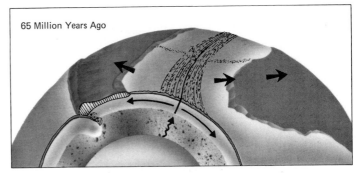

65 Million Years Ago

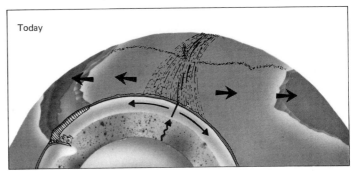

Today

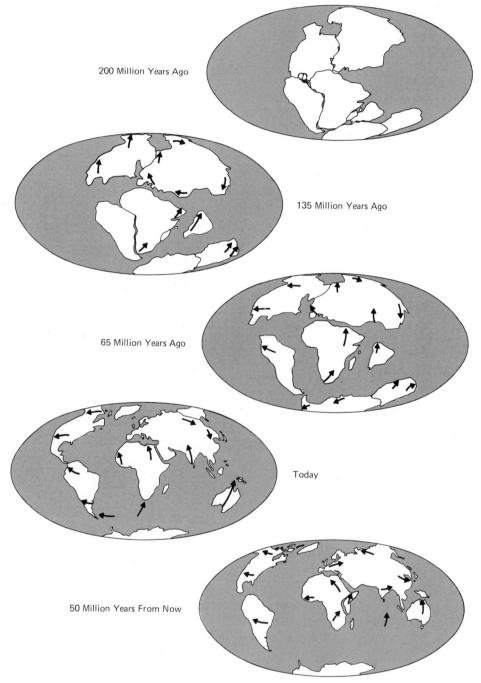

200 Million Years Ago

135 Million Years Ago

65 Million Years Ago

Today

50 Million Years From Now

**Figure 6–24.** *The Drifting Continents.* This series of maps shows the changes in the appearance of the earth over a quarter of a billion years.

porting evidence is that Pangaea did exist over 200 million years ago. This single huge land mass was surrounded by a vast ocean called *Panthalassa*. About 200 million years ago, Pangaea began dividing into two super-continents, *Laurasia* and *Gondwanaland*. Laurasia and Gondwanaland were initially separated by a body of water called the *Tethys Sea*. Small pieces of the Tethys Sea survive today as the Mediterranian, Caspian, and Black Seas. A few million years later South America and Africa became detached as a unit from the rest of Gondwanaland. Indeed, 135 million years ago, the split between Laurasia and Gondwanaland was complete and India had also broken off and was heading on a collision course toward Eurasia.

By 65 million years ago, South America and Africa had drifted apart giving rise to the South Atlantic. Laurasia was breaking up into North America and Eurasia. The Indian Ocean, with India as a large island, had also developed. But Australia and Antarctica were still attached to each other.

Within the past few dozen million years, North America and Eurasia became completely separate. India careened into Asia, thrusting up the Himalaya Mountains, while Australia and Antarctica drifted apart.

If present trends continue, Australia will plow further northward away from Antarctica. The Atlantic and Indian Oceans will continue to widen while the Pacific Ocean becomes narrower. California will split off from the rest of the United States and sections of East Africa will separate from the rest of the African continent. All these changes in the appearance of the earth over a quarter billion years are displayed in a series of maps in Figure 6–24.

In trying to understand the processes involved in the shifting earth's crust, geologists have come to realize that the continents are merely passengers riding on huge plates. These studies have resulted in a new field in geology called *plate tectonics*. There are seven major plates and a large number of smaller plates. All of these plates are comparatively rigid and constitute the *lithosphere*. The continents and the earth's crust are the uppermost layers of the lithosphere. The lithosphere is literally a shell of hard rock 50 to 100 kilometers thick.

The rigid lithosphere floats on a plastic layer of denser rock called the *asthenosphere*. The asthenosphere is the upper part of the mantle and because of its high temperature (just below the melting point of the rock) the asthenosphere exhibits plastic behavior. Like a pot of thick, hot soup on a kitchen stove, there are *convection currents* in the asthenosphere. In certain locations, hot material is rising upward, while in other locations, cooler material is sinking downward. The lithospheric plates are pushed around by these convection currents as shown in Figure 6–25.

In the *rifts* between adjacent plates the hot asthenosphere is rising upward. Liquid rock, or *magma,* surges up through the rift and solidifies. In this way, new crust is formed as the ocean floors spread. Long submarine mountain ranges in the Atlantic and Pacific Oceans mark the locations of major rifts.

At other locations, called *subduction zones,* cooler material in the asthenosphere is sinking downward. Subduction zones also occur along plate boundaries and as the crust at the edge of the plate is pulled downward into the earth, deep oceanic *trenches* are formed. Deep trenches off the coast of Chile, Japan, and the Aleutian

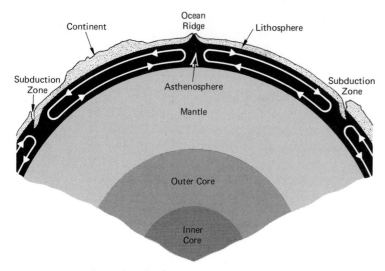

**Figure 6-25.** *The Lithosphere and the Asthenosphere.* The astheno-sphere is a layer of nearly molten rock that lies below the cool, rigid lithosphere. Convection currents in the asthenosphere are responsible for pushing around lithospheric plates to which the continents are attached.

**Figure 6-26.** *Rifts and Subduction Zones.* As a result of convection currents in the asthenosphere, magma surges up through rifts, thereby forming a new crust and causing the ocean floor to spread. Downward currents at the leading edge of a lithospheric plate form deep trenches.

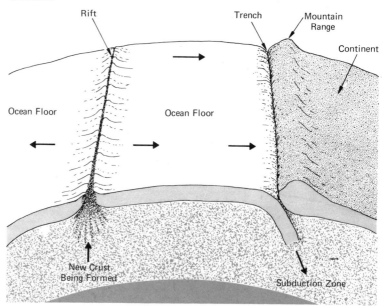

Islands are excellent examples of subduction zones. As shown in Figure 6–26, subduction zones and rifts mark the edges of lithospheric plates.

As might be expected, the boundaries between the lithospheric plates are the locations of violent geological activity. Almost all earthquakes occur along plate boundaries. Volcanoes are found accompanying earthquakes where hot magma wells upward from the asthenosphere along a rift. Along subduction zones, which mark the leading edges of plates, two plates are colliding. As one plate dives under the other, mountain ranges are formed. The Andes in South America and the Himalayas between India and China are excellent examples. Simply by mapping the locations of earthquakes over several years, geologists can deduce the locations of various plate boundaries. Figure 6–27 shows the locations of earthquakes from the mid–1950s to the mid–1970s, while Figure 6–28 shows the major plates that have been identified.

Plate tectonics is one of the major processes actively shaping and changing the surface of our planet. The spreading ocean floor and the collision of huge plates in the lithosphere are largely responsible for some of the most dramatic and violent events in geology. The formation of vast mountain ranges and the occurrence of devastating earthquakes are a direct result of these processes. Indeed, most earthquakes occur along *faults* in the lithosphere where two plates collide or slip past each other.

While plate tectonics often result in sudden changes, other processes are more gradually shaping and reforming the earth. These dramatic processes involve the atmosphere and the oceans.

Life is possible here on earth in part because of the air we breath. Our atmosphere is a mixture of gases, mostly nitrogen ($N_2$) and oxygen ($O_2$), whose chemical composition is shown in Figure 6–29. Most of the atmosphere is restricted to a thin layer extending only a few miles above the earth.

Perhaps the best way of displaying the distribution of the atmosphere above the earth involves *atmospheric pressure*. Since we live at the bottom of a layer of air, we are subjected to the weight of all the gases above us. This pressure amounts to 14.7 pounds per square inch at sea level. The higher we go in the atmosphere, the less air there is above us and hence the atmospheric pressure falls. Atmospheric pressure is usually expressed in pounds per square inch or in *millibars* or in *atmospheres*. Millibars and atmospheres are both units of pressure; 14.7 pounds per square inch equals 1,013 millibars, which also equals one atmosphere. The variation of atmospheric pressure as a function of altitude is shown in Figure 6–30.

While atmospheric pressure declines steadily with increasing altitude, the temperature of the atmosphere varies widely. In the lower atmosphere, temperature declines at a rate of about $6\frac{1}{2}$°C per kilometer (4°F per 1,000 feet). At an altitude of 5 kilometers (3 miles) above sea level, the temperature is down to roughly −20°C while the pressure has dropped to half an atmosphere. The temperature continues to decline at higher altitudes. At 11 kilometers (7 miles) above the ground, the air pressure has dropped to one-quarter atmosphere and the temperature is about −55°C. Seventy-five per cent of the earth's atmosphere lies below this altitude. This level constitutes the top of the *troposphere*.

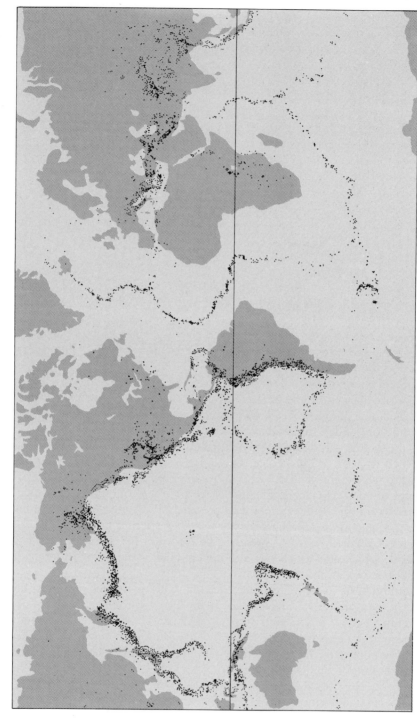

**Figure 6-27.** *Locations of Earthquakes.* Each black dot indicates the location of an earthquake that occurred during a 20-year period from the 1950s to the 1970s. Most earthquakes occur along plate boundaries.

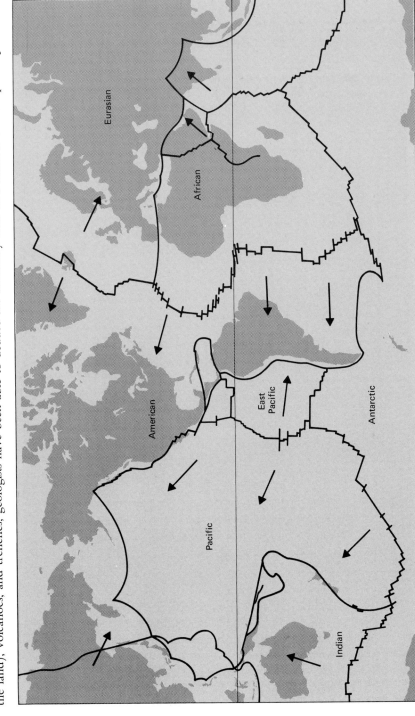

**Figure 6-28.** *The Major Plates.* From studying the locations of earthquakes, mountain ranges (both under the oceans and on the land), volcanoes, and trenches, geologists have been able to deduce the boundary lines of a number of lithospheric plates.

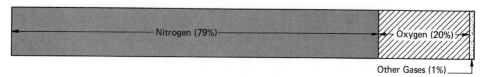

**Figure 6–29.** ***The Composition of Air.*** The chemical composition of dry air at ground level is shown here. Almost four fifths of the atmosphere is nitrogen. Animal life critically depends on the one fifth of the air consisting of oxygen.

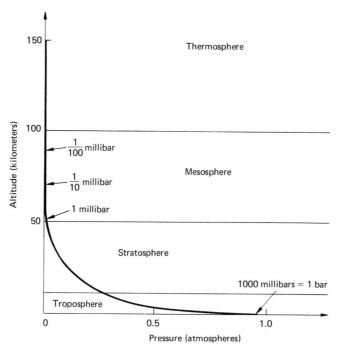

**Figure 6–30.** ***Atmospheric Pressure.*** Atmospheric pressure decreases with increasing altitude above the earth's surface. "One atmosphere" is the pressure at sea level ($= 14.7$ pounds per square inch).

The temperature remains constant for a few miles above the troposphere and then begins to rise. At an altitude of 50 kilometers (30 miles), the temperature reaches a maximum of $10\,°C$ and then begins to decline once again. This altitude at which the temperature reaches a maximum marks the boundary between the *stratosphere* and the *mesosphere*. Through the mesosphere the temperature continues to decline, reaching a minimum of $-76\,°C$ at 80 kilometers (50 miles) at the base of the *thermosphere*. The thermosphere extends upward to 600 kilometers (370

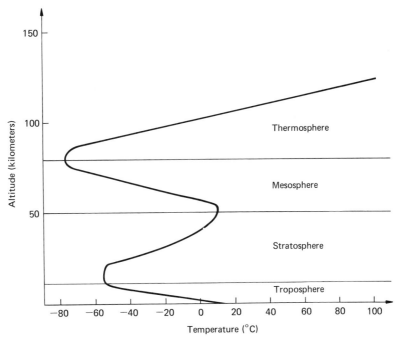

**Figure 6-31.** *Atmospheric Temperature.* In the first 100 kilometers above the ground, the average temperature in the earth's atmosphere varies widely. Maxima and minima in the temperature mark the boundaries of various layers in the atmosphere.

miles) where the temperature levels off to 1,000 °C in the daytime. The outermost layer of the atmosphere, the so-called *exosphere,* gradually fades off into interstellar space. The variation of temperature with altitude is shown in Figure 6–31.

In spite of the extent of the atmosphere, most of the interesting phenomena that affect the earth's surface and our daily lives occur in the lowest few kilometers of the troposphere. Powered by sunlight, horizontal movements of the air, called *winds,* circulate across the face of our planet. Winds blow from regions of high atmospheric pressure, called *highs,* to regions of low atmospheric pressure, called *lows.* These pressure differences can be traced in one way or another to an uneven heating of the earth's surface by sunlight. Warm regions heat the air causing it to rise upward while cool air sinks in those parts of the atmosphere where the temperature is lower. In this way, *convection currents* are set up. The upward motion of the air in a warm region leaves behind a low-pressure area into which the air flows from adjacent high-pressure areas cooled by air descending from higher altitudes.

If the earth did not rotate, air would always be rising from the warm equatorial regions and descending back down in the cooler polar regions. The general, overall circulation pattern for a nonrotating earth is shown in Figure 6–32. But the earth

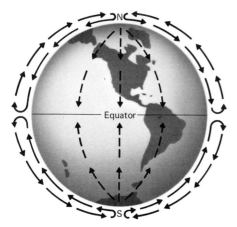

**Figure 6–32.** *Hypothetical Circulation Pattern.* If the earth did *not* rotate, warm air from the equatorial regions would rise upward, flow toward the poles, and sink downward in the cool arctic and antarctic regions. Prevailing winds on the earth's surface would always be from the poles to the equator as indicated by the dashed arrows.

**Figure 6–33.** *Major Convection Currents in the Earth's Atmosphere.* As a result of the earth's rotation, winds are deflected from straight-line paths. This results in six major convection currents circling the earth. The directions of the surface winds are indicated with dashed arrows.

does rotate. Rotation of the earth causes winds to be deflected from their usual straight-line paths as a result of the so-called *Coriolis effect.* As a result, the huge hypothetical convection currents shown in Figure 6–32 are broken up into six large-scale eddies. These six major convection currents circle the globe, three in the northern hemisphere and three in the southern hemisphere, as shown in Figure 6–33.

Occasionally, warm moisture-laden air encounters a region of cooler air, or vice versa. When this happens, the water content of the moist air is precipitated out in the form of rain or, if the temperature is low enough, snow. Boundaries between warm and cold air masses develop *fronts* (either *warm fronts* or *cold fronts,* depending on which way the air is moving), which usually mark the locations of rain showers, thunderstorms, and blizzards. As the water falls to the ground, it flows from higher elevations to lower elevations under the force of gravity forming streams, rivers, and lakes. This flowing water erodes and carves the landscape. Gradually, sometimes over decades or centuries, major changes in the earth's surface are wrought by these relentless forces of nature.

The most effective carving of the landscape results from running water. Glaciers, wind, and waves are also important agents in erosion, but their effects are usually confined to comparatively small, selected regions of the earth's surface.

*Stream erosion* begins when run-off water from rain or melting snow flows under the force of gravity from high elevations to low elevations. This results in the formation of a *gully* in the soft material of a hillside. After each successive rainfall, the gully deepens and widens. As the stream cuts into the landscape, the gully develops a characteristic V-shape and evolves into a *river valley*. As the years pass by, the river valley broadens and sediment carried by the flowing water fills the deepest part of the valley. Its V-shape becomes less and less pronounced. Tributaries extend in a tree-like fashion to smaller valleys and gullys on either side of the river. This characteristic tree-like pattern consists of adjacent streams separated by sharp *divides*. Over the years, erosion smooths and eventually obliterates these divides. The final result is a broad, meandering river.

The V-shape of stream-eroded valleys stands in sharp contrast to the shape of valleys cut by glaciers. In cold climates, snow may not completely melt or evaporate during the summer. The resulting build-up of snow from year to year can result in the formation of a *glacier*. The weight of layer after layer of snow causes pressure that transforms the snow into ice. If the ice is very thick, gravity can cause it to move gradually downhill. This movement carves a valley which is U-shaped with steep sides. Like a gigantic bulldozer, the glacier pushes huge quantities of material in front of it. The pile of debris in front of a glacier is called a *moraine*. As the glacier eventually melts and recedes, the moraine appears as in a low ridge containing a mixture of fine and coarse material. The existence of U-shaped glacial valleys and moraines in the northern United States stands in mute testimony to extensive glaciation during ancient *ice ages* when the earth's polar caps must have been much larger than they are today.

**Figure 6-34.** *River and Glacier Valleys.* Flowing water erodes a V-shaped valley. By contrast, the gradual downhill movement of a glacier carves a U-shaped valley.

Glacier Valley

River Valley

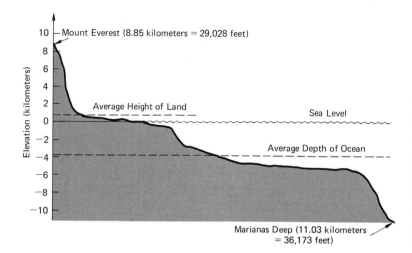

**Figure 6–35.** *Profile of the Earth's Surface.* The average ocean depth is 3.7 kilometers while the average elevation of the land above sea level is only 0.8 kilometers. The highest mountain has an elevation of almost 9 kilometers while the deepest trench descends to a depth of 11 kilometers below sea level.

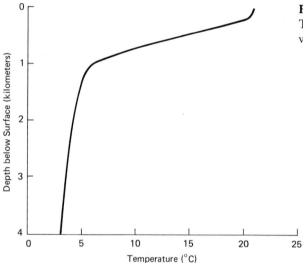

**Figure 6–36 (*left*).** *The Ocean's Temperatures.* The temperature of water in the ocean decreases with increasing depth below the surface.

**Figure 6–37 (*below*).** *Major Ocean Currents.* The circulation of water in the oceans is an important factor in determining the climates on various parts of the earth.

Virtually all the water for rain and snow comes from evaporation off the *oceans*. Most of the water on our planet is located in vast basins surrounding the continents. As shown in Figure 6–35, the average ocean depth is 3.7 kilometers (2.3 miles), while the average height of the continental land masses is 0.8 kilometers (0.5 miles). Because of the presence of minerals dissolved in sea water, the oceans are salty. The total salt content of sea water is about 3.5 per cent by weight. The major constituent of this salt is sodium chloride, in addition to chlorides, sulfates, and bicarbonates of magnesium, calcium, and potassium.

Pressure in the oceans increases rapidly with increasing depth as a result of the great weight of the overlying water. Indeed, at a depth of $9\frac{1}{2}$ kilometers (6 miles) the water pressure is an incredible 8 tons per square inch—about 1,000 atmospheres. Along with this rise in pressure, there is a corresponding decrease in temperature primarily because of the inability of the sun's rays to penetrate very far into oceans. Figure 6–36 shows the variation of temperature with depths below sea level.

Just as sunlight powers convection currents in the earth's atmosphere, there are a number of major *ocean currents* in the waters around the globe. These major currents are shown in Figure 6–37. The circulation of water around the globe has an important effect on the climatic conditions at various locations. For example, the famous Gulf Stream transports warm water from the tropics toward Great Britain. England is therefore comfortably habitable in spite of the fact that its latitude is roughly the same as portions of Hudson Bay and Siberia.

We live on an active, changing planet. Stillness is a remarkable rarity. The earth has been active since its formation $4\frac{1}{2}$ billion years ago and shows no sign of calming down. It is almost as if this planet, teeming with life, is itself alive.

## *The Earth in Space*  6.4

As EVERY boy scout surely knows, the earth has a magnetic field. The earth's magnetic field is said to be *dipolar* in that it has a north magnetic pole and a south magnetic pole, just like a bar magnet. Indeed, the overall shape of the earth's magnetic field is essentially the same as the magnetic field of a hypothetical bar magnet at the earth's center, as shown in Figure 6–38. Just as distance is measured in feet or kilometers, the strength of a magnetic field is measured in *gauss*. On the earth's surface, near the equator, the strength of the earth's magnetic field is about 0.3 gauss. In contrast, a good-quality bar magnet has a field strength of roughly 10,000 gauss, while a child's toy magnet has a strength of several hundred gauss.

The earth's magnetic axis—that is, the line joining the north and south magnetic poles—is not exactly parallel to the earth's rotation axis. Instead, these two axes are inclined to each other by an angle of $11\frac{1}{2}°$, as shown in Figure 6–38. Consequently, compasses do not point to "true" north but rather to a location in

**Figure 6-38.** *The Earth's Magnetic Field.* The geomagnetic field is dipolar; there is a north magnetic pole and a south magnetic pole. The shape of the geomagnetic field is very similar to the magnetic field produced by a bar magnet.

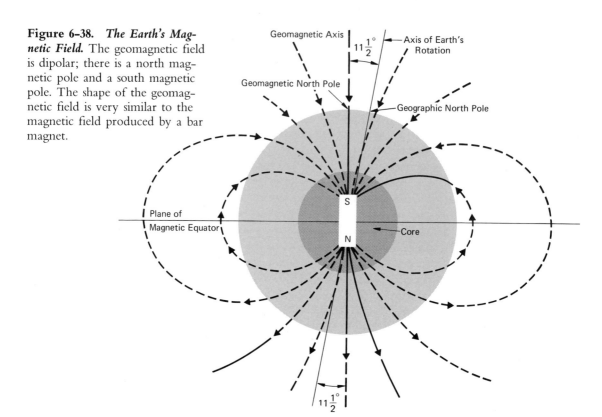

northern Canada at $78\frac{1}{2}°$N, $69°$W. This is the magnetic north pole. The magnetic south pole is exactly on the opposite side of the globe at $78\frac{1}{2}°$S, $111°$E. At the poles, the geomagnetic field strength is about 0.6 gauss, or twice the equatorial strength.

Of course, there isn't a huge bar magnet at the earth's center. Instead, scientists believe that the geomagnetic field is caused by convection currents and motions in the liquid iron core, as shown in Figure 6–39. The circulation of charged particles in the molten core produces the geomagnetic field just as electric current flowing through a wire produces a magnetic field.

For many years, it was strongly suspected that the earth's magnetic field plays an important role in the physics of the outermost layers of the earth's atmosphere. One of the primary motivations for this belief came from observations of *aurorae*. During the night at latitudes near the earth's magnetic poles, it is often possible to see auroral phenomenon from the ground. These phenomena are called *northern lights* (or aurora borealis) in the northern hemisphere and *southern lights* (or aurora australis) in the southern hemisphere. Aurorae appear as faint colored lights—reds, blues, greens, and yellows—forming delicate, constantly changing patterns across the sky. An example is shown in Figure 6–40, although no photograph can truly convey the breath-taking beauty of a spectacular auroral display. Most aurorae are seen over regions of the earth 23° from the magnetic poles called *auroral zones*.

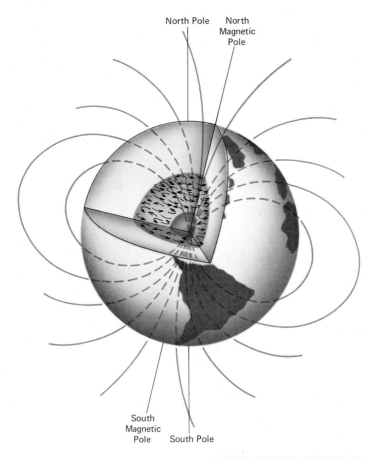

North Pole  North
Magnetic
Pole

South
Magnetic
Pole  South Pole

**Figure 6–39.** *Currents in the
Liquid Core.* It is believed that
the geomagnetic field is caused
by motions of molten iron in
the earth's liquid core.

**Figure 6–40.** *Aurora.* This
aurora was photographed
from an airplane flying over
northern Canada in 1969. No
black-and-white photograph
can do justice to the intricate,
delicate, constantly changing
patterns of colored lights seen
during an auroral display.
(*NASA*)

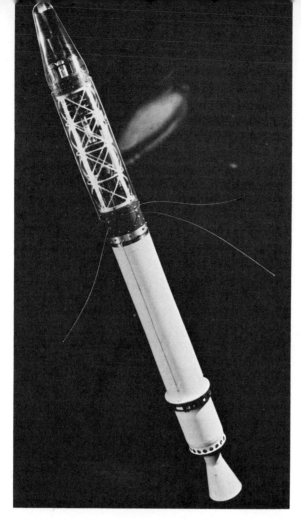

**Figure 6-41.** *Explorer 1.* The United States' first Earth-orbiting satellite, called Explorer 1, weighed 18 pounds and carried a geiger counter. (*NASA*)

Aurorae are a common sight in Alaska, Northern Canada, Scandinavia, and southern Australia and New Zealand.

It had been realized for many years that aurorae have something to do with the sun. Not only does the frequency of aurorae approximately follow the 11-year solar cycle, but violent events on the sun's surface (for example, solar flares) are usually followed a few days later by a spectacular auroral display here on earth. In addition, with the invention of the discharge tube in the mid–1800s (that is, tubes such as those used in neon signs and fluorescent lamps which are filled with gas and glow when attached to a source of electricity) colors similar to those seen in aurorae could be produced in the laboratory. Putting the pieces together, scientists speculated that activity centers on the sun produce streams of charged particles that cause the air to glow and fluoresce as they strike atoms high in the earth's atmosphere. Indeed, most aurorae occur at elevations of roughly 100 kilometers and a delay of two days between a solar flare and an aurora corresponds to a speed of almost 900 kilometers per second for electrons and protons from the sun. Unfortunately, in the absence of any direct observations and experiments in outer space, it was impossible to formulate a detailed understanding of the interaction among particles from the sun, the earth's magnetic field, and aurorae. Phenomena involving the earth and the sun were obviously somehow related, but there simply was not enough information to explain exactly what is going on.

After several disasterous attempts, the United States succeeded in placing a satellite in earth orbit on January 31, 1958. A Jupiter-C rocket managed to get off the launching pad without blowing up and placed a satellite, called Explorer 1, in a highly elliptical orbit about the earth. Perigee (closest distance to the earth) was 360 kilometers and apogee (farthest distance from the earth) was 2,500 kilometers as Explorer 1 circled the globe every 115 minutes. Fortunately, Explorer 1 (Figure 6–41) carried a geiger counter. To everyone's surprise, this geiger counter detected a region of intense radioactivity indicating that there must be large quantities of charged particles trapped above the earth. The existence of numerous charged particles above the earth was confirmed later in 1958 with the flight of Explorer 3. Dr. James Van Allen, who conducted these experiments, realized that the data from his geiger counters could be explained by two huge belts of protons and electrons trapped by the earth's magnetic field. These two belts are today called the *Van Allen radiation belts.*

As shown in the cross-section in Figure 6–42, the inner Van Allen belt starts at

**Figure 6–42.** *The Van Allen Radiation Belts.* Two huge belts of charged particles trapped by the earth's magnetic field completely encircle our planet. The existence of these belts was the first important and surprising discovery to come from man's exploration of outer space.

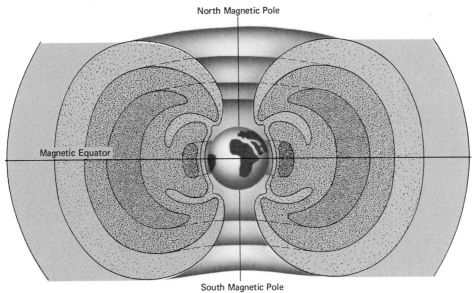

an altitude of about 2,000 kilometers and is approximately 3,000 kilometers thick. It contains mostly protons. In contrast, the outer belt contains mostly electrons and is about 6,000 kilometers thick at an altitude of 16,000 kilometers above the earth. The discovery of these two huge doughnut-shaped radiation belts was so surprising that many scientists initially believed that the geiger counter on Explorer 1 was simply malfunctioning.

While Van Allen and his colleagues were making important experimental discoveries, Dr. Eugene N. Parker at the University of Chicago was making equally revealing theoretical calculations concerning material in the solar corona. The solar corona (see Section 5.1) consists primarily of protons and electrons at a very high temperature. Although electrically neutral (that is, for every negatively charged electron there is a positively charged proton), a substance that is completely ionized behaves very differently than a gas in which the atoms are not ionized. Such a

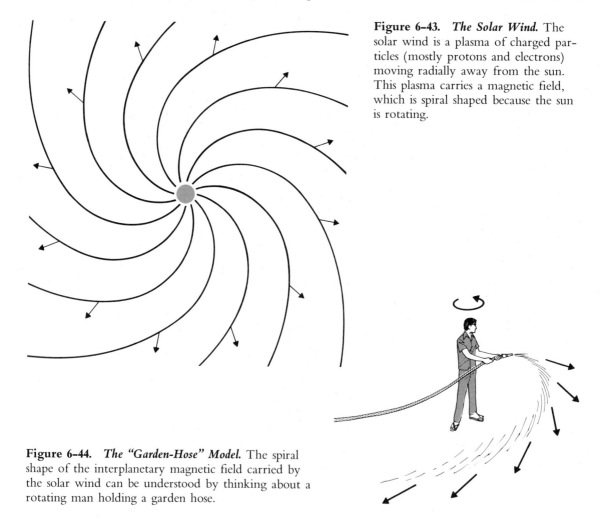

**Figure 6-43.** *The Solar Wind.* The solar wind is a plasma of charged particles (mostly protons and electrons) moving radially away from the sun. This plasma carries a magnetic field, which is spiral shaped because the sun is rotating.

**Figure 6-44.** *The "Garden-Hose" Model.* The spiral shape of the interplanetary magnetic field carried by the solar wind can be understood by thinking about a rotating man holding a garden hose.

substance is called a *plasma* and the branch of science dealing with ionized gases is called plasma physics.

Dr. Parker's calculations proved that the plasma of the solar corona could not be stable and must be continuously expanding. This conclusion supported earlier speculation by Dr. Ludwig F. Biermann who suggested that there must be a continuous outflow of gas from the sun in order to explain the shape and behavior of comet tails. Dr. Parker's work showed that this "gas" is really a plasma of protons and electrons (plus some ions of heavier elements) that is constantly flowing out from the sun. This continuously expanding plasma is called the *solar wind.*

Since the solar wind is a plasma of charged particles, it can conduct electricity and carry a magnetic field. Specifically, a plasma carries with it a "frozen-in" magnetic field. Parker was therefore able to calculate the strength and direction of the interplanetary magnetic field carried along with the solar wind, assuming that this field originated with the sun.

All the charged particles in the solar wind are moving directly away from the sun. But since the sun is rotating, the magnetic field carried along with these particles must have a spiral shape, as shown in Figure 6–43. To understand why this is so, imagine holding a garden hose from which water is streaming. As water pours out of the nozzle in a horizontal stream, imagine that you start rotating. Even though each droplet of water is moving directly away from you, since you are rotating the shape of the entire stream will be a spiral, as shown schematically in Figure 6–44. Parker's ideas about the solar wind and the accompanying magnetic field are often called the *garden-hose model.*

The solar wind was first detected by a Soviet team of scientists under the direction of Dr. K. I. Gringauz with the flight of Luna 3 in 1959. Unfortunately, the Soviet equipment, as well as the apparatus on Explorer 10 flown in 1961, was not specifically designed to examine the solar wind and therefore detailed observations were not possible.

The first extensive and persuasive observations of the solar wind were made during the flight of Mariner 2, shown in Figure 6–45. Mariner 2 was one of the first American probes to explore interplanetary space on its way to Venus. During its long flight, Mariner 2 confirmed the continuous flow of protons and electrons away from the sun. The velocity of the solar wind was found to be very "gusty," varying in speed from 350 to 800 kilometers per second. Mariner 2 detected an average of 9 charged particles per cubic centimeter at a temperature of about 160,000°K.

With the conclusive discovery of the solar wind, scientists realized that a continuous plasma flowing past the earth should dramatically affect the shape of the geomagnetic field. Specifically, the solar wind should "blow" the earth's magnetic field to one side, away from the sun. In addition, as the high-speed solar wind encounters the earth's magnetic field, a *shock wave* results, just as a shock wave occurs when a supersonic aircraft moves through stationary air. Mariner 2 detected this shock wave in 1962. The shock wave is bow shaped, not unlike the shape of the wake of a speed boat moving across still water.

**Figure 6–45.** *Mariner 2.* During its journey to Venus, Mariner 2 conclusively detected and measured the solar wind. (*NASA*)

**Figure 6–46.** *The Earth's Magnetosphere.* As the supersonic solar wind encounters the earth, a huge bow-shaped shock wave is formed. The geomagnetic field is severely distorted by the continuous "blowing" of the solar wind.

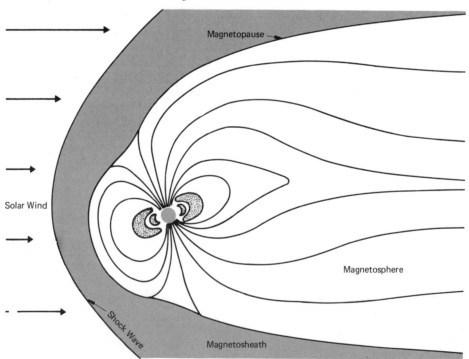

In between the shock wave and the distorted geomagnetic field is an elongated cavity, observed during the flights of Explorer 10, 12, 14, 18, and 21. The outermost boundary of the earth's magnetic field is called the *magnetopause*. The magnetopause forms the inner surface of this elongated cavity while the shock wave constitutes the outer surface. By the early 1960s, scientists could begin putting all the pieces together. The final result is shown in Figure 6–46. The earth's *magneto-sphere,* encased in the magnetopause and shock wave, extends for millions of kilometers away from the sun like the tail of a huge invisible comet.

In conclusion, the aurora commonly seen near the polar regions of the earth involve phenomena far more intricate and complex than anything suspected only two decades ago. High-speed particles from the solar wind are captured, trapped, and accelerated in the earth's magnetosphere. Descending along magnetic field lines, these particles collide with atoms high in the earth's atmosphere causing it to glow and fluoresce. Indeed, the most northerly and southerly "edges" of the Van Allen belts lie nearly overhead of the auroral zones on the earth. All of these discoveries associated with the familiar northern lights were only possible when mankind made its first, hesitant step into outer space.

## *A New Perspective* **6.5**

IN OCTOBER OF 1957, the Soviet Union shocked the world by successfully orbiting a small satellite called Sputnik 1. This was followed a few weeks later with the successful launch of Sputnik 2. Meanwhile, rockets developed in competing programs by the United States Army and Navy were blowing up on the launching pads at Cape Canaveral. The Korean War was fresh in the minds of most Americans. The Russians had successfully detonated their own H-bomb thereby ending a monopoly on thermonuclear weaponry briefly enjoyed by the United States. The Cold War, characterized by distrust and severe lack of communication between the two superpowers to emerge from World War II, had only recently reached its peak during the infamous "McCarthy Era." Alleged "commie" spys were sentenced to death by electrocution. Against this background of national and international turmoil, the Soviet Union had clearly demonstrated its technological superiority by orbiting Sputnik 1 and 2. One result was a period of harsh self-castigation and re-examination encompassing many aspects of American life. Especially severe criticism was leveled at educational and scientific institutions ranging from kinder-garten to the most prestigeous universities. This was soon followed by massive appropriations by the Congress so that America could "catch up" with the Russians. The Space Race was on.

As mentioned in the previous section, the first American earth-orbiting satellite carried a geiger counter. Data returned from this one instrument demonstrated that our planet is surrounded by huge radiation belts filled with high-speed electrons and protons. This discovery was so surprising that scientists began suspecting that future space flights would reveal information about the universe to boggle the mind. The past two decades of space exploration has shown that these suspicions were entirely justified.

Numerous scientific missions have been flown since the late 1950s. The initial programs involved some of the early Explorer, Pioneer, and Discoverer satellites. During the 1960s, as the United States focused its efforts on manned lunar missions, primary efforts in the space program centered on the Orbiter and Ranger series, as well as the Mercury, Gemini, and Apollo flights. In the early 1970s, interest in expensive manned missions declined rapidly. Instead, a wide variety of smaller scientific payloads have been recently launched into both earth and solar orbits. Some of these satellites are examining the details of space near the earth. For example, on June 3, 1974, the Hawkeye scientific satellite was inserted into a highly elliptical polar orbit about the earth. With an apogee of 201,603 kilometers (125,274 miles), Hawkeye is collecting important data concerning the interaction of the solar wind with the earth's magnetosphere.

A second important class of scientific satellites are those that orbit the earth but are designed to make observations of the distant universe. A good example is the

**Figure 6-47.** *An Orbiting Astronomical Observatory.* Major discoveries about the nature of our universe are being made as a result of orbiting telescopes above the earth's atmosphere. (*NASA*)

**Figure 6-48.** *The Space Shuttle.* The Space Shuttle will permit scientists to transport equipment, apparatus, and men from the ground to Earth orbit and back again. This will permit much-improved observations of the universe. (*NASA*)

Orbiting Astronomical Observatory 2 (OAO2). Launched on December 7, 1968, OAO2 weighed 4,400 pounds and carried eleven telescopes. During the first two months of operation, OAO2 transmitted back to earth more information on young stars than had been gathered during the previous fifteen years. Another example is Explorer 42 (*Uhuru*) mentioned in Section 4.2. *Uhuru* was launched into equatorial orbit on December 12, 1970. Its two X-ray telescopes have given astronomers an excellent view of the X-ray sky. After four years of flawless operation, *Uhuru* had identified almost 200 separate X-ray sources including quasars, exploding galaxies, supernova remnants, and even a black hole. The next major advance in astronomy from space will come from the Space Shuttle program. For the first time, scientists will be able to go up with their telescopes, take photographs, and actually bring the photographic film back to earth for further study rather than relying on radio transmissions from unmanned satellites. An anxiously awaited series of observations in this regard involves an all-sky survey in ultraviolet light using telescopes similar to those first developed by Dr. Carruthers (see Figures 4-18, 4-19, and 4-20).

**Figure 6–49a.** *Echo 1.* The first communication satellite was simply a large balloon (shown inflated during a test) that reflected transmissions like a huge mirror in space. (*NASA*)

By far, some of the most exciting discoveries to come from the space program involved vehicles that journeyed to distant objects in the solar system. The moon and all the planets as far as Saturn have been studied during such missions and most of the remainder of this text is devoted to the wealth of information gathered from these historic space flights.

With the inception of the space program, it was soon obvious that the technological capacity to orbit equipment around the earth had numerous commercial applications. The first to come to mind involved world-wide communications. Thousands of miles of copper wire and hundreds of microwave relay stations on mountain tops could be eliminated in favor of a few well-designed communications satellites.

The first communications satellite, Echo 1, was launched on August 12, 1960. It was nothing more than a large balloon 100 feet in diameter made from aluminum-coated plastic. Echo 1 was "passive." Like a huge mirror, it was used to reflect radio waves from one part of the earth to another. A far better idea would be an "active" satellite to receive transmissions, record, amplify, and retransmit them back to the ground. The first active repeater communications satellite was the 500-pound Courier 1B launched on October 4, 1960. This was followed by the much-improved Telstar 1 on July 10, 1962. The 170-pound Telstar 1 was the world's first privately funded satellite, having been built by the Bell System of American Telephone and Telegraph Company.

Up until 1963, communication satellites were placed into orbits fairly near the earth. This means that their orbital periods were only a few hours long. As seen from the ground, sometimes a satellite was overhead and sometimes it wasn't. The next major breakthrough came with *synchronous* satellites. By placing a satellite in a circular orbit exactly 22,300 miles over the equator, the orbital period is precisely

**Figure 6–49b.** *Telstar 1.* The world's first privately funded communications satellite was covered with 3,600 solar cells. Sunlight powered the equipment that relayed transmissions between widely separated parts of the earth. (*NASA*)

**Figure 6–49c.** *Syncom 2.* By placing a satellite in an equatorial, synchronous orbit, the satellite appears to "hover" over a particular point on the earth. Syncom 2 was the first synchronous communications satellite. (*NASA*)

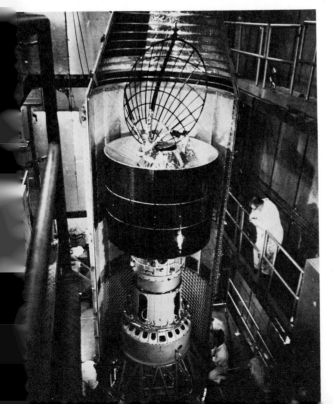

**Figure 6–49d.** *Westar A.* This satellite, owned and operated by Western Union, is in synchronous orbit over the Galapagos Islands in the Pacific Ocean. It is one of three satellites in the Westar series. (*NASA*)

**Figure 6–50a.** *Tiros 1.* Tiros 1 was the first meteorological satellite launched by the United States. In a nearly circular orbit 450 miles above the ground, Tiros 1 transmitted 23,000 cloud-cover photographs back to Earth in less than three months. (*NASA*)

**Figure 6–50b.** *Nimbus 1.* This 830-pound satellite contained visual and infrared television systems permitting continuous 24-hour photographing of the earth. (*NASA*)

228

**Figure 6–50c.** *ESSA 1.* This satellite was the first in a successful series produced by the Environmental Science Service Administration. A total of nine meteorological satellites in the ESSA series were orbited between 1966 and 1969. (*NASA*)

**Figure 6–50d.** *NOAA 1.* This satellite was the first in a series produced by the National Oceanic and Atmospheric Administration. The NOAA program was the successor to the Tiros program. (*NASA*)

24 hours. As seen from the ground, the satellite is always in exactly the same place in the sky, day after day.

The first synchronous communication satellite, Syncom 2, was launched on July 26, 1963. It was then no longer necessary to wait for a satellite to pass overhead in order to broadcast across the Atlantic or Pacific Oceans. Like Syncom 2, all recent communications satellites are in synchronous orbits over the equator. For example, Western Union recently orbited three satellites (Westar A, B, C) to carry domestic transmissions of voice, data, and video throughout the continental United States. Additional synchronous satellites are planned as traffic increases. The National Aeronautics and Space Administration (NASA) also regularly launches satellites designed for international communications. A consortium of 85 nations is responsible for the Intelsat satellites while a joint French–German partnership has resulted in the Symphonie satellites.

Another significant application of space technology involves weather satellites. Meteorologists were quick to point out that photographs of large portions of the earth could be invaluable in improving weather forecasts. The central idea in designing weather satellites is to employ television cameras with reusable photographic recording systems that would transmit views of the earth to receiving stations on a continuous basis.

The first meteorological satellite was the 270-pound Tiros 1 launched on April 1, 1960. Within three months, this satellite transmitted 23,000 cloud-cover photographs back to earth, thereby giving meteorologists their first opportunity to study global weather patterns. During the following three years, seven additional Tiros satellites were launched including Tiros 8 (launched December 21, 1963), which carried the so-called Automatic Picture Transmission (APT) system. This APT system permitted real-time readout of cloud-cover pictures that could be received with relatively inexpensive, portable ground equipment.

As technology improved, the Tiros program was supplanted by the Nimbus program. Nimbus 1 was launched into polar orbit on August 28, 1964, and carried both visual and infrared cameras permitting daytime and nighttime photography. Now weather patterns around the world could be studied continuously during day and night.

Additional major efforts in meteorology from space include the ESSA and NOAA satellites. ESSA 1, operated by the Environmental Science Service Administration was launched on February 3, 1966, while NOAA 1, operated by the National Oceanic and Atmospheric Administration, was launched on December 11, 1970. Both ESSA and NOAA programs have been eminently successful. Cloud-cover pictures from later satellites in the ESSA and NOAA series are regularly seen across the nation every night during news broadcasts by the major television networks.

In studying many of the pictures transmitted from weather satellites, scientists realized that they might be able to take photographs from space to learn a lot more than just where the clouds are. Specifically, different materials on the ground reflect sunlight in very different ways. Of course, dirt is brown and grass is green. While these differences in visible light may seem large, there are still greater differences in reflected infrared light. Figure 6–51 shows the relative intensities of reflected light from four types of ground surfaces in both visible and infrared. Notice, for example, the huge differences in reflected light at 7.5 microns between clay and soybeans. Indeed, the exact amount of light reflected from soybeans critically depends on the health of the crop. A crop that is insect infested or not receiving enough water will reflect infrared light very differently than a healthy crop. The National Aeronautics and Space Administration was therefore strongly motivated to build a spacecraft that could photograph the earth simultaneously in a wide range of visible and infrared wavelengths.

In order to explore the full potential of conducting earth science from space, the Earth Resources Observation System (EROS) was established in 1966 and is operated by the Geological Survey of the Department of the Interior. The primary purpose of EROS is to handle the enormous amount of data pouring in from the Landsat satellites. Landsat 1 (formerly called ERTS 1 for Earth Resources Technology Satellite) was launched on July 23, 1972. Landsat 1, shown in Figure 6–52

**Figure 6–51.** *Spectral Signatures.*
Different materials on the ground reflect light in very different ways. These differences in reflectivity are most pronounced at near-infrared wavelengths.

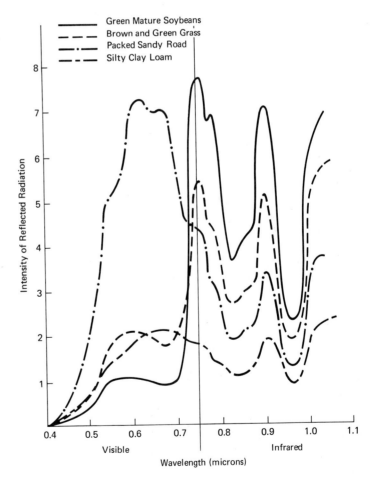

**Figure 6-52.** *Landsat 1.* This sophisticated satellite takes photographs of the earth's surface simultaneously in four separate wavelength ranges. Analysis of the photographs has begun to reveal the enormous amount of information that can be obtained by studying the earth from space. (*NASA*)

**Figure 6-53a (*lower left*).** *San Francisco (green band).* The sediment-laden, shallow portions of the San Francisco Bay show up clearly in this view. (*ERTS; NASA*)

**Figure 6-53b (*below*).** *San Francisco (red band).* Cultural features and urban areas, especially in the San Joaquin Valley, are prominent in this view. (*ERTS; NASA*)

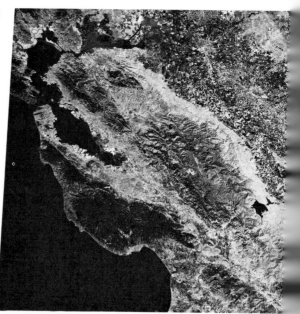

has four cameras that take simultaneous photographs at four separate wavelengths: green band (0.5–0.6 microns), red band (0.6–0.7 microns), near-infrared band (0.7–0.8 microns), and infrared band (0.8–1.1 microns). Each picture from each band is ideally suited to photograph different features and properties of the earth's surface. The green band (0.5–0.6 microns) emphasizes movements of sediment in water and clearly shows areas of shallow water such as shoals and reefs. The red band (0.6–0.7 microns) clearly reveals man-made and cultural features such as roads, buildings, and excavated soil. The near-infrared band (0.7–0.8 microns) emphasizes live vegetation, land forms, and water–land boundaries. And finally, the infrared band (0.8–1.1 microns) gives the best penetration of atmospheric haze, again revealing vegetation and land forms. Figure 6–35 (a, b, c, and d) shows four photographs of the San Francisco Bay area simultaneously taken by the four Landsat 1 cameras.

Each of the four black-and-white photographs from each of Landsat's cameras can be examined separately, or they can be combined to give a *false-color composite*. The resulting color photograph has nothing to do with the real colors seen from space. Instead, the printing of the color film is done in such a way that the differences between the various bands show up in a very vivid and striking fashion. On a false-color composite, healthy vegetation appears red, clear water is black,

**Figure 6–53c (*left*).** *San Francisco (near-infrared band).* Vegetation and land forms, especially mountainous terrain, stand out in this view. (*ERTS; NASA*)

**Figure 6–53d (*right*).** *San Francisco (infrared band).* The boundaries between land and water are emphasized in this view. (*ERTS; NASA*)

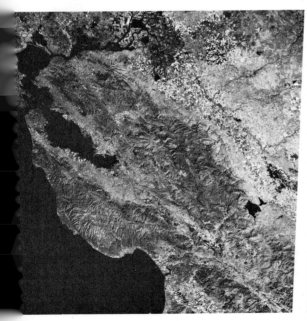

sediment-ladened or polluted water is blue, cities and urban areas are gray. A false-color composite of the San Francisco Bay area is shown in Plate 5. This photograph was made from the four black-and-white views shown in Figure 6–53 (a, b, c, and d). For comparison, a photograph in visible light taken by astronauts on Skylab in the summer of 1973, is shown in Plate 6. Plate 6 shows exactly how San Francisco looks in visible light on a clear day. A similar pair of photographs, one false-color composite from Landsat 1 and one true-color photograph from Skylab, is shown in Plates 7 and 8 for the New York City area.

Landsat 2 was launched on January 22, 1975, into a circular orbit 920 kilometers (570 miles) above the earth's surface. It orbits the globe every 103 minutes or about 14 times a day. Each Landsat completely photographs the entire earth (except the poles) every 18 days. From examining the photographs, scientists now realize that they can evaluate the health and vigor of crops. They can accurately estimate the extent and depth of snow covers as well as the moisture content of soil. Is there too much livestock grazing on a particular range in Wyoming? Is there enough winter snow in the Sierras to meet the water requirements for Los Angeles in the summer? Is a forest in Oregon insect infested or diseased? All of these questions can be answered by an analysis of Landsat photographs.

As the level of technology and industry increases along with a constantly growing population, mankind must turn to sophisticated techniques to cope successfully with the demands made on our environment. Just being able to feed 7 billion people in the year 2000 poses monumental problems. If we are to live in a stable world-community, man must learn to make use of his planet in the most efficient and ecologically sound manner. Information from satellites like Landsat with its applications in land management, pollution control, marine resources, geological and hydrological engineering clearly point in a very constructive direction.

## Questions and Exercises

1. What is the difference between a rock and a mineral?
2. Why do different minerals have different crystal forms?
3. What is meant by cleavage and how can cleavage aid in the identification of minerals?
4. What is the Mohs hardness scale?
5. What is meant by specific gravity?
6. What is the most common rock-forming mineral?
7. Name one common rock-forming mineral that is not a silicate.
8. Compare and contrast igneous, sedimentary, and metamorphic rocks.
9. Name one common igneous rock.
10. Name one common sedimentary rock.

11. Name one common metamorphic rock.
12. Discuss how rocks can be dated by measuring their content of certain types of isotopes.
13. What is meant by the focus and epicenter of an earthquake?
14. Contrast and compare the three types of seismic waves produced by an earthquake.
15. What is meant by shadow zones?
16. Discuss the structure of the earth deduced from the propagation of P-waves and S-waves through the earth.
17. Present a physical argument explaining why the earth's interior consists of a liquid core sandwiched between a solid mantle and a solid inner core.
18. Who was Alfred Wegener?
19. What were Pangaea, Panthalassa, and Gondwanaland?
20. What are the lithosphere and the asthenosphere?
21. Briefly describe the process of plate tectonics.
22. Briefly describe the variation of temperature and pressure in the earth's atmosphere.
23. Briefly describe the overall circulation of the earth's atmosphere.
24. How do valleys formed from stream erosion differ from those formed by glacial erosion?
25. Compare the average ocean depth to the average height of land above sea level.
26. What is an aurora?
27. What surprising discovery came from the flight of Explorer 1?
28. Briefly describe the Van Allen belts.
29. What is the solar wind?
30. Using the garden-hose model, briefly describe the shape of the interplanetary magnetic field.
31. What is the magnetopause?
32. Briefly describe the overall appearance of the earth's magnetosphere.
33. Briefly discuss why astronomers are interested in making observations of the universe from earth-orbiting satellites.
34. What are Echo, Telstar, Syncom, and Westar?
35. What are Tiros, Nimbus, ESSA, and NOAA?
36. What is Landsat and what kinds of information does it give us about the earth?

# 7 Conquest of the Moon

## 7.1 Earth's Nearest Neighbor

WITHOUT exception, our moon consistently provides the most dramatic astronomical sight in the nighttime sky. Throughout the course of recorded history, philosophers, scientists, poets—people of all persuasions—have watched as the silvery moon runs through its phases every four weeks. Great works of art, literature, music, and poetry have been inspired by the earth's nearest celestial neighbor. Most recently, the prospect of successfully landing men on the moon gave rise to one of the most ambitious technological projects of all time. Indeed, Isaac Asimov has argued that if the earth did not have a moon, if the only sight regularly seen in the nighttime sky were tiny pinpoints of starlight, ancient man would not have been very motivated to take up the study of astronomy. The fact that the earth possesses a satellite, so large and near that some of its features can be seen with the naked eye, is certainly in part responsible for mankind's preoccupation with the heavens over the ages.

**Figure 7-1 (*below*).** ***The Appearance of the Moon.*** This series of photographs shows the moon at various times during the lunar month. (*Lick Observatory photograph*)

**Figure 7-2.** ***The Moon.*** Through a small telescope, such as that used by Galileo, it is possible to see craters, mountains, valleys, and plains. (*Lick Observatory photograph*)

**237**

In ancient times, the gradations in gray seen across the lunar surface with the naked eye were thought to be color variations rather than topographical features. With the invention of the telescope around 1600, however, Galileo made the remarkable discovery that the moon was actually covered with craters, mountains, valleys, and vast planes. These features can be easily seen with any small telescope or a good pair of binoculars. The resulting view is similar to that shown in Figure 7–2.

The largest features on the moon's surface are relatively flat planes called *maria*. This term comes from the Latin word meaning "seas" (*mare* = "sea") and dates back to the seventeenth century when astronomers believed that there were oceans on the moon. Of course, we today know that there is absolutely no air or water on the moon. The moon is literally bone-dry. Nevertheless, this grossly inaccurate term has stuck with us.

In examining the moon with telescopes, early astronomers gave very fanciful names to the maria they saw. In keeping with the scholarly tradition of the times, all the names were in Latin. Mare Imbrium (the "Sea of Showers") is the largest of the fourteen lunar maria. It measures about 1100 kilometers (700 miles) across and

**Figure 7–3.** *Mare Imbrium.* This earth-based photograph is centered on the largest of the 14 lunar maria. This "sea" is actually dry land pockmarked with a few small craters. (*Hale Observatories*)

**Figure 7–4.** *Mare Tranquilitatis.* As Apollo 8 astronauts circled the moon, they took this oblique photograph of Mare Tranquilitatis. (*NASA*)

is shown in Figure 7–3. Other prominent "seas" include Mare Nubium (the "Sea of Clouds"), Mare Nectaris (the "Sea of Nectar"), Mare Tranquilitatis (the "Sea of Tranquility"), and Mare Serenitatis (the "Sea of Serenity"). A close–up view of Mare Tranquilitatis is shown in Figure 7–4, while Mare Serenitatis is shown in the lower half of the earth-based photograph in Figure 7–7.

Perhaps the most characteristic type of feature seen on the moon are lunar *craters*. From the earth, 30,000 craters can be seen ranging in size from 1 to more than 100 kilometers in diameter. These circular depressions are named after famous scientists and philosophers of antiquity. The largest craters are Clavius and Grimaldi, each measuring almost 240 kilometers (150 miles) across. An earth-based photograph of the heavily cratered southern part of the moon (with Clavius near the bottom) is shown in Figure 7–5.

In general, craters are found over all parts of the moon except for those regions covered with maria. In sharp contrast to the relatively smooth maria, cratered terrain is usually very rugged. Craters often overlap and are piled on top of one another.

**Figure 7-5.** *Lunar Craters.*
Nearly 30,000 craters can be seen from the earth. This Earth-based photograph shows a heavily cratered region near the moon's south pole. (*Hale Observatories*)

**Figure 7–6a.** *Copernicus from Earth.* This is an excellent Earth–based view of the crater Copernicus. (*Lick Observatory photograph*)

**Figure 7–6b (center left). Oblique View of Copernicus.** On the last manned mission to the moon, Apollo 17 astronauts took this photograph of Copernicus. Copernicus is 100 kilometers (60 miles) in diameter. (*NASA*)

**Figure 7–6c (bottom left).** **Copernicus Nearly Edge On.** While in lunar orbit in December 1972, Apollo 17 astronauts took this photograph of Copernicus on the horizon. (*NASA*)

**Figure 7–6d (below).** **Telephoto View of Copernicus.** Lunar Orbiter 2 transmitted this telephoto view of Copernicus' central peak and far wall back to earth in 1968. The spacecraft was only 28 miles above the moon's surface when this photograph was taken. (*NASA*)

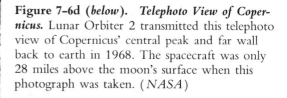

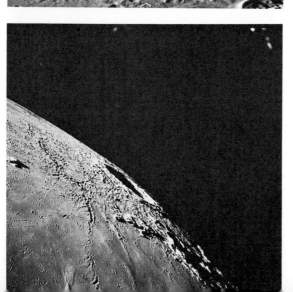

**Figure 7-7.** *Northern Region of the Moon.* Two mountain ranges, the Apennine and the Caucasus Mountains, are shown extending upward from the lower-left corner of this Earth-based view. Mare Serenitatis dominates the lower half of the photograph. The Alpine Valley is the straight feature just above left center, near the terminator. ( *Hale Observatories* )

In most cases, the floor of a crater lies below the surrounding ground level. The inside crater walls are almost always steeper than the outside walls. These walls sometimes rise thousands of feet above the crater floor. In addition, many large craters have prominent central peaks. For example, the crater Copernicus, shown in Figure 7–6 (a, b, c, d) has a central peak rising to an elevation of 1000 feet above the flat crater floor.

In addition to craters and maria, there are also *mountain ranges* on the moon. These lunar mountain ranges have been named after terrestrial mountain ranges: Alps, Apennines, Carpathians, Caucasus, and so forth. Except for their names, lunar mountains bear little resemblance to their terrestrial counterparts. With no water, air, rain, and snow, lunar mountains are devoid of drainage and erosion features that are so very obvious in the case of mountains here on earth.

Many years ago, astronomers realized that the heights of lunar mountains could be deduced from measuring the lengths of the shadows they cast. Such observations reveal that lunar mountain ranges have elevations up to 25,000 feet or more— comparable to the Himalayas between India and China. Two mountain ranges, the Apennines and Caucasus mountains are shown in the lower-left portion of Figure 7–7.

**Figure 7–8 (*left*).** ***The Alpine Valley.*** This view of the Alpine Valley was transmitted back to earth from Lunar Orbiter 5 in August of 1967. The spacecraft was 82 miles above the lunar surface when the photograph was taken. (*NASA*)

**Figure 7–9 (*center*).** ***Rilles.*** This photograph, taken by Apollo 8 astronauts in December of 1968, shows a series of rilles crisscrossing the crater Goclenius. Goclenius is about 64 kilometers (40 miles) in diameter. (*NASA*)

There are many other types of features seen across the lunar surface besides maria, craters, and mountains. For example, there are *valleys* on the moon. The small, straight feature in the above left-center portion in Figure 7–7, near the *terminator* (that is, near the dividing line between day and night) is the famous Alpine Valley. This cigar-shaped depression is 120 kilometers (74 miles) long and up to 10 kilometers (6 miles) wide. It is located in the Alps near the edge of Mare Imbrium. An excellent, close-up view, shown in Figure 7–8, was transmitted back to earth from Lunar Orbiter 5 in 1967.

There are numerous crevasses or clefts on the lunar surface. These features are called *rilles*. A sinuous rille can be seen down the middle of the Alpine Valley in Figure 7–8. Figure 7–9, a photograph taken by Apollo 8 astronauts orbiting the moon, shows a series of prominent rilles crisscrossing the crater Goclenius.

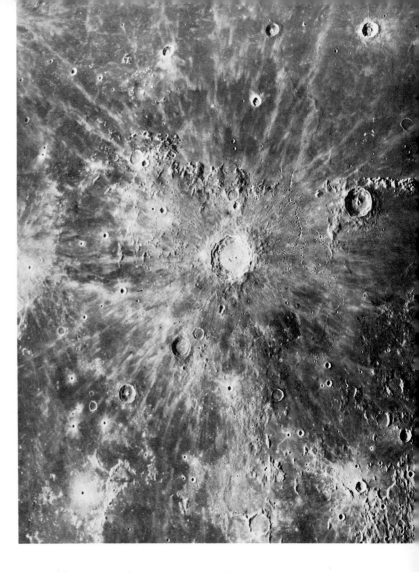

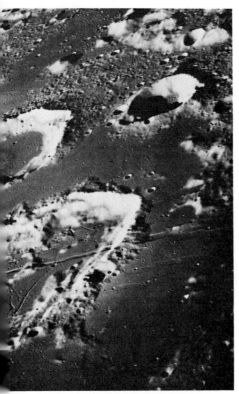

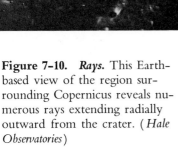

**Figure 7-10.** *Rays.* This Earth-based view of the region surrounding Copernicus reveals numerous rays extending radially outward from the crater. (*Hale Observatories*)

Finally, another type of feature commonly seen on the moon are *rays*. Rays are bright streaks radiating from certain craters such as Tycho and Copernicus. Rays are almost perfectly straight. They have widths ranging from 5 to 10 kilometers and extend for hundreds of kilometers across the lunar surface. An earth-based photograph of the region surrounding Copernicus is shown in Figure 7–10. Numerous rays can be seen extending radially away from the center of the crater. This system of rays strongly suggests that Copernicus was formed by an explosive event, probably by the impact of a large meteoroid on the lunar surface.

Many of these lunar features—maria, craters, mountains, valleys, rilles, and rays—have been observed by astronomers over the past several centuries. The primary reason for this, of course, is the invention of the telescope and the fact that the moon is so very near the earth. At an average distance of 384,400 kilometers

**243**

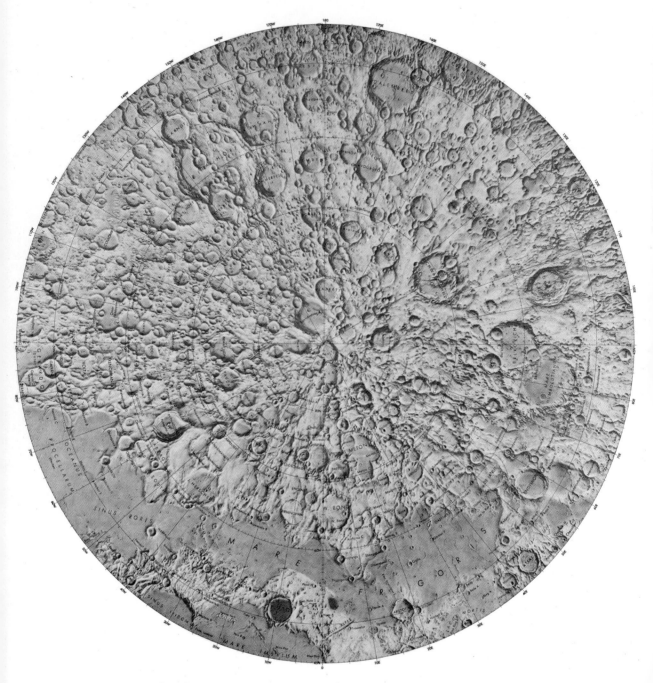

**Map 1.** *Lunar North Pole Map* (*NASA Chart LMP-3*)

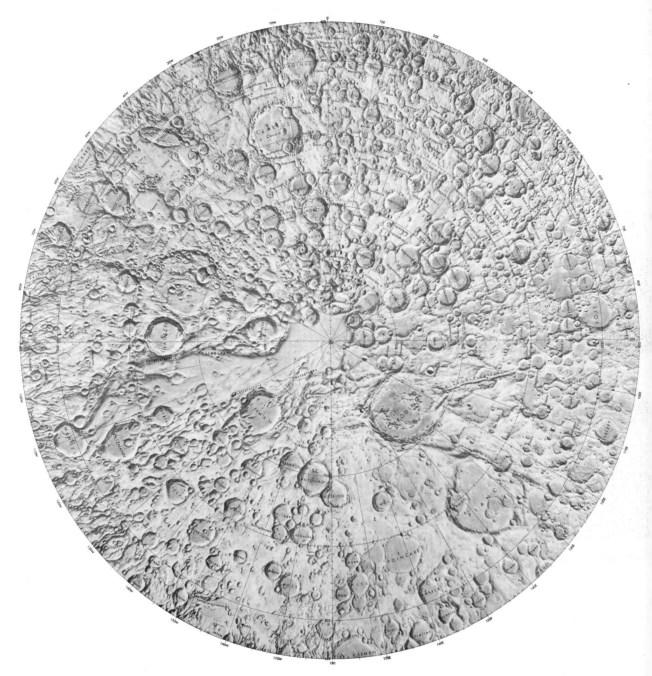

**Map 4.** *Lunar South Pole Map* (*NASA Chart LMP-2*)

**Table 7–1**

| Facts About the Moon | |
|---|---|
| Average Earth–Moon Distance | 384,400 kilometers |
| Sidereal Period | 27.322 days |
| Diameter of Moon | 3,476 kilometers |
| Mass of Moon | 0.0123 of Earth mass |
| Density of Moon | 3.34 grams per cubic centimeter |
| Escape Velocity | 2.38 kilometers per second |
| Surface Gravity | 0.165 of Earth surface gravity |
| Orbital Eccentricity | 0.0549 |
| Inclination of Orbital Plane to Ecliptic | 5°9′ |
| Inclination of Lunar Equator to Ecliptic | 1°32′ |
| Temperature Range | −189°C to +117°C |

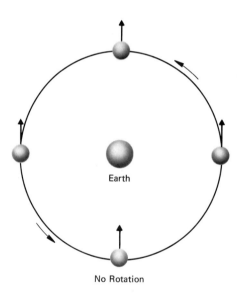

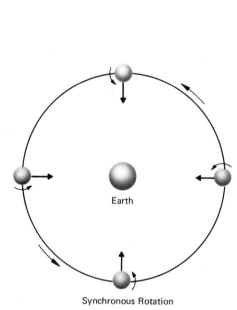

**Figure 7–11.** *Rotation of the Moon.* If the moon did not rotate, all sides of the moon would be seen from the earth, as shown on the left side of this diagram. However, the moon always keeps one side facing the earth, as shown on the right side of the diagram. The moon's orbital period therefore equals its rotational period.

(about one quarter million miles), lunar features as small as 1 kilometer in size can be resolved with high-quality earth-based telescopes.

The diameter of the moon is 3,476 kilometers (2,160 miles), slightly smaller than the distance between New York and Los Angeles. By applying Newtonian mechanics to the observed orbit of the moon about the earth, the moon's mass can be deduced. The mass of the moon is very nearly $\frac{1}{81}$ the mass of the earth. After learning both the mass and size of the moon, it is easy to calculate quantities such as the moon's average density, escape velocity, and surface gravity. For example, the average density of the moon is simply its mass divided by its volume. This comes out to 3.34 grams per cubic centimeter, somewhat less than the earth's average density (5.5 grams per cubic centimeter). Similar calculations reveal that the lunar surface gravity is about $\frac{1}{6}$ as large as the earth's surface gravity. A person standing on the moon would weigh $\frac{1}{6}$ what he does on the earth. Likewise, the escape velocity from the moon is 2.4 kilometers/second (1.5 miles per second), far too low to retain any atmosphere. By comparison, the escape velocity from the earth is almost five times larger (11.2 kilometers per second = about 7 miles per second). Table 7–1 summarizes some significant data about our moon.

One of the well-known properties of our moon is that it always keeps the same side facing the earth. The earthside of the moon may be dark (during new moon phase) or may be fully illuminated (during full moon phase). Nevertheless, the same familiar lunar features always face the earth (refer back to Figure 7–1). This does *not* mean that the moon does not rotate about its axis. Very much to the contrary, *if* the moon did not rotate, then it would be possible to see the entire lunar surface from the earth every four weeks, as shown on the left side of Figure 7–11. Instead, the moon rotates once about its axis during each orbit of the earth, as shown schematically on the right side of Figure 7–11. This phenomenon is called *synchronous rotation* because the rotational period of the moon exactly equals its orbital period.

In spite of the fact that the moon always keeps one side facing us, it is possible to see slightly more than 50 per cent of the lunar surface from earth. The moon's axis of rotation is not exactly perpendicular to its orbital plane. Consequently, we can sometimes see a little over the lunar north pole or a little over the lunar south pole. In addition, although the moon rotates at a constant rate, the moon's orbit is an ellipse. From Kepler's second law, it follows that the speed of the moon around the earth varies slightly; the rates of rotation and revolution are sometimes not exactly in step with each other. Consequently, we can sometimes see slightly around the east limb or the west limb. Finally, although it is usually said that "the moon orbits the earth," in reality, the moon and the earth each orbit their common *center of mass*. These three effects mean that we see the moon from a constantly changing vantage point. As seen from the earth, the moon therefore appears to "wobble" slightly, thereby permitting us to view a little more than half the lunar surface. This phenomenon is called *libration,* illustrated in contrasting views in Figure 7–12. As a result of libration, about 59 per cent of the moon's surface can be seen from earth.

In spite of libration, large quantities of the lunar surface are eternally hidden from earth-based view. In addition, features near the limb or visible edge of the

**Figure 7-12.** *Libration.* As seen from the earth, the moon appears to "wobble" slightly in its orbit about the earth. As a result of these librations, 59 per cent of the lunar surface can be seen from earth. ( *Yerkes Observatory photograph* )

moon are viewed at such an oblique angle that their exact shapes and sizes are only poorly revealed. And finally, objects less than $\frac{1}{2}$ kilometer in size cannot be resolved even with the finest earth-based telescopes. The only viable solution to these limitations involves space flight to the moon. This alternative became technologically feasible in the mid-1900s and resulted in one of the most ambitious scientific and engineering endeavors of the human race since the building of the Pyramids in ancient Egypt almost 5000 years ago.

## 7.2 Rangers, Surveyors, and Orbiters

EFFECTIVE lunar exploration began in the early 1960s with the *Project Ranger*. The purpose of the Ranger spacecraft was to make a "hard" landing on the moon. During descent, television cameras would transmit close-up views of the lunar surface back to earth. After impact, a special scientific package—specifically designed to withstand the crash—would make important scientific measurements of conditions on the moon. For example, the package was planned to contain a seismograph to detect *moonquakes*.

The first Ranger, Ranger 1, was launched into earth orbit on August 23, 1961, primarily to test the spacecraft systems and make measurements of the earth's magnetosphere. Rangers 2, 3, 4, 5, and 6 were launched between November 1961 and January 1964. All failed to perform properly as a result of some sort of malfunction. It was during this very difficult time that an important decision was made that dramatically affected the entire course of lunar exploration. In an unforgettable speech, President Kennedy announced that the goal of the United States space program was to land Americans on the moon and return them safely before the end of the decade. The effect of this decision was to scrap all scientific aspects of future missions that would not directly support a manned landing. For example, it was not critically important to know about moonquakes in order to land men on the moon. Yet, it was important to know exactly what the lunar surface looks like. Engineers had to know how rugged the moon's surface is in order to design the landing vehicle properly. Consequently, the small scientific package was dropped from the Ranger Program and much greater emphasis was on visual *imaging,* that is, on the television systems to send back pictures.

The first successful Ranger, Ranger 7, was launched on July 28, 1964. This was followed by the equally successful flights of Rangers 8 and 9, launched on February 17 and March 21, 1965, respectively. All of these spacecrafts had the same basic design, as shown in Figure 7–13. Spacecraft power was delivered from two solar panels. The $24\frac{1}{2}$ square feet of solar cells produced 200 watts of power. Each panel had 4896 solar cells.

**Figure 7–13.** *Basic Features of the Rangers.* All of the three successful Rangers (7, 8, 9) were essentially the same. They all carried six television cameras (two wide angle and four narrow angle) and the primary purpose of the missions was close-up lunar photography. This diagram shows the basic features of the Ranger spacecraft.

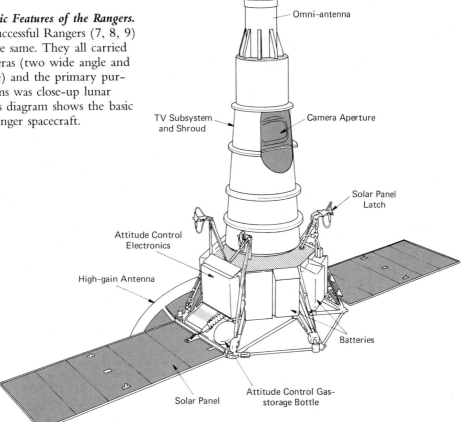

Omni-antenna

TV Subsystem and Shroud

Camera Aperture

Solar Panel Latch

Attitude Control Electronics

High-gain Antenna

Batteries

Solar Panel

Attitude Control Gas-storage Bottle

**Figure 7-14.** *Impact Locations.* The target locations of the three successful Ranger missions are shown against this composite photograph of the moon. All three spacecrafts impacted less than 15 miles from their planned targets. A total of more than 17,000 photographs of the lunar surface were transmitted back to Earth. (*Lick Observatory photograph*)

Communication with the spacecraft was achieved with two antennas. The omnidirectional antenna was employed during launch and at the time of mid-course corrections. The disk-shaped, high-grain directional antenna was used just before impact to transmit television pictures back to earth.

Each Ranger carried six television cameras: two wide-angle cameras and four narrow-angle cameras. The entire television system on each spacecraft weighed 375 pounds and included—in addition to the six cameras—a camera sequencer, a video combiner, a telemetry system, transmitters, and batteries. The total weight of each Ranger was 804 pounds. The television systems were designed at RCA in Princeton, New Jersey, and the entire spacecrafts were built at the Jet Propulsion Laboratory in Pasadena, California. All of the design and engineering work that went into the Rangers set the stage for all future unmanned spacecrafts. It is no accident that, for example, the Mariner spacecrafts that went to Mars, Venus, and Mercury look very much like the Rangers.

The landing sites for Rangers 7, 8, and 9 are shown in Figure 7-14. Both Rangers 7 and 8 impacted in maria. Ranger 7 landed only 9 miles from its target between Mare Cognitum and Mare Nubium while Ranger 8 was about 15 miles off target in Mare Tranquilitatis. In contrast, Ranger 9 landed only 3 miles from its target on the floor of the crater Alphonsus.

**Figure 7–15a (*left*).** *Ranger 7.* Engineers at the Jet Propulsion Laboratory check out Ranger 7. This spacecraft carried six television cameras: two wide angle and four narrow angle. It was successfully launched on July 28, 1964, and sent back 4,316 high resolution pictures of the lunar surface. The total flight time to the moon was 68 hours, 36 minutes. (*NASA*)

**Figure 7–15b (*right*).** *Ranger 8.* This artist's drawing shows Ranger 8 just before impact on the lunar surface. The spacecraft was successfully launched on February 17, 1965, and sent back 7,137 close-up television pictures of the moon. Total flight time to the moon was 64 hours, 53 minutes. (*NASA*)

**Figure 7–15c.** *Ranger 9.* This artist's drawing shows the final Ranger mission to the moon. While the first two successful Rangers (Rangers 7 and 8) "hard landed" in maria, Ranger 9 impacted only 3 miles from the planned target on the eastern floor of the crater Alphonsus. Ranger 9 was successfully launched on March 21, 1965 and sent back 5,814 pictures. Total flight time to the moon was 64 hours, 31 minutes. (*NASA*)

**Figure 7-16.** *View from Ranger 7.*
Ranger 7 sent back this photograph from
an altitude of 470 miles above the lunar
surface. The large crater in the upper
right-hand corner is Guericke, located between Mare Nubium and Mare Cognitum.
The area shown in the photograph is
roughly 75 miles by 65 miles. The smallest
craters seen are about 800 feet across.
(*NASA*)

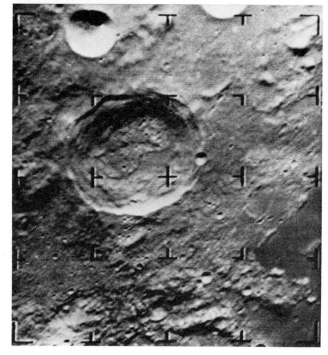

**Figure 7-17.** *View from Ranger 8.* At an
altitude of 470 miles above the lunar surface, Ranger 8 sent back this photograph
approximately 7 minutes before impact in
Mare Tranquilitatis. The area shown is
roughly 90 miles by 70 miles. The large
crater near the center of the view is Delambre, which is 32 miles in diameter.
(*NASA*)

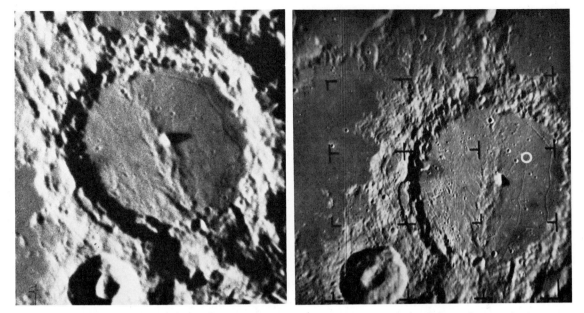

**Figure 7-18.** *Contrasting Views.* A dramatic increase in resolution is clearly illustrated in these contrasting views of the crater Alphonsus. The left photograph was taken with an excellent Earth-based telescope while the right photograph was transmitted back from Ranger 9. The small white circle indicates the impact target of Ranger 9. ( *NASA* )

Typical views from Rangers 7 and 8 are shown in Figures 7-16 and 7-17, respectively. Craters as small as 800 feet in diameter on the mare floor were seen when Ranger 7 was at an altitude of 470 miles above the lunar surface. Figure 7-17, also taken at an altitude of 470 miles, shows the crater Delambre near the southwest corner of Mare Tranquilitatis.

Rangers 7 and 8 sent back a total of 13,453 pictures. The final pictures, transmitted from altitudes of less than half a mile, show details as small as 3 feet in size. This dramatic increase in resolution over earth-based observations is clearly shown in Figure 7-18. The view on the left represents just about the very best that can be done from earth a quarter of a million miles away. Nothing smaller than half a mile in size can be distinguished. The view on the right is from Ranger 9 roughly three minutes before impact. Notice how much sharper and clearer the image from Ranger 9 is compared with the earth-based view.

A sequence of photographs taken during the descent of Ranger 9 are shown in Figure 7-19 (a, b, c). Notice how the resolution increases as the spacecraft approached the lunar surface. Ranger 9 sent back 5,814 pictures before crashing to the floor of Alphonsus in March of 1965. These photographs provided scientists with their first close-up view of the inside of a crater.

**Figure 7-19a.** *View from Ranger 9 (9 minutes, 18 seconds before impact).* This photograph from Camera B was taken at an altitude of 775 miles above the lunar surface. The total area covered is 147 miles by 123 miles. Portions of three craters are seen. Ptolemasus is at the top, Alphonsus is on the left, and Albategnius is on the right. The central peak in Albategnius rises 4,500 feet above the crater floor. (*NASA*)

**Figure 7-19b.** *View from Ranger 9 (2 minutes, 50 seconds before impact).* This photograph from Camera A was taken at an altitude of 258 miles. The total area covered is 121 miles by 109 miles. Alphonsus fills the right half of the picture. The eastern-most portion of Mare Nubium is seen on the left side. Notice the intricate pattern of rilles and ridges on the floor of Alphonsus. (*NASA*)

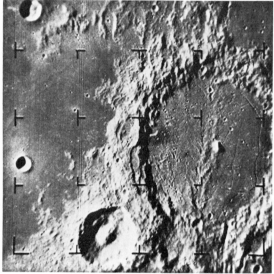

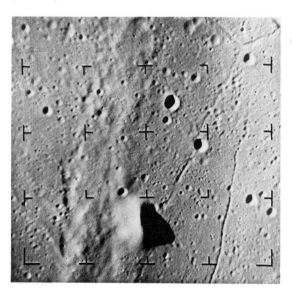

**Figure 7-19c.** *View from Ranger 9 (39 seconds before impact).* At an altitude of 58 miles, Camera A sent back this picture of the central peak in Alphonsus. The total area covered in this view is 28 miles by 26 miles. The central peak in Alphonsus rises to an elevation of 3,300 feet above the crater floor. Features only a few feet in size can be distinguished. Ranger 9 sent back only seven more pictures before impacting on the floor of Alphonsus. (*NASA*)

The obvious next step in preparing for a manned landing is to soft land an instrument package on the lunar surface. This was the purpose of the *Surveyor Program*. In order to achieve a soft landing, the spacecraft had to be equipped with a "retro rocket" to slow the vehicle's descent to the lunar surface along with "legs" and "foot pads" so that the spacecraft can remain upright after landing.

The overall design of the Surveyors is shown in Figure 7–20. Electric power was supplied by a solar panel mounted above the vehicle along with the high-gain antenna. Thermally-controlled compartments contained communications and spacecraft control devices. Using radar, the altitude and descent speed of the spacecraft were measured during the lunar landing. Information from a radar altimeter and velocity sensor controlled the retro rocket to insure a smooth landing. In addition to a television camera, the Surveyors carried a small scoop on an extendable mechanical arm for digging into the lunar soil. A small box, called an alpha-scattering device (see Figure 7–21c), was carried on later Surveyor missions. It contained a radioactive source lowered to the surface to aid in analysing the

**Figure 7–20. *Basic Features of the Surveyors.*** Descent to the lunar surface was slowed by a retro-rocket controlled by information from the "radar altimeter-velocity sensor." In addition to a television camera, Surveyor carried a scoop (the "soil mechanics surface sampler"), which dug small trenches in the lunar soil.

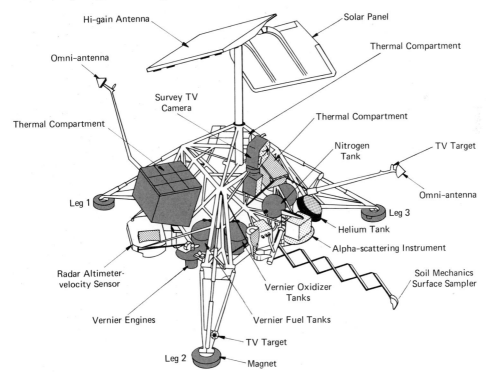

**Figure 7–21a (*left*).   *Surveyor 1.*** This full-scale model of Surveyor 1 shows how the spacecraft looked after its successful lunar landing in June of 1966. Surveyor 1 sent back a total of 11,237 pictures. (*NASA*)

**Figure 7–21b (*right*).   *Surveyor Prior to Launch.*** A scientist is shown working on a Surveyor prior to launch. A total of more than 87,000 photographs of the lunar surface were sent back to Earth during the five successful Surveyor missions. (*NASA*)

chemical composition of lunar soil. These later missions also carried a simple magnet on which metallic particles could be seen by the television camera. Various views of the Surveyors are shown in Figure 7–21 (a, b, c, d).

   The flight of Surveyor 1 was an immediate and unprecedented success. It landed in Oceanus Procellarium in June of 1966. The landing sites of the successful missions are indicated in Figure 7–22 (rockets on Surveyors 2 and 4 malfunctioned and these missions failed).

   As a result of the five successful Surveyor missions, scientists were able to make their first detailed study of the lunar soil. Prior to Surveyor 1, some scientists believed that the moon was covered with a layer of dust so thick that future astronauts would simply sink out of sight upon landing. Other scientists hypothe-

**Figure 7-21c (*left*).  *The Alpha-Scattering Instrument.*** This device was carried on Surveyors 5 and 7. It contained a radioactive source used to obtain a chemical analysis of the lunar soil. (*NASA*)

**Figure 7-21d (*right*).  *The Last Surveyor.*** This full-scale experimental model of Surveyor 7 is being examined by engineers. Unlike the previous four successful Surveyors, which landed near the lunar equator, Surveyor 7 landed on January 9, 1968 near the crater Tycho in the moon's southern hemisphere. (*NASA*)

**Figure 7-22.  *Landing Sites.*** The landing sites of the five successful Surveyors are shown against this composite photograph of the moon. The first Surveyor landed in June 1966, while the last Surveyor landed in January 1968. (*Lick Observatory photograph*)

**Figure 7-23.** *A Moon Rock.* Surveyor 1 sent back this picture of a rock on the moon. The rock is 20 inches long and is located about 15 feet from the spacecraft. (*NASA*)

**Figure 7-24.** *A Lunar Panorama from Surveyor 6.* Five pictures from Surveyor 6 were used in producing this view of the Sinus Medii region of the moon. A low ridge, typical of ridges on lunar maria, is seen near the horizon. (*NASA*)

sized that constant bombardment by the solar wind would have given "moondust" an electric charge so great that this dust would cling all over spacecrafts and astronauts like lint on socks in an electric clothes dryer. These fears were immediately dispelled.

The Surveyors showed that the moon surface resembled sand and gravel peppered with innumerable small depressions. Because the term "soil" is usually associated with growing plants here on earth, scientists refer to the granulated material on the lunar surface as the *regolith*. Nevertheless, the terms *regolith* and *soil* are often used interchangeably in referring to this unconsolidated material on the lunar surface. A typical moon rock and a panorama spanning almost 30° of the lunar surface are shown in Figures 7–23 and 7–24.

The main results from the Surveyor Program were:

1. The upper layers of the regolith are compressible but strong enough to support spacecrafts and astronauts.
2. The soil has an average density of roughly 1.5 grams per cubic centimeter consisting of particles ranging in size from 2 to 60 microns—similar to terrestrial clay.
3. Soil and rocks are uniformly medium gray.
4. Large objects, up to a meter in size, are clearly rocks while small objects, millimeters in size, are sometimes clumps or clods of loosely compressed material.
5. There are innumerable craters, many as small as a few millimeters in diameter, probably caused by bombardment of tiny meteoroids.
6. The chemical composition often resembles that of basalts found here on earth.
7. The regolith is being constantly churned up or "tilled" by meteoritic impacts.

The first four successful Surveyor landings were accomplished in maria regions near the moon's equator. The last Surveyor, launched on January 7, 1968, successfully soft-landed near the crater Tycho, a rugged region in the moon's southern hemisphere. These spacecrafts were clearly so well designed that landings did not have to be confined to the "safe," flat areas on the moon.

Shortly after the successful landing of Surveyor 1, another equally important program in lunar exploration neared completion. In order to insure the best possible selection of landing sites for manned missions to the moon, a high-quality photographic survey of the lunar surface was required. This was the purpose of the *Lunar Orbiter Missions.*

Each of the five Lunar Orbiters was a complete, self-contained photographic laboratory (see Figures 7–25 and 7–26). Unlike the Ranger and Surveyor spacecrafts, which relied on traditional television cameras for imaging, the Orbiters used photographic film. Each spacecraft carried 260 feet of black-and-white 70-millimeter film. As shown in Figure 7–27, the pictures were developed in a "processor" and then scanned by a "scanner." Data from the scanner were transmitted back to earth where the photographs were reconstructed. The scanner, involving an electronically controlled beam of light, examined each photograph in strips $\frac{1}{10}$-inch wide and $2\frac{1}{4}$ inches long. Each strip was transmitted back to earth separately where the entire photograph was reconstructed simply by placing the strips alongside each other. For this reason, all the Orbiter photographs have a characteristic striped appearance.

Lunar Orbiter 1 was successfully launched on August 10, 1966, and went into a low orbit around the moon's equator. A total of 207 photographs were taken, including mankind's first view of the earth as seen from the moon (Figure 7–29).

As Lunar Orbiters 2 and 3 followed in November of 1966 and February of 1967, respectively, an important property of the lunar surface became apparent.

**Figure 7-25.** *The Orbiter Spacecraft.* Five lunar orbiters were successfully sent to the moon. A total of 1,950 high-resolution photographs were transmitted back to Earth. These photographs covered slightly more than 99½ per cent of the entire lunar surface. (*NASA*)

**Figure 7-26.** *Basic Features of the Orbiters.* Each Orbiter was a complete photographic laboratory. Electric power was supplied by four solar panels and small rockets ("attitude-control jets") were used in aiming the camera. The developed photographic film was scanned by a "scanner" inside the spacecraft and the resulting picture was transmitted back to earth by the "directional antenna." (*NASA diagram*)

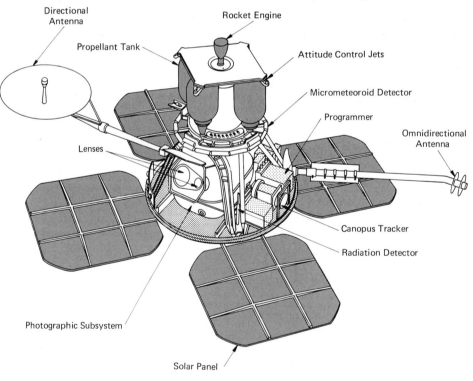

Directional Antenna

Propellant Tank

Rocket Engine

Attitude Control Jets

Micrometeoroid Detector

Programmer

Omnidirectional Antenna

Lenses

Canopus Tracker

Radiation Detector

Photographic Subsystem

Solar Panel

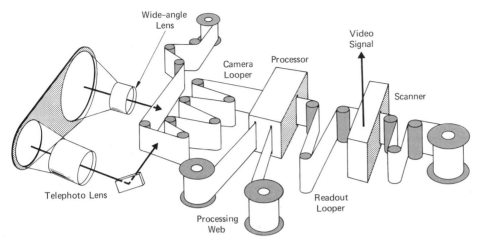

Figure 7-27. *The Photographic Subsystem.* Each Orbiter carried 260 feet of 70-millimeter photographic film, which was developed and scanned on board the spacecraft. An Orbiter could take either wide-angle or telephoto pictures. (*Adapted from NASA diagram*)

Figure 7-28 (*right*). *Orbiter's Camera.* An engineer examines the camera lenses on Orbiter 5. The wide-angle lens is behind the large aperature while the small aperature contains the telephoto lens. (*NASA*)

Figure 7-29. *Our Planet.* Lunar Orbiter 1 took this dramatic photograph of the earth when the spacecraft was on its sixteenth orbit of the moon, August 23, 1966. (*NASA*)

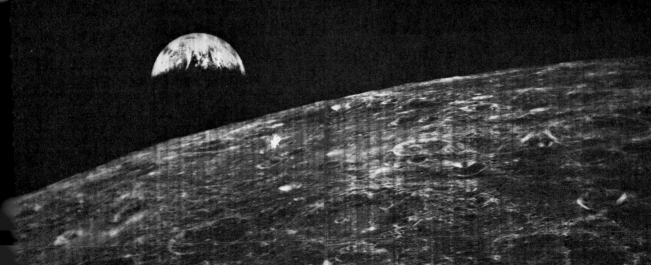

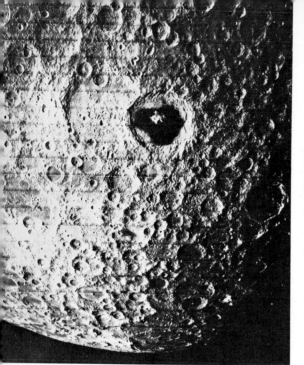
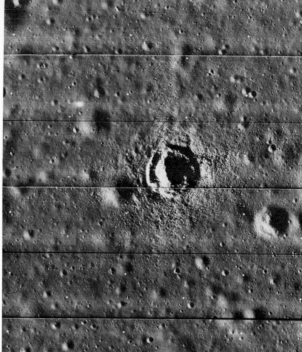

**Figure 7-30 (*left*).** *The Hidden Side of the Moon.* This photograph from Lunar Orbiter 3 shows a typical view of the moon's hidden side. Unlike the earth side, there are no large maria. This photograph was taken when the spacecraft was approximately 900 miles above the lunar equator. The prominent large crater, named Tsiolkovsky, is about 150 miles in diameter. (*NASA*)

**Figure 7-31 (*right*).** *A Young Crater.* This photograph from Lunar Orbiter 3 shows an area measuring 2,800 feet by 3,600 feet. The recently formed crater near the center of the photograph is 500 feet in diameter. A landslide is probably responsible for its double-walled appearance. Boulders a few feet in size can be seen. (*NASA*)

The earth side of the moon is dominated by large maria. With the success of the Orbiter Program, however, scientists now had the ability to see details on the hidden side of the moon. Much to their surprise, there are *no* large maria on the hidden side at all! Almost without exception, the entire hidden side is heavily cratered. A typical photograph is shown in Figure 7-30.

The first three Orbiters were so successful in photographing potential manned landing sites near the lunar equator that the remaining two spacecrafts were sent into near-polar orbit about the moon. A drawing showing the orbits and photographic coverage of the five missions is shown in Figure 7-32.

Near the end of its mission, Lunar Orbiter 4 sent back a remarkable photograph of one of the largest features on the lunar surface. Called the Oriental Basin, this feature is located on the extreme western edge of the moon as seen from earth. Consequently, this region can be seen only poorly at a highly oblique angle from earth. Figure 7-33 looks almost straight down on the Oriental Basin from an

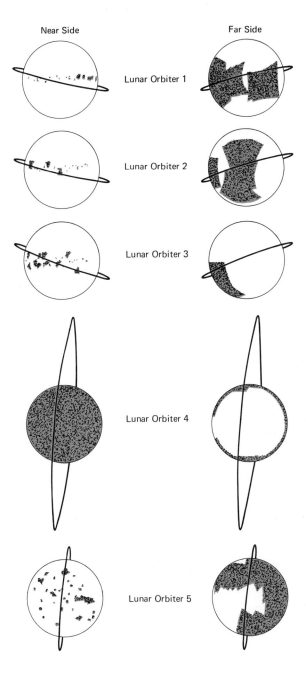

Near Side        Far Side

Lunar Orbiter 1

Lunar Orbiter 2

Lunar Orbiter 3

Lunar Orbiter 4

Lunar Orbiter 5

**Figure 7-32.** *The Five Orbiter Missions.* The first three Lunar Orbiters were in low equatorial orbits about the moon. The last two Lunar Orbiters were in near-polar orbit. More than 99½ per cent of the moon's surface was photographed. (*Adapted from NASA diagram*)

**Figure 7-33.** *Oriental Basin.* This view of Oriental Basin was taken on May 25, 1967, from Lunar Orbiter 4 at an altitude of 1690 miles. The largest ring in this "bull's eye" feature is 600 miles in diameter. (*NASA*)

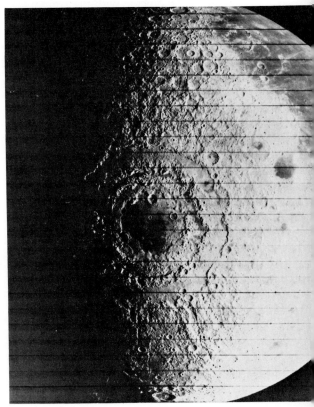

altitude of 1690 miles. The largest ring in this "bull's eye" feature is 600 miles in diameter and consists of the Cordillera Mountains. These are among the highest mountains on the moon, rising some 20,000 feet above the surrounding surface.

The final mission in this highly successful series was launched on August 1, 1967: A total of 212 photographs were received from Lunar Orbiter 5, thereby completing a photographic survey of the entire moon. Resolution varied from one mission to the next, depending on the altitude of the spacecraft. At best, objects as small as 1 meter (3 feet) in size could be seen. This is almost 1000 times better than the best earth-based observations.

One of the surprises to come from the Lunar Orbiter Program had nothing at all to do with photographs. While examining the details of the observed orbits of the spacecrafts, Dr. Paul M. Muller at the Jet Propulsion Laboratory noticed some unexpected departures from the anticipated orbits. These departures were caused by accelerations resulting from gravitational perturbations. This effect has been crudely likened to the motions of a "rollercoaster," but on a much smaller scale. Using mathematics, Drs. Muller and Sjogren at J. P. L. showed that these perturbations must be caused by unusually high levels of gravity over certain selected regions on the earth side of the moon. These regions of higher-than-average gravity are called *mass concentrations* or *mascons* for short. Surprisingly, the locations of these mascons coincide with the large maria and basins, including Imbrium, Oriental, Serenitatis, Humorum, Nectaris, and Crisium, as shown in Figure 7–34.

**Figure 7-34.** *Mascons.* The locations of major mascons are shown against this Earth-based photograph of the full moon. Notice that mascons are associated only with some maria and not with others. The largest mascon, producing the greatest gravity anomaly, is in Mare Imbrium. (*Hale Observatories*)

The nature and origin of these concentrations of matter below the lunar maria are still subjects of intense debate. Some scientists feel that mascons might be due to large nickel-iron meteorites buried below the lunar surface. Others feel that mascons were created as dense lava flowed into the basins shortly after their creation. None of the explanations are fully satisfactory.

The Ranger, Surveyor, and Orbiter missions constituted a major series of successes for the United States space program. They set the stage for one of the most dramatic technological achievements in the history of the human race: the landing of a man on the lunar surface in the afternoon of July 20, 1969.

## The Apollo Program 7.3

THE decision to transport human beings to the lunar surface was primarily political. American technology was suffering from a severe inferiority complex. After all, the Soviet Union had been first in successfully launching an earth-orbiting satellite (Sputnik 1 on October 4, 1957), first to orbit a man around the earth (Yuri Gagarin on April 12, 1961), and first to succeed in an unmanned photographic mission to the moon (Luna 3 on October 6, 1959). The possibility of seeing an American place an American flag on the lunar surface excited politicians and business men into embracing the manned space program with gusto and vigor. With perhaps the exception of the work of Dr. Jack Schmitt, a professional geologist and astronaut on the final Apollo mission, almost everything accomplished by human beings on the moon's surface could have been done years earlier at a greatly reduced cost by using robots. Nevertheless, at that point in history in the early 1960s, committing technology and industry to manned lunar exploration was something the United States simply "had to do."

In view of the value placed on human life, the steps toward safely landing men on the moon were long and tedious. The first major success was the suborbital flight of Alan B. Shepard on board Freedom 7 on May 5, 1961. Shepard simply went up and came back down in a 15-minute flight that landed him in the Atlantic Ocean. Ten months later, on February 20, 1962, John H. Glenn became the first American to orbit the earth. His three earth orbits on board Friendship 7 lasted 4 hours, 55 minutes.

With improvements in spacecraft and booster design, engineers in the mid-1960s were ready to attempt the orbiting of two-man crews. The first successful manned mission in the Gemini series, Gemini 3, occurred on March 23, 1965, with three earth orbits lasting 4 hours, 53 minutes by Virgil I. Grissom and John W. Young. Grissom had previously made the second 15-minute suborbit flight (Liberty Bell 7 on July 21, 1961) and was destined to die in a tragic spacecraft fire on January 27, 1967, which also took the lives of two other astronauts.

The first successful three-man Apollo mission to the vicinity of the moon

**Figure 7–35a.** *Freedom 7.* The first successful manned sub-orbit flight by the United States was made with this spacecraft. On May 5, 1961, Alan B. Shepard spent 15 minutes above the earth's surface in a flight from Florida to the middle of the Atlantic Ocean. (*NASA*)

**Figure 7–35b** (*lower left*). *Friendship 7.* Using another spacecraft in the Mercury series, John H. Glenn became the first American launched into Earth orbit. His three orbits lasted for 4 hours, 55 minutes. (*NASA*)

**Figure 7–35c** (*center*). *Gemini 3.* While the Mercury spacecrafts were designed for only one man, the vehicles in the Gemini series carried a two-man crew. Virgil I. Grissom and John W. Young flew the first manned Gemini mission, completing three Earth orbits in 4 hours, 53 minutes. (*NASA*)

**Figure 7–35d** (*lower right*). *Apollo 8.* In order to accomplish a manned lunar landing, a three-man crew is necessary. Frank Borman, James A. Lovell, and William A. Anders made ten orbits of the moon in Apollo 8 in December of 1968. (*NASA*)

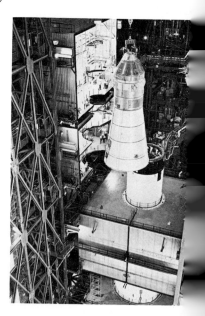

occurred in December of 1968. Astronauts Frank Borman, James A. Lovell, and
William A. Anders spent 147 hours testing the spacecraft in preparation for a future
manned landing. This flight of Apollo 8 made ten orbits of the moon and returned
excellent photographs of the lunar surface. An example is shown in Figure 7–36.
Because the film was brought back to earth for processing, the photographs were far
superior to the views transmitted back from previous unmanned missions.

Developmental work leading up to a manned lunar landing entailed six
one-man flights (the Mercury Project; from May 1961 through May 1963), ten
two-man flights (the Gemini Series; from March 1965 through November 1966),

**Figure 7-36.** *The Moon from Apollo 8.* This nearly vertical view of the moon's heavily cratered far side was taken by Apollo astronauts in lunar orbit in December 1968. The area covered in this photograph is approximately 20 miles on a side. (*NASA*)

**Figure 7-37.** *The Apollo Landing Sites.* The landing sites of the six successful manned lunar missions are shown on this composite photograph of the moon. Each site was carefully chosen for geological interest and safety of the astronauts. (*Lick Observatory photograph*)

**Table 7-2.** *Six Lunar Landings*

| Mission Name and Launch Date | Astronauts | Landing Site | Some Geological Features of Landing Site |
|---|---|---|---|
| APOLLO 11<br>July 16, 1969 | Neil A. Armstrong<br>Michael Collins<br>Edwin E. Aldrin, Jr. | Mare Tranquilitatis | Old Mare |
| APOLLO 12<br>November 14, 1969 | Charles Conrad, Jr.<br>Richard F. Gordon, Jr.<br>Alan L. Bean | Oceanus Procellarum | Younger Mare |
| APOLLO 14<br>January 31, 1971 | Alan B. Shepard, Jr.<br>Stuart Roosa<br>Edgar D. Mitchell | Fra Mauro | Hilly Upland |
| APOLLO 15<br>July 26, 1971 | David R. Scott<br>Alfred M. Worden<br>James B. Irwin | Hadley–Apennine | Mountain Front, Rille, Mare |
| APOLLO 16<br>April 16, 1972 | John W. Young<br>Thomas K. Mattingly<br>Charles M. Duke, Jr. | Descartes | Highland Hills, and Plains |
| APOLLO 17<br>December 7, 1972 | Eugene A. Cernan<br>Ronald E. Evans<br>Harrison H. Schmitt | Littro–Taurus | Highland Massifs and Dark Mantle |

and four three-man flights (Apollo 7, 8, 9, and 10; from October 1968 through May 1969). A three-man crew was ultimately necessary so that one astronaut could stay behind in lunar orbit in the "Command and Service Module" while the other two astronauts descended in the "Lunar Module." Following completion of their activities on the moon, these two astronauts would then return to lunar orbit, "dock" with the Command Module and then transfer themselves, their equipment, and moon rocks to the Command Module. The Lunar Module would then be "jettisoned" for the trip back to earth.

Few people will ever forget that historic afternoon on July 20, 1969, when word was received from Neil A. Armstrong that "the Eagle has landed." Six and a half hours later, Armstrong emerged from the Apollo 11 Lunar Module and stepped out on the southwest section of Mare Tranquilitatis. For more than two hours, he and Edwin E. Aldrin, Jr., collected rocks, drilled into the regolith, and set up scientific apparatus. After returning to lunar orbit, they joined Michael Collins in the Command and Service Module for the 60-hour trip back to earth. They brought back 22 kilograms (49 pounds) of moon rocks in tightly sealed containers.

A total of six successful manned lunar landings were accomplished between July 1969 and December 1972. Each of the landing sites (see Figure 7–37) was carefully chosen on the basis of geological interest and the safety of the astronauts. The first two missions were to maria. The Apollo 11 site on Mare Tranquilitatis was believed to be covered with a thick (3 to 6 meters) regolith, while the Apollo 12 target on Oceanus Procellarum was on a ray from the crater Copernicus. The hilly terrain of the Apollo 14 site presented a navigational challenge at the moment of touchdown (the earlier Apollo 13 mission was aborted because of a malfunction onboard the spacecraft), and the remaining three missions were in or near the mountainous lunar "highlands." Some historical data on these six successful lunar missions are included in Table 7–2.

In the first three missions (Apollo 11, 12, 14), astronauts explored the lunar surface entirely on foot. On the remaining three missions (Apollo 15, 16, 17), a "lunar roving vehicle" or "Rover" was taken along (see Figure 7–38). This device

**Figure 7-38.** *The "Lunar Roving Vehicle."* The last three lunar landings (Apollo 15, 16, 17) included a "Rover." This car-like device permitted the astronauts to cover a large area surrounding the landing site. An excellent view of the regolith is seen in the foreground of this photograph. (*NASA*)

permitted the astronauts to cover a much larger area in the vicinity of the landing sites and collect a wider range of lunar samples. Improved mobility is reflected in Figure 7–39, a bar graph that shows a comparison of lunar activities on the six successful missions.

Apollo 11 landed in a relatively "safe" region near the edge of Mare Tranquilitatis. Similar areas of this mare had been investigated by Ranger 8 and Surveyor 5. The landing site was level and covered with numerous small, rimless craters and depressions. In addition to collecting rocks, the astronauts set up a variety of scientific instruments, called ALSEP or "Apollo lunar surface experiments package." These instruments included a seismograph for detecting moonquakes (Figure

**Figure 7–39.** *Comparison of Lunar Surface Activities.* This bar diagram compares the six lunar landings in terms of equipment delivered, time spent and distance covered during surface activities, and weight of lunar samples returned. (*Adapted from NASA diagram*)

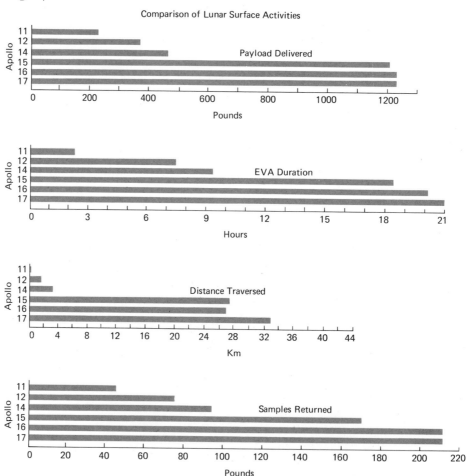

7–40a) and a sheet of aluminum foil (Figure 7–40b) that was exposed to the solar wind and later brought back to earth for analysis.

Apollo 12 came down almost exactly on target in Oceanus Procellarum, a short distance from Surveyor 3, which had landed two and a half years earlier. Indeed, the astronauts visited Surveyor 3 from which they retrieved some parts for examination on earth to learn about short-term exposure to the lunar environment. In addition to collecting moon rocks (see Figure 7–41), the astronauts set up numerous experiments and apparatus. These instruments were similar to those carried on Apollo 11 but also included a magnetometer to measure the moon's magnetic field and a solar wind spectrometer to measure the composition of particles from the sun.

The Apollo 13 mission, launched on April 11, 1970, was aborted because of

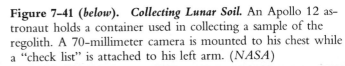

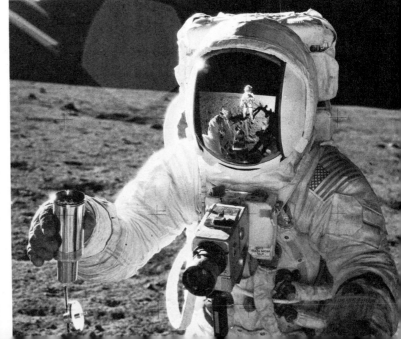

**Figure 7–40a (left).** **Seismic Apparatus.** Astronaut Aldrin is shown standing near seismic apparatus at "Tranquility Base," with the Apollo 11 Lunar Module in the background. This apparatus detected moonquakes and transmitted lunar seismic data back to earth. (*NASA*)

**Figure 7–40b (below left).** **Solar Wind Experiment.** During their stay on the lunar surface, the Apollo 11 Astronauts exposed this sheet of aluminum foil to the solar wind. Prior to leaving "Tranquility Base," they rolled up the foil and brought it back to earth for chemical analysis. (*NASA*)

**Figure 7–41 (below).** **Collecting Lunar Soil.** An Apollo 12 astronaut holds a container used in collecting a sample of the regolith. A 70-millimeter camera is mounted to his chest while a "check list" is attached to his left arm. (*NASA*)

**Figure 7–42a** (*above*). *A Field of Boulders.* This is one of several photographs taken by Apollo 14 astronauts just northwest of their landing site. (*NASA*)

**Figure 7–42b** (*below*). *A Field of Boulders.* This is one of several photographs taken by Apollo 14 astronauts just northwest of their landing site. (*NASA*)

**Figure 7–42c** (*lower left*). *A Field of Boulders.* This is one of several photographs taken by Apollo 14 astronauts just northwest of their landing site. (*NASA*)

**Figure 7–42d** (*center*). *An Interesting Boulder.* This unusual specimen was found in the boulder field just northwest of the Apollo 14 landing site. Notice the difference in color and structure between the upper and lower parts of the rock. (*NASA*)

**Figure 7–42e** (*right*). *Close-up of a Large Boulder.* Many of the boulders near the Apollo 14 site were probably ejected from a few meters below the lunar surface during the formation of craters. (*NASA*)

severe technical problems in the Service and Command Module. Fortunately, all three astronauts were returned safely and Apollo 14 was launched nine months later.

Apollo 14 landed in a hilly region near Cone Crater in the so-called Fra Mauro Formation. This region of the moon consists of rolling hills covered with boulders as large as 3 to 5 meters on a side. Figure 7–42 (a through h) shows numerous views of this boulder-strewn region of the lunar surface. It is reasonable to assume that most of these boulders were dug up and tossed out of Cone Crater during its formation. The term *ejecta* is used to denote material excavated or "ejected" in this fashion. In examining this ejecta, the astronauts therefore had access to material usually buried well below the regolith. Among the instruments and apparatus carried by the Apollo 14 astronauts was a "thumper," a portable device that strikes the lunar surface with a predetermined force. This device produced near-surface seismic waves detected by a series of geophones, or seismic detectors, buried in the regolith by the astronauts. This experiment revealed that the regolith near Fra Mauro was about 9 meters thick, and below it is a layer of rubbly debris 20 to 70 meters thick.

**Figure 7–42f** (*left*).   *Close-up of a Large Boulder.* This large boulder is located in a field of boulders just northwest of the Apollo 14 landing site. (*NASA*)

**Figure 7–42g** (*center*).   *Astronaut Standing near a Boulder.* Astronaut Shepard stands beside a boulder on the lunar surface. Notice the lunar dust clinging to the right boot and leg of his spacesuit. (*NASA*)

**Figure 7–42h** (*right*).   *Several Large Boulders.* A geological hammer and sample bag are shown on a lunar boulder to give some indication of its size. (*NASA*)

**Figure 7-43.** *The Apollo 15 Landing Site.* Apollo 15 landed at the base of the Apennine Mountains near Hadley Rille. The landing site is indicated by the arrow on this Earth-based photograph. Mare Serenitatis is seen on the right while the three craters in the upper left are Aristillus, Archimedes, and Autolycus. (*NASA*)

Apollo 15 was the first mission to include a lunar "Rover." With this battery-operated vehicle, the astronauts were able to cover a total of 28 kilometers (17 miles) on the lunar surface and explore both the Apennine Front (that is, the base of the Apennine Mountains) and the Hadley Rille. An excellent earth-based photograph of this region is shown in Figure 7-43. Among the most important discoveries from Apollo 15 was the recognition of an apparent layering or *stratification* in the exposed mountains or *massif*. Close examination of Mount Hadley (Figure 7-44a) reveals this lineation, which may have been caused by successive lava flows. The Apollo 15 ALSEP (Figure 7-44c and d) consisted largely of the most fruitful devices carried by the Apollo 11, 12, and 14 astronauts.

Apollo 16 landed about 60 kilometers north of the 50-kilometer-wide crater Descartes on April 20, 1972. This was the only Apollo site well within the lunar highlands. Using the Rover, the astronauts covered 20 kilometers (12 miles) and spent a total of more than 20 hours exploring the lunar surface. A total of 95 kilograms (209 pounds) of rocks and soil were collected. Some samples were as large as footballs. Among the surprises on the Apollo 16 mission was the absence of volcanic rock that geologists had expected from examination of preflight photographs.

Some of the apparatus in the Apollo 16 ALSEP is shown in Figure 7-45 (c and d). The heat-flow experiment, designed to measure the amount of heat coming from below the lunar surface, detected a higher-than-expected temperature inside the

**Figure 7–44a (*above*).** *Mount Hadley.* This mountain, northeast of the Apollo 15 landing site, rises about $3\frac{1}{2}$ kilometers (2 miles) above the surrounding plains. (*NASA*)

**Figure 7–44b (*top right*).** *Hadley Rille.* The Hadley Delta forms the background of this scenic view, looking almost due south. The Rover is parked near Hadley Rille seen in the right-center of the photograph. (*NASA*)

**Figure 7–44c (*bottom right*).** *Experiments at the Apollo 15 Site.* An astronaut is seen drilling into the regolith. The solar wind composition experiment is in the foreground. (*NASA*)

**Figure 7–44d (*below*).** *ALSEP at the Apollo 15 Site.* The Apollo Lunar Surface Experiments Package (ALSEP) is shown deployed near the Apollo 15 landing site. (*NASA*)

**Figure 7–45a (*left*). *A Boulder at the Apollo 16 Site.*** The surface of a large boulder is being examined by an astronaut not far from the Apollo 16 landing site in the lunar highlands. (*NASA*)

**Figure 7–45b (*below left*). *"Plum Crater."*** Many craters visited by Apollo astronauts received unofficial names. This crater is about 40 meters (130 feet) in diameter, seen with the Apollo 16 Rover in the background. (*NASA*)

**Figure 7–45c (*above right*). *"Stone Mountain" and ALSEP.*** The heat flow experiment is seen in the foreground. "Stone Mountain" lies behind the Apollo 16 Rover in this view. (*NASA*)

**Figure 7–45d (*right*). *Apollo 16 ALSEP.*** Some of the ALSEP equipment is shown here. The passive seismic experiment is in the foreground, the central station is in the center background with the radioisotope thermoelectric generator to the upper left. (*NASA*)

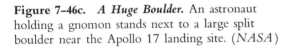

**Figure 7–46a** (*above*). *Near the Apollo 17 Landing Site.* The desolate lunar landscape is seen in this view of astronaut Schmitt standing near the Rover. (*NASA*)

**Figure 7–46b** (*above right*). *"Orange Soil" and Gnomon.* The tripod-like object, called a gnomon, is used to determine sun angle, scale, and lunar color. The surrounding regolith has an orange color. (*NASA*)

**Figure 7–46c.** *A Huge Boulder.* An astronaut holding a gnomon stands next to a large split boulder near the Apollo 17 landing site. (*NASA*)

**Figure 7–46d.** *A Fractured Boulder.* This boulder was probably ejected during the formation of a crater and fractured upon impact with the lunar surface. (*NASA*)

moon. Magnetic and seismic experiments indicate that the moon must be "cold," that is, without a molten core. Thus, the heat-flow experiment is in contradiction with magnetic and seismic experiments. Perhaps the moon's interior is partially molten, or perhaps the excess heat comes from long-lived radioactive isotopes just below the lunar surface. Electric power for these experiments was produced by a "radioisotope thermoelectric generator" channeled through the "central station;" both are shown in Figure 7–45d.

The final manned lunar landing, Apollo 17, occurred in December of 1972. The astronauts landed in a dark-floored valley between two massive hills in the Taurus Mountains, 35 kilometers (21 miles) south-southwest of the crater Littrow. This region constitutes the highlands bordering Mare Serenitatis. Rugged mountains, called *massifs,* rise 2 to 3 kilometers (almost 10,000 feet) above the dark valley floor. Four views are shown in Figure 7–46 (a, b, c, d). Considerable excitement was generated by the discovery of "orange soil" (see Figure 7–46b), which seemed to suggest recent volcanic activity on the moon. This would mean that the lunar interior is hot and molten, like the earth. Chemical analysis of returned samples, however, proved that this unusual color is due to a high concentration of titanium oxide ($TiO_2$) rather than sulphides, arsenides, and iron-bearing rocks associated with volcanic vents on earth.

Almost 800 pounds of moon rocks and soil were brought back to earth during the six Apollo missions. In addition, two remotely controlled, unmanned Soviet spacecrafts (Luna 16 and Luna 20) brought back samples from Mare Fecunditatis in 1970 and 1971. All of these samples have been intensely studied by teams of geologists around the world. Their analysis, along with the enormous amount of data from the six ALSEP stations, has brought mankind much nearer to understanding our satellite.

# 7.4 *Lunar Geology*

THE 800 pounds of lunar rock brought back to earth by the Apollo astronauts have opened up one of the most exciting fields of study in the science of geology. Quite literally, these lunar samples have permitted geologists to examine specimens from another body in the solar system. It is almost as if an entirely new planet were suddenly available for detailed scrutiny.

Prior to the Apollo Program, there were three competing theories as to the origin of the moon. One theory holds that the moon "split off" or *fissioned* from the earth. A second idea is that the moon, initially orbiting the sun on its own, was "captured" by the earth. And finally, the third and generally accepted view, is that the moon "accreted" or grew from the gradual gravitational accumulation of tiny grains of dust and rock that surrounded the earth billions of years ago. Although analysis of moon rocks has not been able to decide once and for all which of these ideas is correct, some important clues have come to light.

From studying meteorites, it was realized that the relative abundances of various isotopes of oxygen (that is, $^{16}O$, $^{17}O$, $^{18}O$) are very sensitive indicators of where the meteorites came from. Meteorites originating in different parts of the solar system have very different proportions of $^{16}O$, $^{17}O$, and $^{18}O$. From analysing the moon rocks, it was found that the relative proportions of these oxygen isotopes are exactly the same as the relative abundances found in earth rocks. It therefore logically follows that the moon was formed at roughly the same distance from the sun as the earth. The moon could not have come from some very distant part of the solar system such as Mercury or from among the Jovian planets.

Another important discovery to come from chemical analysis of moon rocks is that they are systematically depleted in volatile elements. A volatile element is one which has a low *condensation temperature,* in other words, volatile elements condense from a gaseous state at comparatively low temperatures. In lunar rocks, there is an unusually low abundance of elements with low condensation temperatures (for example, bismuth, thallium, cadmium, and so forth) compared to earth rocks. This effect is exhibited in Figure 7–47. Notice that moon rocks contain about twice as

**Figure 7–47.** *Chemical Analysis of Basalts.* The ratio of the abundances of elements in lunar and terrestrial basalts are plotted against their condensation temperatures. Notice that the moon rocks are exceptionally underabundant in the more volatile elements. (*From "The Moon" by John A. Wood, Copyright © 1975 by Scientific American, Inc. All rights reserved*)

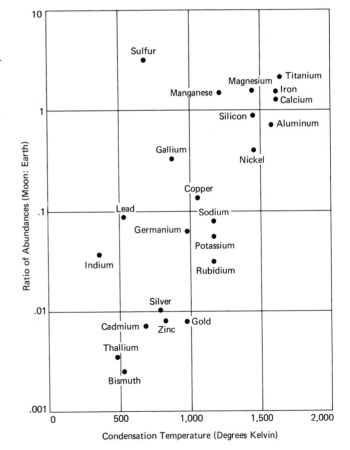

much titanium (which has a high condensation temperature) as earth rocks. In contrast, a moon rock typically contains less than 1 per cent of the amount of bismuth, thallium, and cadmium found in earth rocks.

It is generally believed that the planets formed from the accretion of tiny particles that condensed out of the *primordial solar nebula*. During the formation of the sun, the solar system was presumably filled with large quantities of gas and dust. As this cloud or primordial nebula about the sun cooled, elements with high condensation temperatures would have precipitated out first. Only later, at much lower temperatures, would the more volatile elements have solidified. This process of *fractionization* suggests that the first objects to form in the solar system would be exceptionally "rich" in elements with high condensation temperatures but "poor" in the more volatile elements.

This difference in the abundances of volatile elements between lunar and terrestrial samples is generally interpreted as a severe blow to those who believe that the moon fissioned from the earth. On the other hand, those scientists who prefer the intact-capture theory of the moon by the earth are faced with severe problems in dynamics. Unless some highly unusual process occurred to slow down the moon in its orbit about the sun, it seems quite impossible to understand how the earth could have captured the moon.

While it is extremely difficult to gravitationally capture one very massive object, it is much easier to capture numerous smaller particles and rocks. The early solar system must have been filled with smaller objects orbiting the sun as evidenced by the numerous craters seen on the lunar surface (and also on Mercury and Mars, as we shall learn in later chapters). Presumably the earth could have captured large quantities of these smaller rocks that condensed out of the solar nebula during the very earliest history of the solar system. The gas in the primordial solar nebula, while insufficient to slow down a huge object like an entire moon, would have been adequate to decelerate much smaller particles. In addition, collisions between rocks

**Figure 7–48.  *A Typical Moon Rock.*** At first glance, a moon rock looks like any ordinary greyish rock found on Earth. Close examination of samples from the lunar surface reveal that they are frequently covered with tiny glass craters. (*NASA*)

approaching the earth from different directions would also have substantially decreased the speeds of these rocks. In the final stages of the moon's creation, the mutual gravitational attraction between all these small rocks would cause them to accrete into a larger object. Calculations reveal that the entire moon could have accreted from a disk of particles orbiting the earth in roughly a thousand years.

From examination of the data supplied by the moon rocks and the Apollo experiments, it seems as though the moon went through six major stages in its evolution. The first stage, obviously, was the initial formation of the moon, as described above. It is reasonable to suppose that this accretion process was very violent. The high-speed impacts of numerous rocks striking the moon during its creation would have dissipated large quantities of heat and energy. This energy would have been sufficient to liquify all the accreting matter. Immediately after its creation, therefore, the moon must have been molten. The second major stage in the evolution of the moon must have been a cooling of its outer layers or *crust*.

In order to understand these later stages in the moon's evolution, it is necessary to investigate the properties of moon rocks. At first glance, a typical moon rock (see Figure 7–48) picked up from the lunar surface looks like any ordinary greyish rock found on Earth. Closer examination reveals that the moon rock appears to "sparkle" in the sunlight. This is due to the fact that the "upper" surface of the rock (that is, the part of the rock that was exposed to outer space) is often covered with tiny glass craters and splatterings of glass. A microscope is necessary to see these tiny *zap craters* clearly. Two examples are shown in Figure 7–49 (a and b). Even today,

**Figure 7–49a (left).** *A Zap Crater (side view).* This photograph taken with an electron microscope reveals a tiny crater on the surface of a moon rock. (*NASA*)

**Figure 7–49b (right).** *A Zap Crater (face on view).* Zap craters are formed by the high-speed impact of tiny meteoroids striking the exposed surface of a moon rock. (*NASA*)

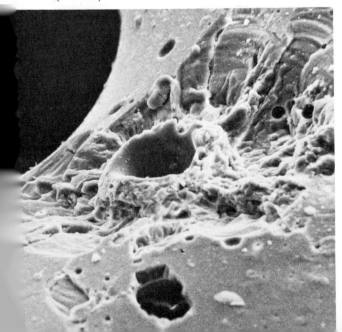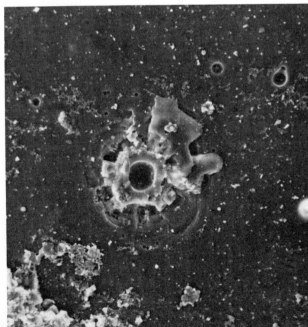

**Figure 7-50 (left).** *A Vesicular Basalt.* This volcanic rock was brought back to Earth by Apollo 15. It is sample number 15016 (the 16th rock in the Apollo 15 collection). Its vesicular appearance suggests that the rock bubbled and frothed, releasing gases as it cooled. This sample weighs 924 grams (about 2 pounds) and measures about 13 centimeters (5 inches) from end to end. It is a typical mare basalt. (*NASA*)

**Figure 7-51 (right).** *A Vuggy Basalt.* This volcanic rock was round by Apollo 15 astronauts near the rim of Hadley Rille. It is sample number 15555 (the 555th rock in the Apollo 15 collection). His huge sample weighs 9614 grams (roughly 20 pounds) and measures about 25 centimeters (10 inches) from end to end. (*NASA*)

**Figure 7-52.** *Limestone Breccia.* A breccia is a rock that consists of smaller jagged fragments that have been cemented together. This terrestrial rock is a limestone breccia. (*Ward's Natural Science Establishment, Inc.*)

long after the formation of all the major craters, the moon is constantly being bombarded by tiny meteoroids. When a micrometeorite strikes the surface of a moon rock, the energy of impact causes a small portion of the rock to melt. Upon resolidification, the liquified rock has turned to glass, often preserving the imprint or crater formed by the micrometeorite.

Since there is no water or air on the moon, it is totally inconceivable that the Apollo astronauts would bring back any sedimentary rocks. Sedimentary rocks, which are so prevalent here on Earth, are totally absent on the moon. In addition, no lunar samples are representative of the primordial composition of the solar system. In view of the violent birth of the moon, all of the lunar rocks were formed after the molten moon cooled about four billion years ago. Of course, by Earth standards, moon rocks are indeed very old. For example, from measuring the isotopic abundances of rubidium and strontium in rock number 15555 (that is, the 555th rock brought back by Apollo 15, shown in Figure 7–51), its age was discovered to be about $3\frac{1}{2}$ billion years. While this is very old, it is certainly not representative of the primordial matter created during the earliest stages of the solar system. All of the moon rocks had their "radioactive clocks" reset when the moon was molten or even later during stages of volcanic activity.

When Apollo 11 returned to Earth, it was found that all the lunar specimens could be arbitrarily classified according to appearance into one of four basic categories. First of all, there are fine-crystalline vesicular basalts ("vesicular" means that the rock is covered with numerous, small spherical openings). A good example of this "Type A" rock is shown in Figure 7–50. The second type of rock, "Type B," is medium-crystalline vuggy basalts ("vuggy" means that the rock has a few irregularly shaped openings), an example of which is seen in Figure 7–51. The third type of specimen, "Type C," is *breccias*. A breccia consists of smaller jaggered rock fragments that have been cemented together. Breccias are found here on Earth (a terrestrial limestone breccia is shown in Figure 7–52). But often wind and water erosion on Earth smooth and round the rock fragments before they have a chance to be cemented together. The resulting rock is then called a *conglomerate*. Of course, no similar erosion processes occur on the moon and therefore no lunar rocks are conglomerates. All cemented lunar rock fragments result in breccias. Just as conglomerates are common here on Earth, breccias are common on the moon. A typical lunar breccia is shown in Figure 7–53. It is believed that most lunar breccias are formed by compression during meteorite impacts on the lunar surface.

**Figure 7-53.** *A Breccia from the Moon.* Many of the rocks brought back by the Apollo astronauts are breccias. This sample (rock number 15450) was collected near the Apollo 15 landing site and measures roughly 25 centimeters (10 inches) in length. (*NASA*)

**Figure 7-54.** *Lunar Soil.* The lunar soil consists of small rock fragments and often contains numerous tiny glass beads. This close-up of the lunar soil was taken by the Apollo 11 astronauts. (*National Space Science Data Center*)

The final type of specimen, "Type D," in this crude system of classification is loose "soil." This includes small particles and rock fragments which make up the regolith (see Figure 7-54). Often these samples contain glass beads. Presumably these tiny glass spheres are formed by the same processes which produce zap craters. During meteorite impact, droplets of molten rock solidify into tiny glass balls.

A more detailed examination and analysis of lunar specimens have led to the recognition of three important types of lunar rock, each of which provided important clues to the evolution of the moon. These three types are: mare basalt, KREEP norite, and anorthositic rock. As their name suggests, more basalts are volcanic rock found on the dark lunar mare. The non-mare lunar "highlands" consist predominantly of the remaining two types of igneous rock. KREEP norite was named for its unusually high content of potassium (K), rare earth elements (REE) and phosphorous (P). Anorthositic rock is by far the most abundant type of rock on the moon. Anorthositic rock is characterized by a great abundance of plagioclase feldspar ($CaAl_2Si_2O_8$) with smaller amounts of other minerals such as pyroxenes and olivine.

Each of these three important types of lunar rock was formed at a different stage in the moon's history. Both mare basalt and KREEP norite have a much lower

**Figure 7-55.** *Anorthosite.* Anorthositic rock is very abundant on the moon. This type of rock has a high melting temperature and therefore was one of the first substances to solidify as the primordial moon cooled. This particular sample (15415) was brought back by the Apollo 15 astronauts and weighs 269.4 grams (0.6 pound). (*National Space Science Data Center*)

**Figure 7-56.** *KREEP Norite.* This specimen brought back by the Apollo 15 astronauts is unusually rich in potassium (K), rare earth elements (REE), and phosphorus (P). The ionized atoms of these elements are large and consequently they were most easily "sweated out" of the more ancient lunar rocks during an early epoch of volcanic activity. At the end of this epoch, the newer rocks that solidified retained a high abundance of KREEP elements. (*National Space Science Data Center*)

melting temperature than anorthositic rocks. Consequently, anorthositic rocks would have formed first as the moon cooled. As mentioned earlier, the second major stage in the evolution of the moon was the cooling of its crust. As the hellish, white-hot ocean of molten rock cooled, minerals rich in heavy elements (such as iron and magnesium) sank, thereby forcing plagioclase feldspar upward. This process of fractionization of the cooling melt produced a crust of anorthositic rock covering the entire moon to a depth of 50 to 100 kilometers. This crust separation occurred very early in the moon's history and therefore anorthositic rocks are among the most ancient found on the moon. The so-called "Genesis Rock" (rock number 15415 found by the Apollo 15 astronauts near the Apennine Front) is a good example of an anorthosite; these typically have ages of about four billion years.

The third stage in the moon's history must have involved an early epoch of vulcanism. It was at this time that KREEP norite was formed. The primary characteristic of KREEP norite is the unusually high abundance of elements such as potassium, barium, uranium, thorium, and phosphorous. All of these elements have very large ions, which are not easily incorporated into the crystalline structure of the minerals in anorthositic rock. Consequently, upon a remelting of anorthositic rock during a period of vulcanism, the first portions of the rocks to liquify would be exceptionally overabundant in these large-ion elements. These elements would therefore have been preferentially "sweated out" of the parent rocks. This accounts for their high abundance in norites that were formed during this third stage.

**Figure 7-57.** *Mare Basalt.* One of the final stages in the series of events that shaped our moon involved extensive lava flows. The maria were formed as lava from the moon's interior flooded huge basins that had been excavated by meteoroids. An example of a mare basalt is shown here. This large sample (15058) was brought back by Apollo 15 astronauts and weighs 2,672.5 grams (5.9 pounds). (*National Space Science Data Center*)

The fourth stage in the evolution of the moon saw numerous impacts of large meteoroids and planetesimals. It was during this barrage that many of the familiar craters were formed. The heavily cratered lunar highlands and the abundance of breccias throughout these highlands bear witness to this period of intense bombardment.

Following the abatement of this bombardment, great quantities of lava surged up from the lunar interior to flood the huge basins excavated by the impacts of the largest planetesimals. During this fifth epoch in the moon's history, the large maria were formed. Over a period of almost a billion years, successive lava flows produced vast flat darkish plains composed of mare basalt.

The sixth and final stage in the evolution of the moon is characterized by a gradual decrease in activity down to the present stage of relative quiescence. The youngest craters (for example, Copernicus and Tycho) were formed during this time. Extensive systems of rays extending from these craters stand in mute testimony to their comparatively recent creation. Actually, rays are not a thin covering of light-colored material, as was originally believed prior to the space program. Instead, rays consist of numerous small craters formed as ejecta from the parent crater fell back onto the lunar surface.

While analysis of lunar samples has resulted in an overall picture of the history of our moon, data from the numerous experiments set up on the lunar surface have provided geologists with an understanding of the moon's interior. Since the magnetometers in the ALSEPs failed to detect any appreciable lunar magnetic field, it is clear that the moon does not have a molten nickel–iron core like the earth. Nevertheless, while the moon was molten, rocks with high specific gravities would have sunk toward the moon's center leaving lower density material near the surface. On the basis of differences in composition and specific gravity, one can speak of a *core, mantle,* and *crust* in discussing the structure of the moon. As shown in Figure 7-58, the core is roughly 1000 kilometers in diameter while the crust is only

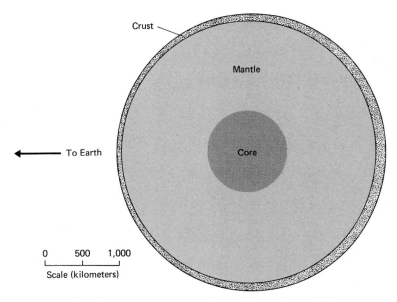

**Figure 7-58.** *The Lunar Core, Mantle, and Crust.* From differences in the composition and specific gravity of rocks, it is possible to distinguish between a core, mantle, and crust of the moon. The core is about 1000 kilometers (600 miles) in diameter while the crust is only 50 to 100 kilometers (30 to 60 miles) thick. (*From "The Moon" by John A. Wood, Copyright © 1975 by Scientific American, Inc. All rights reserved*)

50 to 100 kilometers thick. The crust is probably thicker on the far side of the moon than on the earth side. After all, during the fifth stage in the moon's history, molten lava from the mantle was able to well up and fill the basins excavated on the moon's earth side. Such extensive lava flows were not possible through the thicker crust on the moon's far side, which is almost entirely covered with craters. In addition, the core is probably off-centered slightly toward the earth resulting in the moon's synchronous rotation rate.

From the behavior of seismic waves detected by the ALSEP's seismographs, it is possible to differentiate between a *lithosphere* and *asthenosphere* on the moon. The rigid lithosphere which transmits both S-waves and P-waves is about 1000 kilometers thick. The plastic asthenosphere, in which S-waves are severely dampened, occupies the inner regions of the moon, as shown in Figure 7-59. In spite of the existence of a lithosphere and asthenosphere, there is no tectonic activity on the moon. Plate tectonics here on earth is responsible for mountain-building and earthquakes. All surface features on the moon were formed long ago as the molten moon cooled or as it was bombarded by planetesimals. Moonquakes, which involve far less energy than earthquakes (typically 1 to 3 on the Richter scale), occur preferentially when the moon is at apogee (that is, when the moon is nearest to the

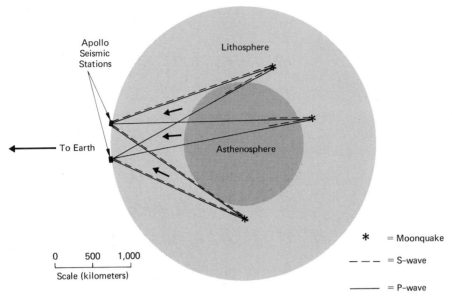

Apollo
Seismic
Stations

Lithosphere

To Earth

Asthenosphere

0    500    1,000
Scale (kilometers)

✳  = Moonquake

— — —  = S-wave

———  = P-wave

**Figure 7-59.  *The Lithosphere and Asthenosphere.*** From data concerning moon-quakes, it is possible to distinguish between a lithosphere and asthenosphere of the moon. Both S- and P-waves easily travel though the rigid lithosphere, which is about 1000 kilometers (600 miles) thick. Evidence for a plastic asthenosphere comes from the fact that S-waves are attenuated by a nonrigid medium. (*From "The Moon" by John A. Wood, Copyright © 1975 by Scientific American, Inc. All rights reserved*)

earth). Thus, while earthquakes are caused by the slippage of tectonic plates, moonquakes are often caused by the tidal forces of the earth on the moon which are largest at lunar apogee.

During the brief period of manned lunar exploration, during the 78 hours when men walked on the lunar surface, samples and data were collected that have proved invaluable in understanding the origin, nature, and evolution of our nearest neighbor in space. The picture is vastly different from earth. Our earth is a very active planet, one on which geological processes are occurring on a vast scale as huge tectonic plates roam across its surface. The moon, in contrast, was geologically active only in the distant past. Once the scene of violent processes over its entire surface, our moon is now desolate, barren, and totally lifeless.

1. How big is the moon?
2. How far away is the moon?
3. What are maria?
4. How many maria are on the moon?
5. Roughly how many craters can be seen through earth-based telescopes?
6. What is the size of the smallest lunar features which can be resolved through the finest earth-based telescopes?
7. Besides maria and craters, what other kinds of features are seen on the lunar surface?
8. Compare the average densities, surface gravities, and escape velocities of the earth and moon.
9. What is meant by synchronous rotation of the moon?
10. What is meant by libration?
11. Briefly describe the Ranger Project. How many successful missions were accomplished?
12. Briefly describe the Surveyor Program. How many successful missions were accomplished?
13. Name three discoveries to come from the Surveyor Program.
14. Briefly describe the Lunar Orbiter Program. How many of the five missions were successful?
15. Compare and contrast the appearances of the near and far sides of the moon.
16. How much of the moon's surface was photographed by the Lunar Orbiters?
17. What is the Oriental Basin?
18. How were mascons discovered?
19. Who was the first Soviet cosmonaut to orbit the earth? Who was the first American astronaut to orbit the earth? When did these historic flights occur?
20. What was the Gemini Program?
21. How many men have walked on the surface of the moon?
22. Briefly describe the successful Apollo lunar landings.
23. What is ALSEP?
24. What is the Lunar Rover and how was it used by the astronauts? Which missions carried Rovers?
25. What is meant by ejecta?
26. What, approximately, is the total weight of moon rocks and soil brought back by the astronauts?
27. Briefly describe and contrast the six Apollo landing sites. What unusual geological features did each one offer?
28. List some of the equipment contained in the ALSEPs.
29. List and briefly discuss the three major competing theories of the moon's creation.
30. What can be concluded from the relative abundances of oxygen isotopes in moon rocks?

31. Compare and contrast the abundances of volatile elements in moon rocks. What does this tell us?
32. List the six major stages in the evolution of the moon.
33. What kinds of rocks are found on the moon?
34. What is a breccia?
35. What is a zap crater?
36. At what stage during the moon's evolution was anorthositic rock formed? Why?
37. What is KREEP norite and when during the evolution of the moon was it formed?
38. What was going on in the moon's evolution when mare basalt was formed?
39. Compare and contrast the core, mantle, and crust of the earth and the moon.
40. Compare and contrast the lithosphere and asthenosphere of the earth and the moon.
41. Discuss some of the differences between moonquakes and earthquakes.

# Journeying to Mars 8

## *Facts and Myths of Mars* 8.1

FOR thousands of years people have looked to the skies for clues to their origin and destiny. From the Pyramids to Palomar, many recurring themes are found in this preoccupation with the heavens. Especially during the last few centuries, these themes have evolved into specific questions. Is life a common phenomenon in the universe? Are we alone and unique or are there other inhabited planets scattered through space? What is the probability that we shall ever communicate with intelligent extraterrestrial creatures?

Questions of this type are in part responsible for the astronomer's fascination with the planet Mars. Of all the planets in the solar system, Mars would seem to be the obvious candidate in the search for extraterrestrial life. The outer planets are so far from the sun that life-forms might have difficulty surviving in their cold, thick atmospheres. Mercury is baked by intense solar heat and Venus is eternally cloud-covered, rendering observation of its surface impossible. Mars is indeed the only planet in the solar system whose solid surface can be examined by earth-based astronomers.

Observing details on the Martian surface is not an easy task. Mars is a small planet. With a diameter of 6,800 kilometers (4,200 miles), Mars is the second smallest planet in the solar system. By comparison, Earth is almost two times larger. In addition, the average distance between the sun and Mars is about 1½ AU or 228 million kilometers (141 million miles). The size and distance to Mars means that

**Figure 8-1.** *Mars.* This excellent earth-based photograph shows the Martian polar cap as well as dark markings that appear greenish against the rusty-red surface of the planet. (*Lick Observatory photograph*)

the planet always appears smaller than one-half minute of arc in diameter. That is roughly the same apparent size of a modest crater on the moon.

While Earth's orbit about the sun is very nearly circular, the orbit of Mars is somewhat more elliptical. Only Mercury and Pluto have orbits which are more eccentric. This eccentricity of Mars' orbit about the sun means that at certain locations the distance between the orbits of Mars and Earth is relatively small, while at other locations this distance is considerably larger. A drawing of the orbits of Earth and Mars is shown in Figure 8–2. Near the perihelion of Mars (that is, where the orbit is nearest the sun), the distance between the two orbits is 56 million kilometers (35 million miles). Mars can never be closer to Earth than 35 million miles. By contrast, near the aphelion of Mars (that is, where the orbit is farthest from the sun), the distance between the orbits is 99 million kilometers (63 million miles).

At roughly two-year intervals, Earth passes between Mars and the sun. As we learned in Section 1.2, this is called an *opposition*. If an opposition occurs when Mars is near its perihelion, the distance between the two planets is only 35 million miles. When oppositions occur far from the perihelion, the distance between the two planets is considerably larger. Obviously, when the distance between the two

**Figure 8–2.** *The Orbits of Earth and Mars.* Earth's orbit is very nearly circular while Mars' orbit is somewhat more elliptical. The distances between the two orbits therefore varies between 35 and 63 million miles. Details on the Martian surface are most clearly seen during oppositions when Mars is near its perihelion.

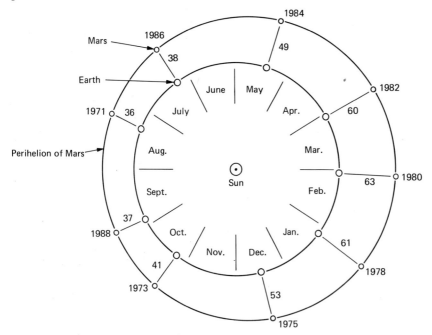

planets is small, Mars appears comparatively big and bright in the sky. An opposition occurring near Martian perihelion is termed "favorable" because, at a distance of 35 million miles, the apparent diameter of Mars is 25 seconds of arc. At such times, the best earth-based views of the Martian surface are obtained. At all other times, Mars appears much smaller through a telescope and it is therefore more difficult to see surface features.

The synodic period of Mars is 780 days and therefore 780 days elapse between each opposition. Favorable oppositions occur only once every 15 or 17 years when Mars is seen shining brightly, high in the late-summer sky around midnight. The last favorable opposition occurred in August 1971. The next is not due until September 1988.

The first telescopic observations of Mars were made by Galileo in the early seventeenth century. A few years later, in 1636, Francisco Fontana made some sketches of the Martian surface. Unfortunately, the telescopes of both of these early Italian astronomers were so poor that no true details on Mars could be seen.

The first reliable observations of the Martian surface were made by the Dutch physicist Christiaan Huygens in October 1659. Huygens drew sketches of a darkish triangular area on Mars now called Syrtis Major. As with early telescopic observations of the moon, astronomers believed that these dark areas were oceans of water and referred to them as *maria,* the Latin word for "seas."

By 1666, G. D. Cassini and others discovered that the rotation period of Mars was nearly 24 hours by observing markings on the Martian surface. Actually, Mars rotates a little slower than Earth; a Martian "day" lasts 24 hours, 37 minutes, 23 seconds. It was also noticed that the two polar caps changed size depending on the Martian seasons. During winter in one of the hemispheres, the polar cap was large and easily visible. With the coming of the Martian summer, the polar cap shrunk and virtually disappeared from view. Just as in the case of Earth, the seasons of Mars are the result of the fact that the planet's equator is inclined to its orbit. As we learned in Section 2.1, the angle between Earth's equator and its orbit is $23\frac{1}{2}°$. In the case of Mars, the angle is 24°. It is, of course, pure coincidence that both the rotation periods and inclination angles of Mars and Earth are very nearly the same. Nevertheless, these coincidences may have played a subconscious role in psychologically motivating early astronomers to believe that Mars, like Earth, was teeming with life.

By the early eighteenth century, telescopes had improved to such an extent that very fine drawings of the Martian surface could be made. In 1719, Giacomo Maraldi noticed changes on Mars which he thought could involve clouds. He also noticed a darkish region bordering the polar caps. In the late 1700s, the English astronomer William Herschel speculated that the dark band surrounding the polar caps might be caused by the melting of ice and snow. It was, incidentally, Herschel who first deduced that the inclination of Mars' axis to its orbital plane was nearly the same as the Earth's.

The first global maps of Mars were made in 1840 by two German astronomers, Wilhelm Beer and Johann Mädler. This was followed in 1863 with the first color sketches by the Italian priest Pietro Angelo Secchi. In 1869, Father Secchi noticed

**Table 8-1.** *Terms Used in the Naming of Martian Surface Features*

| Latin | English |
|-------|---------|
| fretum | strait or channel |
| lacus | lake |
| lucus | grove or wood |
| mare | sea |
| palus | swamp |
| sinus | bay or gulf |

some linear features on Mars which he called "canali." This term was probably chosen because the dark areas were named after bodies of water. These "canali" were destined to receive great notoriety.

By the mid-1800s, so many features on the Martian surface were sufficiently well-known that they began receiving names. In 1871, the English astronomer Richard Proctor published a map in which a number of prominent features were named after British astronomers. This bias did not please the Italian astronomers and in 1877, Giovanni Schiaparelli, the director of an observatory in Milan, devised his own nomenclature. Schiaparelli's names, most of which are from classical and mythical literature, are still used today. Of course, the names were in Latin and Table 8-1 lists some translations. Each of these features was in turn named after Egyptian gods, biblical lands, Greek muses and sirens, and Christian hell. Dark areas were named for bodies of water, while light areas received names of terrestrial lands.

**Figure 8-3.** *A Photograph and Drawing of Mars.* Both views of Mars were obtained at the same time during the opposition of 1926. The drawing suggests "canali," which some observers claim to see on the Martian surface. (*Lick Observatory photographs*)

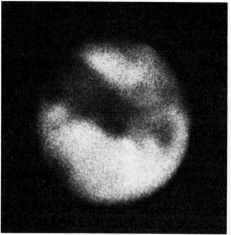 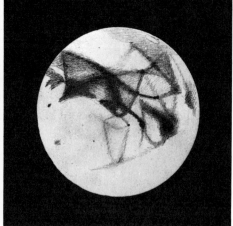

The summer of 1877 was crucial in the history of earth-based observations of Mars. It was during the favorable opposition of this year that Schiaparelli mapped the Martian surface giving over a hundred fanciful names to the features that can be seen. In addition, he claimed to have seen numerous "canali" crisscrossing the Martian surface. A photograph and drawing of Mars revealing some "canali" are shown in Figure 8–3.

This same year marked the first significant contribution by an American astronomer. Working at the United States Naval Observatory, Asaph Hall discovered two small satellites in orbit about Mars. They were named Phobos and Deimos (Greek for "fear" and "panic") after the two mythical horses that drew the chariot of the god of war.

Phobos and Deimos are very dim. Phobos is never brighter than tenth magnitude while Deimos is never brighter than eleventh. From their faintness it was obvious that they are mere chunks of rock, perhaps only ten miles in size.

Both satellites are very near the planet and therefore have short periods of revolution. At a distance of 23,400 kilometers (14,500 miles) from the center of Mars, Deimos takes 1 day, 6 hours, 18 minutes to orbit the planet. But Phobos is only 9,350 kilometers (5,800 miles) from the center of Mars and therefore orbits the planet once every 7 hours, 39 minutes. In view of this short orbital period, Phobos appears to rise in the west and set in the east *twice* each day, as seen from the surface of Mars.

Of all of the discoveries about Mars during the nineteenth century, the "canali" were by far the most notorious and controversial. By 1886, some observers in Europe and the United States were also claiming to see "canali" while others were not. For example, the famous American astronomer E. E. Barnard working at Lick Observatory in 1894 wrote: "To save my soul I can't believe in the canals as Schiaparelli draws them." * Of course, by now the word "canali" had been translated into English as *canals,* implying some sort of irrigation ditches dug by intelligent creatures.

The skepticism of cautious astronomers like Barnard was drowned out by more fanciful pronouncements of other observers such as Percival Lowell. Lowell, a member of a wealthy Boston family, established an observatory near Flagstaff, Arizona for the purpose of studying Mars. By the end of the nineteenth century, Lowell had "discovered" four times as many canals as had been observed by Schiaparelli. Lowell was convinced that these canals were a vast planet-wide network of irrigation ditches used by the Martians to transport water across the surface of their dead, dry planet. Inspired by these observations, the British novelist H. G. Wells wrote *The War of the Worlds,* which depicted an invasion of Earth by Martians abandoning their dying planet in search of a better world. Indeed, a host of monster movies had their origin in the "canali" first observed by Father Secchi in 1869.

The coming of the twentieth century saw the development of photographic techniques and by the mid-1920s, the first good photographs of Mars were taken.

*In a letter to Simon Newcomb dated September 11, 1894.

One might suspect that photography would settle the issue of canals once and for all. But this was not the case. In order to photograph Mars, the astronomers must expose photographic film for a period of time at the focus of a telescope. While taking such a "time-exposure," instabilities and currents in the earth's atmosphere cause the image to wiggle about. The result of this shimmering and quivering is a slightly blurred image. Observers who claimed to see canals did so only during those brief moments when the earth's atmosphere is unusually steady. No one has ever obtained a photograph of a canal.

During the oppositions of 1924 and 1926, astronomers at both Lowell and Lick Observatories succeeded in obtaining photographs of Mars in various colors. It was well known that different wavelengths of light have different abilities to penetrate a planet's atmosphere. Long wavelength red or infrared light easily passes directly through an atmosphere while short wavelength blue or violet light is severely scattered around. Indeed, this is why the daytime sky is blue. These differences in atmospheric penetration are shown in Figure 8–4. The lower half of the illustration shows two views of San Jose taken from Mount Hamilton (on which Lick Observatory is located). The one on the left was taken in violet light, which is so

**Figure 8–4.** *Photographs in Different Colors.* The two views on the left (one of Mars, one looking toward San Jose) were taken in violet light, which is severely scattered by the atmosphere. The two views on the right were taken in infrared light, which easily penetrates the atmosphere. (*Lick Observatory photographs*)

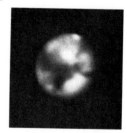

**Views in Violet Light**     **Views in Infrared Light**

Mars Approaching ➡️

The Sun ➡️

Mars Receding ➡️

**Figure 8–5.** *Martian Spectrum.* The pairs of dark absorbtion lines are caused by oxygen molecules in the earth's atmosphere. The upper spectrum was taken while Mars was approaching Earth at a speed of 13.75 kilometers per second (30,760 miles per hour). The lower spectrum was taken while Mars was receding from Earth at a speed of 12.42 kilometers per second (27,780 miles per hour). The middle spectrum is of the sun. The fact that all three spectra are virtually identical means that there is no detectable oxygen in the Martian atmosphere. (*Yerkes Observatory photograph*)

completely scattered that only the immediate foreground is visible. The view on the right was taken in near infrared light, which easily penetrates the atmosphere revealing the landscape many miles away. Similarly with the photographs of Mars, views taken with violet filters show only the upper Martian atmosphere. In contrast, photographs taken with infrared filters transmit the long wavelength radiation that easily penetrates the Martian atmosphere, revealing surface details on the planet.

If Mars is an abode of extraterrestrial life, then spectra of light reflected from the Martian surface might reveal the effects of living organisms. For example, plant life on Earth requires water and produces oxygen. During the oppositions of 1934, 1937, and 1941, W. S. Adams and T. Dunham used the 100-inch telescope at Mount Wilson to obtain spectra of light reflected from Mars. No spectral lines from either $O_2$ or $H_2O$ were found. This result was later confirmed by G. P. Kuiper at McDonald Observatory, who concluded that oxygen was virtually absent and water is extremely scarce on Mars.

There was a momentary flurry of excitement in 1956 when the American astronomer W. H. Sinton announced his discovery of spectral lines in the infrared spectrum of Mars near 3.4 and 3.5 microns. He attributed these spectral features to molecules such as chlorophyll. But a few years later, Sinton and his colleagues correctly identified these features as caused by "heavy water" (molecules of $H_2O$ in which the nucleus of one hydrogen atom contains an extra neutron) in the earth's atmosphere through which they were observing. By the 1960s, the case for life on Mars was looking dimmer and dimmer.

In spite of more than three centuries of telescopic observation, Mars remained shrouded in mystery. The central problem was that Mars is small and far away. Additionally, quality observations could only be made once every 15 to 17 years during a favorable opposition. With the advent of space exploration, however, all this changed. Very early in the history of the space program, Mars was recognized as a prime target. Only by traveling to this tiny planet could the questions raised during the preceding centuries be properly answered.

# 8.2 *The First Martian Probes*

THE first successful mission to Mars was launched on the morning of November 28, 1964 from Cape Kennedy. A small, unmanned spacecraft called Mariner 4 was inserted into an interplanetary trajectory that would ultimately pass within 9,790 kilometers (6,084 miles) of the mysterious red planet. Mariner 3, launched on a similar mission three weeks earlier, suffered severe mechanical failure (Mariners 1 and 2 were directed toward Venus).

Mariner 4 carried a series of instruments to measure interplanetary particles and fields during its eight-month journey to Mars. But by far the most important and surprising discoveries came from a single television camera on board the spacecraft.

Three centuries of earth-based observations of Mars were shattered when Mariner 4 coasted past the red planet on July 14, 1965. (For the flight path of Mariner 4, see Figure 1–28.) Twenty-two pictures were taken which showed

**Figure 8-6.** *Mariner 4.* This 575-pound spacecraft journeyed to Mars and took 22 close-up pictures on July 14, 1965. (*NASA*)

**Figure 8-7.** *Mars from Mariner 4.* This is one of the best views sent back from Mariner 4. A large flat-bottomed crater 175 kilometers (110 miles) in diameter is seen. (*NASA*)

several hundred flat-bottomed craters. The Martian surface was distinctly more like the moon than the Earth. A typical view from Mariner 4 is shown in Figure 8–7. Prior to Mariner 4, a few astronomers had speculated about the possibility of craters on Mars but their prophetic conjectures were almost totally ignored. In contrast, not one single canal was observed by Mariner 4.

After taking this historic series of photographs, Mariner 4 intentionally journeyed behind Mars, as seen from Earth. During this Martian occultation, radio signals from the spacecraft were passed through the Martian atmosphere. Changes in the radio signals allowed scientists to deduce properties of the Martian atmosphere. Specifically, the atmospheric pressure on Mars was found to be extremely low, ranging between 5 to 10 millibars. (Recall from Section 6.3 that the air pressure at sea level on Earth is approximately 1000 millibars.) In addition, the average molecular weight of the gases in the Martian atmosphere was found to be 40. This is consistent with the atmosphere being composed almost entirely of carbon dioxide. (Recall that the atomic weight of carbon is 12 while oxygen is 16, so that $CO_2 = 12 + 16 + 16 = 40$.) This confirmed earlier earth-based spectroscopic observations by G. P. Kuiper in 1947. Additionally, Mariner 4 did not detect any magnetic field about Mars and the maximum surface temperature (daytime in the equatorial regions) was estimated to be roughly 300°K (about 80°F).

With all systems functioning perfectly, Mariner 4 photographed 1,554,000 square kilometers (600,000 square miles) of Mars. This is only 1 per cent of the entire Martian surface. Yet, data from this tiny portion of Mars had a profound impact on our ideas about the planet. Indeed, the mission was so successful that NASA promptly authorized two more flights to the red planet.

**297**

Mariner 6 was launched toward Mars on February 24, 1969. This was followed by an identical spacecraft, Mariner 7, launched a month later on March 27, 1969 (Mariner 5 was directed toward Venus). As with the previously successful Martian mission, Mariners 6 and 7 were programmed to achieve a "fly-by" of the planet, thereby affording a brief period during which the surface of the planet would be studied. They would encounter the planet 5 days apart in the summer of 1969.

Unlike Mariner 4, which carried only one television camera, both Mariners 6 and 7 each had two television cameras, one for wide-angle viewing, the other equipped with a telephoto lens. In addition, each spacecraft (see Figures 8–8 and 8–9) had ultraviolet and infrared spectrometers for examining the Martian atmosphere as well as an infrared radiometer for measuring the surface temperature on Mars.

Mariner 6 began photographing Mars on July 29, 1969 while at a distance of over one million kilometers from the planet. As the spacecraft coasted toward Mars, 50 "far-encounter" photographs were taken. On the following day, the spacecraft flew past the planet taking 25 "near-encounter" photographs. Three days later, on August 2, Mariner 7 began taking 91 far-encounter pictures starting at a distance of almost two million kilometers from the planet. Two days later, on August 4, Mariner 7 successfully acquired 33 near-encounter photographs of the Martian surface.

**Figure 8–8.** *The Design of Mariners 6 and 7.* Each spacecraft weighed 840 pounds and contained two television cameras along with equipment to measure surface and atmospheric conditions on Mars. (*NASA*)

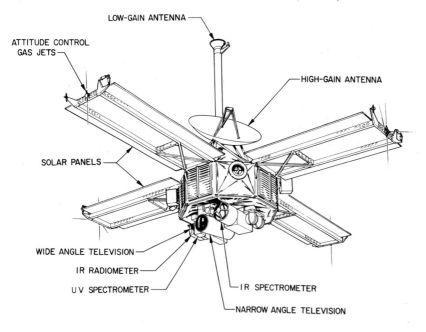

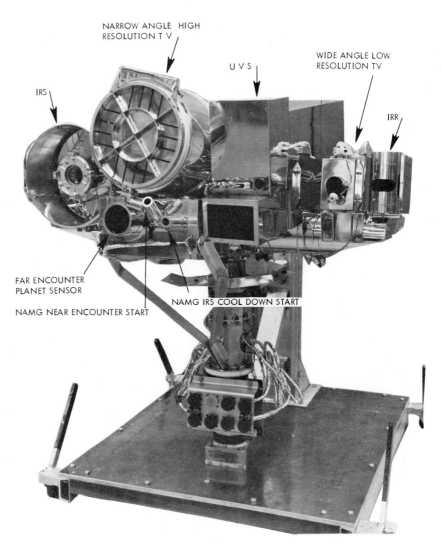

NARROW ANGLE HIGH RESOLUTION T V

U V S

WIDE ANGLE LOW RESOLUTION TV

IRS

IRR

FAR ENCOUNTER PLANET SENSOR

NAMG NEAR ENCOUNTER START

NAMG IRS COOL DOWN START

MARINER MARS 1969 SCIENCE INSTRUMENTS

**Figure 8-9.** *Instruments On Board Mariners 6 and 7.* Details of the instrument packages on board Mariners 6 and 7 are shown. In addition to wide-angle and narrow-angle television cameras, each spacecraft carried spectrometers (IRS and UVS) for examining details of the Martian atmosphere and a radiometer (IRR) for measuring the temperature on the planet's surface. (*NASA*)

The purpose of the far-encounter photographs was to provide the "missing link" between earth-based photography and the close-up views of Mariner 4. An excellent example is shown in Figure 8–12. Although far superior to any view ever seen from earth, not one single canal showed up. Canals are a complete fiction, imagined by over-zealous astronomers. Perhaps the sequence of small, dark features seen near the middle of the view in Figure 8–12 blurred together to give the impression of a straight line. Indeed the alleged "Coprates canal" should have appeared in this region.

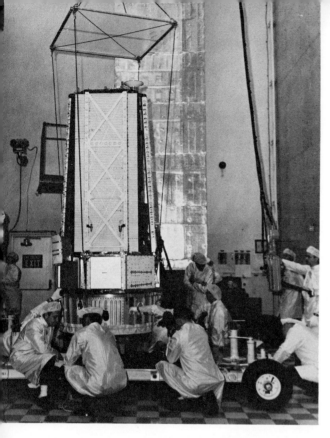

**Figure 8–10.** *Mariner 6 Prior to Launch.* Engineers make a last-minute check of Mariner 6 prior to launch by an Atlas–Centaur rocket. Once in space, the four solar panels unfold and supplied the spacecraft with power from sunlight. (*NASA*)

**Figure 8–11.** *The Flight-Paths of Mariners 6 and 7.* Although Mariners 6 and 7 were launched a month apart, they arrived at Mars within five days of each other because Mariner 7 followed a shorter "inside track."

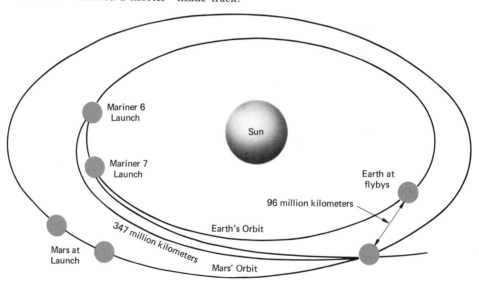

**Figure 8-12.** *Mars from Half a Million Kilometers.* This photograph was taken by Mariner 7 at a distance of 514,811 kilometers (about 320,000 miles). The south polar cap is at the bottom. No canals are seen. (*NASA*)

During the fly-bys, both cameras on both spacecrafts succeeded in taking 58 near-encounter photographs. At closest approach the wide-angle cameras photographed areas approximately the size of Alaska (1000 kilometers by 1000 kilometers) and could reveal details as small as 3 kilometers (2 miles) across. The narrow-angle cameras photographed areas roughly the size of the Los Angeles basin (100 kilometers by 100 kilometers) revealing features 300 meters (1000 feet) across.

A mosaic of seven wide-angle views from Mariner 6 is shown in Figure 8-13. The area shown covers roughly 4000 kilometers by 2000 kilometers and includes

**Figure 8-13.** *A Mosaic of Wide-angle Pictures.* Seven wide-angle photographs taken by Mariner 6 were pieced together in this mosaic. The area covered is roughly comparable to the size of the United States. (*NASA*)

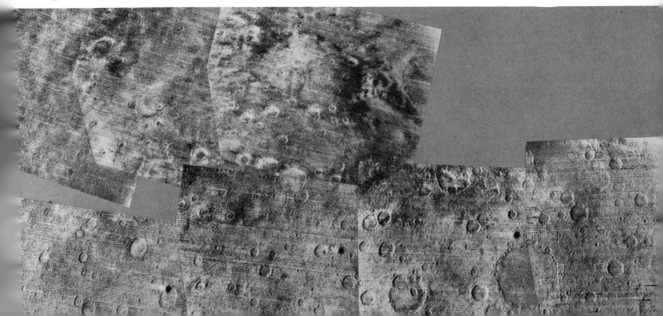

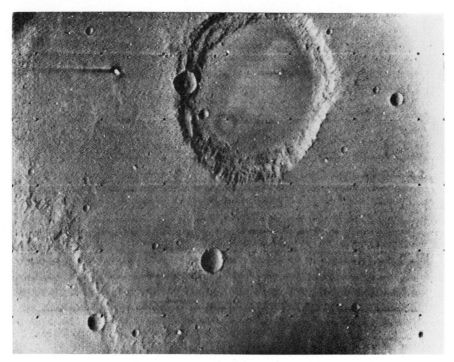

**Figure 8–14.** *A Narrow-angle View from Mariner 6.* The area covered in this picture measures 105 kilometers by 73 kilometers (65 miles by 45 miles). The spacecraft was at a distance of 3746 kilometers (nearly 2330 miles) at the time the photograph was taken. (*NASA*)

the feature Schiaparelli called Meridiani Sinus. An abundance of flat-bottomed craters are seen. A narrow-angle view revealing details of a few of these craters is shown in Figure 8–14. The area covered in this narrow-angle view measures 105 kilometers by 73 kilometers.

From these views it was clear that Martian craters are very different from lunar craters. Unlike lunar craters, craters on Mars experience erosion. For example, Figure 8–14 shows several small, young craters. Their youthfulness is evidenced by their rounded crater floors. In the upper half of the picture is a larger, flat-bottomed crater measuring roughly 25 miles in diameter. In the lower-left corner is the faint suggestion of a still more ancient crater that has been almost completely obliterated.

Over the past few centuries, earth-based observers have occasionally watched as virtually all the surface features on Mars disappeared from view for several weeks. During these times, Mars presents a featureless red-orange disk, as seen through a telescope. It is believed that winds in the Martian atmosphere pick up tiny dust particles producing a planet-wide *dust storm*. These dust storms result in the erosion of surface features on the planet and are probably responsible for the flat-bottomed appearance of older craters which have gradually been filled in.

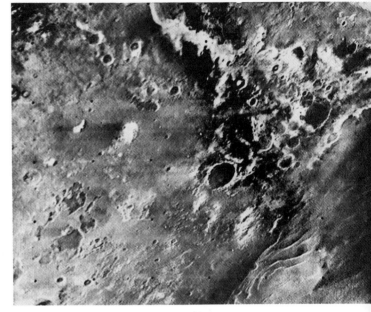

**Figure 8-15 (*above*).** *The South Polar Cap.* This mosaic of five wide-angle views from Mariner 7 shows the southern polar cap of Mars. The spacecraft was roughly 5000 kilometers from the polar cap when the pictures were taken. (*NASA*)

**Figure 8-16.** *The South Pole of Mars.* The south pole of Mars is located in the lower right corner of this wide-angle view from Mariner 7. The area covered in this photograph measures 1549 kilometers by 1394 kilometers (roughly 960 miles by 870 miles). (*NASA*)

It is interesting to note that the areas between craters in Figure 8–14 are smooth, flat, and almost completely devoid of features. Indeed, Mariner 7 succeeded in photographing expansive areas that were completely devoid of any features, most noticeably in the region Schiaparelli called Hellas. Even close-up telephoto views from the narrow-angle camera failed to reveal any features whatsoever.

In addition to cratered terrain and featureless terrain, a third type of Martian topography was identified during the 1969 missions. In certain selected regions, the narrow-angle camera on board Mariner 6 succeeded in photographing jumbled, hilly areas. These regions were termed "chaotic terrain."

While Mariner 6 photographed mainly equatorial regions of Mars, Mariner 7 succeeded in obtaining pictures of the south polar cap. A mosaic of several wide-angle views is shown in Figure 8–15. An enlargement of one of these views appears in Figure 8–16, which includes Mars' south pole. It was early spring in the southern

**303**

hemisphere of Mars at the time of these photographs. Notice that the rims of some of the craters seem to be covered with "frost." Also notice the unusual "layered" features very near the Martian south pole in the lower-right corner of the photograph. Measurements of surface temperature and atmospheric pressure on Mars are consistent with the idea that the "snow" or "frost" seen in the polar regions actually consists of frozen carbon dioxide. Thus, the polar caps on Mars are probably made of dry ice rather than frozen water as is the case here on Earth.

While the television cameras on Mariners 6 and 7 were producing dramatic pictures of the Martian surface, additional equipment on board the spacecrafts were measuring atmospheric conditions around the planet. While Earth's atmosphere consists of almost 80 per cent nitrogen, the ultraviolet spectrometers did not succeed in detecting any of the gas on Mars. In 1976, however, the Viking spacecraft that landed on the planet measured a nitrogen content of 3 per cent and an argon content of $1\frac{1}{2}$ per cent. Very tiny amounts of hydrogen and oxygen are also present, but by far the major constituent of the Martian atmosphere is carbon dioxide. The infrared spectrometers detected clouds of carbon dioxide along with some water ice in the vicinity of the south pole. Additionally, spectral lines in the infrared caused by solid carbon dioxide in the south polar cap were observed at wavelengths from 2 to 4 microns.

The infrared radiometers measured the surface temperature on Mars along the path viewed by the television cameras. Daytime temperatures as high as 280 to 290°K (roughly 10°C = 50°F) in the equatorial regions were recorded. In contrast, temperatures of 160 to 170°K (roughly −110°C = −170°F) were found on the south polar cap. This temperature is approximately as cold as frozen carbon dioxide would be in the prevailing conditions on Mars.

As was the case with Mariner 4, both of the 1969 Mariners passed behind Mars as viewed from Earth. Observing changes in radio transmissions through the Martian atmosphere permitted scientists to measure the atmospheric pressure on Mars. Specifically, at four widely separated locations (one for entering and one for leaving the "shadow zone" for each of the two spacecrafts) the temperature, atmospheric pressure, and planet's radius were measured. The results of these experiments are given in Table 8–2.

This confirmed earlier measurements by Mariner 4, which found an atmospheric pressure in the range of 5 to 10 millibars. Recalling that the air pressure at

**Table 8-2.** *Results of Mariners 6 and 7 Occultation Experiments*

| Location | Planet Radius | | Pressure | Temperature | |
| --- | --- | --- | --- | --- | --- |
| | (kilometers) | (miles) | (millibars) | (°K) | (°F) |
| Meridiani Sinus | 3393 | 2108 | 6.6 | 276 | 37 |
| North Polar | 3373 | 2096 | 6.4 | 160 | −172 |
| Hellespontus | 3383 | 2102 | 3.8 | 211 | −79 |
| Amazonis | 3378 | 2099 | 7.0 | 200 | −99 |

sea-level on Earth is roughly 1000 millibars, we see that the atmospheric pressure on Mars is extremely low.

The flights of Mariners 6 and 7 presented a new and fascinating picture of Mars. Yet, many questions remained unanswered. Although the entire planet had been photographed in far-encounter pictures, the close-up views covered only a small percentage of the Martian surface. The 200 pictures showed that Mars has its own unique character, unlike the moon or Earth. Unfortunately, data could be collected only during a brief span of eight days. The obvious next step was to place a spacecraft in orbit about the planet. Only then could scientists examine Mars over a period of many months. This was the next major goal in unmanned space exploration.

## *Mariner 9 in Orbit* **8.3**

ON MAY 30, 1971, Mariner 9 was carried aloft from Cape Kennedy by an Atlas-Centaur rocket. This spacecraft was destined to become the first man-made satellite of a distant planet. Unfortunately, Mariner 8, launched two weeks earlier on an accompanying mission, was destroyed because of a vehicle failure.

Recall that the first close-up pictures of Mars were sent back in 1965 during the

**Figure 8-17.** *Mariner 9.* The spacecraft is shown here during the final stages of construction at the Jet Propulsion Laboratory. Four solar panels would provide Mariner 9 with power from sunlight. The "retro-rocket," shown at the top of the spacecraft would burn 900 pounds of fuel to slow the spacecraft as it approaches Mars. (*NASA*)

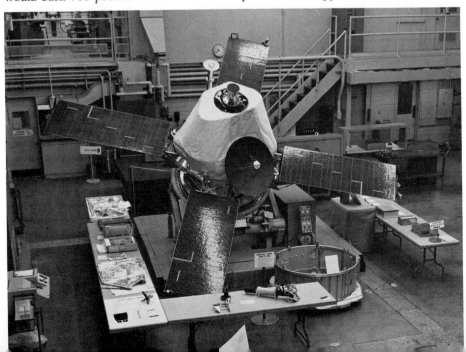

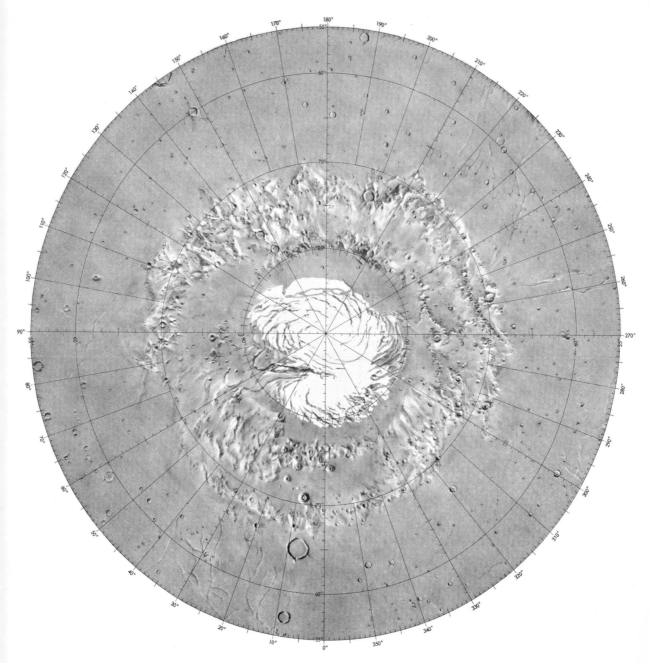

**Map 5.** *Mars North Pole Map*

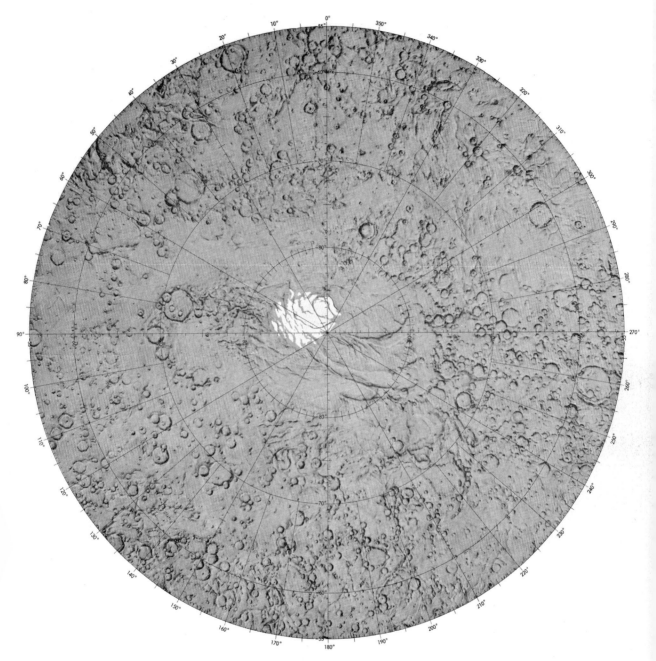

**Map 8.** *Mars South Pole Map*

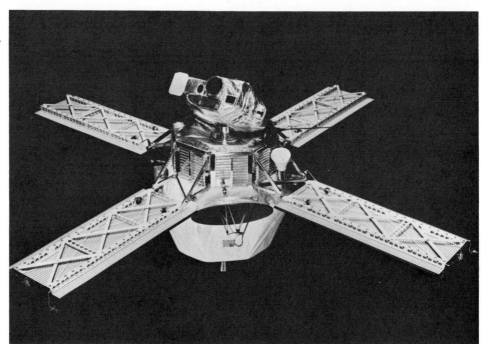

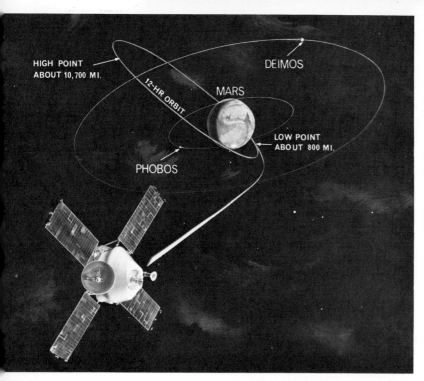

HIGH POINT
ABOUT 10,700 MI.

12-HR ORBIT

DEIMOS

MARS

LOW POINT
ABOUT 800 MI.

PHOBOS

**Figure 8–18 (*above*).** *Mariner 9.*
As with its predecessors, Mariner 9 carried television cameras, infrared and ultraviolet spectrometers, and an infrared radiometer. At launch the spacecraft weighed 2200 pounds. But after arriving at Mars, the "burn-out" weight was 1200 pounds. (*NASA*)

**Figure 8–19.** *The Orbit of Mariner 9.* Mariner 9 orbits Mars twice each day. The high point in the orbit was 17,100 kilometers of the Martian surface while the low point was at a distance of 1,650 kilometers, resulting in an orbital period of 11.98 hours. (*NASA*)

fly-by of Mariner 4. These views revealed a bleak cratered surface reminiscent of the moon. Four years later, in 1969, Mariner 6 and Mariner 7 discovered that the Martian surface was not uniformly cratered. Instead there were hilly, jumbled areas termed chaotic terrain and expansive, flat, featureless regions such as the "desert" called Hellas. During its first year in orbit about the mysterious red planet, Mariner 9 would send back over a hundred times more data than had been accumulated during the previous three successful missions and 350 years of earth-based observations. Scientists would be amazed with incredible views of immense volcanoes, huge canyons, and strange features that look suspiciously like dried-up river beds. Not one hint of any of these Martian features had been observed by any of the previous Mariners.

On September 22, 1971, nearly four months after launch, yet still eight weeks before Mariner 9 completed its 287-million-mile interplanetary journey, earth-based observers noticed a small whitish cloud appear within a few hours over the Noachis region on Mars. Five days later, the growing storm covered two thousand kilometers, from Noachis on the west to Hellas on the east. Syrtis Major to the north was rapidly becoming obscured. Within two weeks of the onset of the storm, a zone 10,000 kilometers long had been covered by the growing dust storm, which advanced across the planet at roughly 35 kilometers per hour (25 miles per hour). By the time Mariner 9 arrived at Mars on November 14, the entire planet was embroiled in one of the most extensive planet-wide storms ever observed in $3\frac{1}{2}$ centuries. With the exception of four hazy spots in the northern hemisphere, not one single surface feature could be seen from Mariner 9. These three spots later turned out to be the summits of four huge volcanoes (Olympus Mons, Ascraeus Mons, Pavonis Mons, and Arsia Mons) protruding through the dust pall.

The immediate effect of the "Great Dust Storm of 1971" was the cancellation of most of the far-encounter picture-taking program. The entire planet looked like a featureless orange blob. Systematic picture-taking was delayed for almost three months as the storm raged on.

As the storm began to subside, scientists focused their attention on some of the features that protruded through the clouds. Scientists were amazed by what they saw: familiar-shaped craters just like those found on the tops of the volcanoes in the Hawaiian Islands. Figure 8–20 shows a photograph of the summit of a Martian volcano taken on November 28, 1971. Years earlier, earth-based observers had named this feature Nodus Gordii (Latin for "The Gordian Knot"). But as the dust clouds parted, the central crater or *caldera* of a huge volcano was revealed. This volcano has since been renamed Arsia Mons; it was the southernmost hazy spot seen in the far-encounter photographs taken several weeks earlier.

As the dust storm continued to clear, greater portions of the sides of the volcanoes were exposed to view. Figure 8–21 shows a wide angle view of Olympus Mons (previously called Nix Olympica, Latin for "The Snows of Olympus") along with a telephoto view of one of its flanks. Careful examination reveals a long fissure and two lava flows running downhill. The summit of the volcano consists of a series of collapsed volcanic craters. The entire caldera is about 65 kilometers (40 miles) in diameter.

**Figure 8-20 (*right*). *The Summit of Arsia Mons.*** The central crater or caldera at the top of the volcano Arsia Mons was seen as the dust storm began to subside in late November 1971. The caldera measures 140 kilometers (90 miles) in diameter. (*NASA*)

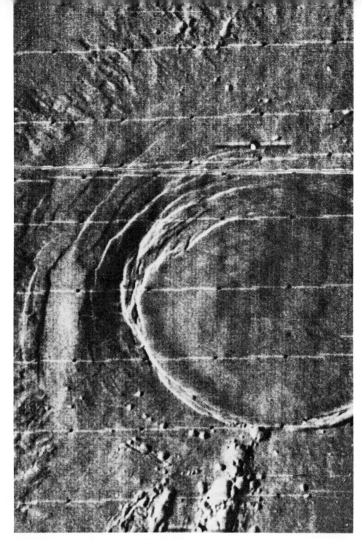

**Figure 8-21. *The Summit and Flank of Olympus Mons.*** The wide-angle view (left) shows the caldera at the summit of Olympus Mons. The telephoto view (right) reveals details of the volcano's flank. (*NASA*)

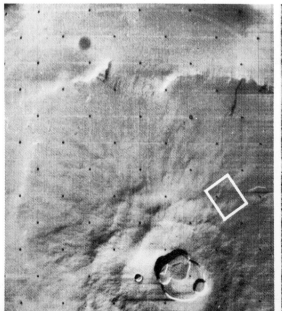

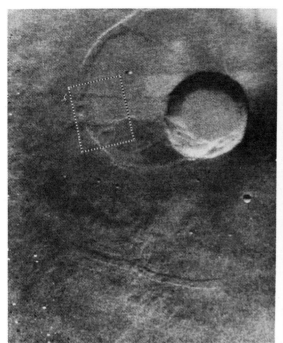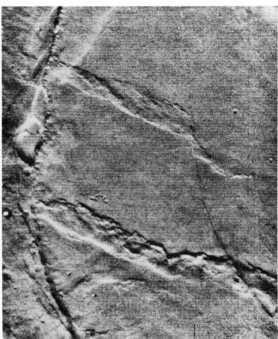

**Figure 8–22** (*above*).
*The Summit and Flank of Pavonis Mons.* The wide-angle view (left) shows the caldera at the summit of Pavonis Mons. The telephoto view (right) reveals lava flows on the volcano's flank. (*NASA*)

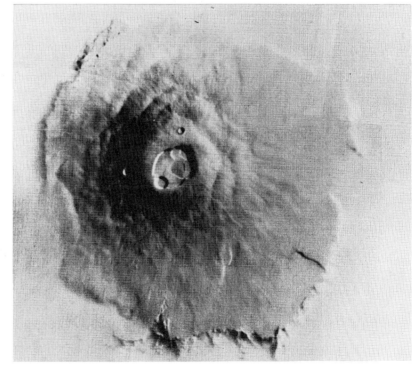

**Figure 8–23.** *Olympus Mons.* This immense volcano is 500 kilometers across at the base and rises to a height of 25 kilometers (80,000 feet) above the surrounding Martian plane. By comparison, the summit of Mount Everest is only 29,000 feet above sea level. (*NASA*)

A similar pair of wide angle and telephoto views of Pavonis Mons (previously called Pavonis Lacus) is shown in Figure 8–22. In this case the central caldera is 45 kilometers (28 miles) in diameter while the telephoto view shows sinuous structures similar to sites of lava flows found on the moon.

Finally, by late January 1972, the dust storm ended and scientists could see the enormous extent of these volcanoes. Olympus Mons is by far the largest. Set on a great plane, the volcano is 500 kilometers (300 miles) across at the base. An excellent mosaic constructed from several wide-angle views is shown in Figure 8–23. If this volcano were on Earth it would cover the whole state of Missouri. Alternatively, it would just barely fit between Boston and Baltimore if it were on the east coast or between San Francisco and Los Angeles on the west coast. From data supplied by the ultraviolet spectrometer, scientists concluded that Olympus Mons rises to an astonishing altitude of 25 kilometers (15 miles = 80,000 feet) above the surrounding Martian surface. By comparison, Mount Everest rises to an altitude of only 29,000 feet above sea level here on Earth. The largest volcano on Earth is Mauna Loa in the Hawaiian Islands. Measured from the floor of the Pacific Ocean, its altitude is only 10 kilometers (6 miles = 30,000 feet). Olympus Mons is

**Figure 8-24.** *Valles Marineris.* This mosaic of several hundred photographs covers one sixth of the surface of Mars along its equator. Olympus Mons is in the upper left-hand corner. Valles Marineris is located horizontally in the middle third of the mosaic. (*NASA*)

2½ times taller and wider. At its base (on the ocean floor) Mauna Loa, the most massive volcanic structure on Earth, measures 225 kilometers (140 miles) across.

One of the features that might be expected to accompany volcanoes is patterns of fractures and faults. In spite of this "reasonable expectation," scientists were shocked by what they saw as the dust settled from the lower parts of the atmosphere. To the east of Arsia Mons there was supposed to be a feature called Coprates or the "Coprates Canal" by earth-based observers. (This feature was discussed briefly in the preceding section in connection with a far-encounter photograph, Figure 8–12, taken by Mariner 7.) By the spring of 1972, it was apparent that Coprates is actually a vast canyon almost 4,000 kilometers (2,500 miles) long! This canyon has been renamed Valles Marineris in honor of the Mariner program. Valles Marineris stretches more than one sixth of the way around Mars. If this feature were on Earth, it would extend all the way from New York to Los Angeles, completely across the United States!

A mosaic of several hundred individual photographs revealing the enormous size of Valles Marineris is shown in Figure 8–24. Eleven million square miles, about one fifth of the entire Martian surface, is covered in this mosaic. Olympus Mons is in the upper left-hand corner. The equator of Mars bisects the mosaic horizontally. Valles Marineris lies slightly south of the equator, running generally east-west in the middle third of the mosaic.

A small section of Valles Marineris is shown in Figure 8–25. At its widest points, the canyon measures almost 200 kilometers (120 miles) across and its depth is as

**Figure 8-25.** *A Section of Valles Marineris.* This section of Valles Marineris is located 490 kilometers (300 miles) south of the Martian equator and about 2,000 kilometers (1300 miles) directly east of the volcano Arsia Mons. The photograph was taken from an altitude of 1,977 kilometers (1225 miles) and covers an area 376 kilometers by 480 kilometers (235 miles by 300 miles). This region was formerly called Tithonius Lacus by Earth-based observers. (*NASA*)

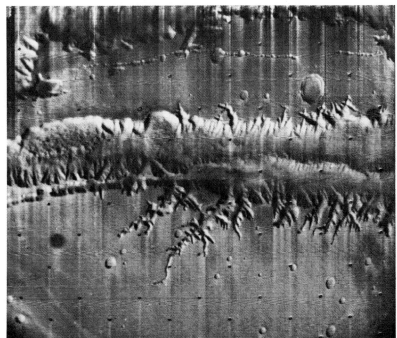

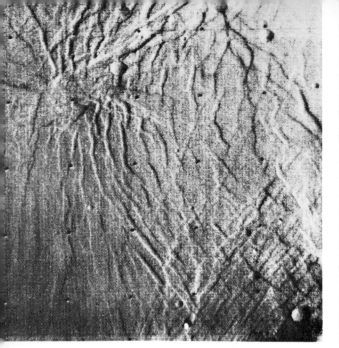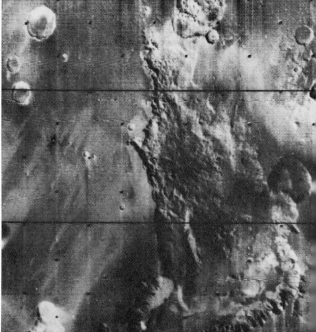

**Figure 8-26 (*left*).** *The "Elephant Skin" Fracture Complex.* This heavily fractured plateau lies near the west end of Valles Marineris. The fault valleys are about 2 kilometers (1½ miles) wide. This region was called Phoenicis Lacus by Earth-based observers. (*NASA*)

**Figure 8-27 (*right*).** *"Chaotic Terrain."* The eastern end of Valles Marineris merges into a jumbled, hilly area termed "chaotic terrain." The view of chaotic terrain shown here includes the feature called Juventae Fons by Earth-based observers. (*NASA*)

much as 6 kilometers (nearly 4 miles = 20,000 feet). By comparison, the Grand Canyon in the western United States is only 150 kilometers long, 6 to 28 kilometers wide and at most 2 kilometers (less than 7,000 feet) deep. Earth's Grand Canyon would appear as only one tiny tributary alongside this vast chasm on Mars.

The west end of Valles Marineris consists of a heavily fractured plateau covered with cracks and faults, as shown in Figure 8–26. At its eastern end, Valles Marineris merges into a jumbled hilly area in the region previously called Aurorae Sinus. These craggy lowlands, termed "chaotic terrain," were first observed by Mariners 6 and 7. An excellent view of chaotic terrain from Mariner 9 is seen in Figure 8–27.

As if massive volcanoes and vast canyons were not enough, early in 1972 Mariner 9 began sending back astonishing pictures of channels that look very much like dried riverbeds. An excellent example is shown in Figure 8–28. This channel, now called Vallis Mangala has a length of 350 kilometers (220 miles) and looks exactly like features on Earth formed by a sudden flood of sediment-laden water after a torrential rain storm. These channels, incidentally, are *not* related to the

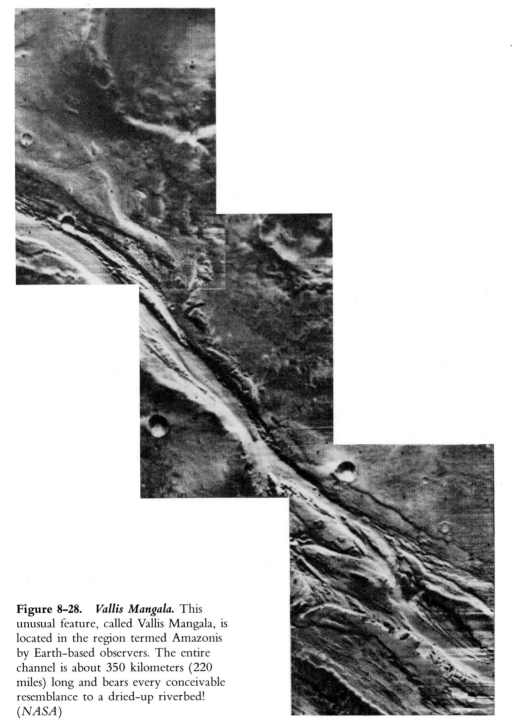

**Figure 8–28.** *Vallis Mangala.* This unusual feature, called Vallis Mangala, is located in the region termed Amazonis by Earth-based observers. The entire channel is about 350 kilometers (220 miles) long and bears every conceivable resemblance to a dried-up riverbed! (*NASA*)

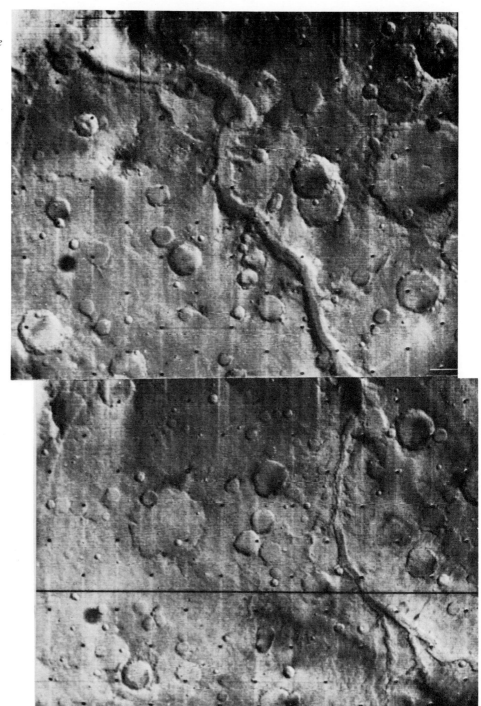

**Figure 8–29 (opposite).** *Vallis Maadion.* This channel is located in the heavily cratered region immediately south of the area called Zephyria by Earth-based observers. The length of the channel seen in these two photographs is 700 kilometers (435 miles). (*NASA*)

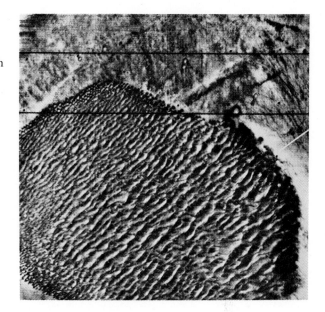

**Figure 8–30.** *A Dune Field.* Sand dunes of loose material on the floor of a crater in the Hellaspontus region of Mars were photographed by the telephoto camera on Mariner 9. The dunes are spaced about a mile apart and the diameter of the crater is 150 kilometers (93 miles). (*NASA*)

"canals" observed by Schiaparelli and Lowell. Another example of these mysterious channels is shown in Figure 8–29. This channel, called Vallis Maadim, is about 700 kilometers (435 miles) long.

These channels were perhaps the most enigmatic features discovered on Mars. The only reasonable explanation for their appearance seems to involve vast quantities of water. But where is the water? As with the earlier Mariner flights, instruments on board Mariner 9 confirmed that Mars is extremely dry. The total amount of water on Mars would barely fill a single good-sized lake here on Earth. Furthermore, the atmospheric pressure on Mars (6.1 millibars) is so low that *if* water existed on Mars, it would promptly boil (in the equatorial regions) or freeze (near the poles). Scientists are at a loss to explain these channels, which look exactly like ancient riverbeds. The only major form of erosion now occurring on Mars involves dust storms. Indeed, Mariner 9 had no trouble locating areas covered with sand dunes such as those shown in Figure 8–30. But there are certainly no lakes or oceans. In fact, if *all* the water in the Martian atmosphere were precipitated out in the form of rain, it would make a layer of water less than a twentieth of a millimeter deep on the planet.

As scientists struggled with the problem of the "missing water" on Mars, attention again focused on the chaotic terrain. Geologists R. Sharp, L. Soderblom, B. Murray, and J. Cutts at the California Institute of Technology suggested that subsurface deposits of water ice or *permafrost* might be responsible for this rugged terrain. If, for some reason (volcanic activity; seasonal or long-term climatic changes) ice and frost under the Martian surface melt, then the terrain collapses and produces a region of jumbled hills. The melting of a subsurface permafrost might be partly responsible for "fretted terrain" just north of the area called Ismenius Lacus

**315**

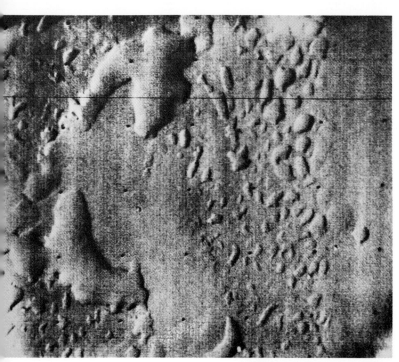

**Figure 8-31** (*left*). *Pitted Terrain.*
Wind excavation probably produced this heavily pitted terrain only 800 kilometers (500 miles) from the south pole. The two large basins on the left are about 16 kilometers (10 miles) across. The smaller pits are 1½ to 3 kilometers (1 to 2 miles) in size. These features were photographed from a range of 3,343 kilometers (2,072 miles) by the telephoto camera on Mariner 9. (*NASA*)

**Figure 8-32.** *Laminated Terrain.*
This oval, layered terrain is near the Martain south pole. This telephoto view was taken from a range of 3,422 kilometers (2,121 miles) and shows an area which measures 60 kilometers by 47 kilometers (37 miles by 29 miles). (*NASA*)

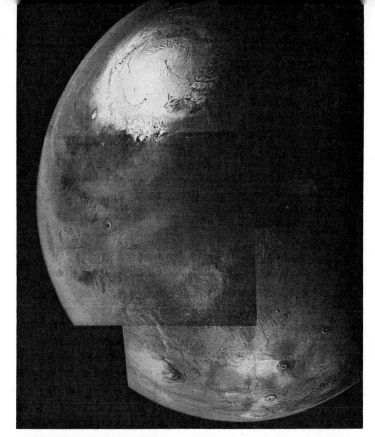

**Figure 8–33.** *The Northern Hemisphere of Mars.* Three wide-angle views taken from an average range of 13,700 kilometers (8,500 miles) on August 7, 1972, were used to construct this mosaic of Mars' northern hemisphere. Details of the northern polar cap, shrinking during the late Martian spring, are seen. The volcano Olympus Mons appears near the lower left. (*NASA*)

in the northern hemisphere and the "pitted terrain" in the south polar region. As the permafrost melts or evaporates, the Martian winds can blow away the loosened soil. This phenomena of wind excavation is believed to have caused the extraordinary series of pits and hollows seen in Figure 8–31. This region of pitted terrain is 800 kilometers (500 miles) from the Martian south pole.

Wind erosion and deposits may also be responsible for the layered or laminated terrain in the Martian polar regions. As shown in Figure 8–32, Mariner 9 photographed features that look like stacks of poker chips or saucers of various sizes. Each layer is roughly 30 meters (100 feet) thick. Perhaps fine dust and volcanic ash carried during dust storms are deposited in the polar regions. As the underlying ice evaporates, a layer of material is left behind. Similar laminations are easily seen around the Martian north pole, shown in Figure 8–33.

Further clues to the mystery of the "missing water" were supplied as scientists watched the melting of the Martian polar caps. With the coming of spring in a Martian hemisphere, the polar cap recedes rapidly. This rate of shrinkage is consistent with the rapid evaporation of a layer of carbon dioxide "frost." However, views from Mariner 9 conclusively established the existence of a permanent, residual polar cap at the south pole. Observations over six weeks, early in 1972, showed that this residual polar cap shrank very slowly—too slowly to be explained by dry ice. This strongly suggested that the residual polar cap is made of ordinary water ice, which evaporates at a much slower rate than solid carbon dioxide.

During June, July, and August of 1972, scientists watched the approach of spring in the northern hemisphere of Mars. Initially, the rate at which the polar cap receded was in agreement with calculations for the rapid rate of the evaporation of dry ice made by R. Leighton and B. Murray at the California Institute of Technology. This was followed by an abrupt halt in the retreat of the polar cap. The relative stability of the remaining, residual northern polar cap again suggests the

**Figure 8–34 (*left*).** *Phobos.* Mars' largest satellite measures roughly 28 kilometers by 20 kilometers (17 miles by 14 miles). This photograph was taken at a distance of 5,540 kilometers (3,444 miles). (*NASA*)

**Figure 8–35 (*right*).** *Deimos.* Mars' smaller satellite, Deimos is about 12 kilometers (7 miles) in diameter. Some faint crater-like markings can be seen with difficulty. (*NASA*)

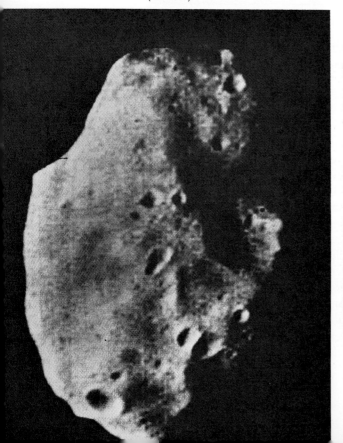
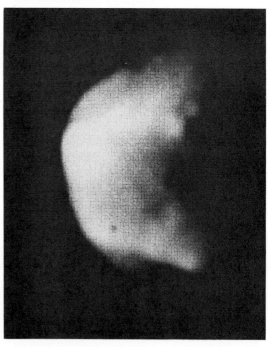

**Table 8-3.** *Sizes of the Martian Satellites*

| | Phobos | | Deimos | |
|---|---|---|---|---|
| | kilometers | miles | kilometers | miles |
| Longest Diameter | 28 | 17 | 16 | 10 |
| Shortest Diameter | 20 | 12 | 10 | 6 |
| Average Diameter | 23 | 14 | 12 | 7 |

existence of ordinary water ice. Temperature measurements from the infrared radiometer support this conclusion. Evidently, the Martian polar caps have a dual composition: dry ice covers a large section of the polar regions in the Martian winters but completely evaporates with the coming of summer, while an underlying, small, permanent cap of ordinary water ice persists throughout the seasons.

Finally, Mariner 9 succeeded in photographing both of the Martian moons. As shown in Figures 8–34 and 8–35, both satellites are mere chunks of rock pitted with a few craters. Indeed, they have been called "flying mountains." From the Mariner 9 photographs, the dimensions of Phobos and Deimos (see Table 8–3) were obtained.

During nine months of operation, Mariner 9 sent back 7,273 photographs revealing details on the surface of Mars never before imagined. Years would be spent analyzing the enormous amount of data collected during this highly successful mission. Unfortunately, yet as expected, Mariner 9 did not provide any clues to the question of life on the planet. Scientists had to wait for the completion of the Viking program in 1976 when two spacecrafts would land on the Martian surface and sample its soil.

## The Nature of Mars 8.4

THE Mariner missions of 1965, 1969, and 1971–72, have shown that Mars has its own unique and distinctive character, unlike either Earth or the moon. Mariner 4 photographed about 1 per cent of the Martian surface showing a cratered, lunar-like landscape. Two years later, Mariners 6 and 7 increased close-up photographic coverage to about 10 per cent of the planet's surface. Expansive plains and jumbled hilly regions were found, proving that only certain restricted areas of Mars are heavily cratered. Finally, in the early 1970s, Mariner 9 succeeded in photographing the entire Martian surface revealing huge volcanoes, vast canyons, and numerous other intriguing topographical features. In examining the overall picture of Mars, scientists immediately noticed that northern and southern hemispheres are very different. The southern hemisphere is more densely cratered, somewhat like the lunar highlands with very few volcanic features. In sharp contrast, there are comparatively few craters north of the Martian equator. The northern hemisphere, however, contains almost all of the huge Martian volcanoes. The reason for this difference between the northern and southern hemispheres is not known.

It is perhaps instructive to compare many of the basic features of Mars with our own planet. Table 8–4 lists some well-known data involving these two planets.

Because Mars is further from the sun, the surface temperatures on Mars are lower than those on Earth. Infrared radiometer measurements reveal that the noontime temperatures near the Martian equator are as high as 290°K 17°C = 62°F) and drop to about 200°K ( −73°C = −100°F) during the night. The tenuous Martian atmosphere does not retain heat very well, resulting in a large

**Table 8–4.** *A Comparison of Earth and Mars*

|  | Earth | Mars |
|---|---|---|
| Average Distance from Sun: |  |  |
| (millions of kilometers) | 150 | 230 |
| (millions of miles) | 93 | 140 |
| (AU) | 1.00 | 1.52 |
| Orbital Period |  |  |
| (Earth days) | 365 | 687 |
| (Earth years) | 1.00 | 1.88 |
| Diameter |  |  |
| (kilometers) | 12,700 | 6,800 |
| (miles) | 7,900 | 4,200 |
| (Earth = 1) | 1.00 | 0.53 |
| Mass |  |  |
| (Earth = 1) | 1.00 | 0.11 |
| Average Density |  |  |
| (grams per cubic centimeter) | 5.5 | 3.9 |
| Period of Rotation |  |  |
| (hours, minutes, seconds) | 23$^h$ 56$^m$ 04$^s$ | 24$^h$ 37$^m$ 23$^s$ |
| Inclination of Equator to Orbit |  |  |
| (degrees, minutes of arc) | 23°27′ | 24°11′ |
| Surface Gravity |  |  |
| (Earth = 1) | 1.00 | 0.38 |
| Escape Velocity |  |  |
| (kilometers per second) | 11.2 | 5.1 |
| (miles per hour) | 24,300 | 11,000 |
| Average Atmospheric Pressure |  |  |
| (millibars) | 1,016 | 6 |
| (pounds per square inch) | 14.7 | 0.09 |
| Atmospheric Composition |  |  |
| (most abundant gases) | nitrogen & oxygen | carbon dioxide |

temperature fluctuation over a 24-hour period. Temperatures in the polar regions of Mars are as low as 150°K ( −120°C = −180°F).

The thin Martian atmosphere is composed almost entirely of carbon dioxide ($CO_2$). Only very tiny amounts of water vapor ($H_2O$), carbon monoxide (CO), hydrogen (H), nitrogen (N), argon (A), and oxygen (O) have been detected. No other gases (such as methane and ammonia, which are important constituents of the atmospheres of the outer planets) have been discovered. The tenuousness of the Martian atmosphere as well as the low surface gravity on the planet results in a very low atmospheric pressure. The average atmospheric pressure on Mars is only 6 millibars (about 0.09 pounds per square inch), compared to 1,016 millibars (about 14.7 pounds per square inch) for air pressure at sea level here on Earth.

While the magnetometers carried to Mars failed to detect a magnetic field around the planet, measurements of the solar wind by particle detectors on board the Mariners suggest that Mars may have a very small magnetic field. At most, the strength of Mars' magnetic field is about 0.2 per cent of Earth's magnetic field. With such a weak magnetic field, no radiation belts have been able to form around Mars.

The fact that Mars possesses almost no magnetic field means that the planet cannot have a molten nickel-iron core like Earth. Recall that the average density of Earth is 5.5 grams per cubic centimeter, in spite of the fact that the density of typical rocks in the earth's crust is only about 2.7 grams per cubic centimeter. This high average density of our planet is caused by existence of a massive nickel–iron core in the earth's interior. The average density of Mars is only 3.9 grams per cubic centimeter, which is more like the average density of the moon (3.3 grams per cubic centimeter). Thus, the interior of Mars is more like the moon than the earth.

Another important way in which Mars is probably different from Earth is that the Martian crust is not divided into moving plates. As suggested by Michael H. Carr of the U.S. Geological Survey, the crust on Mars may be so thick that internal convective motions are insufficient to cause major lateral displacements of the surface. Recall that on Earth, most vulcanism occurs at the intersections of huge tectonic plates that move across our planet's surface. The plates diverge along the mid-oceanic ridges where hot magma wells up from the earth's interior to form new crust. The plates collide along subduction zones characterized by chains of young mountains and deep oceanic trenches. Much of the volcanic activity on Earth is confined to these two types of regions at the plate boundaries. There is, however, at least one major obvious exception. The volcanoes that created the Hawaiian Islands lie at one end of a long line of extinct volcanoes called the Hawaiian-Emperor chain that stretches for thousands of kilometers under the Pacific Ocean. It is believed that there is a *hot spot* in the upper mantle directly below Hawaii. Molten rock forcing its way up from the hot spot is responsible for the volcanic activity often observed on these islands. As the Pacific plate moves toward the northwest, new volcanoes erupt to the southeast, leaving behind a chain of extinct volcanoes that trace the path of the plate over the underlying hot spot.

The equivalent of the volcanoes along the mid-Atlantic ridge is not seen on Mars. Neither are there long chains of young mountains and volcanoes typifying

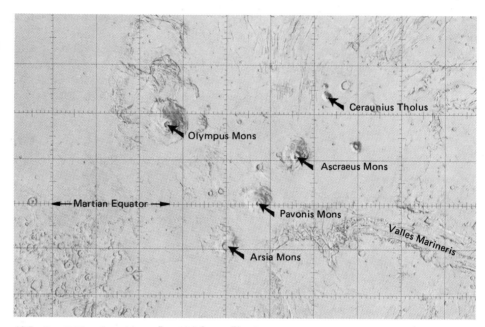

**Figure 8–36.** *The Tharsis Region.* The Tharsis region is the largest and most prominent area of volcanic activity on Mars. The volcanoes are located on a broad bulge in the Martian crust measuring 5,000 kilometers (3,000 miles) in diameter and rising up to 7 kilometers (4 miles) above the average height of the Martian surface. This bulge is called the Syria Rise because its highpoint occurs in Syria Planum just beyond the west end of Valles Marineris. (*U.S. Geological Survey*)

the junctions of colliding plates such as those surrounding most of the Pacific Ocean. But in many respects, the huge volcanoes on Mars do resemble the Hawaiian volcanoes. Perhaps there are hot spots inside Mars, preferentially located in the northern hemisphere, that supply the volcanoes with molten lava.

There are two major areas of volcanic activity on Mars: the Tharsis region (which includes Olympus Mons, Ascraeus Mons, Pavonis Mons, Arsia Mons, and Ceraunius Tholus) and the Elysium region (which includes Elysium Mons, Hecates Tholus, and Albor Tholus). One obvious way in which these volcanoes differ from the Hawaiian volcanoes is their size. This also can be understood from the absence of plate tectonics on Mars. The Hawaiian volcanoes are very short lived. They grow only as long as they are located over the hot spot in the mantle but become extinct as soon as the motion of the Pacific plate carries them away. On Mars, however, a volcano remains stationary over the same hot spot for hundreds of millions of years. This continuous supply of lava along with the planet's low surface gravity results in the Martian volcanoes building up to enormous heights.

In principle, it should be possible to infer the ages of the Martian volcanoes from the density of impact craters in the surrounding areas. All scientists agree that there must have been episodes of impact cratering by planetesimals and meteoroids during the history of Mars. Exactly when such bombardment occurred is, however, a matter of debate. Laurence A. Soderblom of the U.S. Geological Survey argues that the bombardment history of Mars is similar to that of the moon. We know the time span of crater formation on the lunar surface from isotopic age-dating of moon rocks brought back by the Apollo astronauts. Assuming that the rates of crater formation on Mars and the moon are similar, it is possible to deduce the ages of the Martian volcanoes. The results of such dating of the most prominent Martian volcanoes are given in Table 8-5.

As expected, none of the Mariner missions shed any light on one of the most intriguing questions about Mars: Is there life on the planet's surface? The answer could only come from a "soft-landing" followed by an analysis of the Martian soil. Two such missions, Viking 1 and Viking 2, were planned for the summer of 1976. These two missions—essentially Martian versions of the lunar Surveyors ten years

**Figure 8-37.** *The Elysium Region.* The Elysium region is the second largest area of volcanic activity on Mars. It is also located on a broad bulge in the Martian crust. This bulge is smaller than the Syria Rise and measures 2,000 kilometers (1,200 miles) in diameter with a height of only 2 to 3 kilometers (roughly 2 miles). (*U.S. Geological Survey*)

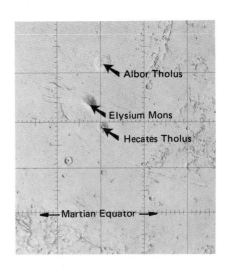

**Table 8-5.** *Sizes and Ages of the Largest Martian Volcanoes*

| Name | Height kilometers | Height miles | Diameter kilometers | Diameter miles | Age (millions of years) |
|------|------------------|--------------|---------------------|----------------|-------------------------|
| Olympus Mons | 26 | 16 | 550 | 340 | 200 |
| Ascraeus Mons | 19 | 12 | 400 | 250 | 400 |
| Pavonis Mons | 19 | 12 | 400 | 250 | 400 |
| Arsia Mons | 19 | 12 | 400 | 250 | 800 |
| Ceraunius Tholus | 3 | 2 | 150 | 90 | 800 |
| Elysium Mons | 15 | 9 | 250 | 150 | 1,500 |
| Hecates Tholus | 7 | 4 | 200 | 120 | 1,500 |
| Albor Tholus | (Unknown) | | 150 | 90 | 1,500 |

earlier—would be the highpoint of space exploration of the mid-1970s. Estimates for finding life ranged from the extremely pessimistic (one chance in a million according to the author of this text) to a highly optimistic 50–50 by Dr. Gerald A. Soffen, chairman of the Viking Project Office at the Langley Research Center. The fact of the matter was: that scientists simply did not know what to expect. In any case, it was clear that possible life forms on Mars must be able to survive under extremely hostile conditions. The planet is very dry and cold; its atmosphere is thin and composed almost entirely of carbon dioxide. In addition, without an ozone ($O_3$) layer or Van Allen belts to shield the planet, deadly ultraviolet radiation and particles from the solar wind impinge on the Martian surface. It is extremely difficult to find life forms on Earth that could survive in such a hostile environment.

## Questions and Exercises

1. What is the minimum distance between Earth and Mars?
2. What is meant by a favorable opposition of Mars?
3. Compare the rotation periods of Earth and Mars.
4. Why do seasons occur on Mars?
5. When were the first global maps of Mars drawn?
6. Who was Giovanni Schiaparelli?
7. Who first used the term "canali" to describe linear features apparently see on the Martian surface?
8. When were the satellites of Mars discovered? By whom?
9. Briefly describe the controversy about Martian canals.
10. If you wanted to photograph surface features on Mars, would you use a violet filter or an infrared filter at your telescope? Why?
11. Briefly discuss the implications of the absence of spectral lines of oxygen and water in spectra of Mars with regard to the possibility of life on the planet.
12. Name one interesting result of the flight of Mariner 4 past Mars.
13. Compare the chemical composition of the atmospheres of Earth and Mars.
14. Compare the atmospheric pressure at the surfaces of Mars and Earth.
15. Briefly compare and contrast the Mariner 4 spacecraft with the Mariner 6 and 7 spacecrafts.
16. Briefly describe the three major types of Martian terrain identified from photographs taken by Mariners 6 and 7.
17. Briefly compare and contrast lunar craters and Martian craters.
18. Describe the temperature ranges on Mars.
19. How do the Martian polar caps differ from the polar caps on Earth?
20. In what important way did the Mariner 9 mission differ from the earlier Mariner flights to Mars?

21. List and briefly describe the important discoveries about the Martian surface made from Mariner 9 that have been missed in the earlier Mariner flights.
22. Briefly compare and contrast Olympus Mons with Mount Everest and Mauna Loa.
23. What is a caldera?
24. Briefly describe Valles Marineris.
25. Compare and contrast Valles Marineris on Mars with the Grand Canyon in the Western United States.
26. Why are scientists puzzled by features like Vallis Mangala and Vallis Maadim?
27. What evidence suggests the existence of subsurface water ice or permafrost on Mars?
28. Describe the recession of the Martian polar caps. Why does this phenomenon suggest the existence of water ice near the poles?
29. Briefly describe the size and appearance of the two Martian moons.
30. Briefly contrast the northern and southern hemispheres of Mars with regard to the placement of volcanoes and craters.
31. Contrast and compare the atmospheric composition, atmospheric pressure, and surface temperature of Mars and Earth.
32. What can be inferred about the interior of Mars from the absence of a magnetic field about the planet.
33. Cite evidence suggesting that the crust of Mars is not divided up into moving plates.
34. Contrast and compare the Hawaiian–Emperor chain of volcanoes on Earth with the volcanoes on Mars.
35. In what way can the ages of Martian volcanoes be deduced?

# 9 Landing on the Red Planet

## 9.1 *The Viking Missions*

IN THE summer of 1975, the United States embarked upon the most ambitious and rewarding unmanned missions in the history of space flight. On August 20, a Titan-Centaur rocket lifted off from Cape Canaveral and successfully inserted Viking 1 into an orbit that would ultimately intersect with Mars. Three weeks later, on September 9, a repeat performance sent Viking 2 toward the red planet. Both spacecraft, identical in virtually every respect, would spend nearly a full year coasting across almost 290 million kilometers (180 million miles) of interplanetary space. Upon their arrival in the summer of 1976, they would soft-land on the planet and begin one of the most exciting and far-reaching quests in the history of science: the search for extraterrestrial life.

Each Viking spacecraft consisted of two parts: an *orbiter* and a *lander*. The orbiters were similar to Mariner 9 while the landers resembled the Surveyor spacecrafts that soft-landed on the moon in the 1960s. As shown in Figures 9–1 and 9–2, the lander and orbiter of each of the Viking spacecrafts were securely attached to each other for the duration of the flight to Mars. After arriving at Mars, the orbiters' rockets were turned on to slow the spacecrafts so that they would go into orbit about the planet. Viking 1 went into orbit about Mars on June 19, 1976, almost exactly ten months after it left Earth. Viking 2 took eleven months to get to Mars and was inserted into Martian orbit on August 7, 1976. Only after final landing site selection had been made, would the landers separate from the orbiters and begin the descent through the thin Martian atmosphere.

The initial purpose of the Viking orbiters was to guide the spacecrafts into orbit about Mars. In addition, the orbiters contained an array of scientific equipment to study the planet. Most importantly, each orbiter carried two television cameras, which surveyed the Martian surface for suitable landing sites. Each camera contained six color filters so that Mars could be photographed at various visible wavelengths. By superimposing views taken at different wavelengths, color pictures of the planet were obtained. Plate 9 shows a planet-wide color view of Mars taken by Viking 1 during its final approach on June 18, 1976. The large impact basin called Argyre along with many craters can be seen near the terminator. Just north

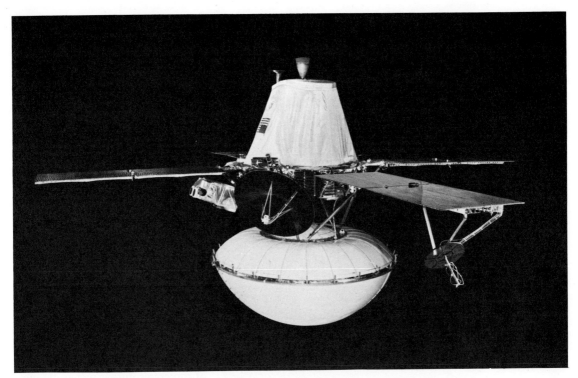

**Figure 9-1.** *The Viking Spacecraft.* Each Viking spacecraft consisted of an orbiter and a lander. During the flight to Mars, the lander remained stowed in a dome-shaped fiberglass case. The total weight of each Viking spacecraft (lander and orbiter with fuel) was nearly four tons. (*NASA*)

**Figure 9-2.** *Orbiter-Lander Configuration.* The Viking 1 and 2 orbiters were similar to Mariner 9, which was successfully orbited about Mars in 1971. After landing sites were selected, the landers separated from the orbiters and began the descent through the thin Martian atmosphere. With fuel, each orbiter weighed 5,125 pounds.

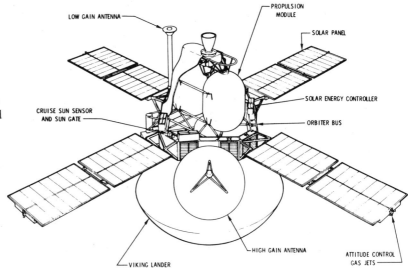

LOW GAIN ANTENNA

PROPULSION MODULE

SOLAR PANEL

CRUISE SUN SENSOR AND SUN GATE

SOLAR ENERGY CONTROLLER

ORBITER BUS

VIKING LANDER

HIGH GAIN ANTENNA

ATTITUDE CONTROL GAS JETS

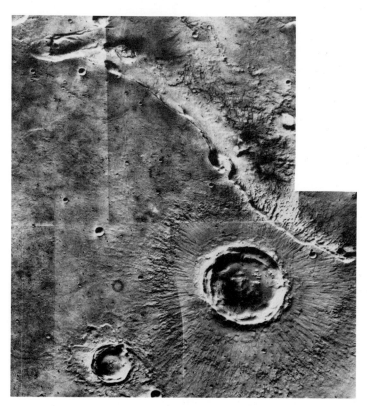

**Figure 9-3.** *A Young Crater.*
While searching for a potential landing site, Viking 1 sent back this photograph of a very young crater. The crater is about 30 kilometers (18 miles) in diameter and is surrounded by fresh ejecta blown out of the crater during meteoritic impact. ( *JPL; NASA* )

of Argyre is a portion of Valles Marineris, the "Grand Canyon" of Mars. An extraordinarily fine color view of Valles Marineris is shown in Plate 10. This picture is a mosaic of several color views taken in September of 1976 by Viking orbiter 1 and covers an area twice the size of Alaska.

As soon as Viking 1 arrived at Mars, it began searching for a potential landing site. In doing so, the cameras on the orbiter sent back some remarkably detailed pictures. For example, views of impact craters are seen in Figures 9–3, 9–4, and 9–5. The crater shown in Figure 9–3 exhibits fresh ejecta patterns, revealing that the crater was only recently formed. However, the ejecta patterns surrounding the craters called Yuty and Arandas (Figures 9–4 and 9–5) seemed to suggest the flow of material in a semi-liquid state. This is especially true in the case of Arandas where the pattern of ejected material is very reminiscent of the flow around experimental craters made by scientists in waterlogged ground here on Earth. Perhaps, Yuty and Arandas were formed in locations containing substantial quantities of subsurface ice. The impact would have melted the ice, thereby giving rise to the observed flow patterns.

As photographic reconnaisance of the Martian surface continued from orbit, evidence began accumulating that substantial water erosion has occurred on Mars.

Specifically, the search for a suitable landing site for Viking lander 1 soon centered about Chryse Planitia (the "Golden Planes"), which is a relatively flat area in the northern hemisphere. The elevation at Chryse is about 2½ kilometers (1½ miles) below the average level of the Martian surface. It was immediately obvious that this huge basin, which measures nearly 1,600 kilometers (1,000 miles) across, was the site of several extensive fluvial episodes. Excellent examples of the resulting water erosion are shown in Figures 9-6, 9-7, and 9-8. Dried riverbeds, flow patterns around up-raised crater rims, and even a natural dam are clearly seen.

The erosion patterns seen at Chryse and elsewhere on Mars seem to have all the tell-tail signs of flash floods. In other words, at sporadic intervals large quantities of water briefly flowed across the Martian surface. These brief episodes of wetness must have been very short-lived and restricted to certain locations, according to Dr. Harold Masursky of the U.S. Geological Survey. Mars never had substantial

**Figure 9-4 (left).** *The Crater Yuty.* Viking 1 photographed this crater, called Yuty, in June 1976, at a range of 1,877 kilometers (1,165 miles). Yuty is 18 kilometers (11 miles) in diameter and has a prominent central peak. The "flows" surrounding the crater consist of layers of broken rock ejected from the crater following impact. The leading edges of the debris flows are similar to ridges formed by huge avalanches here on Earth. (*JPL; NASA*)

**Figure 9-5 (right).** *The Crater Arandas.* After the Viking lander 1 had safely descended to the Martian surface, Viking orbiter 1 started searching for a potential landing site for Viking 2. In doing so, Viking orbiter 1 sent back this view of a remarkable crater called Arandas. Arandas is about 25 kilometers (15 miles) in diameter. Surprisingly, the ejecta seem to have flowed along the surface rather than having been blasted out of the crater. (*JPL; NASA*)

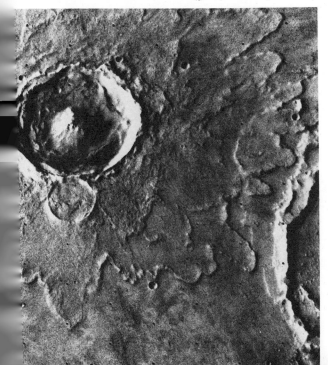

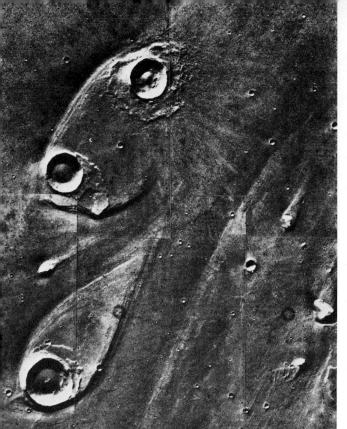

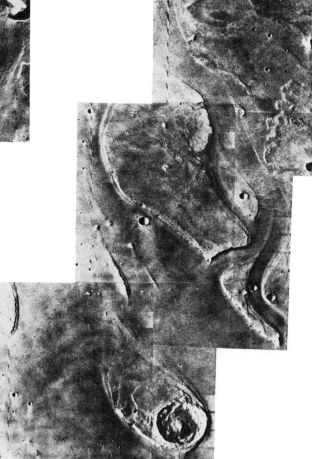

Figure 9-6. *Water Erosion in Eastern Chryse.* Erosion features seen here strongly suggest that portions of the Martian surface were subjected to sudden flooding by large quantities of water. This photograph was taken by Viking 1 at a range of 1,600 kilometers (990 miles). (*JPL; NASA*)

Figure 9-7. *Channels in Northeastern Chryse.* Meandering, intertwining channels are vividly displayed in this view from Viking 1. It is believed that these channels were cut by running water on the Martian surface. (*JPL; NASA*)

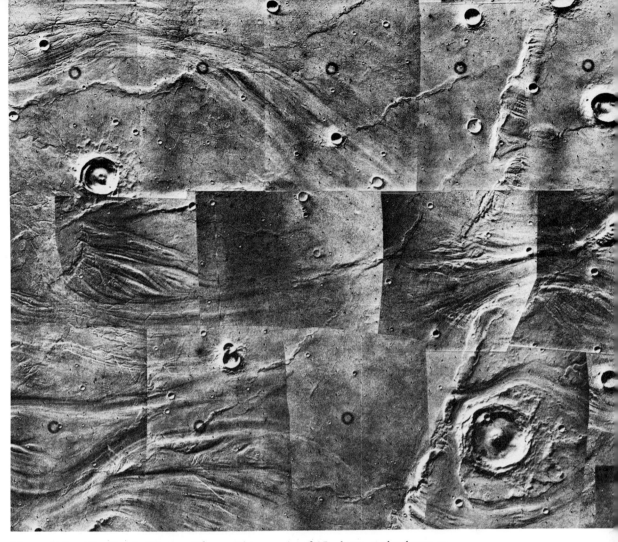

**Figure 9–8.** *Flooding in Western Chryse.* This mosaic of 15 photographs shows an area measuring 250 by 200 kilometers (155 by 120 miles). Note the ridge near the right side of the photograph. It seems reasonable to suppose that a large quantity of water (flowing from left to right in this view) piled up against the ridge that formed a natural dam. Eventually, the resulting lake became so large that the dam broke. Breaks in the natural dam are clearly seen. (*JPL; NASA*)

quantities of water on a global scale like the lakes and oceans here on Earth. Instead, it seems entirely reasonable to suppose that large quantities of ice were trapped beneath the Martian surface. Then, during periods of volcanic activity, the ice deposits were suddenly melted, thereby producing a flash flood.

In light of this idea about subsurface ice reservoirs, we can now understand why the "mouth" of Valles Marineris (the Martian "Grand Canyon") is located very near the Tharsis Plateau (see Figure 8–36 for a map of this region). The Tharsis Plateau contains four of the largest and youngest volcanoes on Mars. The same volcanic activity that produced Olympus Mons and its neighbors also would have melted nearby reservoirs of ice trapped beneath the Martian surface. As the ice turned to water, the overlaying rocks and soil would have collapsed into the subsurface pools, thereby forcing the water up to the surface. The resulting valley floor, caused by the collapse or downfaulting of the Martian surface, could then take on the appearance of the "chaotic terrain" that had mystified scientists after the flights of Mariners 6 and 7.

An excellent example of this scenario proposed by Masursky is shown in Figure 9–9. This mosaic (oriented with south toward the top) covers 90,000 square

**Figure 9–9.** *A Rille in the Capri Plateau.* The melting of subsurface ice and the ensuing collapse or downfaulting of the Martian surface easily explain the features seen in this mosaic taken by Viking 1 on July 3, 1976 from a range of 2,300 kilometers (1,400 miles). Water liberated by these geologic processes then flowed eastward (to the left) across the nearby plains. The area shown in this photograph measures 300 by 300 kilometers (180 by 180 miles). (*JPL; NASA*)

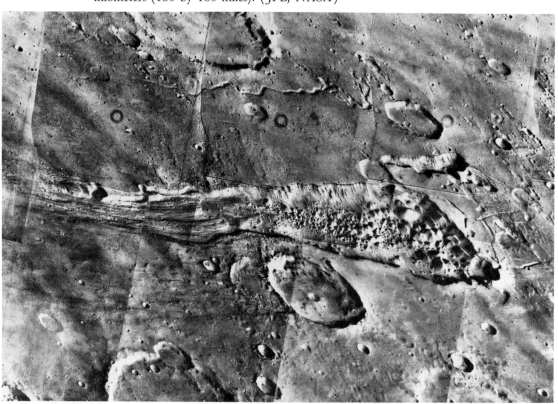

kilometers (32,000 square miles) in the Capri plateau near the Martian equator at the eastern end of Valles Marineris. Note the obvious downfaulting of the Martian surface at the right (western) side of the picture. Hummochs on the floor of the valley look like chaotic terrain while the walls of the valley show clear evidence of lanslides. Also note the water erosion on the left (eastern) side of the photograph, strongly suggesting that the collapse of the Martian surface in this region liberated a large quantity of water, which then flooded the nearby planes.

An even more impressive photograph of down-faulting and landslides (perhaps caused by the melting of subsurface ice) is shown in Figure 9–10. This mosaic shows a small portion of the southern wall of Valles Marineris. Perhaps this entire region was once the site of a vast underground ice reservoir, which was melted by volcanic activity in the nearby Tharsis Plateau. Processes of this type may be (in part, at least) the solution to the "mystery of the missing water" that had baffled scientists since the flight of Marine 9 in 1971.

In addition to the television cameras, each Viking orbiter also carried an infrared thermal mapper and a water vapor mapper. Both devices operate at infrared wavelengths. The thermal mapper focuses reflected light onto antimony-bismuth detectors that measure the total amount of incoming infrared light. The amount of reflected infrared light in various wavelength ranges is directly correlated with the

**Figure 9–10.** *A Huge Landslide in Valles Marineris.* Viking 1 took this remarkable picutre of the south wall of Valles Marineris on July 3, 1976, from a range of 2,000 kilometers (1,200 miles). The wall of the canyon obviously collapsed resulting in a huge landslide that flowed down and across the canyon floor. The area shown here measures 70 by 150 kilometers (43 by 94 miles) and therefore encompasses only a tiny fraction of the entire canyon, which extends for 4,000 kilometers (2,500 miles) along the Martian equator. (*JPL; NASA*)

temperature of the planet's surface. The water vapor mapper is an infrared spectrometer designed to detect spectral lines caused by water molecules. The "strengths" of these spectral lines are related to the amount of water vapor in the Martian atmosphere. As the Viking 1 and Viking 2 missions progressed, both of these instruments began revealing evidence for surprisingly large quantities of water vapor and water ice in the Martian polar regions.

As emphasized in the previous chapter, Mars is very dry by terrestrial standards. There is so little water on Mars that, by Martian standards, Death Valley in California looks like a Louisiana swamp. But there was a time in the recent past that many scientists believed Mars was totally dry. Consequently, finding any appreciable amount of water would constitute a major discovery. Of course, this water must be in the form of either vapor or ice. The atmospheric pressure on Mars is so low that liquid water could not exist for very long on the planet's surface.

As mentioned in the previous chapter, scientists carefully watched the recession of the northern polar cap of Mars during 1972 with the aid of Mariner 9. As spring came to the northern hemisphere, the polar cap initially shrank rapidly in size in agreement with the expected evaporation of dry ($CO_2$) ice. But as summer approached, the shrinkage stopped, suggesting that there might be a permanent polar cap made of ordinary ($H_2O$) water ice. Thermal mapping from the Viking orbiters clearly support this idea. When the Vikings arrived at Mars, it was summer in the northern hemisphere and winter in the southern hemisphere. Temperatures at the winter polar cap are very low, about $150°K$ ($-120°C = -190°F$) or colder. At such brutally cold temperatures, carbon dioxide in the Martian atmosphere freezes and covers the ground like snow. But in September of 1976, Viking Orbiter 2 measured the surface temperature around the northern polar cap that was experiencing the Martian summer. Dark regions in the polar cap had a temperature of $235°K$ ($-38°C = -36°F$) while the white regions had a temperature of $205°K$ ($-68°C = -90°F$). This is indeed cold but not nearly cold enough to support frozen carbon dioxide. Dry ice would evaporate well below these temperatures. This conclusively means that the *residual polar caps* that survive through the summer months must be made *entirely of water ice*. It furthermore means that there can be no permanent trapped reservoirs of carbon dioxide on the planet, as some scientists had theorized. And finally, since the residual caps at both poles are now known to be made entirely of water ice, the amount of water trapped at the poles could be 1,000 to 100,000 times greater than the total amount of water currently in the Martian atmosphere.

Confirmation of the large water-ice content of the northern polar cap came from the water vapor mappers, which detected surprisingly high quantities of water vapor in the atmosphere around the pole. Although very dry by terrestrial standards, the Martian atmosphere was shown to be unexpectedly humid at northern latitudes. This means that the clouds photographed shortly after the arrival of Viking 1 must be made of water-ice crystals. An excellent example of these clouds is seen in Figure 9–11. This mosaic shows an oblique view of the summit of Olympus Mons, the largest of the Martian volcanoes. As the prevailing Martian winds blow moisture up the slopes of the volcano, the water vapor condenses into

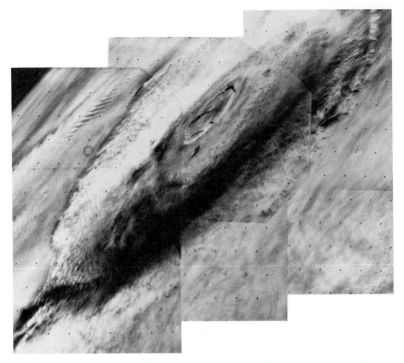

**Figure 9-11.** *Clouds Around Olympus Mons.* The largest of the Martian volcanoes was photographed by Viking orbiter 1 on July 31, 1976, from a range of 8,000 kilometers (5,000 miles). The summit of the volcano consists of a caldera 80 kilometers (50 miles) across and rises into the stratosphere 24 kilometers (15 miles) above the surrounding plains. Clouds form around the volcano as moisture in the atmosphere cools and condenses while moving up the slopes. (*JPL; NASA*)

ice crystals. The resulting clouds, which start forming in the mid-morning, are mostly concentrated on the western side of the volcano. In the Martian afternoon, the cloud-cover has developed sufficiently that it can be seen from Earth. This phenomenon of clouds around Olympus Mons is seasonal, largely limited to spring and summer in the northern hemisphere.

The water vapor mappers on the Viking orbiters also showed higher concentrations of humidity at the lowest elevations. By contrast, the higher elevations remain extremely dry. This indicates that the "ground fog" observed in low-lying areas just after sunrise on Mars must be water vapor. Examples of this early morning fog are clearly seen in Figure 9-12. The two photographs were taken one half hour apart shortly after Martian dawn. The earlier photograph (upper view) shows bright patches of fog which had disappeared only 30 minutes later, as shown in the lower view. These fog patches are the first visible evidence that water

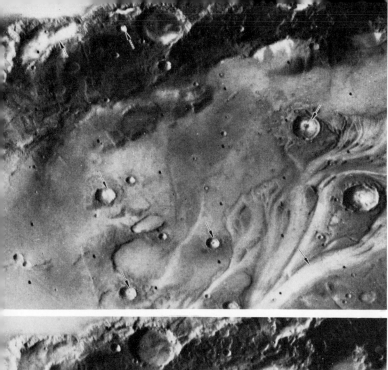

**Figure 9-12.** *Morning Fog on Mars.*
These two pictures were taken one
half hour apart shortly after Martian
dawn. In the earlier (upper) photograph, bright patches of fog are seen
hovering close to the ground inside
craters. Half an hour later, the fog
patches have disappeared as shown in
the lower photograph. (*JPL; NASA*)

exchange occurs between the Martian surface and atmosphere. This was a tremen-
dously exciting discovery for scientists at the Jet Propulsion Laboratory because the
ultimate goal of the Viking missions was to search for extraterrestrial life. All life
here on Earth requires water. If Mars had turned out to be as arid as some scientists
had expected, then any form of life (as we know it) would be impossible. To the
contrary, there is *some* water vapor in the Martian atmosphere. Hopes were
therefore high as the Viking landers descended to the Martian surface.

**336**

**Plate 1.** *Earth.* Most of Africa and portions of Europe and Asia can be seen in this spectacular photograph taken by the Apollo 11 astronauts on their way to the moon. (*NASA*)

**Plate 2.** *Mars.* This photograph of Mars was taken during the Spring of 1967. The North Polar Cap and "greenish" markings can be seen. (*NASA; Lunar and Planetary Observatory*)

**Plate 3.** *Jupiter.* This photograph of the largest planet in the solar system was taken in January 1967. (*NASA; Lunar and Planetary Observatory*)

**Plate 4.** *Saturn.* This photograph of Saturn was taken in October 1968. Faint surface markings as well as a major "gap" in the rings can be seen. (*NASA; Lunar and Planetary Laboratory*)

**Plate 5.** *San Francisco (false color composite).* This composite was constructed from four black-and-white photographs in visible and infrared light. Healthy vegetation appears red. (*NASA; EROS*)

**Plate 6.** *San Francisco (true color).* This view taken by the Skylab astronauts shows the true appearance of San Francisco in visible light. (*NASA*)

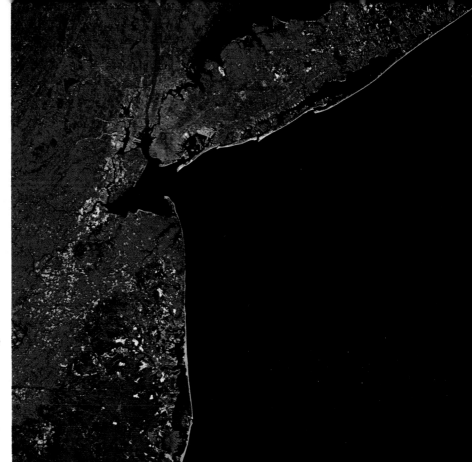

**Plate 7.  *New York City (false color composite).*** This false color view from Landsat 1 covers an area 185 kilometers (115 miles) on each side. Urban areas appear bluish-gray. (*NASA; EROS*)

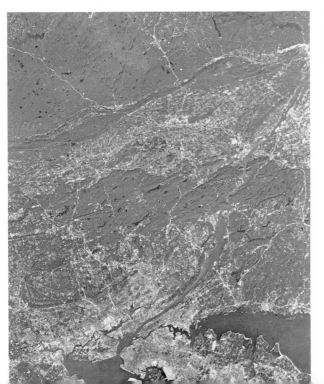

**Plate 8.  *New York City (true color).*** This view of New York City and northern New Jersey was taken by the Skylab astronauts. The area covered is 144 kilometers (90 miles) on each side. (*NASA*)

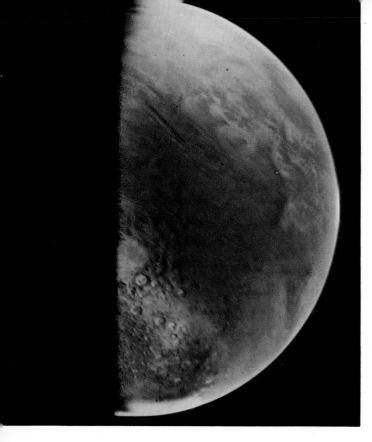

**Plate 9.** *Mars.* Viking 1 took this photograph of Mars on June 18, 1976, one day before the spacecraft went into orbit about the planet. Just below the center of the picture and near the morning terminator is the large impact basin called Argyre. North of Argyre is a portion of Valles Marineris, the "Grand Canyon" of Mars. (*NASA*)

**Plate 10.** *Valles Marineris.* This vast canyon near the Martian equator is 5,000 kilometers (3,000 miles) long. The area covered in this view taken by Viking orbiter 1 measures 1,800 by 2,000 kilometers (1,100 by 1,200 miles). This color mosaic was constructed from 15 black-and-white photographs taken through three color filters (violet, green, and red) in the orbiter's camera. (*NASA*)

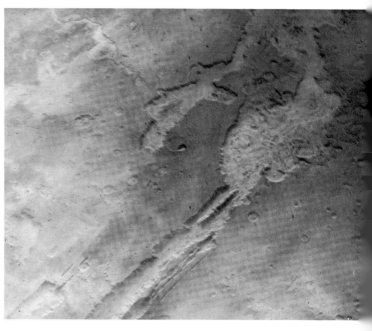

**Plate 11.** *Viking 1 Scene.* This color picture of the Martian surface was taken on July 21, 1976, one day after Viking 1 successfully landed on the planet. The reddish color of the rocks and regolith is perhaps caused by a form of rust called limonite (hydrated ferric oxide). The reddish color of the sky is caused by tiny dust particles suspended in the Martian atmosphere. (*NASA*)

**Plate 12.** *Viking 1 Panorama.* Trenches dug by the Viking's sampler scoop can be seen to the left of this panorama of the Viking 1 site. The dark, coarse-grained rock to the right of the trenches is about 25 centimeters (10 inches) across. Large boulders measuring one to two meters (three to six feet) in size can be seen near the horizon. Viking's sampler scoop is in the "parked position" at the center of the photograph. (*NASA*)

**Plate 13.** *Sand Dunes at the Viking 1 Site.* A series of beautiful rolling sand dunes are located northeast of Viking Lander 1. The largest boulder in this view (called "Big Joe") is only 8 meters (25 feet) away from the spacecraft. Big Joe is 1 meter high and 3 meters wide (3 feet high and 10 feet wide). Note that the upper surface of this large, dark boulder is covered with reddish Martian soil. (*NASA*)

**Plate 14.** *Chryse at Sunset.* This photograph of the Viking Lander 1 site was taken about 15 minutes before sunset on August 21, 1976. The low elevation of the sun accentuates the circular depression just above one of the Lander's legs. Beyond the depression are several rocks measuring 30 centimeters (1 foot) across. Near the horizon, several bright patches of bedrock are seen. (*NASA*)

**Plate 15.** *Viking 2 Scene.* Viking 2 successfully landed on Mars on September 3, 1976, and sent back this view, looking northward across the Utopia plain. Rocks in the foreground are noticeably vesicular, clearly indicating their volcanic origin. As the result of an 8° tilt of the spacecraft (one leg must be resting on a rock), the distant horizon appears to be tilted. (*NASA*)

**Plate 16.** *Viking 2 Panorama.* This mosaic of three color views taken in early September 1976 shows almost 200° of the Martian surface at the Viking Lander 2 site. A computer was used to remove the effects of the 8° tilt of the spacecraft. The flat Martian horizon is about 3 kilometers (1.8 miles) from the spacecraft. (*NASA*)

**Plate 17. *Jupiter.*** This photograph taken with the 200-inch Palomar telescope is an excellent example of the best Earth-based view of this giant planet. The Great Red Spot is shown faintly on the upper left-hand portion of the planet. (*Copyright by the California Institute of Technology and Carnegie Institute of Washington. Reproduced by permission of the Hale Observatories*)

**Plate 18. *Jupiter from Pioneer 10.*** This was the view of Jupiter from Pioneer 10 when it was 1,550,000 miles from the planet. The shadow of Io, one of Jupiter's four largest satellites, is seen on the planet's surface. (*NASA*)

**Plate 19.** *Jupiter from Pioneer 11.* Jupiter's Great Red Spot appears prominently in this photograph taken by Pioneer 11 when the spacecraft was 660,000 miles from the planet. (*NASA*)

**Plate 20.** *The North Pole of Jupiter.* On December 12, 1974, Pioneer 11 passed over the north pole of Jupiter. This photograph was taken when the spacecraft was 750,000 miles from the huge planet. (*NASA*)

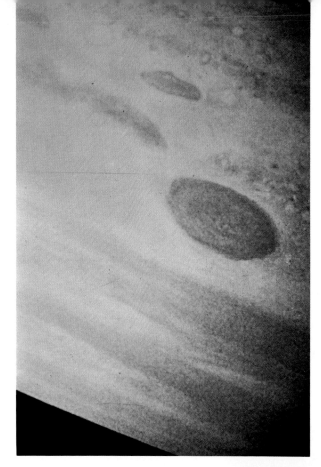

**Plate 21.** *The Great Red Spot.* This photograph of Jupiter's Great Red Spot was taken by Pioneer 11 at a distance of 545,000 kilometers (338,000 miles). This remarkable feature on Jupiter has been seen for hundreds of years by Earth-based astronomers, although never with the clarity shown here. (*NASA*)

**Plate 22.** *The Little Red Spot.* The photograph of Jupiter's "Little Red Spot" was taken by Pioneer 10 at a distance of 1,153,000 kilometers (715,000 miles). This temporary feature on Jupiter was observed for only several months in the early 1970s. (*NASA*)

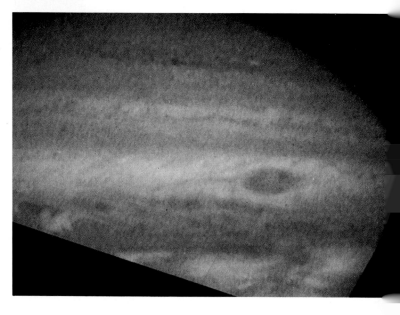

THE central idea behind the Viking missions was that after suitable landing sites had been selected, the landers would separate from the orbiters and descend to the Martian surface. Prior to this separation, each lander was stowed in a fiberglass "bioshield," which measures nearly 3.8 meters (12 feet) in diameter and 1.9 meters (6⅓ feet) from top to bottom. During separation, the base of the bioshield is jettisoned, thereby exposing the aluminum "aeroshell" that encases the lander. The base of the aeroshell consists of a heat shield along with small rockets and scientific equipment that measures properties of the upper Martian atmosphere during descent. The top of the aeroshell contains a parachute that would be deployed during the final stages of the descent. The arrangement of the bioshield, aeroshell, and lander is shown in Figure 9–13.

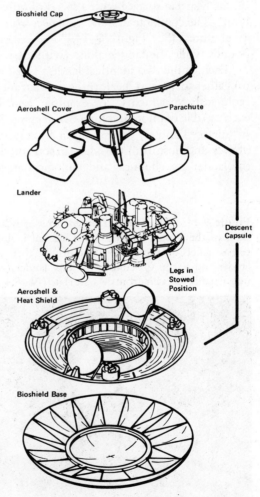

**Figure 9–13.** *The Descent Capsule.* Prior to separation from the orbiter, the descent capsule is stowed in a fiberglass bioshield. During descent, the lander remains encased in the aeroshell, which contains rockets, scientific instruments, and a parachute. During the final stages of the descent, the aeroshield is jettisoned, thereby exposing the lander. (*Martin Marietta Aerospace*)

The aeroshell contains eight "deorbit" rocket engines, which are ignited immediately after separation. These engines slow the descent capsule during the initial stages of entry so that it can be pulled by gravity toward the planet's surface. During entry, the heat shield protects the lander from the heat that develops from friction with the upper Martian atmosphere. During this period, instruments such as a spectrometer as well as pressure and temperature sensors measure conditions in the upper atmosphere.

At 5.8 kilometers (19,000 feet) above the Martian surface, the descent capsule has been slowed to a speed of about 600 miles per hour from its initial velocity of over 10,000 miles per hour. At this elevation, the heat shield is jettisoned and the parachute is deployed. The parachute continues to slow the lander's descent to 140 miles per hour and at an elevation of 1.4 kilometers (4,600 feet) the parachute is jettisoned and a series of eighteen "lander engines" at the base of the lander are ignited. These small rockets further slow the descent so that the lander settles down on the Martian surface with a speed of only 8 feet per second (about $5\frac{1}{2}$ miles per hour). The complete descent sequence, from separation to touch-down, takes about 10 minutes. (See Figure 9–14.) The entire descent capsule weighs about 2,400 pounds while the lander alone (without rocket fuel) weighs only 1,260 pounds.

Each of the two identical landers is packed with scientific equipment designed to gather data and perform a wide range of geological, meteorological, and biological experiments. For example, each lander is equipped with two cameras, each capable of photographing overlapping 342° panoramas of the Martian surface. A "surface sampler," consisting of a long mechanical arm and a scoop, collects bits of rock and dust from the Martian soil and delivers the specimens to instruments in the lander for chemical and biological analyses and experiments. A seismometer

**Figure 9–14.** *The Viking Descent.* This artist's rendition shows some of the basic features of the Viking landing. Final descent is slowed by means of a parachute and retro-rockets on the lander. The orbiter then serves as a "relay station" between the lander and Earth. Of course, the tiny lander (roughly the size of an automobile) cannot actually be seen from Mars orbit. (*NASA*)

listens for Marsquakes while sensitive apparatus on the "meteorology boom" measures temperature, pressure, and wind velocity in the Martian atmosphere. A photograph of the lander is shown in Figure 9–15, and important components are identified in Figure 9–16.

By mid-July of 1976, a landing site had been selected for Viking 1 in Chryse. Great care was taken in selecting this site because if a lander attempted to touch down in a boulder-strewn region, the spacecraft could have easily been toppled or punctured. Landing on a boulder the size of a household garbage can would have resulted in failure of the mission.

Viking 1 safely landed on Mars on July 20, 1976, almost exactly one month after the spacecraft went into orbit about the planet.

By mid-August, the search for a suitable landing site for Viking 2 centered about a region called Utopia Planitia. Both Viking landing sites were chosen in the northern hemisphere of Mars because one of the primary goals of the mission was to search for life. Based on water vapor mapping as well as observations of erosion features and morning ground fog, scientists believed that there might be slightly more water in the atmosphere north of the Martian equator. Since life, as we know it, depends critically on water, the decision was made to land in the northern hemisphere.

The landing site for Viking 1 in Chryse, northeast of Valles Marineris is at a latitude 22.4° north of the Martian equator. Utopia, like Chryse, is a large basin which appeared especially attractive because the amount of water in the atmosphere was found to be at least three times greater than at the Viking 1 site. Utopia is located nearer to the Martian north pole than Chryse but on the opposite side of the planet. While the Viking 1 site at Chryse is not far from the Tharsis plateau, the most active volcanic region on Mars, Utopia is adjacent to Elysium Planitia, the second largest volcanic region on the planet. The final selection of a site for Viking 2 in Utopia placed the spacecraft 47.9° north of the Martian equator, 7,500 kilometers (4,600 miles) from the Viking 1 site.

In spite of a few hair-raising moments when communications with the spacecraft were temporarily lost, Viking 2 safely landed on Mars on September 3, 1976, 45 days after the landing of Viking 1.

One of the first tasks initiated immediately upon arrival at the planet's surface was to turn on the cameras. Figures 9–17 and 9–18 show the first pictures sent back from Vikings 1 and 2, respectively. It was immediately obvious that the two sites were different. Rocks at the Viking 1 site had flat, angular facets but rocks at the Viking 2 site were generally more rounded and remarkably vesicular. These vesicles or holes indicate that the rocks had a volcanic origin. As the molten Martian lava cooled, bubbles of gas escaping from the seething rock left numerous porous holes.

After a few initial pictures had been taken, the cameras on each of the landers were slowly slewed around the horizon to produce panoramas. The views were breathtaking: rocks and boulders strewn over the Martian landscape as far as the eye could see. The first panoramas from the Viking 1 and Viking 2 sites are shown in Figures 9–19 and 9–20, respectively.

**Figure 9-15 (left).** *The Viking Lander.* Each lander weighs about 1,300 pounds and is roughly the size of a small automobile. The spacecraft rests on the Martian surface by means of three "legs." The "surface sampler" is shown extended in front of the spacecraft while the "meteorology boom' is seen to the upper right of the lander. A parabolic "high-gain antenna" at the top of the spacecraft sends data to the orbiter, which relays the information back to Earth. (*NASA*)

**Figure 9-16.** *Details of the Lander.* Basic components of the Viking lander are identified in this diagram (compare with Figure 9–15). The scientific equipment on board the lander was designed to take pictures, collect data, and perform a wide range of experiments on the Martian surface. (*Martin Marietta Aerospace*)

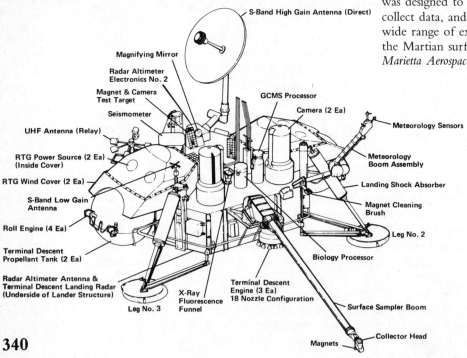

S-Band High Gain Antenna (Direct)

Magnifying Mirror

Radar Altimeter Electronics No. 2

Magnet & Camera Test Target

Seismometer

GCMS Processor

Camera (2 Ea)

UHF Antenna (Relay)

Meteorology Sensors

RTG Power Source (2 Ea) (Inside Cover)

Meteorology Boom Assembly

RTG Wind Cover (2 Ea)

Landing Shock Absorber

S-Band Low Gain Antenna

Magnet Cleaning Brush

Roll Engine (4 Ea)

Leg No. 2

Terminal Descent Propellant Tank (2 Ea)

Biology Processor

Radar Altimeter Antenna & Terminal Descent Landing Radar (Underside of Lander Structure)

X-Ray Fluorescence Funnel

Terminal Descent Engine (3 Ea) 18 Nozzle Configuration

Surface Sampler Boom

Leg No. 3

Collector Head

Magnets

**Figure 9–17.** *First Photograph from the Viking 1 Site.* After landing in Chryse on July 20, 1976, Viking lander 1 sent back this picture of the Martian surface. The center of the image is 1.4 meters (5 feet) from the camera. The large rock near the upper middle of this view is about 10 centimeters (4 inches) across and shows three rough facets. One of the lander's footpads is seen at the lower right. (*JPL; NASA*)

**Figure 9–18.** *First Photograph from the Viking 2 Site.* After landing in Utopia on September 3, 1976, Viking lander 2 sent back this picture of the Martian surface. The center of the image is 1.4 meters (5 feet) from the camera. Several small boulders measuring 10 to 20 centimeters (4 to 8 inches) are visible in the field of view. One of the lander's footpads is seen at the lower right. (*JPL; NASA*)

**Figure 9–19 (above).** *Viking 1 (Camera #2) Panorama.* This 300° panorama of the Chryse plains was taken by Viking lander 1 on July 20, 1976, the day of the successful landing. The view faces southward with east near the left and north at the right. The horizon is about 3 kilometers (2 miles) from the spacecraft. Large boulders are strewn near the uneven horizon. (*JPL; NASA*)

**Figure 9–20 (below).** *Viking 2 (Camera #2) Panorama.* This 330° panorama of the Utopia plains was taken by Viking lander 2 on September 3, 1976, the day of the successful landing. The view faces eastward and covers all but one tenth of the scene from the spacecraft. The horizon (computer corrected to remove the effects of an 8° tilt of the lander) is much flatter than at the Viking 1 site. (*JPL; NASA*)

The panorama from Viking 1, taken on the day of the landing, showed a barren, rock-strewn scene. The terrain represented a mixture of processes including impact crater debris, old volcanic material, wind erosion, and possibly the effects of water. Water mixing, which may have occurred long ago at the Viking 1 site, brought together fine materials as well as large rocks without any sorting.

While Viking lander 1 was essentially level (there was a slight 3° tilt of the spacecraft toward the west), Viking lander 2 was tilted by 8.2° to the west, as though one of the spacecraft's legs is resting on a rock. As a result, the Utopia horizon appeared lop-sided in the photographs. The panorama sent back from Viking 2 on the day of its successful landing (Figure 9–20) was therefore "com-

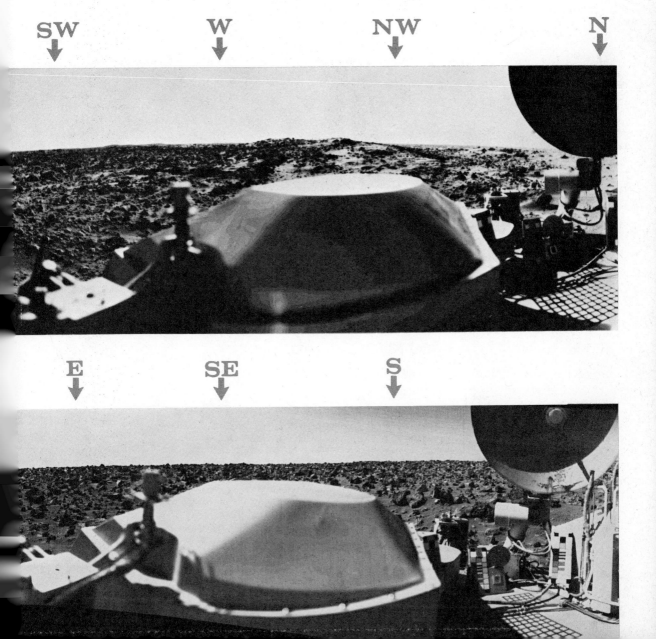

**Figure 9–21 (*above*).** ***Viking 1 (Camera #1) Panorama.*** During the fourth day on Mars, the other camera (Camera #1) on Viking lander 1 was turned on and photographed this 300° panorama. The view is northward with west near the left and south at the right. This view includes the 60° slice not visible from Camera #2. Drifts of sand and a large boulder (called "Big Joe") were revealed in this second panorama. (*JPL; NASA*)

**Figure 9–22 (*below*).** ***Early Morning at Chryse.*** This 100° panorama was taken two hours after sunrise at the Viking 1 site, on August 3, 1976. The view faces northward and is centered on the drifts of sand seen from the lander's Camera #1. Sharp crests on the drifts indicate that recent wind storms moved sand in the general direction from upper left to lower right in the photograph. (*JPL; NASA*)

puter corrected" to eliminate the effects of the spacecraft's tilt. As with the Viking 1 panorama, the scene was again barren and rock-strewn. More vesicular rocks were observed which were smooth and cut with grooves as if subjected to a sand-blasting process. None of the expected sand dunes were seen.

Both the Viking 1 and Viking 2 sites clearly resembled rocky deserts similar to those found in the southwestern United States. The important difference immediately apparent to Viking scientists is there are no large life forms at either site. Nothing resembling sage-brush or cactus could be seen at Chryse or Utopia. While this was not a surprising development, it confirmed earlier suspicions that Martian life forms, if they exist at all, must be microscopic.

**NE** **E** **SE** **S**

**E**

During the fourth day on Mars, the second camera on Viking lander 1 was activated and photographed a second panorama (see Figure 9–21). This second panorama covered the 60° slice not visible in earlier photographs of the Chryse plains. A series of beautiful sand dunes were revealed along with a huge boulder (christened "Big Joe") only 8 meters (25 feet) from the spacecraft.

This scene was so interesting that another panorama was taken at higher magnification and is shown in Figure 9–22. While the previous panorama from Camera #1 on Viking lander 1 was taken in the Martian afternoon, the second panorama was photographed two hours after sunrise at Chryse in order to accentuate details in the sand drifts. The drifts are remarkably similar to those found in California and Mexico. Sharp dune crests indicate that recent wind storms moved the sand over the dunes. Small deposits on the downwind side of the rocks confirm that the wind direction was generally from upper right to lower left in the photograph. The large boulder, "Big Joe," can be seen even more clearly.

In general, many of the rocks at the Viking lander 1 site are sharply faceted. Drifts of fine sand or silt–like material can be seen among the rocks at many locations. Some of this fine material covers the upper surfaces of the larger rocks

**Figure 9–23.   *Rocks at the Viking Lander 1 Site.*** Many surface rocks photographed by Viking lander 1 are sharply faceted. Fine-grained sandy material covers the ground between the rocks. Light-colored bedrock is seen near the top of the picture. (*JPL; NASA*)

**Figure 9-24.** *Rocks at the Viking Lander 2 Site.* Many surface rocks photographed by Viking lander 2 are rounded and very vasicular. As at the Viking lander 1 site, fine-grained sandy material covers the ground between the rocks. The object lying on the ground is the aluminum shroud, which was ejected from the collector head of the surface sampler. The shroud is 12 inches long and 4½ inches in diameter (*JPL; NASA*)

and boulders. For example, "Big Joe" has a covering of red Martian sand as seen in Plate 13. Plates 11, 12, and 14 show other color views around the Viking 1 site. Occasional light-colored outcroppings of bedrock are visible as seen in Figure 9–23.

In contrast to the Viking 1 site, rocks photographed by Viking 2 were generally more rounded and very vesicular, as shown in Figures 9–24, 9–25, and 9–26. Once again, fine-grained sandy material covered the ground between the rocks. But unlike those at the Viking 1 site, the larger boulders in the distance did not have coverings of reddish Martian soil. Excellent high-resolution views, such as that shown in Figure 9–26, allowed geologists to identify several types of igneous rocks as well as breccias. The color view shown in Plate 15 covers much of the area contained in Figure 9–26. Plate 16, a 200° mosaic of the Viking lander 2 site, reveals a larger portion of the Utopia plains in color.

In addition to visual examination of the Martian terrain, geologists were able to discover the chemical composition of the Martian surface by means of a "X ray-

**Figure 9-25.** *Vesicular Rocks in Utopia.* This view shows a few square meters (yards) near Viking lander 2. This location was later chosen as a site for collecting samples by the lander's trenching scoop. Notice that some of the areas on the ground are lighter than others. This suggests the presence of two kinds of fine-grained sandy material. (*JPL; NASA*)

**Figure 9-26.** *A Mosaic View of Utopia.* Several high-resolution photographs were combined to produce this scene looking northeast from Viking lander 2. Note that the bottom third is the same as Figure 9–25. The large rock at the center of the picture is 60 centimeters long and 30 centimeters high (2 feet long and 1 foot high). (*JPL; NASA*)

**Table 9–1.** *Chemical Composition of
Martian Soil*

| Element | Abundance (range in per cent) |
|---|---|
| Magnesium | 2.5 to 5.5 |
| Aluminum | 2.0 to 5.0 |
| Silicon | 18.5 to 24.0 |
| Sulfur | 2.5 to 5.0 |
| Cerium | 0.1 to 0.6 |
| Potassium | 0 to 0.8 |
| Calcium | 3.0 to 4.5 |
| Titanium | 0.3 to 0.5 |
| Iron | 12.5 to 15.0 |

fluorescence spectrometer" carried on each lander. Specifically, the scoop on the lander's mechanical arm was used to deposit a small sample of Martian soil into the spectrometer. By radiating the soil with X rays and observing how the X rays are reemitted from the atoms in the soil, geologists could ascertain an approximate chemical composition of the specimen. It was found that silicon and iron are among the most abundant elements. Table 9–1 lists the results of this chemical analysis.

Magnetic properties of the Martian soil were detected by means of small magnets attached to the lander's scoop. As the scoop dug trenches on the Martian surface, bits of rock and dust clung to these magnets indicating an abundance of 3 to 4 per cent of magnetic material.

All of these physical, chemical, and magnetic analyses suggest that common minerals on Mars include hematite and magnetite. Other less well-known iron-rich minerals (for example, maghemite purpurite and montmorillonite clay) may also be present. This high abundance of iron, which gives the soil its reddish color, suggests that Mars is *not* as strongly chemically differentiated as Earth. In the case of our planet, most of the iron sank to the core, leaving comparatively little in the surface rocks. On Mars, the iron is more uniformly spread out through the planet as evidenced by the high iron content of the surface samples.

While geologists were examining views of the Martian surface, meteorologists were collecting data sent back by meteorological instruments on board the Viking spacecrafts. Prior to the actual landings, instruments in the aeroshell analysed the upper Martian atmosphere. During descent, ionized atoms and molecules were detected starting at 350 kilometers (220 miles) above the planet's surface. While carbon dioxide is the most abundant gas in the lower atmosphere, oxygen and nitrogen are important constituents of the upper atmosphere. For example, at an altitude of 140 kilometers (90 miles) there is roughly six times more singly ionized oxygen molecules ($O_2^+$) than singly ionized carbon dioxide molecules ($CO_2^+$). In addition, nitrogen was detected for the first time. While 95 per cent of the lower atmosphere is composed of carbon dioxide, the percentage composition of carbon

**Table 9-2.** *Chemical Composition of the Martian
Atmosphere*

| Gas | Abundance (in per cent) | |
| --- | --- | --- |
| | Mars | Earth |
| Carbon Dioxide ($CO_2$) | 95 | 0.03 |
| Nitrogen ($N_2$) | 2 to 3 | 78 |
| Oxygen ($O_2$) | 0.1 | 21 |
| Argon (Ar) | 1 to 2 | 0.9 |

dioxide decreases rapidly above the planet, yielding to a 50 per cent constituency of nitrogen above 100 kilometers (60 miles).

After landing on the Martian surface, a so-called gas chromatograph mass spectrometer on each of the Viking spacecrafts was utilized to determine the chemical composition of the lower atmosphere. As expected from previous Mariner flights, 95 per cent of the Martian atmosphere is carbon dioxide. But in addition, nitrogen, oxygen, and argon were also detected. The final results are listed in Table 9–2, which also includes the composition of Earth's atmosphere for comparison.

In addition to determining the composition of the Martian atmosphere, meteorologists were able to measure a variety of conditions (for example, pressure, temperature, wind velocity) with sensitive instruments on the Viking's meteorology

**Figure 9-27.** *The Meteorology Boom.* Meteorology sensors are mounted atop an extended boom, about four feet above the ground. These sensors measure temperature, pressure, and wind velocity in the Martian atmosphere. (*JPL; NASA*)

boom. This boom, shown in Figure 9–27, extends to one side above the Viking lander about four feet off the ground. Atmospheric pressure is measured by means of a thin diaphragm stretched over an evacuated chamber. As the atmospheric pressure changes, the diaphragm flexes slightly and by measuring the amount of movement or flexing of the diaphragm, meteorologists deduce the pressure of the Martian atmosphere. Atmospheric temperature is measured by thermocouples whose electrical properties are sensitive to changes in temperature. And finally, wind velocity is measured with a hot film anemometer. This device consists of a heated aluminum oxide cylinder. The central idea is that as the wind blows against the exposed cylinder, one side becomes cooled and wind speed is deduced from knowing how much electrical power is necessary to keep the device at a constant temperature. Indeed, the anemometer operates in much the same way that a boy scout might hold a moistened finger in the air to determine wind speed and direction.

As anticipated from earlier Mariner flights, atmospheric pressure on Mars is about 7 millibars (recall that atmospheric pressure on Earth is roughly 1,000 millibars = 1 bar). But data collected over several weeks revealed that the atmospheric pressure is gradually decreasing, as shown in Figure 9–28. The reasonable explanation of this gradual decrease involves the fact that at the time the data were

**Figure 9–28.** *Atmospheric Pressure.* Air pressure on the surface of Mars is typically in the range of 6 to 8 millibars (recall, for comparison, that air pressure at sea level on Earth is nearly 1,000 millibars). Both Viking landers recorded a gradual decrease in atmospheric pressure during the weeks following their successful landings. The logical explanation for this decrease involves the fact that carbon dioxide was being frozen out of the Martian atmosphere at the planet's frigid south pole. The data shown here is from Viking lander 1.

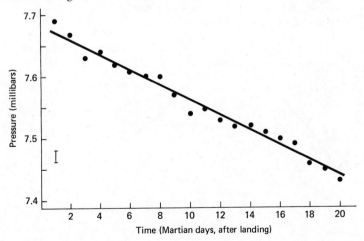

taken it was winter in the southern Martian hemisphere. Presumably, carbon dioxide over the frigid south polar regions becomes frozen and covers the ground like snow or frost. As the carbon dioxide is frozen out of the atmosphere, the air pressure must decrease. In 1977, as springtime came to the southern hemisphere, the carbon dioxide frost vaporized and the atmospheric pressure once again started to rise.

Temperature at the Viking landing sites varied with the time of day much in the same way one would naturally expect from heating by sunlight. Lowest temperatures at both Viking sites (around $190°K = -85°C = -120°F$) were recorded just before sunrise while highest temperatures (around $240°K = -30°C = -20°F$) occur in the early afternoon. Figure 9–29 shows the variation of air temperature over the course of a Martian day. For comparison, a graph showing the daily variation in air temperature measured in California is also included.

Wind velocity averaged 8 miles per hour on a typical day at Chryse with maximum speeds around 20 miles per hour and an occasional gust at about 35 miles

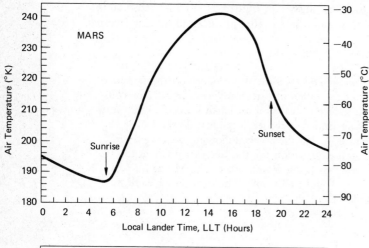

**Figure 9–29.** *Daily Temperature Variation.* The upper graph shows the typical temperature variation on Mars during a Martian day. Lowest temperatures ($190°K = -85°C = -120°F$) are recorded just before sunrise while highest temperatures ($240°K = -30°C = -20°F$) occur in the early afternoon. The lower graph shows typical temperature variations on Earth, as measured in the desert in California.

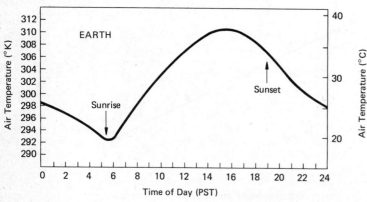

per hour. Lowest wind speeds (around 3 miles per hour) occur just before midnight. At Utopia, the average wind speed is in the range of 4 to 6 miles per hour with occasional gusts between 20 to 40 miles per hour. Minimum velocities range from calm to about 2 miles per hour.

In short, the Martian weather is boring: the same thing day after day. A typical Martian weather report, like those issued by Viking Meteorologist Seymour Hess, reads simply. "Light winds at 10 miles per hour, shifting as any sensible wind is supposed to do. Temperatures ranging from a low of $-120°F$ to an afternoon high of $-20°F$. Air pressure is 7 millibars. No rain in sight." No rain indeed! It has not rained on Mars for eons.

## The Search for Life  9.3

THE third planet from the sun is teeming with life. Although all life forms on Earth have fundamentally the same molecular structure in the sense that they are all based on the double-helix DNA molecule, the full range of possible species is absolutely incredible. Reptiles including tiny lizards and huge dinosaurs have roamed across the face of our planet. Mammals, from the smallest rodents to the largest whales, populate the continents and the oceans. Microscopic one-celled creatures are found everywhere.

In addition to assuming an unbelievable variety of sizes and shapes, the life-cycles of terrestrial life forms cover an equally immense range. Certain insects may live for only a few hours or days while trees such as the Sequoia and bristle-cone pine survive for hundreds and even thousands of years. And most interesting of all, an unusual mammal fully capable of logic and rational thought has recently evolved on Earth. While earlier creatures insured their survival in a hostile environment by means of hard shells, tough skins, and sharp claws and teeth, this recent mammal is blessed only with a complex, highly sophisticated central nervous system. As a result, this creature insures its survival by methods very different from those employed by any other life forms with which he shares his planet. For an ample supply of food, this creature invented agriculture and animal husbandry. For protection against the weather, he developed clothing and architecture. To ward off attacks and hostile invasions by other creatures, he constructed weapons at the macroscopic level (for example, clubs, spears, and guns), while at the microscopic level he discovered vaccines and penicillin. Clearly, it was only a matter of time before this remarkable creature would wonder if his was the only planet on which life had evolved. It was only a matter of time before this creature would build machines and vehicles that he would send to other planets and start exploring the universe for extraterrestrial life. Indeed, this was the central motivation behind the Viking missions to Mars.

Two spacecrafts successfully soft-landed on the Martian surface on July 20, 1976

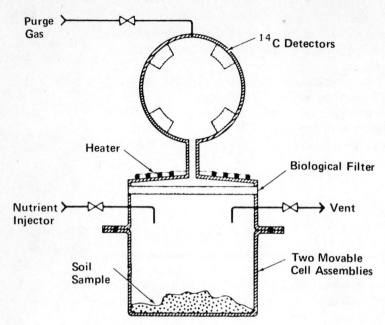

Purge Gas

$^{14}$C Detectors

Heater

Biological Filter

Nutrient Injector

Vent

Soil Sample

Two Movable Cell Assemblies

**Figure 9–30 (*left*). *The Labeled-Release Experiment.*** Scientists searched for signs of metabolism by extraterrestrial microorganisms on Mars by moistening a soil sample with a nutrient containing radioactive carbon-14. If microorganisms are present, they would presumably "eat" the nutrient and give off gases containing some of the radioactive carbon-14. Detectors above the soil sample measure the amount of carbon-14 released. (*Martin Marietta Aerospace*)

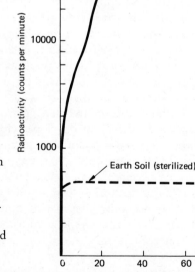

**Figure 9–31 (*right*). *Labeled Release from Terrestrial Soil.*** Soil from Earth containing bacteria released large quantities of gases containing radioactive carbon-14 when moistened with nutrient. But when the terrestrial soil was first sterilized and then "fed" with nutrient, a much smaller amount of carbon-14 was released.

and September 3, 1976. While no macroscopic life-forms were seen at either landing site, apparatus on board the two Viking landers was activated to examine the Martian soil for microscopic organisms. In particular, three different devices performed three types of experiments to search for extraterrestrial life. The *labeled release experiment* looked for signs of metabolism. The *pyrolytic release experiment* tried to detect microorganisms that function by photosynthesis or chemotrophy. And the *gas exchange experiment* searched for respiration by organisms by measuring changes in the gases in a closed container.

The central idea behind all three experiments was very simple: living organisms alter the environment they inhabit. Consider, for example, the so-called labeled release experiment. A small sample (roughly $\frac{1}{2}$ cubic centimeter) of Martian soil is loaded into a container by means of Viking's mechanical arm. The container, shown in Figure 9–30, is then sealed and the sample is moistened with some nutrient. This nutrient is rich in vitamins and amino acids containing radioactive carbon-14 ($^{14}C$). Scientists nicknamed the nutrient "chicken soup" since no sensible terrestrial organism could possibly resist this delicious substance. If microorganisms are present in the sample of Martian soil, they would presumably "eat" or metabolize the nutrient. In doing so, the organisms would release gases. Since the nutrient contains radioactive carbon-14, the gases produced by metabolism should contain carbon-14. Consequently, after incubating the soil sample for a few days, carbon-14 detectors located above the sample would measure for any radioactive carbon in gases (for example, $^{14}CO$, $^{14}CO_2$, $^{14}CH_4$) released from the sample.

Of course, certain chemical reactions might mimic biological processes. Perhaps the Martian soil might contain a tiny amount of very active chemicals that, when moistened with the nutrient, might fizz and bubble thereby releasing gases containing carbon-14 into the atmosphere above the sample. This would fool scientists into thinking that they had discovered extraterrestrial organisms when, in fact, they had merely detected the effects of some interesting chemistry. To avoid this possible confusion, the Viking landers were equipped with the ability to sterilize the Martian soil with heat. By heating the sample, scientists hoped to be able to kill any Martian organisms yet leave the chemicals unaffected. Thus, if an unsterilized sample gave a positive response (that is, lots of carbon-14 released) while repeating the experiment with a sterilized sample gave a negative response (that is, little or no carbon-14 released), scientists could be in a position to conclude that they had discovered life on Mars. But if both sterilized and unsterilized samples gave essentially the same positive responses, then chemical reactions would be the obvious explanation. Of course, negative responses with all samples would indicate no biology and no unusual chemistry.

To insure the effectiveness of their apparatus, scientists performed the labeled release experiment many times with terrestrial and lunar soil prior to the Viking landings. Figure 9–31 shows the results of a terrestrial soil sample loaded with bacteria. When the unsterilized sample was moistened with nutrient, relatively large amounts of gas containing radioactive carbon-14 were released as evidenced by the large number of counts per minute measured by carbon-14 detectors. But

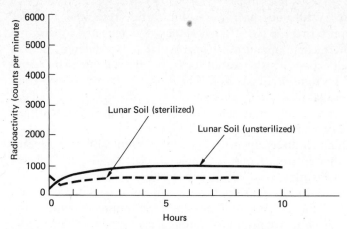

**Figure 9–32.** *Labeled Release from Lunar Soil.* Since the moon is totally devoid of any life forms, both cycles of the labeled-release experiment (unsterilized and sterilized) gave the same negative response. Virtually no gases containing radioactive carbon-14 were released.

when this experiment was repeated with a sterilized terrestrial soil sample, far fewer counts per minute were recorded indicating that the bacteria had been killed.

Figure 9–32 shows the results of performing this experiment with lunar soil. We know, of course, that the moon is totally devoid of any life forms. Thus, both unsterilized and sterilized soil samples gave essentially the same negative response. Virtually no gases containing radioactive carbon-14 were released from either sample. It was generally agreed that a ten-fold difference in the counts per minute between unsterilized and sterilized soil samples would be positive evidence for life.

After the Viking spacecrafts had landed on Mars, the labeled release experiment was performed several times with both unsterilized and sterilized samples of Martian soil. In all experiments at both Chryse and Utopia the results were essentially the same: large quantities of gas containing carbon-14 were released from unsterilized samples, while sterilized samples released much smaller quantities of radioactive gas. Typical data from the experiments are shown in Figure 9–33. Viking scientists were shocked and amazed by the large quantity of gas containing carbon-14 rapidly released from the unsterilized samples. Indeed, to quote Dr. Gerald Soffen, "It was too much, too soon!" The unsterilized sample was behaving as though active chemical processes were at work. Perhaps the gases were released because the soil sample fizzes and froths when moistened with nutrient. But a chemical interpretation of the data did not seem quite right. After all, the sterilized samples released significantly less radioactive gas. This dramatic difference between the behavior of sterilized and unsterilized Martian soil samples is clearly seen in Figure 9–34.

In the Viking 1 labeled release experiments, the samples were sterilized by heating the soil to 160°C (320°F) for three hours. Perhaps this was too hot. Perhaps

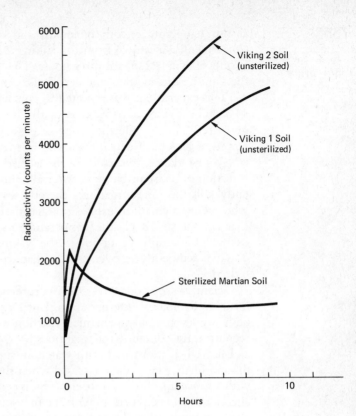

**Figure 9-33.** *Labeled Release from Martian Soil.* Unsterilized samples of Martian soil released a substantial amount of gases containing radioactive carbon-14 after being "fed" with nutrient. Far less radioactive gas was released from sterilized samples.

**Figure 9-34.** *Long-term Incubation.* The dramatic difference between sterilized and unsterilized Martian soil samples in the labeled-release experiment is clearly shown on this graph. Significantly more gas containing carbon-14 was released from the unsterilized sample compared to the sterilized sample.

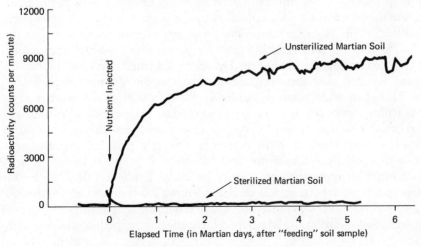

at this temperature both biological and chemical agents would be destroyed. Consequently, a sample from the Viking 2 site was "cold sterilized" by heating it to only 50°C (122°F) for three hours. The "cold sterilized" sample gave essentially the same results as the "hot sterilized" samples: virtually no gas released.

This experiment was repeated in January, 1977, with cold sterilization at only 44°C (111°F) instead of 50°C. As with all other sterilized samples, very little gas was given off. Thus, while the range of possible chemical reactions was greatly narrowed, a biological explanation was not reduced.

At first glance, the results of the labeled release experiment clearly point in the direction of Martian biology. After all, almost any amount of sterilization apparently kills the microorganisms thereby severely inhibiting the reaction. Anyone who prefers a chemical explanation has the difficult task of thinking of a chemical that is *stable* at 18°C (the temperature to which the unsterilized samples were exposed by simply being inside the Viking landers) but is *unstable* and breaks down at temperatures over 44°C. That would be a very extraordinary chemical indeed.

Scientists are at odds over the interpretation of the labeled release results. Some favor the biological interpretation. But Dr. Norman Horowitz of Caltech vigorously argues that only a chemical reaction was observed. He feels that the Martian regolith must contain large quantities of "peroxides" or "superoxides" that froth and fizz when moistened with water. In support of his position, Horowitz noted that no further gases were released when the unsterilized samples were moistened with a *second* injection of nutrient. This is consistent with the idea that all the active chemicals had been used up with the first squirt of nutrient. If Martian microorganisms had been growing and thriving as a result of the first injection of nutrient, a second injection should have produced still more gas. But no additional gas was released.

In addition to the three biological experiments, the gas chromatograph mass spectrometer on the Viking landers should have been able to shed some light on the question of biology vs. chemistry. This device was designed to detect organic compounds in the Martian soil. A small soil sample is packed into a tube in an oven which heats the specimen. As the temperature rises, any organic compounds present in the soil are driven off. The resulting gases are passed into a spectrometer which measures and detects organic chemicals. Dr. Klaus Biermann and the molecular analysis team detected only three organic chemicals in the Viking 1 experiment. They were methyl chloride, acetone, and freon. All of these chemicals were used to clean the machine before it was launched toward Mars. All of these chemicals are clearly terrestrial residues. In the Viking 2 experiment, several more organic molecules of terrestrial origin were detected. The fact that Biermann's team identified organic molecules used to clean the device prior to launch clearly demonstrated that his apparatus was working properly. Evidently, the oven in the gas chromatograph mass spectrometer on board Viking 2 was not as clean as that on Viking 1 prior to blast-off. Of course, the whole point is that absolutely *no* organic molecules were discovered in the Martian soil. By contrast, experiments done with antarctic soil here on Earth release all sorts of organic molecules that were obviously not used as solvents to clean the apparatus.

What is the meaning of these conflicting results? On the one hand, the labeled release experiment gave indications of biology. If microorganisms are indeed the cause of the positive data in the labeled release experiments, then the Martian soil should be loaded with extraterrestrial microbes (both dead and alive). But on the other hand, the gas chromatograph mass spectrometer failed to detect any organic (biological) molecules whatsoever. Dr. Harold P. Klein, leader of the Viking biology team, pointed out that in order for the gas chromatograph mass spectrometer to detect Martian organic chemicals there must be more than a million microbe cells (either dead or alive) in the analysed soil sample. If there were significantly fewer than one million cells present in the soil sample, then the device, which has a limited sensitivity, would probably fail to detect them. In addition, Dr. Klein pointed out that the results of the labeled release experiment could be explained by the action of only one million living cells. Thus, perhaps the explanation of the data could be that while there were just enough cells to give positive results in the labeled release experiment, there were not quite enough cells to be detected in the gas chromatograph mass spectrometer analysis. But that doesn't sound quite right either. Consider, for example, a stable community of one million living creatures. If this million-member community had been in existence for any reasonable length of time, then there should be an enormous number of corpses lying around. The gas chromatograph mass spectrometer should detect any organic substances, regardless of whether they come from living or dead cells while the labeled release experiment responds to the action of living cells only. If the results of the labeled release experiment were the result of a million living cells, why didn't the gas chromatograph mass spectrometer discover any trace of the billions of corpses left behind by their ancestors? Perhaps Martian microorganisms are cannibals! After all, conditions on Mars are very harsh and there certainly is not an ample supply of food around for any living creatures. Perhaps Martian organisms have adapted to these conditions of scarcity by eating their dead ancestors.

This hypothesis might have been tenable if the remaining two biological experiments on board the Viking landers had given positive data like the labeled release experiment. Unfortunately this was not the case.

In the pyrolytic release experiment, a small sample ($\frac{1}{4}$ cubic centimeter) of Martian soil is placed in a sealed container, shown in Figure 9–35. The container is then filled with gases containing radioactive carbon-14 (for example, $^{14}CO$ and $^{14}CO_2$) and illuminated with a light simulating the sun seen from Mars, minus the deadly ultraviolet waves. This experiment could be performed with the light either on or off. The central idea is that here on Earth, plants depend on photosynthesis. In this process, they remove carbon dioxide from the air and, using the energy in sunlight, incorporate the carbon into their cells while releasing oxygen back into the atmosphere. Perhaps Martian organisms do the same. If so, some of the radioactive carbon in the gases in the container would end up in the cells of the organisms in the soil sample. Consequently, after incubating the sample for several days, the

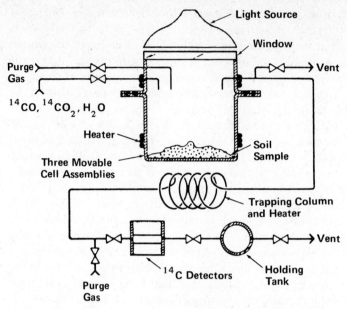

**Figure 9–35.** *The Pyrolytic Release Experiment.* Scientists searched for signs of photosynthesis by extraterrestrial organisms in the Martian soil by exposing a sample to light in a chamber containing radioactive carbon-14 gas. If microorganisms are present, they should absorb carbon monoxide and/or carbon dioxide and incorporate some of the radioactive carbon-14 in their cells. By flushing the chamber with nonradioactive purge gas and then baking the soil sample, the carbon-14 would be driven off and detected. (*Martin Marietta Aerospace*)

container is flushed with a clean purge gas containing no radioactive carbon-14. Any carbon-14 not ingested by the organisms would thereby be removed from the chamber. The sample is then baked at 625°C (1,060°F). At this temperature, any carbon-14 assimilated by Martian cells would be driven off and detected.

The results of performing the pyrolytic release experiment with three terrestrial soil samples are shown in Table 9–3. In all cases, the incubation time is 22 hours. So-called "peak 1" is obtained as the unused radioactive gases are flushed from the chamber. Peak 1 is simply a measure of how much carbon-14 went into the experiment. "Peak 2" is a measure of how much carbon-14 was released after baking the soil sample. Evidence for life depends on the relative sizes of peak 1 and peak 2, both of which measure the amount of carbon-14 in counts per minute. If peak 2 is large compared to peak 1 (for example, peak 2 = 100 and peak 1 = 10,000), then lots of radioactive carbon-14 has been assimilated by microorganisms in the soil. If peak 2 is quite small (for example, a dozen counts per minute), then very little carbon-14 had been assimilated and no photosynthesis has

| Soil Sample | Algae | Bacteria | Light or Dark | Peak 1 | Peak 2 |
|---|---|---|---|---|---|
| A609 (unsterilized) | 500 | 5,000 | light | 26,400 | 605 |
| A609 (unsterilized) | 500 | 5,000 | dark | 61,600 | 286 |
| A609 (sterilized) | 0 | 0 | light | 30,800 | 11 |
| A641 (unsterilized) | 0 | 12,500 | light | 77,000 | 341 |
| A641 (unsterilized) | 0 | 12,500 | dark | 136,000 | 187 |
| A641 (sterilized) | 0 | 0 | light | 107,800 | 40 |
| A638 (unsterilized) | 33 | 1,500 | light | 8,600 | 106 |
| A638 (sterilized) | 0 | 0 | light | 13,000 | 18 |

occurred. Note that a typical positive response indicating life would be a peak 2 of 100 counts per minute while peak 1 was 7,000 or 8,000 counts per minute. A peak 2 of less than 15 counts per minute (compared to a peak 1 of more than 7,000 counts per minute) is a negative result.

The first time this experiment was run on Viking lander 1, peak 1 measured 7,400 counts per minute while peak 2 measured 96 counts per minute. As with the unsterilized terrestrial sample A638, this is a positive result!

In the second experiment by Viking lander 1, the soil sample was sterilized by baking at 160°C for three hours before the experiment was started. Peak 2 was then measured at only 15 counts per minute, again in keeping with terrestrial sample A638. But when the experiment was repeated a third time with an unsterilized sample, peak 2 was measured at only 27 counts per minute. That's borderline. With new unsterilized samples at Viking 1, the initial positive response could not be repeated!

Data from Viking lander 2 were equally ambiguous. On the first run of the experiment with an unsterilized sample, peak 2 was 23 counts per minute while peak 1 was around the usual 7,000 counts per minute. Another borderline case. So someone came up with the idea that perhaps the Martian organisms needed a little water. After all, plants that depend on photosynthesis here on Earth need water. Consequently, during the second run of the experiment by Viking lander 2 with an unsterilized sample, a small amount of water vapor (only 100 micrograms) was injected into the chamber along with lots of gas containing radioactive carbon-14. Peak 1 was measured at a whopping 12,500 counts per minute indicating that a plentiful amount of $^{14}CO$ and $^{14}CO_2$ went into the chamber. But peak 2 was only 3 counts per minute. Clearly a negative result! This was a big surprise. Evidently the injection of water vapor stops the Martian soil from doing any photosynthesis-like processes.

The labeled release experiment looked for signs of metabolism. The pyrolytic release experiment looked for signs of photosynthesis. And finally, the gas exchange experiment looked for signs of respiration. As living organisms breath, they alter the

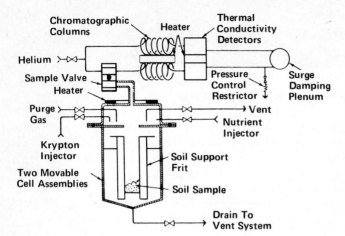

**Figure 9-36.** *The Gas Exchange Experiment.* Scientists searched for signs of respiration by extraterrestrial organisms in the Martian soil by measuring changes in the chemical composition of gases above a soil sample in a sealed container. This experiment could be performed either dry, humid, or wet, depending on how much nutrient was added. (*Martin Marietta Aerospace*)

chemical composition of the atmosphere in which they live. To check for similar processes by any Martian organisms, a soil sample was placed in a container, shown in Figure 9–36, along with a controlled atmosphere of helium, krypton, and carbon dioxide. Sensitive instruments periodically analyse the gases in the container searching for chemicals usually produced during life processes. This experiment can be run either dry, humid, or wet depending on how much nutrient (more "chicken soup") is injected into the chamber.

On various runs of the gas exchange experiment by both Viking landers 1 and 2, the chemical composition of the gases in the chamber did indeed change. Various amounts of nitrogen, oxygen, and carbon dioxide were given off depending on the amount of nutrient used. But in all cases, the released gases could be explained in terms of chemical processes. There was no clear suggestion of biology at all.

The whole business turned out to be rather frustrating. Some experimental results were positive, some were negative, and some were half way in between. In desperation, Viking scientists started looking under rocks. Someone came up with the idea that perhaps Martian organisms might be living under rocks that shield them from the sun's ultraviolet rays. As shown in Figure 9–37, Viking's mechanical arm was used to push aside surface rocks so that shielded samples could be obtained and analyzed. Once again, however, the results were as inconclusive as before.

The Viking missions have not provided a definitive answer concerning the question of life on Mars. Since some results are positive while others are negative, the question is still open. Nevertheless, we know a lot more than we did prior to the Viking landings. As the weeks and months passed by during the fall of 1976, Viking biologists began wishing that they could land a third spacecraft on the planet. This new spacecraft would carry a new set of experiments, redesigned on the basis of our recent discoveries, to decide once and for all between biology and chemistry in the Martian soil. Meanwhile, Viking geologists looking at the surface photographs were asking incredibly obvious questions like "I wonder what's behind

**Figure 9-37.** *Pushing Rocks.* Using Viking's mechanical arm, scientists obtained soil samples located under Martian rocks for biological analysis. Some scientists believed that if life forms exist on Mars, they might use rocks as shelter from the sun's deadly ultraviolet rays.

those sand dunes? . . . or over that hill? . . . or beyond the horizon?" They, too, were envisioning a third soft-landing on Mars, this time by a vehicle equipped with wheels!

At this point it is clear that we must send a Viking 3 to Mars. But certainly the issue will be raised as to whether it would be worth it. After all, Viking 1 and 2 each cost $660 million ($244 million for each orbiter and $416 million for each lander). With inflation, a Viking 3 would probably run about one billion dollars. And that sounds like a lot of money!

In 1976, Americans spent $569 million on pickles. In that same period, they spent $800 million on chewing gum. They are spending over a billion dollars a year in bowling alleys, almost ten billion dollars a year on cigarettes, and twenty billion on alcoholic beverages. Thus, for roughly the same amount of money that Americans spend on booze in three weeks, we could land Viking 3 on Mars.

1. Briefly describe the Viking missions.
2. Briefly describe the evidence for water erosion found by the Viking orbiters.
3. Describe the evidence that leads scientists to believe that reservoirs of water ice were trapped beneath the Martian surface.
4. Descibe one way in which the water in an underground ice reservoir on Mars might be liberated to produce a flood.
5. Why do some scientists believe that it is significant that one end of Valles Marineris is located near the Tharsis Plateau?
6. On what basis did scientists conclude that the residual polar caps must be composed entirely of water ice?
7. Describe the clouds seen around Olympus Mons.
8. Describe the morning fog seen in low-lying regions shortly after the Martian dawn.
9. Why can there be no permanent reservoirs of carbon dioxide dry ice on Mars?
10. Briefly describe the Viking landers and the sequence of events which transpired immediately prior to landing.
11. Why were the Viking landing sites selected in the northern hemisphere?
12. What does the Martian surface look like?
13. Compare and contrast the Martian surface at Utopia and Chryse.
14. What are the most abundant elements in Martian rocks and soil?
15. What is the chemical composition of the Martian atmosphere? How does this compare with the composition of Earth's atmosphere?
16. Briefly discuss a possible explanation for the gradual decrease in atmospheric pressure detected by the Viking landers.
17. Briefly describe the variation in temperature over a Martian day at the planet's surface.
18. What are typical wind speeds on Mars?
19. What experiments were carried onboard the Viking landers to search for microscopic life forms in the Martian soil?
20. Describe the labeled-release experiment.
21. What were the results of the labeled-release experiment with terrestrial, lunar, and Martian soil?
22. What were the results of organic analysis of the Martian soil by Vikings' gas chromatograph mass spectrometer?
23. Describe the pyrolytic release experiment.
24. Why were the results of the pyrolytic release experiment on Mars ambiguous?
25. Describe the gas exchange experiment.
26. Why were the results of the gas exchange experiment not indicative of biological processes in the Martian soil?
27. If you were a member of the Viking team, how sure would you feel about concluding whether there is or is not life on Mars?

# Cloud-covered Venus 10

## *A Mysterious Planet* 10.1

EXCEPT for the sun and the moon, Venus is often the brightest object regularly seen in the sky. Since Venus is an inferior planet (that is, its orbit lies closer to the sun than Earth's), it is always seen near the sun. During its synodic period, Venus appears to swing back and forth from one side of the sun to the other. When Venus is east of the sun in the sky, the planet sets after sunset. At such times, Venus dominates the evening sky and is sometimes called the "evening star." When Venus is west of the sun in the sky, the planet rises before sunrise and is sometimes called the "morning star." At maximum brilliancy, Venus has an apparent magnitude of −4 which means that it is ten times brighter than Sirius, the brightest-appearing star in the night sky.

Ancient peoples thought that the morning and evening "stars" were two separate objects. Indeed, the earliest Greek astronomers called these two bright objects *Phosphorus* and *Hesperus*. In the sixth century B.C., however, Pythagoras realized that Phosphorus and Hesperus were, in fact, the same planet.

Up until the 1950s, Venus was thought to be Earth's "twin." This misconception was the result of the fact that—at first glance—Venus has many characteristics similar to our own planet. As shown in Table 10–1, quantities such as the mass, size, and density of Venus are more like Earth's than any other planet's in the solar system. These similarities were so striking that many people in the nineteenth century imagined that Venus might be teeming with life, just like our own planet.

Nothing could be farther from the truth. The environment on Venus is incredibly hostile and all similarities with Earth are entirely superficial. One of the reasons for these misconceptions is the fact that Venus is perpetually covered with a thick cloud-cover beneath which the Venusian surface is eternally hidden from view. Of course, this cloud-cover is responsible for Venus' high albedo; the planet reflects more sunlight than any other object in the solar system. Nevertheless, this massive cloud-cover has shrouded Venus in mystery ever since it was first observed by Galileo in the early 1600s.

The first person to observe Venus through a telescope was Galileo Galilei. As

**365**

**Table 10-1.** *Venus-Earth Similarities*

| Property | Venus | Earth |
|---|---|---|
| Mass of Planet (Earth = 1) | 0.82 | 1.00 |
| Diameter of Planet (Earth = 1) | 0.95 | 1.00 |
| (kilometers) | 12,112 | 12,742 |
| (miles) | 7,526 | 7,920 |
| Average Density (grams per cubic centimeter) | 5.5 | 5.3 |
| Surface Gravity (Earth = 1) | 0.91 | 1.00 |
| Escape Velocity (kilometers per second) | 10.3 | 11.2 |

noted in Section 1.3, Galileo discovered that Venus exhibits phases just like our moon. This discovery was instrumental in the overthrow of the geocentric Ptolemaic system, since it proved that Venus must be in orbit about the sun rather than the earth.

An excellent earth-based view of the crescent phase of Venus is shown in Figure 10–1. Of course, Venus cannot be seen at full phase because at such times the planet is at superior conjunction on the opposite side of the sun from Earth. Neither can Venus be seen very well at new phase, which occurs at the time of inferior conjunction.

At superior conjunction, Venus is at its greatest distance from Earth and—if Venus could be seen—would have an angular diameter of only 10 seconds of arc. On the other hand, at inferior conjunction, Venus is nearest Earth and has an angular diameter of 64 seconds of arc. Although Venus appears largest near the time of inferior conjunction, the planet's apparent brightness is quite low because Earth-based observers are seeing mostly the dark (night) side of the planet. Similarly, Venus is dim near the time of inferior conjunction. Although Earth-based observers see a nearly full-phase view of Venus near inferior conjunction, the planet is so far away that its apparent magnitude is again quite low. Maximum brilliancy therefore amounts to a trade-off between the percentage of the illuminated (daytime) side of Venus that can be seen from Earth and the angular size of Venus. Maximum brilliancy occurs when Venus has an elongation of about 39° from the sun in the sky. This takes place 36 days before and after inferior conjunction. For comparison, the greatest elongation of Venus is 47°. Venus can never appear more than 47° away from the sun, as shown in Figure 10–2.

Since Venus is an inferior planet, it might be thought that Venus could be seen as a black dot against the sun at the time of inferior conjunction. Such a phenomenon is called a *transit* since Venus would appear to move across the bright solar disk. Transits of Venus are, however, extremely rare because the orbit of the planet is inclined to the ecliptic by slightly more than 3°. Since the angular diameter of the sun seen from Earth is only $\frac{1}{2}$°, this means that at inferior conjunction Venus usually lies above or below the sun.

**Figure 10-1.** *Venus.* Near the time of maximum brilliancy, Venus exhibits a pronounced crescent phase as shown in this Earth-based photograph. (*Hale Observatories*)

**Figure 10-2.** *The Configurations of Venus.*
As an inferior planet, Venus is never more
than 47° away from the sun. Maximum
brilliancy occurs 36 days before and after
inferior conjunction when the planet's
elongation is 39°.

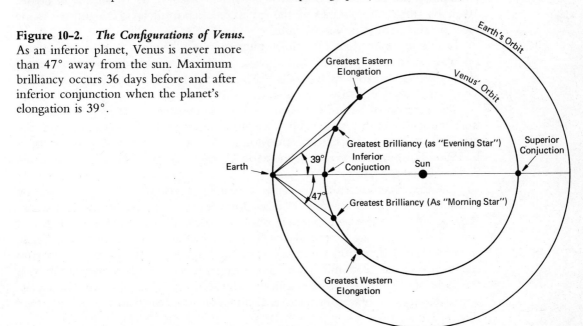

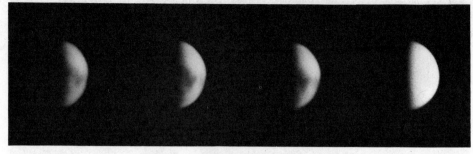

Ultraviolet Light                  Visible Light

**Figure 10-3.** *Views of Venus.* These four photographs were taken in June 1967. The three views on the left show faint markings that reveal the motion in Venus' upper atmosphere. The view on the right was made in visible light and shows no features. (*New Mexico State University Observatory*)

Transits of Venus occur in pairs separated by eight-year intervals. The most recent Venusian transits occurred in 1874 and 1882. The next pair of Venusian transits are scheduled for June 8, 2004 and June 6, 2012.

Observations in visible light of Venus through a telescope reveal no discernible markings whatsoever. Only a shimmering, featureless cloud-cover is seen. At ultraviolet wavelengths, however, some indistinct patterns in the planet's upper atmosphere can be seen. Several excellent Earth-based views are shown in Figure 10-3. The three photographs on the left were taken in ultraviolet light, while the view on the right is in visible light.

Examining successive ultraviolet photographs of Venus over several nights reveals a remarkable property of the planet's rotation. Unlike the other terrestrial planets, Venus' rotation is backward or *retrograde*! In other words, as seen from Venus, the sun rises in the west and sets in the east.

The rotation of Venus had been a subject of controversy among astronomers since the seventeenth century. The first hint of retrograde motion came from the work of Robert S. Richardson at Mount Wilson and Palomar Observatories in 1956. This was confirmed in the early 1960s by the two French astronomers B. Guinot and C. Boyer.

The first extensive study of Venus at various wavelengths had been made by Frank E. Ross in 1928 at Mount Wilson Observatory. Stimulated by Boyer's work, Ross again took numerous photographs in the near-ultraviolet in the 1960s and succeeded in detecting the motion of features in Venus' upper atmosphere. These motions suggested that the clouds in Venus' atmosphere have a speed of up to 100 meters per second (230 miles per hour), retrograde! Final confirmation came with the flight of Mariner 10 (discussed in the next section), which showed that Venus' upper atmosphere rotates in a retrograde direction about the planet once every four days.

Of course, all these observations of Venus' rotation rate are restricted to motions of the planet's upper atmosphere. They do not necessarily tell us anything about the rotation of the planet itself. In order to discover the planet's actual rotation rate, astronomers had to make observations that somehow penetrate the thick Venusian cloud-cover. This was first accomplished in the early 1960s using *radar*. Radar waves easily pass through the cloud-cover and are reflected off of the planet's rocky surface. By bouncing radar signals off of Venus, astronomers discovered that the planet itself does indeed rotate retrograde, but the rate of rotation is very slow. Venus rotates once about its axis every 243 days. Thus, the rapid rotation rate of Venus' upper atmosphere must be the result of high-speed winds extending over large areas.

This interval of 243 days for Venus' rotation period is measured with respect to the stars; a sidereal day on Venus lasts for 243 Earth-days. But Venus also orbits the sun, and as a result a solar day on the planet is much shorter. The time from one "high noon" on Venus to the next is 116.8 days. A "day" on Venus lasts 116.8 Earth-days. But recall that the synodic period of Venus (that is, the time between inferior conjunctions) is 584 Earth-days. That is *exactly* 5 Venus-days. In other words, at every inferior conjunction, the same side of Venus faces Earth. Some scientists therefore believe that Venus' rotation has been "captured" by tidal forces exerted by Earth. In order for this to have occurred, however, the mass of Venus must be asymmetrically distributed about the planet. Instead of being perfectly spherical, the planet must have a lump or bulge on one side. Careful analysis of the orbits of spacecrafts passing near Venus has not detected any such asymmetry. The details of Venus' rotation rate therefore remain unexplained.

One of the most important tools that the astronomer has at his disposal is the spectrograph. Identification of spectral lines in the spectra of light from a celestial object permits the astronomer to discover the chemical composition of the source of light. Like all planets, Venus shines by reflected light from the sun. Therefore, any spectral lines already present in the sunlight falling on the planet will also appear in a spectral analysis of light from Venus—nothing in the Venusian atmosphere could possibly "erase" spectral lines originating in the sun's atmosphere. But other chemicals not present on the sun would be expected to add new spectral lines to the light reflected from Venus.

In the early 1930s, Walter S. Adams and Theodore Dunham, Jr., at Mount Wilson Observatory succeeded in identifying spectral lines of carbon dioxide ($CO_2$) in the near-infrared spectrum of light from Venus. One of these early spectra is shown in Figure 10–4. The upper spectrum simply shows ordinary sunlight. The lower spectrum also exhibits all the usual solar spectral lines, but in addition there are two series of molecular lines beginning at 7820 Å and 7883 Å. Both are caused by carbon dioxide. Recent space flights have shown that carbon dioxide constitutes 97 per cent of the Venusian atmosphere.

In addition to the chemical composition of Venus' atmosphere, the temperature of the clouds can also be measured from Earth. Venus is closer to the sun than Earth and therefore should have a higher temperature than our own planet. Indeed, Venus receives about twice as much sunlight than Earth. Infrared measurements made at Mount Wilson and Palomar Observatories indicate that the planet has a

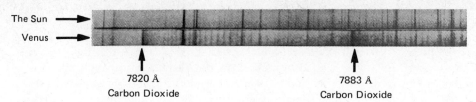

The Sun ➔

Venus ➔

7820 Å
Carbon Dioxide

7883 Å
Carbon Dioxide

**Figure 10–4.** *Carbon Dioxide on Venus.* Substantial amounts of carbon dioxide in the Venusian atmosphere was discovered by analysing reflected sunlight from the planet. The upper spectrum is of the sun; no molecular lines are seen. The lower spectrum is of reflected sunlight from Venus. Two series of lines caused by carbon dioxide are seen. (*Yerkes Observatory photograph*)

rather uniform temperature on both sunlit and night sides ranging from 230 to 240°K (−27 to −45°F), which is below the freezing point of water. Infrared radiation, however, originates at the top of the cloud-cover surrounding the planet. While the cloud tops are obviously cool, these observations gave no clues to the surface temperature of the planet. Once again astronomers had to turn to radio telescopes to penetrate the thick Venusian atmosphere. Surprisingly, radio observations beginning in the mid-1950s at a wavelength of several centimeters indicated a surface temperature of 700°K (900°F). This result was so incredibly high that many scientists believed that the measurements were erroneous. Recent space flights have, however, confirmed this high temperature and scientists today realize that, except for the sun, Venus is the hottest object in the solar system.

# 10.2  *Probing the Atmosphere*

AS IN the case of the moon and Mars, many intriguing questions about Venus could only be answered by actually traveling to the planet. But unlike the missions to Mars or the moon, which have little or no atmosphere, pictures of Venus were not expected to be the most fruitful source of data about the cloud-covered planet. Even from a few thousand kilometers, television cameras would only be able to photograph the cloud tops of the thick Venusian atmosphere. Scientists therefore concentrated their efforts on other types of equipment (for example, radiometers to measure temperature and magnetometers to detect magnetic fields) to probe the environment of Venus. Indeed, up until the mid-1970s, 11 missions were launched to Venus by the United States and the Soviet Union (only 7 of the flights returned usable data) but not one picture was taken.

The first successful mission to Venus was the flight of Mariner 2, launched on August 26, 1962. Two earlier spacecrafts, Mariner 1 and Venera 1, had malfunctioned. These Mariner spacecrafts were constructed at the Jet Propulsion Laboratory and were patterned after the successful Ranger series used in lunar exploration. At first glance Mariner 2 (see Figure 10–5) looks entirely like one of the later Ranger spacecrafts. But while the purpose of the Rangers was to produce close-up photographs of the lunar surface, Mariner 2 was designed to measure temperatures, magnetic fields, and particles in interplanetary space and around Venus.

As mentioned in Section 6.4, one of the important results to come from the flight of Mariner 2 was the conclusive discovery of the solar wind. The solar plasma spectrometer detected charged particles from the sun and found that the solar wind moves outward from the sun at speeds ranging from 300 to 800 kilometers per second (roughly 700,000 to 2 million miles per hour). During its four-month flight to Venus, the spectrometer on board Mariner 2 measured the density of the solar wind. The extreme tenuousity of the solar wind is almost impossible to comprehend: only 10 to 20 protons and electrons per cubic inch (about 1 particle per cubic centimeter).

Mariner 2 passed within 35,000 kilometers (22,000 miles) of Venus on December 14, 1962, making numerous measurements of temperatures and magnetic fields. The microwave radiometer detected temperatures of 400°K (260°F) deep within the Venusian atmosphere while the infrared radiometer measured temperatures of 240°K (−45°F) near the cloud tops. The magnetometer on board Mariner 2 failed to detect any magnetic field about the planet. Because the average density

**Figure 10–5.** *Mariner 2.* This 450-pound spacecraft flew past Venus in December 1962. While cruising to Venus, Mariner 2 confirmed the existence of the solar wind. At Venus, Mariner 2 measured temperatures in the planet's cloud-cover and found that Venus has no detectable magnetic field. (*NASA*)

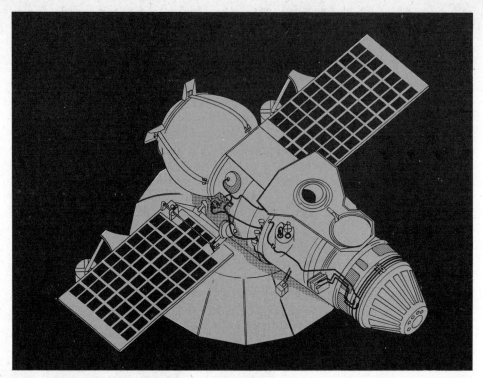

**Figure 10-6.** *The Venera Spacecraft.* Veneras 4, 5, 6, 7, and 8 looked very similar. Upon arriving at Venus, the spherical device at the back of the spacecraft separated from the rest of the vehicle and descended through the Venusian atmosphere by means of parachutes.

of Venus (5.5 grams per cubic centimeter) is similar to the average density of Earth (5.3 grams per cubic centimeter) it seems reasonable to suppose that Venus—like Earth—possesses a dense nickel–iron core. But unlike Earth, Venus does not have a magnetic field. The absence of such a field is perhaps caused by the slowness of Venus' rotation.

While the United States focused its efforts during the 1960s on manned lunar exploration and unmanned missions to Mars, scientists in the Soviet Union directed much of their attention to Venus. The first two Soviet missions (Venera 1 and 2), launched in 1961 and 1965, flew past Venus but failed to return data. Venera 3 landed on Venus in March of 1966 but again failed to return any data. The first successful Soviet flight, Venera 4, was launched on June 12, 1967. Four months later, October 18, 1967, the bulky Russian spacecraft encountered the planet and descended by means of parachutes into the thick Venusian atmosphere. The probe transmitted data concerning the pressure, temperature, and chemical composition for 94 minutes during the parachute descent. Radio contact with Venera 4 was lost while the spacecraft was at an altitude of about 25 kilometers (roughly 15 miles) above the Venusian surface.

**Figure 10-7.** *Mariner 5.* This 542-pound spacecraft was launched toward Venus two days after the Russians launched Venera 4. Combined data from both spacecrafts proved that Venus' atmosphere is extremely hot and dense. (*NASA*)

Two days after the launch of Venera 4, the United States embarked on its second unmanned mission to Venus. Mariner 5 lifted off from Cape Canaveral on June 14, 1967 and passed within 4,100 kilometers (2,500 miles) of Venus on October 19, 1967, only one day after the successful descent of Venera 4. Unlike the Soviet mission, however, American scientists chose to examine Venus as their spacecraft flew past the planet. Much of the equipment and spacecraft design was patterned after Mariner 4, which had been successfully sent to Mars two years earlier.

The flight of Mariner 5 permitted an accurate determination of the mass and size of Venus. The mass of Venus was found to be 0.8149988 Earth-masses and the diameter of the planet was measured at 12,106 kilometers (7,523 miles). This means that the top of the Venusian cloud-cover is located at an altitude of between 65 and 70 kilometers (roughly 40 miles) above the planet's surface. As with the previous Mariner mission to Venus, the magnetometer on board Mariner 5 did not detect any magnetic field surrounding the planet. Consequently, the planet is not surrounded by any radiation belts such as the Van Allen belts encircling the earth. One of the most important experiments conducted during the flight of Mariner

5 centered about the occultation of the spacecraft by Venus. The orbit of Mariner 5 had been chosen so that the spacecraft would pass behind the planet as seen from Earth. At the beginning and end of the occultation, radio signals from Mariner 5 had to pass through the thick Venusian atmosphere. Detailed examination of changes in the radio signals transmitted through the Venusian atmosphere permitted scientists to calculate the ranges of temperature and pressure in the gases above the planet. Prior to 1967, some scientists had grave doubts about the incredibly high

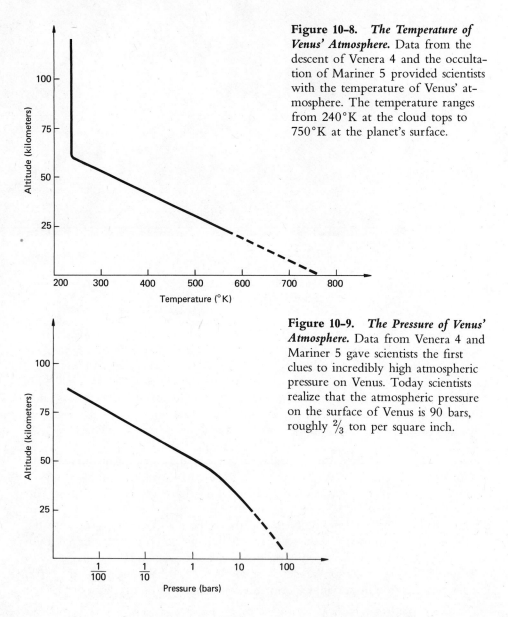

**Figure 10–8.** *The Temperature of Venus' Atmosphere.* Data from the descent of Venera 4 and the occultation of Mariner 5 provided scientists with the temperature of Venus' atmosphere. The temperature ranges from 240°K at the cloud tops to 750°K at the planet's surface.

**Figure 10–9.** *The Pressure of Venus' Atmosphere.* Data from Venera 4 and Mariner 5 gave scientists the first clues to incredibly high atmospheric pressure on Venus. Today scientists realize that the atmospheric pressure on the surface of Venus is 90 bars, roughly ⅔ ton per square inch.

temperatures reported earlier by radio observations from Earth. The descent of Venera 4 and the fly-by of Mariner 5 dispelled all such misgivings. The data from the Soviet and American flights were in complete agreement: Venus is indeed very hot.

The temperature profile of Venus' atmosphere obtained from data returned by Venera 4 and Mariner 5 is shown in Figure 10–8. Confirming earlier earth-based observations, the temperature of the cloud tops was found to be 240°K (−45°F). Once inside the clouds, however, the temperature rises sharply. Although the data stopped at an elevation of about 23 kilometers above the Venusian surface, it was possible to extrapolate the temperature down to the ground (indicated by dashed lines in Figure 10–8). The surface temperature came out to be 760°K. As a result of later Venera flights, particularly Venera 8, which successfully landed on the surface in 1972, this surface temperature was revised downward slightly. The currently accepted surface temperature on Venus is 750°K (480°C = 890°F).

In addition to measuring the temperature, the descent of Venera 4 and the occultation of Mariner 5 permitted scientists to deduce the atmospheric pressure on Venus. Recall that the average atmospheric pressure on Earth at sea level is 1 bar ( = 1,000 millibars = "1 atmosphere"). Surprisingly, scientists found that the atmospheric pressure on Venus is considerably higher. Once again, data were not available below an altitude of about 25 kilometers, but extrapolation down to the ground (dashed lines in Figure 10–9) gave a pressure of 100 bars at the Venusian surface. Later Venera flights that transmitted data from the surface of the planet resulted in a slight downward revision. The currently accepted atmospheric pressure at the surface of Venus is 90 bars (1,300 pounds per square inch). The atmospheric pressure on Venus is ninety times greater than here on Earth.

Finally, direct chemical analysis of Venus' atmosphere during the descent of Venera 4 showed that the content of carbon dioxide is in excess of 90 per cent. Later measurements made by Soviet scientists revealed that 97 per cent of Venus' atmosphere is composed of carbon dioxide.

The extremely high temperature, pressure, and carbon dioxide contents of Venus' atmosphere implies some exotic and fascinating chemical and physical processes. First of all, the high temperature can be understood as a result of the so-called *greenhouse effect*. Think about an automobile parked in the sunshine on a hot summer day with all the windows rolled up. It is common experience that the temperature inside the car can get very high, much higher than the air temperature outside the car. The reason for this is that visible sunlight easily penetrates the car windows. This sunlight is absorbed by objects inside the car that reradiate the light at much longer wavelengths as infrared radiation. This infrared radiation cannot escape back out through the windows and therefore the car heats up. Similarly with Venus, trapped infrared radiation is responsible for the planet's high surface temperature. Of course, most of the sunlight falling on Venus is reflected back out into space by the clouds, which have a high albedo. But the radiation that does penetrate to the surface is reradiated at infrared wavelengths, which cannot escape back through the thick cloud-cover.

The high surface temperature on Venus is directly associated with the chemical

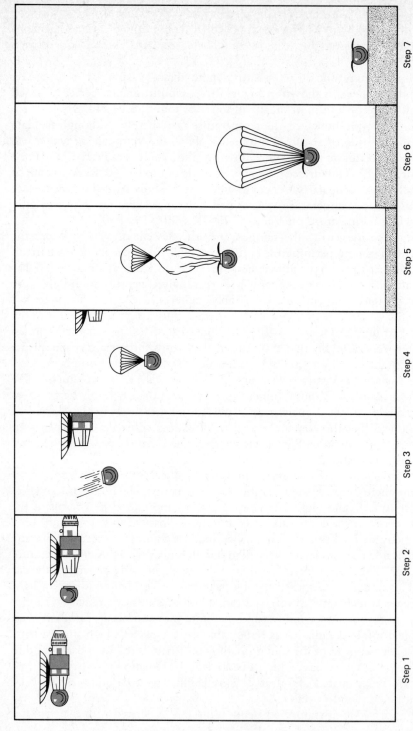

Step 1     Step 2     Step 3     Step 4     Step 5     Step 6     Step 7

**Figure 10–10.** *The Venera Plan.* Upon approaching Venus, the descent capsule separated from the spacecraft. The speed of descent was slowed by means of several parachutes.

composition of its atmosphere. Here on Earth there are vast quantities of carbon dioxide, but almost all of this gas is contained as carbonates in rocks. If Earth were to get as hot as Venus, the temperatures would be so high that the carbon dioxide in the carbonates would be liberated from the rocks. Typical reactions that might occur include:

$$CaCO_3 + SiO_2 \longrightarrow CaSiO_3 + CO_2$$

(calcite)   (quartz)      (wollastonite)

$$MgCO_3 + SiO_2 \longrightarrow MgSiO_3 + CO_2$$

(magnesite)   (quartz)      (entatite)

$$(CaMg)CO_3 + SiO_2 \longrightarrow (CaMg)SiO_3 + CO_2.$$

(dolomite)   (quartz)      (diopside)

The temperature on Venus is so high that all the carbon dioxide simply remains in the planet's atmosphere. Similarly, hydrogen chloride (HCl) and hydrogen fluoride (HF) are "baked" out of the rocks on Venus' hot surface. Spectral lines of both of these highly corrosive gases were discovered by William S. Benedict at the University of Maryland in high resolution earth-based spectra of Venus taken by Pierre and Janine Connes in France. Here on Earth, hydrogen chloride and hydrogen fluoride are "cooked" out of rocks only during volcanic activity. These gases do not survive long in the air; they rapidly combine with material on the ground and are neutralized.

Hydrogen chloride and hydrogen fluoride (which result in hydrochloric acid and hydrofluoric acid when dissolved in water) are minor constituents of the Venusian atmosphere. In sharp contrast, sulfuric acid ($H_2SO_4$) is an extremely important component of the clouds on Venus. The existence of sulfuric acid in the Venusian atmosphere was first proposed by Godfrey Sill at the University of Arizona and Ronald G. Prinn at M.I.T. in the 1970s. Sulfuric acid has a great affinity for water. Under the conditions on Venus, the sulfuric acid "gobbles" up water molecules and then dissociates into two ions $H_3O^+$ and $HSO_4^-$. First of all, this helps explain why Venus is so very dry. Ronald A. Schorn and Edwin S. Barker, both working at the McDonald Observatory, have detected only tiny amounts of water vapor in the upper atmosphere. The relative humidity is somewhat less than 1 per cent. Radio observations have established that there is less than .1 per cent water vapor in the lower atmosphere. Venus is much dryer than the most arid deserts here on Earth.

The proposal of sulfuric acid also helped explain several puzzling features detected in infrared spectra of the planet. In the early 1960s, W. H. Sinton at Lowell Observatory and V. I. Moroz in the Crimea discovered absorption features between 3 and 4 microns. Later, Sinton and J. Strong of Johns Hopkins University found broad absorption features at 11.2 microns. All of these spectral lines remained unexplained for over a decade, until the sulfuric acid proposal. The $H_3O^-$ ion readily absorbs light in the 3 to 4 micron range while the $HSO_4^-$ ion absorbs at 11.2 microns.

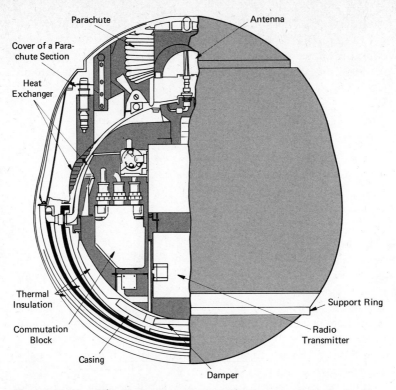

**Figure 10–11.** *The Venera Descent Capsule.* Descent through the Venusian atmosphere involved a spherical device measuring about two feet in diameter. The early Veneras were crushed by the enormous atmospheric pressure encountered before reaching the surface of the planet. Veneras 7 and 8 were much sturdier; both capsules landed safely and transmitted data from the Venusian surface for a few minutes.

While the United States was busy landing men on the moon, the Soviet Union sent several more spacecrafts to Venus. Veneras 5 and 6 were launched in January 1969, and made parachute descents through the Venusian atmosphere in mid-May of that year. As with Venera 4, both spacecrafts ceased functioning before landing on the surface. The descent capsules were evidently crushed by the enormous atmospheric pressure above the planet. Venera 7 was launched on October 17, 1970, and parachuted to Venus on December 15, 1970. It is claimed that Venera 7 survived for 23 minutes on the planet's surface. The currently accepted values of the surface pressure (90 atmospheres) and surface temperature (750°K) first came from this mission. The flight of Venera 8 proved to be even more successful. The massive 1,090-pound spacecraft was launched on March 28, 1972, and descended safely through the Venusian atmosphere on July 22, 1972. Venera 8 detected the base of the cloud cover at an altitude of 35 kilometers (22 miles) above the planet's surface. Wind velocities of 50 meters per second (110 miles per hour) were measured at an altitude of 45 kilometers (28 miles) above the planet. The wind speeds decreased steadily to 2 meters per second (5 miles per hour) at 10 kilometers (6 miles) above the surface.

In the meantime, Michael A. Janseen at the Jet Propulsion Laboratory made Earth-based microwave observations of Venus' atmosphere that indicated the presence of particles or droplets near the base of the Venusian cloud-cover shrouding the planet. In addition, Mariner 5 and Mariner 10 (discussed in the next section), detected a strong absorption of microwaves at altitudes of less than 50 kilometers (30 miles) above the planet. Again, these results could be explained by sulfuric acid. A sulfuric acid solution has a high electrical conductivity (this is the chemical used in automobile batteries) and could account for these effects if large enough droplets are present. Indeed, the current picture of the base of the Venusian cloud-cover involves a hot rain of concentrated sulfuric acid!

If this were not enough, sulfuric acid in Venus' atmosphere would be expected to react with the hydrogen fluoride to produce fluorosulfuric acid ($HSO_3F$). Fluorosulfuric acid is one of the most corrosive substances known to man. It will dissolve lead, tin, mercury, and most rocks. The surface of Venus is a vicious inferno. No wonder Soviet scientists had so much trouble with the early Venera missions.

## *Close-ups and Landings* 10.3

BACK in the 1960s, it was realized that one spacecraft could explore several planets. On rare occasions, the planets are positioned so that the gravitational field of one planet can be used to redirect a spacecraft on to another planet. Initial proposals centered about the "Grand Tour" in which a single spacecraft would pass by Jupiter, Saturn, and Uranus in the late 1970s. Unfortunately, this mission was abandoned for lack of financial support. Instead, another less-ambitious mission to the inner planets was undertaken.

From the fall of 1973 through the spring of 1974, the three planets closest to the sun were positioned in such a way that a single spacecraft launched from Earth would pass close to Venus and then head on to encounter Mercury. This was the goal of Mariner 10 whose flight trajectory is shown in Figure 10–12.

Mariner 10 was patterned after the earlier Mariners (6, 7, and 9) that successfully journeyed to Mars in 1969 and 1971. Like its predecessors, Mariner 10 carried two television cameras, magnetometers, an ultraviolet spectrometer, and an infrared radiometer. Perhaps the most obvious difference in the appearance of Mariner 10 (see Figure 10–13) was the elimination of two of the four solar panels. Since Mariner 10 was to journey toward the sun rather than away, only two solar panels were needed to supply the spacecraft with electrical power.

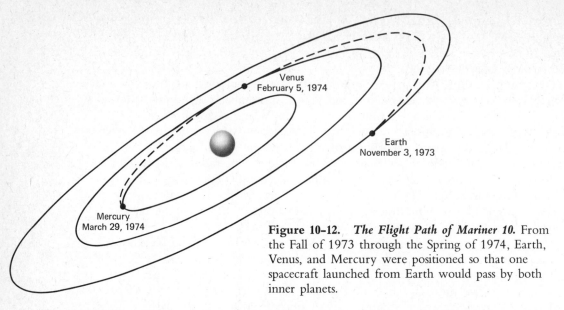

Venus
February 5, 1974

Earth
November 3, 1973

Mercury
March 29, 1974

**Figure 10–12.** *The Flight Path of Mariner 10.* From the Fall of 1973 through the Spring of 1974, Earth, Venus, and Mercury were positioned so that one spacecraft launched from Earth would pass by both inner planets.

**Figure 10–13.** *Mariner 10.* The design of this spacecraft for the Venus–Mercury fly-bys was patterned after the successful Mariners (6, 7, 9) that went to Mars in 1969 and 1971. (*NASA*)

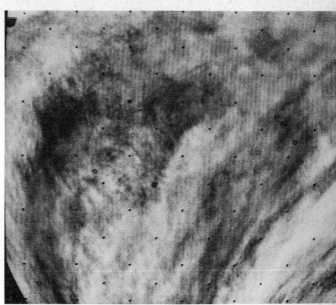

**Figure 10-14 (left).** *Venus from Mariner 10.* Photographs taken in ultraviolet light reveal circulation patterns in Venus' upper atmosphere. This particular view is a composite of many photographs taken at a range of 720,000 kilometers (450,000 miles) in February 1974. (*NASA*)

**Figure 10-15 (right).** *Close-up Details.* Whorls, mottled patterns, and spiral streaks result from convection in Venus' atmosphere. Details as small as 7 kilometers (about 4 miles) in size could be recognized in the ultraviolet photographs transmitted back to earth from Mariner 10. (*NASA*)

Mariner 10 was launched from Cape Canaveral on November 3, 1973. Three months later, on February 5, 1974, the spacecraft coasted past Venus at an altitude of 5,800 kilometers (3,600 miles) above the planet's surface. The most dramatic results came from Mariner's two television cameras. Photographs in visible light were very unrewarding. They showed only a featureless, shimmering cloud-cover. In visible light, Venus has a yellowish color. The precise reason for this yellowish hue is poorly understood; some scientists suggest that this coloring might come from sulfur in the planet's atmosphere.

While photographs in visible light were disappointing, those taken with ultraviolet light were spectacular. Using filters which transmitted light around 3550 Å, Mariner 10 detected light and dark markings (up to 30 per cent difference in brightness) in the upper cloud-cover. An excellent view of the entire planet is shown in Figure 10–14, while a close-up of the clouds appears in Figure 10–15. The dark and light markings must be caused by the uneven distribution of a substance (probably carbon monoxide, CO) in Venus' upper troposphere and lower stratosphere.

**Figure 10-16.** *The Classical "Y."* Earth-based ultraviolet photographs sometimes faintly revealed a "Y" pattern on Venus' cloud-cover. This view was taken four days after Mariner 10 swung past the planet. The distance to the spacecraft was 2,800,000 kilometers (1,700,000 miles). (*NASA*)

Some dark and light markings had been photographed from Earth at near-ultraviolet wavelengths. Astronomers sometimes spoke of the classical "Y" feature that faintly appeared in such photographs. Views from Mariner 10 (see Figure 10–16) showed this feature as dark belts near the planet's equator. All these patterns seen on Venus are caused by large-scale convection currents in the planet's atmosphere resulting from solar heating. A series of photographs taken at various intervals (see Figure 10–17) confirmed the fact that Venus' entire atmosphere rotates in a retrograde direction about the planet every four days. Since the planet itself rotates much slower (recall that a sidereal day on Venus lasts 243 Earth-days), these atmospheric motions must constitute high-speed winds extending over large areas.

As with all previous flights to Venus, Mariner 10 did not detect any magnetic field about the planet. Consequently, Venus does not possess any Van Allen belts or magnetosphere like Earth. The solar wind can, therefore, approach Venus much more closely than it approaches Earth. The overall structure of Venus' exosphere is shown in Figure 10–18. As the solar wind strikes the upper atmosphere, a shock wave is formed. This marks the boundary between supersonic and subsonic speeds of the particles in the solar wind. Inside the shock front is a boundary called the *anemopause,* where the pressure of the solar wind equals the planet's atmospheric pressure. As the solar wind flows around the anemopause, ions in Venus' uppermost atmosphere are swept away. This explains why the Mariner and Venera probes had found that Venus' ionosphere was much weaker than expected.

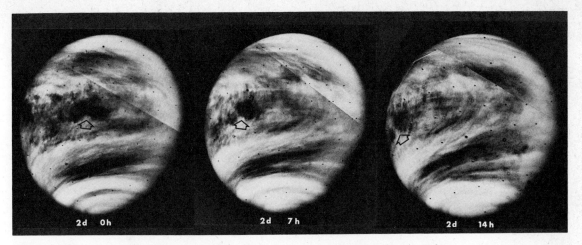

**Figure 10-17.** *Rotation of the Atmosphere.* These three views were taken at 7-hour intervals beginning 2 days after Mariner 10 flew past Venus. They reveal that the planet's upper atmosphere rotates in a retrograde direction about the planet once every 4 Earth-days. The dark feature marked with an arrow is about 1,000 kilometers (600 miles) across. (*NASA*)

**Figure 10-18.** *Venus' Exosphere.* Since Venus has no magnetic field, the solar wind can approach the planet much closer than it approaches the earth. The so-called anemopause marks the boundary where the pressure of the solar wind equals the pressure in the planet's upper atmosphere.

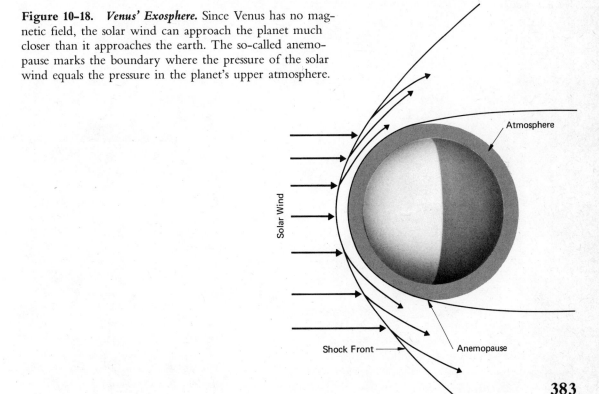

**383**

After a brief but fruitful encounter with Venus, Mariner 10 continued on to its rendezvous with Mercury. The next major advance in the exploration of Venus came as a result of the Soviet space program. Once again photographs were taken, but unlike the American mission, the best photographs came from the planet's surface.

Prior to the flights of Venera 9 and 10 in 1975, observations of Venus' surface were achieved by bouncing radar signals off the planet. Radar waves easily penetrate the thick cloud-cover and analysis of the "echo" permits scientists to get some idea about the planet's surface. A pioneer in this field is Dr. Richard M. Goldstein at the Jet Propulsion Laboratory. Using the huge antenna at the Goldstone Tracking Station in California (see Figure 10–19), Dr. Goldstein and his colleagues sent powerful bursts of radar waves to Venus. When a burst of radar waves hits Venus, several things happen. First of all, waves that hit a mountain peak are reflected back to Earth a little earlier than those that strike the rest of the surface. Similarly, waves

**Figure 10–19.** *The Goldstone Tracking Station.* This antenna in the California desert is 65 meters (210 feet) in diameter. In addition to receiving transmissions from spacecrafts, this antenna has been used to map the surface of Venus. (*JPL*)

that are reflected from the bottom of a valley or depression on Venus will arrive back at Earth a little later than all the other waves. Consequently, as a result of the topology of Venus' surface, the initial burst of radar waves gets spread out in time. The part of the "echo" that gets back to Earth first must have been reflected from high elevations while those parts that get back last must have bounced off of depressions on Venus' surface.

In addition to a spreading out in time, the burst of radar waves also gets spread out in wavelength. Those radar waves reflected from Venus' receding side are slightly redshifted while those reflected from Venus' approaching side are slightly blueshifted. This consequence of Venus' rotation and the Doppler effect permits scientists to tell which side of the planet the valleys and mountains are on.

The final result of radar mapping is a "photograph" in which departures (either depressions or rises) from the planet's average surface level appear as light markings, as shown in Figure 10–20. Recall that Venus' rotation is synchronous; the same side of Venus faces Earth at every inferior conjunction. Since the distance between Earth and Venus is smallest at an inferior confunction, this is the best time to bounce radar signals off the planet. Consequently, all the maps made at inferior conjunction look the same. At inferior conjunction, the same surface features on Venus are exposed to our radar transmissions.

**Figure 10–20.** *A Radar Map of Venus.* Analysis of radar signals bounced off of Venus gives this "map" of the planet's surface. Rough spots (mountains, craters, valleys, and so forth) appear as bright areas on an otherwise generally smooth surface. (*Courtesy of Dr. Goldstein; JPL*)

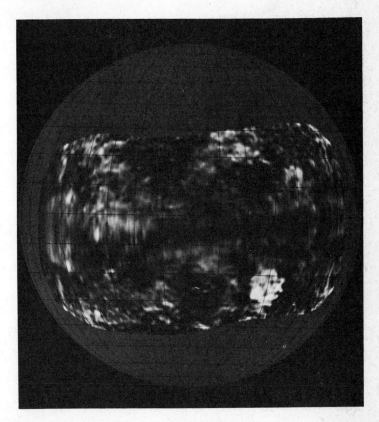

**Figure 10-21.** *Test Model of Veneras 9 and 10.* Two of these huge vehicles successfully landed on the Venusian surface in October 1975. Each lander weighed 3,400 pounds and descended through the cloud-cover by means of parachutes. (*Sovfoto*)

One of the major successes of the Soviet space program centers about the flights of Veneras 9 and 10. In an ambitious dual mission, two huge spacecrafts were launched toward Venus in June 1975. They arrived at the planet three days apart, on October 22 and 25, 1975. Each spacecraft consisted of a "bus" and a "lander" having a total weight of 4,650 kilograms (nearly 10 tons). Upon arriving at Venus, each spacecraft orbited the planet at distances ranging from roughly 1,000 to 110,000 kilometers with orbital periods of nearly 2 days. Following a "cooling maneuver," which lowered the temperatures of the spacecrafts to $-10°C$, the landers separated from the buses to begin their arduous descents through the Venusian atmosphere. A photograph of a test model of the Veneras 9 and 10 landing crafts is shown in Figure 10–21.

Using a system of heat shields and parachutes, each spacecraft took 75 minutes to descend through the Venusian atmosphere. They landed about 2,200 kilometers (1,400 miles) apart on the daylight side of the planet. Venera 9 lasted for 53 minutes on the planet's surface while Venera 10 survived for 65 minutes. At the Venera 9 landing site, the atmospheric pressure was 90 atmospheres (1,320 pounds per square inch) and the temperature was $760°K$ ( $= 485°C = 900°F$). At the

Venera 10 landing site, the atmospheric pressure was a little higher (92 atmospheres = 1,350 pounds per square inch) but the temperature was a little lower
(740°K = 465°C = 870°F).

During the descents of the two spacecrafts, numerous measurements of conditions in the Venusian atmosphere were made. There were no surprises; information
concerning chemical composition, pressure, and temperature confirmed data from
earlier flights. The overall structure of Venus' atmosphere is shown in Figure 10–22
while the chemical composition of the atmosphere is indicated in Figure 10–23.

**Figure 10–22.** *The Structure of
Venus' Atmosphere.* The overall structure of Venus' atmosphere is shown
in this composite graph. Most of the
clouds are located between altitudes
of 26 to 60 kilometers (16 to 37
miles) above the planet's surface. A
hazy layer 20 kilometers (12 miles)
thick exists above the clouds. The
curve indicates the temperature in
the atmosphere (measured on the
bottom axis of the graph) while the
approximate atmospheric pressure is
given on the right.

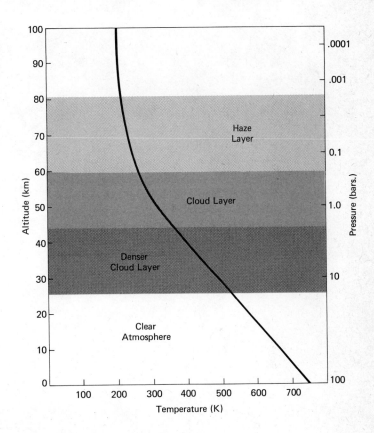

**Figure 10–23.** *The Composition of Venus' Atmosphere.* Venus' atmosphere consists almost entirely of carbon dioxide (97 per cent). Nitrogen
is probably the second most abundant gas, comprising less than 2 per
cent of the Venusian atmosphere. This leaves about 1 per cent for all
other gases (oxygen, water vapor, ammonia, and corrosive gases).

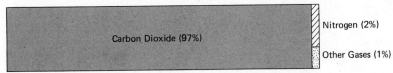

**Figure 10–24 (top).** *The View from Venera 9.* Sharp, angular rocks measuring 12 to 16 inches across are shown in this panorama of Venus' surface. Surface details 500 feet from the spacecraft can be seen. The horizon (upper right) is about 800 feet away. (*Sovfoto*)

**Figure 10–25 (bottom).** *The View from Venera 10.* The Venera 10 landing site was smoother than the Venera 9 site, 1,400 miles away. The area covered in this view ranges from 7 feet across near the spacecraft (at bottom) to 660 feet across near the horizon (top). The arrow points to a density device extending from the spacecraft. (*Sovfoto*)

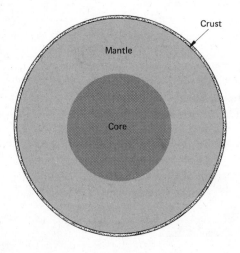

**Figure 10–26.** *The Internal Structure of Venus.* Like Earth, Venus evidently has a crust, mantle, and nickel–iron core. An Earth-like crust is inferred from the similar radioactive content of Venusian and terrestrial rocks. An Earth-like interior is deduced from the similar size, mass, and average density of the two planets.

388

The most impressive results from Venera 9 and 10 came after the spacecrafts had safely landed. Prior to these missions, some scientists believed that the Venusian atmosphere was so thick that no sunlight could penetrate the cloud-cover to illuminate the planet's surface. Much to everyone's surprise, both Veneras 9 and 10 succeeded in taking and transmitting photographs of the Venusian surface back to Earth. A panorama seen from the Venera 9 landing site is shown in Figure 10–24. Rocks measuring 12 to 16 inches across were seen near the base of the spacecraft.

Many scientists were surprised to see sharp, angular rocks revealed in the photographs from Venera 9. Recall that Venus' atmosphere rotates about the planet once every 4 days, but the planet itself rotates much slower, once every 243 days. It was therefore expected that high winds raging across the planet's surface would have severely eroded any rocks, producing fine gravel and sand. To the contrary, the wind speeds on the Venusian surface are quite low. At the Venera 10 landing site wind velocities of 3.5 miles per second (8 miles per hour = 7 knots) were measured. Consequently, the speed of the winds in Venus' atmosphere is slowed considerably at ground level.

A panorama taken from the Venera 10 landing site (see Figure 10–25) revealed a smoother surface than the Venera 9 site. Soviet paleontologist Mikhail Marov aptly described the Venusian surface as a "stoney desert" covered with rocks and boulders and *not* with a finely ground regolith like the moon.

While operating on the Venusian surface, gamma ray spectrometers on board Veneras 9 and 10 measured the content of radioactive elements in the planet's crust. The results were: potassium = 0.3 per cent, thorium = 0.0002 per cent, and uranium = 0.0001 per cent. This is very similar to basalt and granite here on Earth. This strongly suggests that Venus—like Earth—is chemically differentiated. Like Earth, Venus has a crust, mantle, and nickel—iron core, as shown in Figure 10–26. An earth-like interior is inferred from the fact that both Venus and Earth have very similar masses, sizes, and average densities.

While the interior of Venus is perhaps very similar to our own planet, the atmosphere is vastly different. The obvious question is: Why? What happened on Venus to give the planet an atmosphere with the incredible properties our space-crafts detect? Actually, here on Earth there is roughly the same amount of carbon dioxide as is found in Venus' oppressive atmosphere. *But* on Earth, almost all of this carbon dioxide is contained in rocks.

Perhaps, long ago, Venus was very much like our own planet. Eventually, because Venus is closer to the sun than Earth, the Venusian oceans began to boil. Venus then became enshrouded in a thick cloud-cover composed of water vapor. Due to the greenhouse effect discussed in the previous section, the temperature on the planet's surface began to rise dramatically. Finally, the temperature got so high that the carbon dioxide in the rocks was baked out. Molecules of water in the planet's upper atmosphere were then broken apart or *dissociated* by ultraviolet radiation from the sun, thereby allowing the hydrogen to escape off into outer space.

In about five billion years, the structure of our sun will change dramatically. Depletion of hydrogen fuel for the thermonuclear "fires" at the sun's core will signal the end of the sun's life as a main-sequence star. The sun will then expand as it becomes a red giant star. The increased size of the sun will cause temperatures here on Earth to rise hundreds of degrees. The oceans will boil and our planet will become enshrouded in a thick cloud-cover. Eventually, as the temperatures approach 1,000°F, the carbon dioxide in the rocks will be baked out giving our planet a thick, oppressive carbon dioxide atmosphere. Perhaps, in looking at Venus, we see the terrifying fate of our own Earth.

## Questions and Exercises

1. Why were early astronomers misled in believing that conditions on Venus might be very similar to those on Earth?
2. Discuss the telescopic appearance of Venus during one synodic period.
3. When is Venus at maximum brilliancy? Briefly discuss the competing factors which are important in determining when Venus is at maximum brilliancy.
4. Why don't transits of Venus occur at every inferior conjunction? When will the next transits occur?
5. In what way does the rotation of Venus differ from the rotation of all the other inner planets?
6. How was the rotation rate of Venus' surface first determined?
7. How does the rotation rate of Venus' surface differ from that of its upper atmosphere?
8. What is the most abundant chemical in the Venusian atmosphere and how was it first discovered?
9. How was the temperature of Venus' surface first measured? What was the result?
10. Why is it reasonable to suppose that Venus has a dense core, perhaps composed of nickel and iron?
11. What is the strength of Venus' magnetic field?
12. What is the temperature at the top of the Venusian cloud-cover? At what altitude above the Venusian surface are the cloud-tops located?
13. What is the atmospheric pressure and temperature at Venus' surface?
14. What is meant by the greenhouse effect and does it account for Venus' high surface temperature?
15. What percentage of the Venusian atmosphere is carbon dioxide?
16. Give two reasons why sulfuric acid is an important constituent of the Venusian cloud cover. What observations are easily explained by sulfuric acid?
17. What is the altitude above Venus' surface of the bottom of the cloud-cover? Approximately how thick is the cloud-cover?
18. In your own words, briefly summarize the major properties of the atmosphere of Venus.

19. Describe the appearance of the Venusian atmosphere in visible and ultraviolet light as discovered by the flight of Mariner 10.
20. Briefly discuss the interaction of Venus and the solar wind.
21. What is the anemopause?
22. How are radar waves used to "see" Venus' surface?
23. In your own words, describe the appearance of the Venusian surface as photographed by Veneras 9 and 10.
24. What, roughly, is the wind speed on the Venusian surface?
25. Why do scientists believe that rocks on the Venusian surface are similar to rocks found in the earth's crust?
26. Why is the internal structure of Venus probably similar to that of Earth?
27. Why do some scientists believe that our planet may someday look like Venus?

# 11 On to Mercury

## 11.1 Mercury from Earth

MERCURY is one of the brightest objects regularly seen in the sky, exceeded only by the sun, moon, Venus, Mars, and Jupiter. With an apparent magnitude of −1.9 at greatest brilliancy, Mercury gets as bright as Sirius, the brightest-appearing star in the nighttime sky. Yet in spite of its brilliancy, few people have ever seen this planet. According to legend, on his deathbed Nicholas Copernicus lamented that he had never seen Mercury. Even today, many professional astronomers have never personally viewed this innermost planet in our solar system.

Mercury is difficult to see simply because it is so very close to the sun. The average distance between Mercury and the sun is only 0.39 AU. This means that, as seen from Earth, Mercury always appears less than 28° away from the blinding solar disk in the sky (see Figure 11–1). Mercury appears farthest from the sun at greatest eastern and western elongations. At greatest eastern elongation, Mercury is seen at dusk as an "evening star" and sets less than two hours after sunset. At greatest western elongation, Mercury rises less than two hours before sunrise and appears as a "morning star" at dawn. Because of its proximity to the sun, Mercury is never seen clearly when the sky is completely dark. The careful observer must search for it in the twilight sky.

Mercury was well known to ancient astronomers. But as in the case of Venus, these early astronomers did not realize that the planet seen in the morning and evening skies was the same object. For example, the early Greeks called the planet Mercury when it appeared as an "evening star," but referred to it as Apollo when it appeared as a "morning star." The corresponding names in ancient India were Raulineya and Buddha, and in Egypt it was termed Horus and Set. Since the synodic period of Mercury is 116 days, it appears three times in the evening twilight and three times in the morning twilight each year.

Being the innermost planet in the solar system, Mercury takes a very short time to circle the sun. The sidereal period of Mercury is 88 days and its average orbital speed is 48 kilometers per second (roughly 100,000 miles per hour). As Mercury orbits the sun, the distance from the sun to the planet varies from 46 million kilometers (29 million miles) at perihelion to 70 million kilometers (43 million

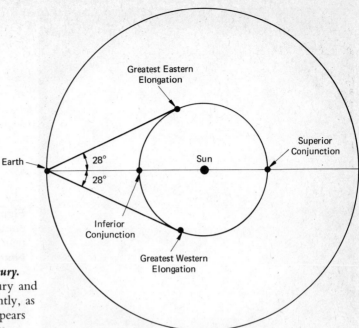

**Figure 11-1.** *Configurations of Mercury.*
The average distance between Mercury and
the sun is about 0.39 AU. Consequently, as
seen from Earth, Mercury always appears
less than 28° from the sun in the sky.

miles) at aphelion. This is because Mercury's orbit is highly elliptical. Most of the
planets have very circular orbits; only Pluto's orbit is more eccentric than Mercury's. In addition, Mercury's orbit is more steeply inclined to the ecliptic than any
other planet except for Pluto. The orbits of most planets lie very nearly in the plane
of the ecliptic, but Mercury's orbit is inclined to the ecliptic by an angle of 7°.

In spite of the steep inclination of Mercury's orbit, transits of this tiny planet are
seen more frequently than those of Venus. Mercury is closer to the sun and more
frequently passes through inferior conjunction than Venus. On the average, 13
transits of Mercury occur each century. These transits are seen only in May and
November because only at such times can Mercury be both at inferior conjunction
*and* in the plane of the ecliptic—obvious requirements for an Earth-based observer
to see the tiny black disk of Mercury passing across the face of the sun. A photograph of a transit is shown in Figure 11-2.

Mercury is the smallest planet in the solar system. With a diameter of 4,880
kilometers (about 3,000 miles) it is slightly larger than the moon and only about
one third as big as Earth. Indeed, Callisto, one of the largest satellites of Jupiter, is
very nearly the same size as Mercury.

Mercury also has the lowest mass of any planet in the solar system. Since
Mercury has no natural satellite, the planet's mass could be determined only by
measuring the size of perturbations on objects passing near the planet. Occasionally,
comets and asteroids pass near Mercury. By observing how much these objects are
deflected from their usual orbits, the mass of Mercury could be estimated. Prior to
the 1970s, one of the best determinations came from the asteroid Icarus, which

**393**

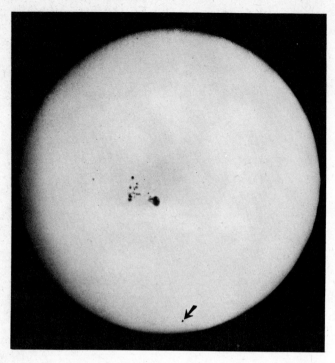

**Figure 11–2. *A Transit of Mercury.*** This photograph shows the transit of Mercury that occurred on November 14, 1907. Note how tiny the planet appears. Transits are seen only in May and November, when Mercury is both at inferior conjunction and in the plane of the ecliptic. This combination of circumstances occurs about 13 times each century. (*Yerkes Observatory photograph*)

passed within 16 million kilometers (10,000 miles) of the planet in 1968. More recently, the spacecraft Mariner 10 passed very near the planet three times in 1974 and 1975. This permitted an extremely accurate determination of Mercury's mass. Mercury's mass is $\frac{1}{18}$ that of Earth.

In view of Mercury's mass and size, the planet's average density turns out to be 5.4 grams per cubic centimeter. Recalling the average density of Earth (5.5 grams per cubic centimeter), this strongly suggests that Mercury must have a dense core, perhaps composed of nickel and iron, just like our own planet.

In addition, Mercury's mass and size result in the planet having a very low escape velocity, only 4.3 kilometers per second (about 10,000 miles per hour). For comparison, recall that the escape velocity from Earth is 11.2 kilometers per second (26,000 miles per hour) and from the moon is 2.4 kilometers per second (5,600 miles per hour). A low escape velocity, coupled with the fact that Mercury is so very near the sun, means that Mercury possesses no atmosphere. Daytime temperatures on the planet range up to 700°K (430°C = 800°F). At such high temperatures, molecules of any atmospheric gas are traveling so fast that they easily escape from Mercury's weak gravitational pull.

Since Mercury is so small and always so very near the sun, it is almost impossible to obtain a clear view of the planet from Earth. Of course, as an inferior planet, Mercury exhibits phases just like Venus. But Mercury's angular size is always very small, roughly 7 seconds of arc at greatest elongation. By comparison, Venus appears three times larger at greatest elongation and still larger at greatest brilliancy.

To make matters worse, astronomers can obtain good photographs of objects only when they are far above the horizon. The unsteadiness of the earth's atmosphere severely degrades the images of objects seen close to the horizon. Consequently, the best observations of Mercury are made when the planet appears high in the sky. But at such times the sun is also high in the sky. Astronomers interested in Earth-based observations of the innermost planet are therefore forced to photograph Mercury during daylight.

An excellent Earth-based photograph of Mercury is seen in Figure 11–3. The view is obviously unrewarding. Experienced observers have reported seeing faint surface markings on the planet. Vague suggestions of maria-like features can be seen in Figure 11–3.

Until recently, astronomers had to rely on the faint maria-like markings on Mercury's surface in order to determine the planet's period of rotation. These markings are so faint and indistinct that for many years astronomers erroneously believed that Mercury's rotational period equalled its orbital period of 88 days. If this were true, Mercury would always keep one side facing the sun, just as one side of our moon always faces Earth.

In 1962, however, a team of radio astronomers at the University of Michigan detected radio waves from Mercury's dark side. This was surprising because, if one side of Mercury always faced the sun, then the dark side of Mercury should be shrouded in a perpetual frigid night. Temperatures on the night side would then be so cold that, according to Planck's radiation law (Section 4.1), very few radio waves would have been emitted from the planet's cold surface. Yet, some radiation was detected, indicating that the temperature of the night side was not as low as expected.

The next major breakthrough occurred in 1965 when Rolf B. Dyce and Gordon H. Pettengill at the Arecibo Observatory bounced radar pulses off of Mercury. By measuring the Doppler shift of the radar waves from the edges of the planet, it was concluded that the true rotation period was roughly 59 days.

The Italian physicist Guiseppe Colombo was quick to realize the significance of a 59-day rotation period. He noted that 59 is almost exactly two thirds of 88, the number of days in Mercury's sidereal period. More precisely, Mercury orbits the

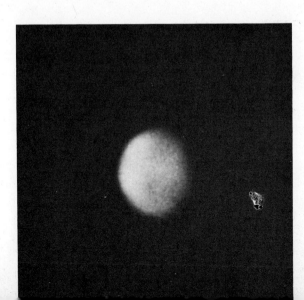

**Figure 11–3. *Mercury.*** Even through the best telescopes on Earth, Mercury is an unrewarding sight. Like Venus and the moon, Mercury exhibits phases. Faint surface markings on the planet are also sometimes seen. (*New Mexico State University Observatory*)

sun once every 87.97 days and $\frac{2}{3} \times 87.97 = 58.65$. Colombo therefore conjec-
tured that Mercury's rotation period is *exactly* 58.65 days. This means that Mercury
rotates three times about its axis while circling the sun twice. This phenomenon is
known as *spin–orbit coupling.* Colombo's brilliant guess has been confirmed with a
high degree of accuracy in recent years, especially as a result of the flight of Mariner
10.

In the late 1960s, several astronomers—particularly Peter Goldreich, Stanton J.
Peale, Irwin I. Shapiro, and Colombo—presented arguments explaining why
Mercury exhibits this spin–orbit coupling. If Mercury is not a perfect sphere, but
rather is a little fatter along one direction in the plane of its equator, then the tidal
forces of the sun acting on this bulge will try to point the bulge toward the sun
every time Mercury is nearest the sun. Of course, this would occur if the same side
of Mercury always faced the sun. But another likely situation is the 2:3 spin–orbit
coupling. At successive perihelions, the bulge alternately points directly at or
directly away from the sun (see Figure 11–4). Actually, Mercury is highly spherical
and thus deviations from a perfect sphere are very tiny. It must have taken billions
of years for the tidal forces from the sun to slow down Mercury's rotation to its
present state.

The rotation period of Mercury has an interesting implication for an adven-
turous astronaut who someday might visit the planet. During most of the Mercurian "year" of 88 days, the sun would be seen to drift slowly across the sky from
east to west, just as here on Earth. But since a Mercurian "day" is so long,
something very unusual would occur near perihelion. At perihelion, Mercury is so
close to the sun that, according to Kepler's second law, the planet is moving very

**Figure 11–4.** *Spin-Orbit Coupling.* The simplest case of spin-orbit coupling is where
the orbital period equals the rotational period. The planet would then keep the
same side always facing the sun. Mercury exhibits 2:3 orbit coupling. Mercury rotates three times about its axis during two complete orbits of the sun.

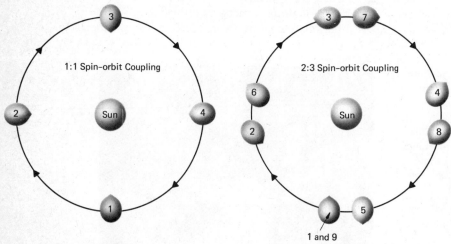

rapidly along its orbit. The orbit velocity is then so high that the apparent movement of the sun across the sky is dominated by motion of Mercury around the sun rather than the rotation of Mercury about its axis. During most of a Mercurian "day," the rotation of the planet causes the sun to appear to move across the sky from east to west. But near perihelion, the sun stops and backs up (that is, goes west to east) for a short period of time before resuming its usual westward course. This interval of retrograde motion lasts for 8 Earth-days, when Mercury's orbital angular speed exceeds its rotational angular speed near perihelion. An astronaut on the daylight side of Mercury near perihelion would actually see the sun "stop in the sky" and go "backward" for 8 Earth-days. At the *terminator* (that is, at the boundary of the daylight and night portions of the planet) this effect would even be more dramatic. At the evening terminator, an astronaut would see the sun sink below the horizon, and then momentarily come back up in the west before finally setting again, thereby plunging that portion of the planet into the long, cold Mercurian night. Similarly, at the morning terminator near perihelion, two sunrises would be seen. The sun would come up briefly, go back down in the east, and then once again rise roughly a week later.

Traditionally, the two inferior planets have posed severe observational problems for earth-based astronomers. Venus is perpetually shrouded in thick clouds while Mercury is always seen very near the blinding sun. As with Venus, important questions concerning the nature of the innermost planet in our solar system could only be answered by interplanetary space flight.

## *Views from Mariner 10* 11.2

FOLLOWING its highly successful encounter with Venus on February 5, 1974, Mariner 10 continued on course to Mercury. The Venus fly-by was specifically chosen so that the gravitational field of Venus would redirect Mariner 10 toward the innermost planet of our solar system, as shown in Figure 11–5. Mercury was the prime target of the mission.

After coasting for 21 weeks across a quarter of a billion miles of interplanetary space, Mariner 10 arrived at Mercury on March 29, 1974. Designed like the previous Mariner, which went to Mars three years earlier, Mariner 10 carried two television cameras, magnetometers, and spectrometers (see Figure 11–6). In addition to taking pictures, these instruments would measure Mercury's magnetic field, its interaction with the solar wind, the surface temperature of the planet and search for traces of an atmosphere.

Almost a week before the encounter, while at a distance of 5 million kilometers (3 million miles), Mariner's television cameras began sending back extraordinary views of Mercury. For the first time, the surface of the planet could be seen clearly. Day after day, teams of scientists stood in front of television monitors as the pictures

**Figure 11-5.** *The Flight of Mariner 10.* Venus' gravitational field was used to redirect the spacecraft toward Mercury. This was the first attempt of a "gravity-assisted" mission whereby the gravitational field of one planet is used as a slingshot to propel a spacecraft toward another planet. (*NASA*)

**Figure 11-6.** *The Mariner 10 Spacecraft.* Mariner 10 weighed 499 kilograms (1,100 pounds) and measured 7.6 meters (26 feet) from one end of the extended solar panels to the other. A total of 8,122 solar cells provided electrical power to operate the television cameras, magnetometers, spectrometers, and telemetry devices.

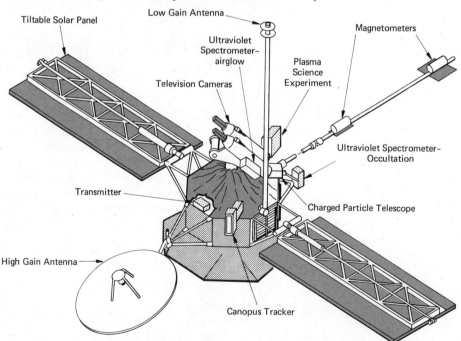

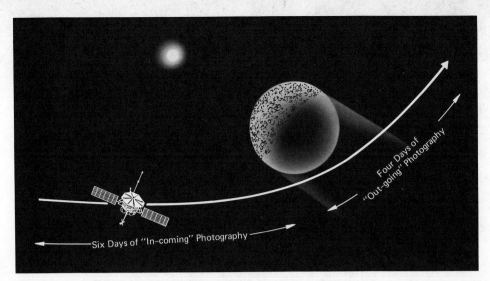

Figure 11-7. *The Fly-by of Mercury.* Mariner 10 passed behind the nighttime side of Mercury on March 29, 1974. Closest approach was 5,800 kilometers (3,600 miles) from the planet's center, or 700 kilometers (400 miles) above its surface. The first in-going photograph was obtained 6 days before closest approach and the last out-going photograph was sent back 4 days after closest approach.

came back. To everyone's surprise, Mercury looked very much like our own moon!

As shown in Figure 11–7, Mariner 10 passed behind the nighttime side of Mercury. Naturally, no pictures could be taken of the planet's unilluminated hemisphere. Consequently, all the photographs of visible surface features came either before encounter ("in-coming") or after encounter ("out-going"). Mosaics of the in-coming and out-going views of Mercury are shown in Figures 11–8 and 11–9. Representative close-up views revealing details of Mercury's lunar-like surface are seen in Figures 11–10, 11–11, and 11–12. At closest approach, the spacecraft was 5,800 kilometers (3,600 miles) from the planet's center, or 700 kilometers (400 miles) above the Mercurian surface. Near the time of closest approach, craters measuring 150 meters (500 feet) across could be easily resolved, as shown in Figure 11–13.

Although the illuminated side of Mercury presented views that could be mistaken for a lunar landscape, there are some significant differences between the innermost planet and our moon. Even the most heavily cratered portions of Mercury exhibit conspicuous plains (see, for example, Figures 11–10 and 11–12). By contrast, the heavily cratered lunar highlands are packed with numerous overlapping craters. The intercrater plains commonly seen on Mercury are very rare on the moon.

An important reason for this difference in the Mercurian and lunar surfaces was suggested by Donald E. Gault at NASA. He noted that the surface gravity on Mercury (39 per cent of Earth) is slightly more than twice the surface gravity on

**Figure 11-8.** *The "In-Coming" View.* Eighteen pictures taken at 42-second intervals as Mariner 10 approached Mercury were fashioned into this photo-mosaic. The spacecraft was 200,000 kilometers (120,000 miles) from the planet and six hours away from closest approach at the time of photography. The largest craters measure roughly 200 kilometers (120 miles) in diameter. (*NASA*)

**Figure 11-9.** *The "Out-Going" View.* Eighteen pictures taken at 42-second intervals as Mariner 10 receded from Mercury were fashioned into this photo-mosaic. Mariner 10 had passed closest approach six hours earlier and was at a distance of 210,000 kilometers (130,000 miles) from the planet at the time of photography. A large circular basin, about 1,300 kilometers (800 miles) in diameter, is partially seen along the terminator. (*NASA*)

**Figure 11-10.** *Mercury's Southwestern Quadrant.* Four hours before closest approach on March 29, 1974, this view of the southwestern portion of Mercury was received from Mariner 10. The spacecraft was almost 150,000 kilometers (about 90,000 miles) from the planet when this photograph was taken. The largest craters seen in this picture are about 100 kilometers (60 miles) across. (*NASA*)

**Figure 11-11.** *Lunar-like Craters.* This photograph was taken from a range of 87,000 kilometers (54,000 miles), just two hours after closest approach on March 29, 1974. The prominent crater with a central peak near the bottom of the photograph is 30 kilometers (20 miles) in diameter (*NASA*)

**Figure 11-12.** *Craters and Plains.* This photograph was taken from a range of 55,000 kilometers (35,000 miles) and shows a portion of Mercury's northern hemisphere. The view is about 500 kilometers (300 miles) wide. Lava flows, such as those that were responsible for maria on our moon, probably created the plains seen in this view. (*NASA*)

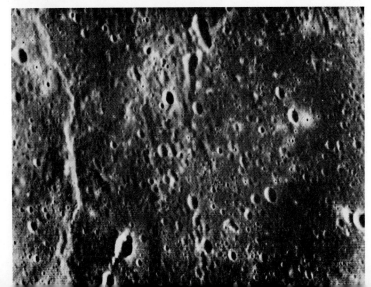

**Figure 11-13.** *Mercury at Closest Approach.* Only minutes after closest approach to Mercury on March 29, 1974, Mariner 10 sent back this high-resolution photograph. The view measures 50 by 40 kilometers (31 by 25 miles) and craters as small as 150 meters (500 feet) in diameter can be seen. (*NASA*)

the moon (16.5 per cent of Earth). Consequently, when a meteoroid hits the suface of Mercury, ejecta from the impact are scattered over a much smaller area than on the moon. Indeed, material ejected during crater formation on Mercury would cover an area only one sixth as large as the area that would have been covered on our moon. Secondary impact craters on Mercury would therefore be more closely clustered around the primary impact crater and hence any underlying plains would not be obliterated as easily as on the moon.

Another way in which the surfaces of Mercury and the moon differ involves the presence of long, shallow cliffs on the Mercurian surface. These cliffs, called *lobate scarps,* run for hundreds of kilometers across the planet's surface. An edge-on view of one such scarp is identified in Figure 11–14. Robert G. Strom of the University of Arizona has suggested that these scarps may have been caused by a wrinkling of the planet's surface. Perhaps, billions of years ago, as the planet cooled, it contracted slightly. A shrinkage in the size of Mercury would have resulted in a wrinkling of the crust, thereby producing the observed cliffs. Similar scarps are not seen on the moon or Mars.

One of the largest features on Mercury was seen from Mariner 10 shortly after

**Figure 11–14.** *A Lobate Scarp.* Close examination of the Mariner 10 pictures revealed that Mercury is covered with long cliffs called lobate scarps. These scarps, such as the one seen near the limb of the planet in this view, were probably formed by a wrinkling in the crust. (*NASA*)

closest approach. In general, the "out-going" view of the planet showed significantly more expansive plains than the "in-coming" view (compare Figures 11–8 and 11–9). One of these flat regions involves a huge impact basin seen along the terminator. This feature, which is partly hidden from view in the Mercurian night, is called the *Caloris Basin* or *Caloris Planitia*. Its name comes from the Latin word for "heat," because it lies at the subsolar point every two Mercurian "years" when the planet is at perihelion. At alternate perihelion passages, when Mercury is closest to the sun, the Caloris Basin is the hottest place on the planet because the nearby sun is almost directly overhead.

The Caloris Basin (see Figure 11–15) is 1,400 kilometers (870 miles) in diameter, which makes it comparable to Mare Imbrium on the moon. The basin is surrounded by mountains that rise as high as 2 kilometers (over 6,000 feet) above the surrounding plains. The floor of the basin is very heavily fractured and wrinkled. The fractures range in width from 8 kilometers (5 miles) down to a few hundred meters (a thousand feet), and perhaps smaller.

**Figure 11-15.** *The Caloris Basin.* One of the largest features on Mercury is this huge impact basin partially hidden from view at the planet's terminator. The basin measures 1,400 kilometers (870 miles) in diameter and its floor is very heavily fractured and wrinkled. (*NASA*)

**Figure 11–16.** *"Weird" Terrain.* Antipodal to the Caloris Basin, exactly on the opposite side of the planet, Mariner 10 photographed this unusual rippled terrain. Scientists believe that shock waves from the Caloris impact billions of years ago may have caused the surface to become rippled in this peculiar fashion. (*NASA*)

Exactly opposite or *antipodal* to the Caloris Basin, Mariner 10 photographed an unusual rippled terrain. As shown in Figure 11–16, numerous small hills and ridges cut across many of the craters and intercrater plains. It has been argued by Donald Gault and Peter Schultz at NASA that shock waves from the violent impact of the huge meteoroid that created the Caloris Basin would have come to a focus on the opposite side of the planet thereby causing the rippled appearance we observe. Although not quite so prominent, similar rippled terrain is found antipodal to the Imbrium and Oriental basins on the moon.

From analyzing the pictures sent back from Mariner 10, scientists divide the history of Mercury into five major stages, each quite similar to the prominent epochs that shaped our moon. Recalling that Mercury's average density implies a nickel–iron core, the planet must have chemically fractionated very shortly after its formation. Thus, during this first stage, the nickel and iron sank toward the planet's center, leaving a silicate and basaltic mantle.

Since there seem to be no surface features that could be interpreted as scars of the initial accretion process from which the planet formed, the second stage may have involved substantial vulcanism. During this epoch, all traces of primordial craters were obliterated. In addition, cooling and shrinking of the planet's core would have caused the crust to wrinkle, thereby forming the lobate scarps.

The third stage must have involved heavy bombardment of the planet, culminating with the violent impact that created the Caloris Basin. This was then followed by another epoch of vulcanism that characterized the fourth stage. Lava flows must have filled in the low-lying areas in much the same way that maria formed on the moon.

During the fifth and final stage, which extends up until the present time, very little has happened. During the past few billion years, occasional light impacts of small meteoroids formed the final generation of craters. These young craters show conspicuous rays. A good example is the crater Kuiper seen slightly above the center of the "in-coming" view in Figure 11–8.

During a few short weeks in the spring of 1974, more was learned about the planet Mercury than the total amount of knowledge that had been accumulated over the entire span of human history. The photographs transmitted back from Mariner 10 showed a remarkable lunar-like surface. Equally important, however, was the fact that while the television cameras were taking dramatic pictures, Mariner's other instruments were collecting data that show that Mercury has a very Earth-like interior.

## An Earth-like, Moon-like Planet 11.3

BACK in the 1960s, when a gravity-assisted space flight to Mercury was first proposed, scientists did not realize how auspicious the orbit of Mariner 10 would be. After launch from Earth (November 3, 1973), the spacecraft would coast past Venus (February 5, 1974) and then head on toward an encounter with Mercury (March 29, 1974). After the Mercury fly-by, Mariner 10 would assume an elliptical orbit about the sun. The unforeseen bonanza involves the fact that the orbital period of Mariner 10 about the sun turned out to be 176 days. That is *exactly* two Mercurian "years." Since the orbital period of Mariner 10 is exactly twice the orbital period of Mercury, the spacecraft was destined to encounter on September 21, 1974 and again on March 16, 1975, as shown in Figure 11–17. Fortunately, the spacecraft carried enough fuel so that the second and third encounters with Mercury would send back usable data. The three historic fly-bys were termed "Mercury I," "Mercury II," and "Mercury III." Table 11–1 lists some significant information about these encounters.

During the first Mercury encounter, only one half of the planet's surface was photographed. It was obviously impossible to see any features on Mercury's dark side. Unfortunately, the later fly-bys could also not provide any views of Mercury's hidden side. Recall that Mercury exhibits a 2:3 spin-orbit coupling; during two revolutions about the sun, the planet rotates exactly three times about its axis. Since Mariner 10 orbits the sun once every 176 days ( = 2 Mercurian "years"), exactly the same view of the planet was exposed to the spacecraft's cameras on each of the fly-bys. Thus, while one half of Mercury has been photographed in great detail, the other half has never been viewed.

Aside from the astounding pictures of Mercury's surface, one of the surprises to come from the flight of Mariner 10 was the discovery that Mercury has a magnetic field. Recall that Earth has a nickel–iron core and a magnetic field. It is believed that Earth's magnetic field is caused by electric currents arising from fluid motions in the core. As Earth rotates once every 24 hours, electric currents in the core produce our planet's magnetic field.

Based on the average density of the planets, scientists had been aware of the fact

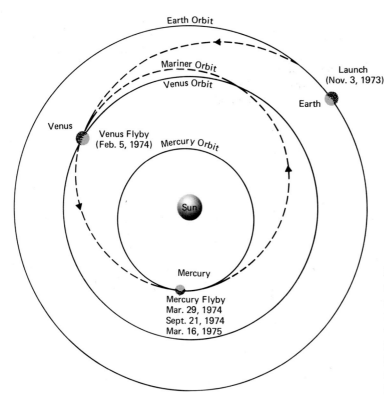

**Figure 11–17.** *The Orbit of Mariner 10.* After encountering Mercury on March 29, 1974, Mariner 10 continued to orbit the sun with a period of 176 days. Since the sidereal period of Mercury is 88 days, Mariner 10 is in an orbit that encounters Mercury every two Mercurian "years."

**Table 11–1.** *The Mercury Encounters*

| Name | Date | Distance* | Resolution† |
|------|------|-----------|-------------|
| Mercury I | March 29, 1974 | 703 kilometers (436 miles) | 450 meters (1,500 feet) |
| Mercury II | September 21, 1974 | 50,000 kilometers (31,000 miles) | 30,000 meters (19 miles) |
| Mercury III | March 16, 1975 | 327 kilometers (203 miles) | 204 meters (670 feet) |

\* Distance above planet's surface at closest approach.
† Size of smallest surface detail visible at closest approach.

that both Venus and Mercury possess nickel–iron cores. The interior structure of Venus was discussed in Section 10.3 (see Figure 10–26) and the interior of Mercury is shown in Figure 11–18. Compared to Earth (see Figure 11–19), a much larger percentage of Mercury's mass is contained in its core. But numerous American and Soviet space flights to Venus have failed to detect any Venusian magnetic field. The absence of a magnetic field about Venus is apparently caused by the planet's slow rotation. Venus rotates once about its axis every 243 days. This is so slow that electric currents in Venus' core cannot produce a planet-wide magnetic field.

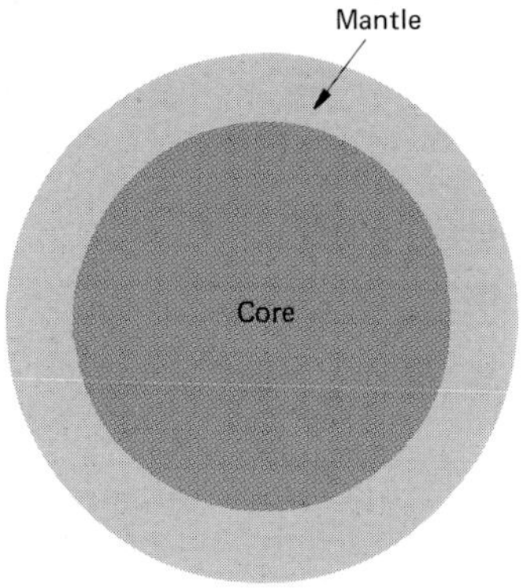

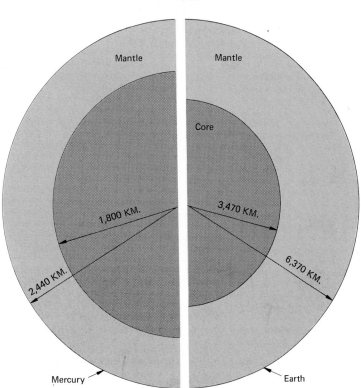

**Figure 11-18.** *Mercury's Interior Structure.* Mercury possesses a very large nickel–iron core. While the diameter of the planet is only 4,880 kilometers (3,030 miles), the diameter of the core is 3,600 kilometers (2,200 miles). The iron core contains 80 per cent of Mercury's mass.

**Figure 11-19.** *Mercury and Earth Compared.* This illustration shows cut-away views of Mercury and Earth drawn to the same scale. Compared to Earth, a much larger percentage of Mercury's mass is contained in its core. (*From "Mercury" by Bruce C. Murray, Copyright © 1975 by Scientific American, Inc. All rights reserved* )

Mercury also rotates very slowly. This innermost planet rotates once about its axis every 59 days. As in the case of Venus, it was assumed that this slow rate of rotation would fail to produce a planet-wide magnetic field about Mercury.

To everyone's surprise, Mariner 10 detected a magnetic field about Mercury during the first fly-by in March 1974. Unfortunately, during the second fly-by, the spacecraft never got closer than 50,000 kilometers (31,000 miles) to the planet. This was too far away to measure the magnetic field. But during the third encounter, Mariner 10 skimmed within 327 kilometers (203 miles) of the planet's surface and confirmed the existence of Mercury's magnetic field.

At Mercury's surface, the strength of the planet's magnetic field is only 1 per cent as strong as the geomagnetic field here at Earth's surface. Although weak by terrestrial standards, Mercury's magnetic field is considerably stronger than that of either Venus or Mars. Specifically, Mercury's magnetic field is strong enough to cause a substantial interaction with the solar wind. As shown in Figure 11–20, supersonic particles in the solar wind impinging on Mercury first encounter a shock wave, just as in the case of Earth. The inner portions of Mercury's *magnetosphere* are

**Figure 11-20.** *Mercury's Magnetosphere.* From data collected during the first and third encounters with Mercury, Mariner 10 confirmed the existence of a magnetosphere about the planet. Mercury's magnetosphere is superficially a weak version of Earth's magnetosphere. Mercury's magnetic field is oriented in the same direction as Earth's.

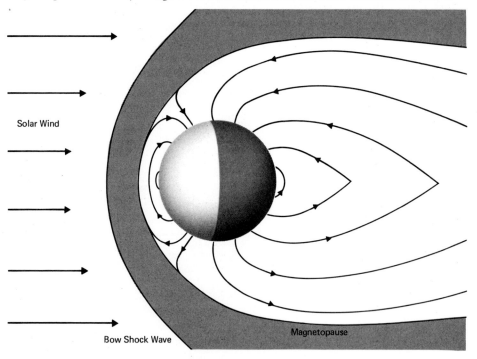

bounded by a *magnetopause* where the strength of the interplanetary magnetic field is counterbalanced by Mercury's magnetic field. Apparently, Mercury's magnetic field is not strong enough to sustain any trapped radiation belts analogous to the Van Allen belts that encircle Earth. Nevertheless, Mercury's field is strong enough to capture some helium nuclei contained in the solar wind. Mariner 10 discovered a very thin envelope of helium gas surrounding the planet. Of course, the planet's comparatively weak gravitational field and high daytime temperatures preclude the retention of a permanent atmosphere. Noontime temperatures on Mercury's equator soar to 700°K (430°C = 800°F). In contrast, temperatures on the planet's dark side drop to only 100°K (−170°C = −280°F) during the long Mercurian night.

Exactly how Mercury's large but slowly rotating core produces the planet's magnetic field remains poorly understood. Nevertheless, there is now considerable agreement concerning the proposal of Gault and Schultz (see Section 11.2) that Mercury's core played an important role in at least one of the planet's surface features. As mentioned in the previous section, one of the largest features on Mercury is the Caloris Basin. Billions of years ago, a comparatively large planetoid—perhaps 10 miles in diameter—must have violently collided with Mercury. Seismic waves from the impact must have shaken the entire planet. Specifically, surface waves traveling through the crust and mantle would have combined with compression waves traversing the core and come to a focus on the opposite side of the planet, as shown in Figure 11–21. Indeed, an unusual rippled terrain is found antipodal to the Caloris Basin on Mercury. Close-up views of the Caloris Basin as well as the antipodal "weird" terrain are seen in Figures 11–22 and 11–23.

Following its third successful encounter with Mercury in March 1975, Mariner 10 exhausted the supply of fuel that kept the spacecraft properly positioned in space. Tumbling helplessly as it orbits the sun every 176 days, Mariner 10 will continue

**Figure 11-21.** *Effects of the Caloris Impact.* Seismic waves from the Caloris impact would have come to a focus on the opposite side of Mercury. This is the likely explanation for the "weird" terrain observed antipodal to the Caloris Basin.

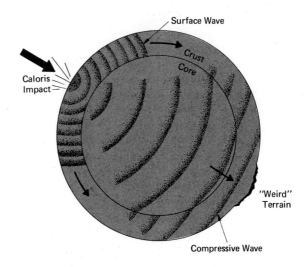

**Figure 11–22.** *The Caloris Basin.* In spite of three encounters with Mercury, no more than one half of the Caloris Basin has ever been viewed from Mariner 10. Lava from the planet's interior must have filled in the basin shortly after impact with a comparatively large planetoid. The interior of the basin was completely flooded while the adjacent lowlands were only partially filled in. (*NASA*)

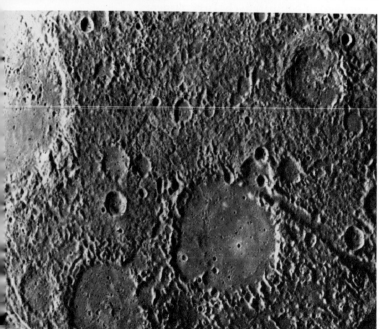

**Figure 11–23.** *"Weird" Terrain.* Details of the unusual rippled terrain antipodal to the Caloris Basin are shown in this close-up. Mariner 10 was at an altitude of 35,000 kilometers (22,000 miles) when this photograph was taken. The picture covers an area of 290 by 220 kilometers (180 by 140 miles). The large valley to the right is 7 kilometers (4 miles) wide and more than 100 kilometers (60 miles) long. The large flat-bottomed crater near the center of the view must have been formed *after* the Caloris impact. This crater is about 80 kilometers (50 miles) in diameter. (*NASA*)

to pass Mercury every six months. Its instruments are silenced; its cameras are blind to the now-familiar features perpetually exposed to the spacecraft's view. Although Mariner 10 is incapable of returning any more data, scientists and engineers are busily preparing follow-up missions. Spacecrafts that would orbit the planet and descend to its surface, like the Viking missions to Mars in 1976, could be one of the most rewarding scientific ventures of the 1980s.

**410**

1. How bright does Mercury get at maximum brilliancy? How does this compare with the brightness of other objects in the sky?
2. Why is it so difficult to obtain a good view of Mercury from Earth?
3. Briefly describe Mercury's orbit about the sun. How do the orbital inclination and eccentricity compare with that of other planets?
4. How often are transits of Mercury seen?
5. Briefly describe the necessary conditions for an Earth-based observer to see a transit of Mercury.
6. Compare Mercury's mass and size to Earth and the moon.
7. Why does Mercury's average density strongly imply that the planet has a dense core?
8. When was the rotation period of Mercury first determined? How?
9. What is the rotation period of Mercury and what exceptional relationship does it have to Mercury's orbital period?
10. What is meant by 2:3 spin-orbit coupling?
11. What familiar object in the sky exhibits 1:1 spin-orbit coupling?
12. Describe the unusual apparent motion of the sun as seen from Mercury near perihelion that results from the planet's orbital eccentricity and its 2:3 spin-orbit coupling.
13. Briefly describe the overall appearance of Mercury's surface.
14. Compare and contrast the surfaces of Mercury and our moon.
15. How much of Mercury's surface has been photographed and mapped?
16. Why and how does the surface gravity of Mercury result in a different distribution of craters than on our moon?
17. What is a lobate scarp? How were lobate scarps probably formed?
18. Describe the Caloris Basin.
19. What kind of unusual terrain is found antipodal to the Caloris Basin?
20. In your own words, briefly describe the five major stages in the history of Mercury.
21. What is meant by "Mercury I," Mercury II," and "Mercury III?"
22. Contrast and compare the interior structure of Mercury and Earth.
23. Why were scientists surprised to learn that Mercury has a magnetic field?
24. Briefly describe Mercury's magnetosphere.
25. Compare and contrast the magnetospheres of Mercury and Earth.
26. Briefly describe the range of surface temperatures on Mercury's surface.
27. Briefly describe how seismic waves from the Caloris impact affected surface features on the opposite side of the planet.

# 12 Voyages Past Jupiter

## 12.1 The Giant Planet

ONE OF the extraordinary accidents in the history of astronomy was the naming of the planets. The planet with the shortest sidereal period was named after Mercury, the fleet-footed messenger of the gods. Venus, sparkling brilliantly in the twilight colors of the morning or evening sky, was named after the goddess of love and beauty. Blood-red Mars took its name from the god of war. And with similar prophetic insight, ancient astronomers named the largest planet after the king of the gods.

Except for the sun, Jupiter is by far the most massive object in the solar system. Jupiter's mass is almost $2\frac{1}{2}$ times larger than the masses of all the other planets combined. Jupiter is 318 times more massive than Earth or, alternatively, about $\frac{1}{1,000}$ the mass of the sun. Additionally, Jupiter is the largest planet in the solar system. Jupiter's diameter is roughly 143,000 kilometers (90,000 miles) or about 11 times larger than Earth's diameter. Consequently, more than 1,300 Earths could fit inside a sphere the size of Jupiter.

In view of Jupiter's mass and size, the planet's average density is 1.33 grams per cubic centimeter. This is considerably less than the average density of the terrestrial planets or of rocks (for example, recall that Earth's average density is 5.52 grams per cubic centimeter). Consequently, Jupiter must be primarily composed of lighter elements such as hydrogen and helium.

The idea that Jupiter could retain even the lightest gases is seen to be plausible once we consider the temperature and surface gravity (or, alternatively, the escape velocity) of the planet. Jupiter's nearly circular orbit is at an average distance of 778 million kilometers (almost half a billion miles) from the sun. At this enormous distance from the sun, the temperature at the top of the Jovian atmosphere should be quite cool, roughly 130°K ($-140$°C $= -220$°F). This means that the average speed of atoms and molecules in Jupiter's atmosphere would be relatively low. In addition, in view of Jupiter's mass and size, the surface gravity is 2.64 times the surface gravity of Earth and the escape velocity is 57.5 kilometers per second (37 miles per second). Thus, a person standing on Jupiter would weight 2.64 times more than he weighs on Earth and in order to escape from Jupiter he would have to be traveling upward at a speed 5 times greater than the escape velocity from Earth (recall that Earth's escape velocity is only 11.2 kilometers per second). The low temperature of the Jovian atmosphere and the high escape velocity work together to insure that no atoms or molecules ever leave the planet. Thus, Jupiter stands in sharp contrast to the inner terrestrial planets. For example, any hydrogen or helium that may have been present in the vicinity of Earth when our planet was formed would have long ago floated off into outer space. Jupiter, on the contrary, would have easily retained these gases. Consequently, the Jovian atmosphere may be representative of the primordial abundance of elements present at the time of the formation of the solar system. Studying Jupiter therefore gives scientists unique opportunities not afforded by any of the terrestrial planets.

The earliest telescopic observations of Jupiter date back to 1610 when Galileo first pointed his telescope toward the sky. While the resolution of Galileo's telescope was so poor that he could not see any surface details, he did discover Jupiter's four largest moons. Each of these four *Galilean satellites* is roughly the same size as our own moon. They were named Io, Europa, Ganymede, and Callisto. Since that time, ten additional small satellites have been discovered in orbit about Jupiter.

In 1660, Giovanni Domenico Cassini, using an improved telescope, noticed a "spot" on Jupiter. Observing this spot over several nights, Cassini discovered that Jupiter's rotation period is about ten hours. Indeed, Jupiter rotates faster than any other planet in the solar system.

As the design of telescopes improved, astronomers found that Jupiter actually has a striped appearance. Dark and light bands parallel to Jupiter's equator cover a large percentage of the planet's surface. By convention, the light bands are called *zones* while the dark bands are termed *belts*. In addition, it was noticed that the "spot" first seen by Cassini has a distinctive reddish color. This persistent feature on Jupiter was therefore christened the *Great Red Spot*. Actually, the Great Red Spot varies slightly in both color and size. Its width is fairly constant at 14,000 kilometers (8,700 miles) but its length ranges from 30,000 to 40,000 kilometers (19,000 to 25,000 miles) over a period of a few years. This Great Red Spot is so large that three earths could fit side by side in this feature, which has been seen on Jupiter for the past 300 years.

An excellent earth-based photograph of Jupiter is seen in Figure 12-1. Belts, zones, and the Great Red Spot are easily distinguished. Figure 12-2 shows a map of Jupiter along with the names of the various belts and zones that have been adopted by astronomers over the years.

By observing various features at different latitudes, astronomers have been able to measure Jupiter's rotation period very accurately. Surprisingly, Jupiter does not rotate like a solid body; different parts of the planet rotate at slightly different speeds. Near the equator, Jupiter rotates with a period of 9 hours, 50 minutes, 30 seconds. At higher latitudes, nearer the poles, the rotation period is slightly longer: 9 hours, 55 minutes, 41 seconds. This is possible because the "surface" of Jupiter seen through a telescope is not solid. Rather, the features seen on the planet are located at the top of a very thick cloud-cover. If Jupiter has a solid surface at all, it must be buried very far below the clouds.

**Figure 12–1.** *Jupiter.* The largest planet in the solar system is 318 times more massive than Earth. Jupiter is roughly 143,000 kilometers (90,000 miles) in diameter, which means that 11 Earths could fit side by side across the giant planet. (*New Mexico State University Observatory*)

**Figure 12–2.** *A Map of Jupiter.* Most of the persistent features on Jupiter have been named. The light-colored bands are called zones and the darker bands are called belts. The exact sizes and colors of these features vary slightly over the years.

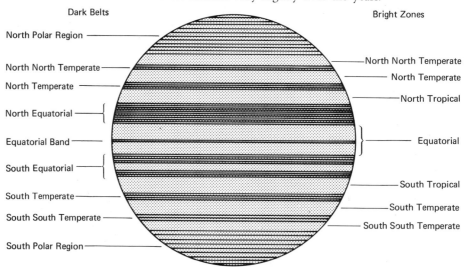

As a result of its high rate of rotation, Jupiter is not spherical. Instead, the planet is substantially flattened at the poles. The equatorial diameter of Jupiter is 142,800 kilometers (88,500 miles) while the diameter through the north and south poles is 134,000 kilometers (83,000 miles). Thus, Jupiter is flattened by almost 9,000 kilometers, which is about two thirds the diameter of Earth. In spite of Jupiter's oblate shape, astronomers have adopted a standard diameter for the planet of 142,744 kilometers (88,700 miles).

Spectroscopic observations of Jupiter extending over many years have easily detected the presence of hydrogen ($H_2$), ammonia ($NH_3$) and methane ($CH_4$). Helium is much more difficult to detect from Earth, but the presence of this gas was confirmed during the fly-bys of the Pioneer 10 and 11 spacecrafts in the 1970s. Examination of the abundances of the various elements in Jupiter's atmosphere reveals that they are present in the same proportions as on the sun. Of course, atoms in Jupiter's atmosphere can combine to form chemicals that could not survive the high temperatures on the sun's surface. Nevertheless, the abundances of elements on Jupiter and the sun are very similar.

Assuming that the composition of Jupiter and the sun is identical, scientists infer that Jupiter must contain a certain percentage of heavy elements. Unlike the case of the sun, however, these heavier elements would have sunk to Jupiter's center resulting in a small, rocky, iron-silicate core.

From both Earth-based observations and data from the flights of Pioneers 10 and 11, John D. Anderson at the Jet Propulsion Laboratory and William B. Hubbard at the University of Arizona have constructed a theoretical model of Jupiter's interior structure. They estimate that Jupiter's rocky core is roughly 25,000 kilometers (15,000 miles) in diameter. In view of the enormous weight of the overlying material, the temperature of this core is thought to be roughly 30,000°K (54,000°F). Just as hydrogen is by far the most abundant element on the sun, most of Jupiter must also be composed of hydrogen. But, unlike the hydrogen on the sun, which is very hot, the hydrogen inside Jupiter can exist in liquid form. From the rocky core out to a distance of 46,000 kilometers (about 29,000 miles) from the planet's center, the pressures are so great that the liquid hydrogen behaves like a metal, perhaps similar to sodium or potassium. *Liquid metallic hydrogen* has never been observed in laboratory experiments simply because it exists only at enormous pressures that cannot be created here on Earth. Nevertheless, scientists feel confident that liquid metallic hydrogen can exist from their theoretical understanding of the properties of hydrogen atoms.

**Figure 12-3.** *A Spectrum of Jupiter.* This portion of Jupiter's spectrum from about 9,000 to 10,000 Å shows numerous molecular lines. Spectral lines such as these are formed primarily by methane and ammonia in the planet's atmosphere. (*Yerkes Observatory photograph*)

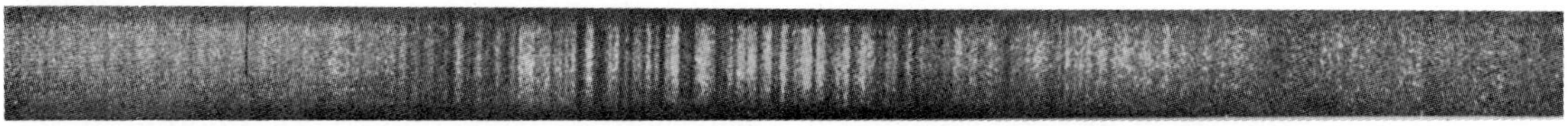

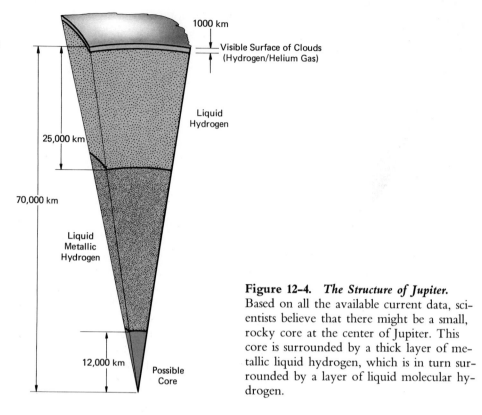

1000 km

Visible Surface of Clouds
(Hydrogen/Helium Gas)

Liquid
Hydrogen

25,000 km

70,000 km

Liquid
Metallic
Hydrogen

12,000 km

Possible
Core

**Figure 12–4.** *The Structure of Jupiter.*
Based on all the available current data, scientists believe that there might be a small, rocky core at the center of Jupiter. This core is surrounded by a thick layer of metallic liquid hydrogen, which is in turn surrounded by a layer of liquid molecular hydrogen.

At a distance of 46,000 kilometers from the planet's center, the pressure becomes sufficiently low that hydrogen can no longer exist in metallic form. As this location inside Jupiter (46,000 kilometers from its center), the pressure has dropped to (only!!) three million earth atmospheres and the temperature is down to 11,000°K. Above this location, hydrogen exists in simple liquid form. This outer layer of liquid molecular hydrogen is about 24,000 kilometers (15,000 miles) thick and extends to the bottom of the Jovian atmosphere. The overall structure of Jupiter's interior is shown in Figure 12–4.

An important discovery about the nature of Jupiter was made in the 1950s when radio astronomers detected radio waves coming from the planet. Radio waves are observed over two broad ranges of wavelengths. At wavelengths of a few tenths of meters, radio radiation is apparently coming from two very different types of sources. Most of this radiation is *thermal;* it simply comes from the planet's surface and is exactly what would be expected from an object emitting heat. A significant portion of this radio emission is, however, *nonthermal.* These emissions are consistent with what would be expected from high-speed electrons spiralling around a huge magnetic field. It was therefore concluded that Jupiter must have a substantial magnetic field, perhaps considerably stronger than Earth's magnetic field. As we shall see in the next section, the existence of a magnetosphere surrounding Jupiter was confirmed by the flights of Pioneers 10 and 11. Jupiter's magnetic field is roughly ten times stronger than Earth's but is oriented in the opposite direction. Compasses on Jupiter would point toward the south pole.

Both the *thermal radiation* from the planet's surface and the *synchrotron radiation* from high-speed electrons in the Jovian magnetosphere produce radio emissions with wavelengths measured in tenths of meters. This is called *decimetric radiation.* At much longer wavelengths, measured in tens of meters, radio emissions from Jupiter are also detected. Unlike the short wavelength radio waves, the source of this longer wavelength radiation or *decametric radiation* is not known. It is sporadic and may be caused by electrical discharges or lightning in the clouds in Jupiter's atmosphere.

In examining radiation from Jupiter, it was discovered that the planet emits roughly $2\frac{1}{2}$ times more energy than it receives from the sun in sunlight. This incredible property of Jupiter was one of the reasons why scientists were anxious to send spacecrafts to the planet. From Earth, we see only the illuminated side of Jupiter. Only with a spacecraft programmed to fly "behind" Jupiter could we examine radiation coming from the planet's dark side. Prior to the flights of Pioneer 10 and 11, it was thought that Jupiter might be gradually shrinking in size. The gravitation contraction of Jupiter would convert gravitational energy into thermal energy, thereby explaining the excess radiation we observe. However, with the recent work of investigators like Anderson and Hubbard, it is now felt that Jupiter's interior is liquid (liquid molecular hydrogen and liquid metallic hydrogen) and it is very difficult to compress liquids. Today it is generally believed that this excess radiation from Jupiter comes from the planet's primordial thermal energy. This energy was stored in the planet as it condensed out of the gases in primordial solar nebula $4\frac{1}{2}$ billion years ago. Thus, as we see Jupiter shining brilliantly in the nighttime sky, some of the radiation striking our eyes actually originated in the ancient past, at the time of the formation of the solar system itself.

## *The Flights of Pioneers 10 and 11* **12.2**

IN THE early evening of March 2, 1972, a huge Atlas–Centaur rocket slowly lifted off its launching pad at Cape Canaveral on the eastern coast of Florida. Steadily building up speed, the vehicle achieved a velocity of over 51,000 kilometers per hour (32,000 miles per hour), well above the escape velocity of Earth. Indeed this record-breaking speed was so high that the payload, Pioneer 10, was eventually catapulted out of the solar system and is now headed toward the star Aldebaran in the constellation of Taurus, the bull.

On April 6, 1973, almost exactly one year after the launch of Pioneer 10, a second identical spacecraft began a follow-up mission to the outer reaches of the solar system. Both spacecrafts, called Pioneer 10 and Pioneer 11, were directed toward the planet Jupiter. Pioneer 10 flew past Jupiter on December 4, 1973, and

**Figure 12-5.** *Pioneer 10.* Pioneer 10 is shown undergoing tests at TRW Systems in California where the spacecraft was constructed. Arrays of compact instruments were arranged behind the spacecraft's 9-foot diameter parabolic antenna. (*NASA*)

Pioneer 11 arrived at Jupiter almost exactly one year later, on December 3, 1974. These two successful missions constituted mankind's first attempt at space exploration beyond the orbit of Mars.

The design of Pioneers 10 and 11 was not unlike other spacecraft that had successfully explored the inner planets. The structure of both Pioneers centered about a 9-foot diameter parabolic antenna, as shown in Figure 12–5. Attitude control and propulsion systems, communication systems, navigational devices, and thermal control units were mounted behind the dish-shaped antenna. In addition, there was an array of compact scientific instruments on each spacecraft, as shown in Figure 12–6. Two magnetometers, one of which was at the end of a $6\frac{1}{2}$ meter (21 feet) boom, measured interplanetary and Jovian magnetic fields. A plasma analyser was used to examine the solar wind while a charged particle detector and cosmic-ray telescope studied the properties of cosmic rays (that is, high–speed particles) from the sun and more distant astronomical sources. A trapped radiation detector was designed to provide important information on Jupiter's magnetosphere.

Since both Pioneers were destined to journey very far from the sun, solar panels would be ineffectual in providing the necessary electrical power for the spacecrafts. Instead, it was decided that electricity would be obtained from two small nuclear generators mounted on each of the Pioneers. These power supplies, called radioisotope thermoelectric generators, had been developed earlier by the Atomic Energy Commission for earth-orbiting meteorological satellites such as Nimbus 3. These generators turned heat from the radioactive decay of plutonium-238 into electricity. In order to reduce the effects of radiation on Pioneer's sensitive equipment, each generator was mounted on an extendable boom. After launch, the booms held the

**Figure 12-6.** *The Design of Pioneers 10 and 11.* Both spacecrafts carried an array of scientific equipment to analyse interplanetary space and conditions near Jupiter. Unlike previous unmanned spacecrafts, there were no solar panels. Electrical power was drawn from two small nuclear generators.

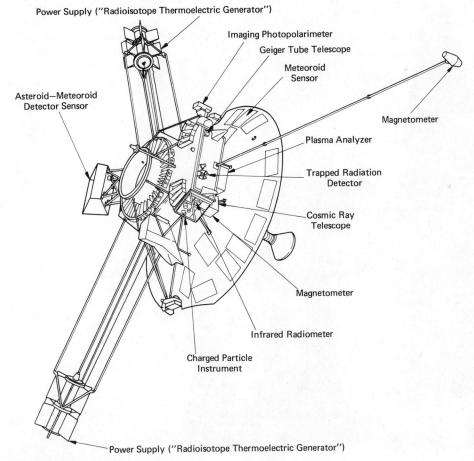

generators at an acceptable distance from the spacecraft. All of this apparatus was designed to fit inside the 3-meter (10-foot) diameter shroud of an Atlas–Centaur rocket, as shown in Figure 12–7.

Since both Pioneers would pass through the asteroid belt between Mars and Jupiter, each spacecraft carried instruments to detect meteoroids and asteroids. Of course, impact with a large meteoroid would have been fatal to the spacecrafts. There was considerable anxiety among scientists as each of the Pioneers traversed the asteroid belt. Fortunately, fears of imminent disaster proved to be unfounded.

When each of the Pioneers arrived at Jupiter, ultraviolet photometers and infrared radiometers made important measurements of conditions in the Jovian clouds at wavelengths far removed from visible light. The presence of helium in

**Figure 12–7.** *Pioneer 11.* Each of the Pioneers weighed 250 kilograms (550 pounds) and were designed to fit inside the 3-meter (10-foot) diameter shroud atop an Atlas–Centaur rocket. Pioneer 11 is shown here undergoing final tests before its launch on April 6, 1973. (*NASA*)

Jupiter's atmosphere could only be detected at ultraviolet wavelengths while measurements of the temperature across the surface of Jupiter as well as the planet's heat output required observations at infrared wavelengths.

Of course, some of the most dramatic results from Pioneers 10 and 11 were the close-up pictures sent back by the television cameras. Actually, these cameras are properly called imaging photopolarimeters. Each consists of a telescope followed by a series of prisms, filters, and mirrors that relay light into one of two detectors that measure the intensity of the incoming radiation. As a result of the arrangement of the two detectors and the red and blue filters, the imaging photopolarimeter simultaneously produced pictures of Jupiter in both red and blue light. At any given instant, the field of view of the telescope encompassed only 0.028 degrees across, which is a very tiny angle. To get a full picture of Jupiter, scientists relied on the fact that the spacecraft was spinning. Each time the spacecraft rotated, the telescope in the imaging photopolarimeter would scan a narrow strip of Jupiter's surface, as shown in Figure 12–8. By placing the resulting strips side by side, a full picture was obtained. As mentioned above, images of Jupiter in red and blue light were produced simultaneously. Representative red and blue pictures from both Pioneers are shown in Figures 12–9 and 12–10.

On their way to Jupiter, both Pioneers sent back a lot of information about conditions in interplanetary space. For example, it was discovered that the strength of the sun's magnetic field, the density of the solar wind, and the number of high-speed particles from the sun all declined steadily with increasing distance from

**Figure 12–8.** *The Imaging Photopolarimeter.* At any one instant, Pioneer's camera could see only a tiny portion of Jupiter. But since the spacecraft was spinning, the telescope would scan a narrow strip across the planet. By moving the telescope, a series of strips were obtained that were then assembled into a full picture.

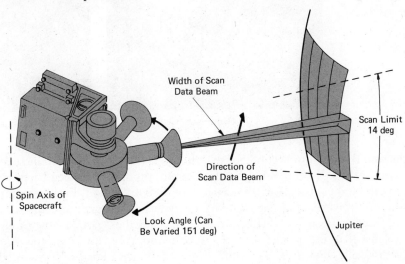

Red Image

Blue Image

**Figure 12-9.** *Views from Pioneer 10.* Five days before closest approach, while at a distance of 5.6 million kilometers (3.5 million miles) from Jupiter, Pioneer 10 sent back these two pictures of the planet. The view on the left is the red image, while the view on the right is in blue light. (*NASA*)

**Figure 12-10.** *Views from Pioneer 11.* Three days before closest approach, while at a distance of 3.6 million kilometers (2.2 million miles) from Jupiter, Pioneer 11 sent back these two pictures of the planet. Notice that the Great Red Spot is prominent in the blue image on the right, but is almost invisible in the red image on the left. (*NASA*)

Red Image

Blue Image

the sun. Recall from Section 5.1 that the paths of the particles in the solar wind are spirals. At the distance of Earth from the sun, the angle between Earth's orbit and a typical spiral path is about 45°, as shown in Figure 12–11. Five times farther from the sun, however, the angle between a typical spiral path and Jupiter's orbit has decreased to about 10°. This is in keeping with the "garden-hose model," which explains that the particles in the solar wind emanate from the rotating sun. Occasionally, hot spots in the sun's corona produce a burst of high-speed particles.

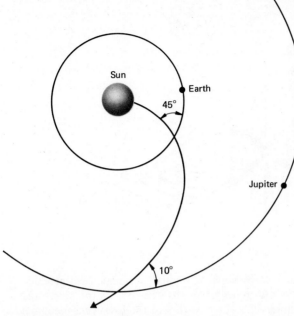

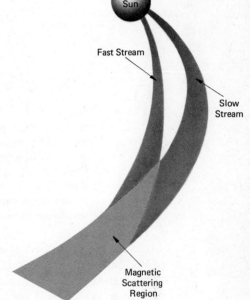

**Figure 12–11.** *The Shape of the Solar Wind.* Particles in the solar wind follow spiral paths away from the sun. At Earth, the angle between Earth's orbit and a typical spiral path is about 45°. At the distance of Jupiter, this angle has decreased to about 10°.

**Figure 12–12.** *Scattering Regions.* Collisions between fast and slow streams of particles in the solar wind produce regions of magnetic scattering. These regions prevent low-speed cosmic rays from entering the solar system.

The spiral paths for these particles are less tightly wound and, as a result, the fast-moving stream catches up with more distant slowly moving streams, as shown in Figure 12–12. Since the solar wind carries the sun's magnetic field along with the streams, collisions between streams moving at different speeds produce regions of *magnetic scattering*. As the solar wind streams collide and rebound, the magnetic field is amplified. Because of these scattering regions, no low-speed cosmic rays can enter the solar system. Low-speed particles from distant astronomical sources are incapable of penetrating these regions of magnetic scattering.

As mentioned earlier, there was considerable concern among scientists about the safety of the spacecrafts as they passed through the asteroid belt. Surprisingly, the Pioneers' meteoroid detectors experienced a fairly constant rate of impacts throughout their trips. For example, the detector on Pioneer 10 experienced 41 puncturing impacts during the four months between launch and June 1972, when

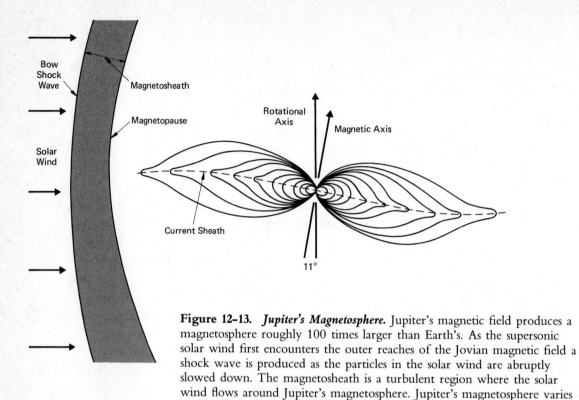

**Figure 12-13.** *Jupiter's Magnetosphere.* Jupiter's magnetic field produces a magnetosphere roughly 100 times larger than Earth's. As the supersonic solar wind first encounters the outer reaches of the Jovian magnetic field a shock wave is produced as the particles in the solar wind are abruptly slowed down. The magnetosheath is a turbulent region where the solar wind flows around Jupiter's magnetosphere. Jupiter's magnetosphere varies in size in response to changes in the solar wind. Charged particles are confined to a sheet near the planet's magnetic equator.

the spacecraft arrived at the asteroid belt. During the next four months in the asteroid belt, an additional 42 impacts were recorded. As far as tiny particles are concerned, it is almost as if the asteroid belt does not exist. Fine particles and dust seem to be scattered rather uniformly through the solar system.

From observations by radio astronomers back in the 1950s and 1960s, it was realized that Jupiter has a substantial magnetic field. It was therefore correctly believed that Jupiter must have a magnetosphere similar to Earth's but on a much larger scale. As they approached Jupiter, both Pioneers passed through a *bow shock wave* at the location at which the supersonic solar wind first encounters the effects of Jupiter's magnetic field. At the bow shock, the high-speed solar wind is abruptly slowed down as it interacts with the outer reaches of Jupiter's magnetic field. Inside the bow shock is the *magnetosheath,* a turbulent region where the solar wind is deflected around Jupiter's *magnetosphere,* as shown in Figure 12–13. The boundary between the magnetosheath and the magnetosphere is called the *magnetopause.* At the magnetopause, the pressure of the solar wind is counterbalanced by the pressure of Jupiter's magnetic field.

Pioneers 10 and 11 reported crossing the magnetopause at distances ranging between 3 to 7 million kilometers (2 to 4 million miles) above Jupiter. For comparison, Earth's magnetopause is located at about 75,000 kilometers (46,000 miles from our planet and thus Jupiter's magnetosphere is about 100 times larger than Earth's. If you could see Jupiter's magnetosphere with your eyes, it would cover an area sixteen times bigger than the full moon in the sky. The fact that the Pioneers crossed the magnetopause (that is, the outer boundary of Jupiter's magnetosphere) at a wide range of distances from the planet strongly suggests that Jupiter's magnetosphere is very "spongy." Evidently Jupiter's magnetosphere is extremely sensitive to fluctuations in the solar wind and expands and contracts by a factor of two, very rapidly and frequently. By comparison, Earth's magnetosphere is much "stiffer"; dramatic changes in the size of our magnetosphere are exceedingly rare.

The strength of Jupiter's magnetic field is about ten times greater than the terrestrial magnetic field. At Jupiter's north pole the strength is about 15 gauss and at the south pole is 11 gauss. Jupiter's magnetic field is oriented opposite to Earth's, and the magnetic axis is inclined to the rotation axis by 11°. Indeed, since Jupiter's surface exhibits differential rotation, best measurements of Jupiter's true rotation period are obtained from radio signals involving the planet's magnetic field, which is alternately pointed toward and away from Earth several times each day. The rotation period obtained in this fashion is 9 hours, 55 minutes, and 30 seconds.

In Earth's magnetosphere, charged particles are largely confined to two belts, the Van Allen belts, which circle our planet like giant doughnuts. In the case of Jupiter, however, the magnetosphere is so huge and rotates so rapidly (once every 10 hours) that the charged particles in the magnetosphere experience very large centrifugal forces. This results in the charged particles being confined to a thin sheet, the so-called *current sheet,* near the plane of Jupiter's magnetic equator. Furthermore, centrifugal forces acting on the particles in the current sheet cause the magnetic field to be stretched out away from the planet, as shown in Figure 12–13.

The properties of Jupiter's magnetosphere are further complicated by the fact that five of Jupiter's fourteen moons have orbits inside the magnetosphere. These five moons (the four Galilean satellites and Amalthea, the innermost moon) sweep the charged particles in the magnetosphere leaving behind a clear corridor. In this process, these inner satellites acquire intense radioactivity. Even tiny Amalthea, measuring only 150 kilometers (93 miles) across, is very effective in reducing the number of charged particles as it passes through the densest regions of the magnetosphere every 12 hours. As a result of this effect of these five satellites, the total radiation near Jupiter can be reduced by a factor of 100 and there is even some evidence that Jupiter's entire magnetosphere empties every 10 hours.

The peak intensity of electrons and protons measured by the Pioneer spacecrafts was thousands of times greater than in Earth's magnetic environment. Fortunately, the spacecrafts were so well designed that sensitive equipment on board continued to function properly as the Pioneers approached the huge planet. Both spacecrafts were therefore successful in obtaining extremely fine photographs of Jupiter's clouds. These photographs, as well as measurements from ultraviolet and infrared detectors, have given scientists a good look at the meteorology of a planet vastly different from Earth.

# 12.3   *The Jovian Atmosphere*

EVERYONE is surely familiar with the fact that predicting weather conditions here on Earth is a tricky business. Every evening, all the major television networks regularly present weather forecasts as an important part of their evening news broadcasts. A meteorologist is usually seen standing in front of a map on which all sorts of data are plotted. Yet, in spite of the fact that these data are gathered from thousands of weather stations and earth-orbiting satellites, the unfortunate meteorologist is often hard pressed to predict tomorrow's weather with any reliable degree of certainty. It is our common experience that, even after listening to this well-qualified meteorologist, we still do not know if the next day will be warm and sunny or cold and rainy. Indeed, it may sometimes seem that the meteorologist is more often wrong than right!

For the average person, weather forecasts may be important only insofar as clothing and wardrobes are concerned. Yet each year, thousands of human lives are lost and billions of dollars of damage are sustained at the apparent whims of atmospheric phenomena such as cyclones, hurricanes, and tornadoes. Of course, important advances in meteorology have been made in recent years, primarily as a result of weather satellites, which constantly monitor atmospheric conditions on a global scale. Yet, there are so many variables and Earth's atmosphere is so complex that forecasting the weather sometimes amounts to making an intelligent guess.

**Figure 12-14 (left).**   *View from Pioneer 10.* This picture was taken when the spacecraft was 1,888,000 kilometers (1,173,000 miles) from the planet. Details in the belts and zones are seen much more clearly and distinctly than is possible from earth-based observations. (*NASA*)

**Figure 12-15 (right)**   *View from Pioneer 11.* This picture was taken when the spacecraft was 1,100,000 kilometers (660,000 miles) from the planet. The Great Red Spot, measuring almost 40,000 kilometers (25,000 miles) in length, dominates this view in blue light. (*NASA*)

One of the important difficulties that traditionally face meteorologists is that their understanding had been based entirely on one planet. Until very recently, the science of meteorology was firmly rooted in the physics of a nitrogen–oxygen atmosphere surrounding one terrestrial planet. It was almost as if biologists were to attempt to understand life by examining only one species of animal or if geologists tried to understand rocks by looking at only one specimen. But with the Venera and Mariner missions to Venus, the Viking landings on Mars, and the flights of Pioneers 10 and 11 to Jupiter, vast quantities of new data are now available. Understanding atmospheric conditions on these other planets will certainly have a profound effect on meteorology. Indeed, unmanned space flights to other planets may be as important for meteorology as Galileo's telescope was for astronomy.

Mankind's first close-up look at Jupiter's atmosphere came from the fly-bys of Pioneers 10 and 11. Prior to this time, the Jovian atmosphere could only be viewed indistinctly from a distance of more than 600 million kilometers (about 400 million miles) when the planet is near opposition. Then, in December of 1973, Pioneer 10 passed within 131,000 kilometers (81,000 miles) of Jupiter and almost exactly a year later Pioneer 11 coasted past Jupiter at a distance of only 46,400 kilometers (28,000 miles). Hundreds of excellent close-up views were suddenly available. Typical black-and-white examples are shown in Figures 12–14 and 12–15. In addition, since images were taken simultaneously through red and blue filters, color photographs could be constructed. Excellent examples are shown in Plates 17 through 22.

Jupiter's atmosphere is about 1,000 kilometers (600 miles) thick and is composed almost entirely of hydrogen and helium, like the sun. At greater depths beneath the cloud-cover, pressures are so high that the gases are liquified. Analysis of the Pioneer data indicates that 82 per cent of the Jovian atmosphere is hydrogen and 17 per cent is helium. This leaves only 1 per cent for all other substances. For every helium atom, there are 15 hydrogen atoms. But for every carbon atom, there are 3,000 hydrogen atoms, and for every nitrogen atom there are 10,000 hydrogen atoms. These abundances are very close to the abundances of the same elements in the sun.

In spite of the relatively low abundances of the heavier elements, chemical compounds involving oxygen, carbon, nitrogen, and sulfur are important solid constituents of the Jovian clouds. According to theoretical calculations by John S. Lewis at the Massachusetts Institute of Technology, the uppermost clouds in Jupiter's atmosphere are composed of ammonia crystals ($NH_3$). Roughly 30 kilometers (20 miles) below the ammonia clouds, there is a layer of ammonium hydrosulfide crystals ($NH_4SH$). And about 40 kilometers (25 miles) deeper in the atmosphere, there is a layer of water-ice crystals ($H_2O$). These substances are suspended in an atmosphere of mostly hydrogen and helium, along with gases such as ammonia ($NH_3$), methane ($CH_4$), and water vapor ($H_2O$).

The arrangement of the cloud layers is shown in Figure 12–16. Also plotted on this graph is the temperature in the Jovian atmosphere, which rises steadily from about 140°K ($-130°C = -210°F$) at the top of the ammonia crystal clouds to almost 300°K ($30°C = 90°F$) at the bottom of the water-ice crystal clouds. This

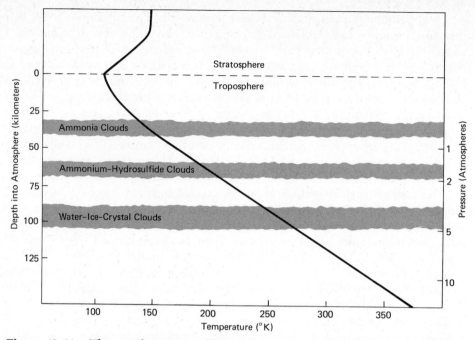

**Figure 12–16.** *Theoretical Structure of Jupiter's Outer Atmosphere.* Clouds of crystalline substances are located at various depths in Jupiter's troposphere. The uppermost layer is composed of ammonia crystals where the temperature is roughly 150°K and the pressure is 0.6 Earth atmospheres. The next layer is composed of ammonium hydrosulfide crystals where the temperature is 200°K and the pressure has risen to about 1.7 Earth atmospheres. At still greater depths, where the pressure is between 4 and 5 Earth atmospheres, there is a layer of water-ice crystals. Beneath the water-ice clouds, the temperature is above 270°K and droplets of water are found.

temperature profile is based on an analysis of infrared data by Glenn S. Orton at the Jet Propulsion Laboratory. Above the ammonia crystal clouds, the temperature reaches a minimum of about 110°K (−160°C = −260°F) and then starts rising again with increasing altitude above the planet. By analogy with Earth's atmosphere, the location of this temperature minimum marks the boundary between the troposphere and stratosphere on Jupiter.

The theoretical picture of Jupiter's atmosphere presented in Figure 12–16 is idealized in the sense that it assumes Jupiter's atmosphere is very uniform. In reality, the situation is complicated by temperature and pressure differences as well as large-scale turbulence. The familiar system of belts and zones clearly indicates such horizontal variations across the planet.

Analysis of infrared data from Pioneers 10 and 11 reveals that the light-colored zones are formed by warm currents rising upward. As material reaches the top of a zone, it cools in the upper atmosphere, spills over the top, and descends into the adjacent dark-colored belts. Zones are therefore regions of high atmospheric

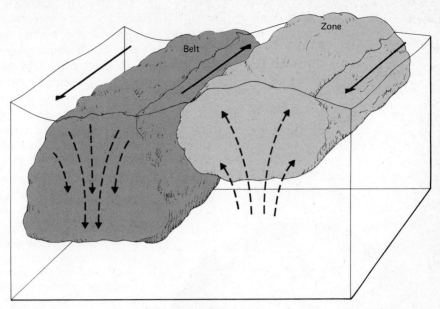

**Figure 12–17.** *Belts and Zones.* The light-colored zones are regions of high atmospheric pressure in which warm gases rise upward. At the top of a zone, the gases cool and spill over into the adjacent low-pressure, dark-colored belts where the gases descend back into Jupiter's atmosphere.

pressure (analogous to "highs" in Earth's atmosphere) where warm gases rise upward, while belts are regions of low atmospheric pressure (analogous to "lows" in Earth's atmosphere) where cool gases descend. Indeed, the tops of the clouds in the zones are substantially higher than the tops of the clouds in the belts, as shown in Figure 12–17.

While data from Pioneers 10 and 11 have shown that belts and zones consist of gases at different altitudes and temperatures, available information does not explain the color variations seen on the planet. It is still not understood exactly why the light-colored zones are primarily white or pale yellow and the dark-colored belts are various shades of reddish-brown. All of the gases mentioned so far are colorless and the crystallized chemicals in the cloud layers are white. Actually, only a small amount of coloring material in the Jovian atmosphere would suffice in explaining the shades, hues, and tints seen on Jupiter. This coloring material might involve sulfur, red phosphorus, or organic molecules. There is even a small chance that simple living organisms might exist in Jupiter's atmosphere because its gases are believed to be similar to those which enveloped Earth billions of years ago when life first appeared on our planet. In any case, a final explanation of the color differences on Jupiter would probably involve light-colored chemicals rising upward in the zones. At the top of the zones, ultraviolet radiation from the sun produces chemical reactions resulting in dark-colored compounds that accumulate in the belts.

An understanding of the horizontal structure, motion, and currents in Jupiter's atmosphere comes primarily from examining the pictures of the planet sent back by the Pioneer spacecrafts. In order to appreciate the resulting discoveries, it is perhaps instructive to review some properties of Earth's atmosphere.

On both Earth and Jupiter, winds blow from regions of high atmospheric pressure toward regions of low atmospheric pressure. In both cases, the exact paths followed by the winds are dramatically influenced by the rotation of the planets. This is caused by the fact that when gases or particles move in a rotating system (such as on the surface of a rotating planet), they are deflected toward one side by so-called *Coriolis forces*. On Earth, in the Northern Hemisphere, winds blow in a counterclockwise direction around a low-pressure region (forming a *cyclone*) and in a clockwise direction around a high-pressure region (forming an *anticyclone*). As shown in Figure 12–18, the situation is reversed in the Southern Hemisphere. In the

**Figure 12–18.** *Circulation in Earth's Atmosphere.* Air spiralling into low-pressure centers produces cyclones and air spiralling out of high pressure centers produces anticyclones. In the Northern Hemisphere, cyclones rotate counterclockwise and anticyclones rotate clockwise. This situation is reversed in the Southern Hemisphere. In the Southern Hemisphere, cyclones rotate clockwise and anticyclones rotate counterclockwise.

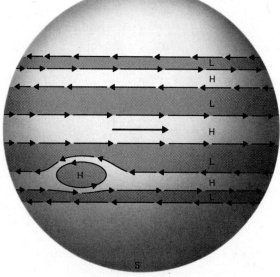

**Figure 12–19.** *Circulation in Jupiter's Atmosphere.* Jupiter is so large and rotating so fast that regions of high and low atmospheric pressure are stretched out all the way around the planet. High-speed winds (jets) flow at the boundaries between the bands. This pattern breaks down in the polar regions.

Southern Hemisphere, the cyclonic winds around a low-pressure region are rotating clockwise while the anticyclonic winds around a high-pressure region are rotating counterclockwise.

On Jupiter, zones are regions of high-atmospheric pressure while belts are regions of low-atmospheric pressure. But Jupiter is so big and is rotating so fast that these low- and high-pressure regions are stretched out all the way around the planet. It is as if you took the cyclones and anticyclones on Earth and stretched them out into long bands. The resulting circulation patterns on Jupiter are shown in Figure 12–19.

The analogy between circulation patterns in the Jovian and terrestrial atmospheres can be seen by comparing Figures 12–18 and 12–19. Notice, for example, that in Earth's Northern Hemisphere, winds north of a cyclone around a low-pressure region are blowing westward (to the left) while winds south of this low-pressure region are blowing eastward (to the right). Similarly on Jupiter, winds at the northern boundary of a low-pressure region in the Northern Hemisphere are blowing westward (to the left) while winds at the southern boundary of this same

**Figure 12–20.** *Turbulence at Polar Latitudes.* In the north and south polar regions of Jupiter, the familiar pattern of belts and zones becomes unstable and breaks down. Numerous swirls and eddys, resembling unorganized hurricanes, are formed. This view was taken in blue light from Pioneer 11 when the spacecraft was 600,000 kilometers (370,000 miles) from the planet. A large portion of Jupiter's north temperate and north polar regions are seen. (*NASA*)

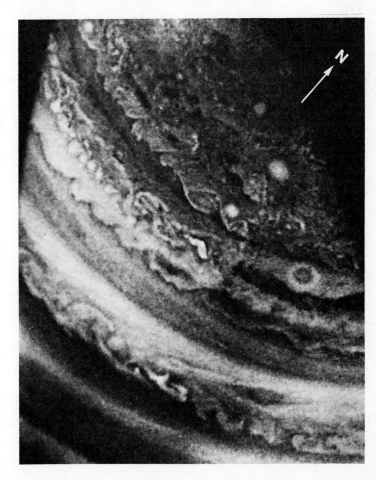

belt are blowing eastward (to the right). Just as on Earth, therefore, the *vorticity* or direction of wind flow on Jupiter is cyclonic in the low-pressure belts and anticyclonic in the high-pressure zones. This pattern breaks down near Jupiter's north and south poles. At latitudes greater than about 45°, the flow between belts and zones becomes unstable and degenerates into numerous swirls and eddys, as shown in Figure 12–20.

One of the most interesting yet puzzling features in the Jovian atmosphere is the Great Red Spot, which has been seen for the past three centuries. Prior to the flights of Pioneers 10 and 11, there was all sorts of speculation concerning the nature of the Great Red Spot, including the erroneous idea that it might be located over a "mountain" buried deep within Jupiter's atmosphere. Today, however, scientists realize that the Great Red Spot is actually a high-pressure center around which anticyclonic winds flow. The Great Red Spot is located in a zone in Jupiter's Southern Hemisphere. The Great Red Spot has a higher atmospheric pressure than the surrounding regions in the zone that already have a comparatively high atmospheric pressure. The Great Red Spot could therefore be called a "superzone." It is an entirely meteorological phenomenon similar to hurricanes here on Earth. Warm gases rise upward in the Great Red Spot and the resulting cloud tops are several kilometers higher than the surrounding cloud layer. Winds circulate in a counterclockwise direction inside the Great Red Spot. Indeed, this feature in Jupiter's atmosphere is rotating like a wheel between two oppositely moving surfaces, as shown in Figure 12–21. The Great Red Spot rotates once every 12 Earth-days. In a series of computer calculations during the early 1970s, Dr. Andrew P. Ingersoll at Caltech found that this rolling wheel configuration is the only wind-flow pattern that resulted in a stable configuration. If no energy were transferred to the surrounding atmosphere, the Great Red Spot should last forever. In reality, some energy is dissipated and the Great Red Spot will someday disappear.

**Figure 12–21.** *Wind Flow in the Great Red Spot.* Winds blow in a counterclockwise direction in the Great Red Spot. According to computer calculations by Dr. Ingersoll, the circulation pattern is like a wheel spinning between two oppositely moving surfaces. (*From "The Meteorology of Jupiter" by Andrew P. Ingersoll, copyright © 1976 by Scientific American, Inc. All rights reserved*)

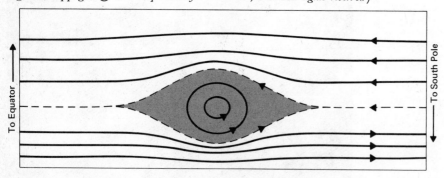

The idea that the Great Red Spot is a temporary (albeit long-lived) hurricane in the Jovian atmosphere received support from observations of a similar feature recently seen in Jupiter's Northern Hemisphere. Early in 1972, a "Little Red Spot" was observed by earth-based astronomers in Jupiter's north tropical zone. The Little Red Spot was photographed in December 1973, by Pioneer 10 (see Plate 22). But when Pioneer 11 flew past Jupiter a year later, the feature was gone. Such reddish-orange spots are not uncommon in zones and typically last for about two years. Only in the case of the Great Red Spot did the Jovian hurricane become so huge that it could survive for several centuries.

A fascinating explanation for the dramatic reddish-orange color of the Great Red Spot comes to mind when we think a little further about the analogy with hurricanes. Vast clouds must be constantly forming as the warm gases rise upward in the Great Red Spot. It is perhaps reasonable to suppose that huge thunderheads are formed and the Great Red Spot is a scene of enormous electrical discharges—thunder and lightning—that rumble across thousands of square miles of the planet's surface. In an interesting experiment, Dr. Cyril Ponnamperuma at the Laboratory of Chemical Evolution of the University of Maryland placed some Jovian-like gases (that is, methane, ammonia, and hydrogen) in a flask. He then threw a switch and electric sparks crackled between electrodes embedded in the flask. After a short time, the colorless gases became cloudy and a reddish material was deposited on the walls of the flask. The color of the thickening material was remarkably similar to the color of the Great Red Spot.

The reddish material produced in Dr. Ponnamperuma's experiment is an organic compound called nitrile. When combined with water, nitrile produces amino acids, which are the building blocks of proteins found in all life forms here on Earth. The implication is that life could have evolved in places like the Great Red Spot on Jupiter. Indeed, many scientists believe that the upper atmosphere on Jupiter is an excellent place to search for extraterrestrial life.

The flights of Pioneers 10 and 11 have unlocked many of the mysteries surrounding Jupiter. But as is often the case in science, even more questions—including the fascinating issue of extraterrestrial life—have been raised. To answer some of these questions, two Voyager spacecrafts, each weighing 750 kilograms (1,650 pounds), were launched toward Jupiter in the late summer of 1977. They are scheduled to arrive in the spring of 1979. This mission is much more ambitious than any other previous fly-by because the spacecrafts will also examine several of Jupiter's moons in addition to the planet's surface before heading on to Saturn. Arriving at Saturn in the early part of 1981, both spacecrafts will (one hopes!) repeat their performance.

Still further in the future, scientists would like to send spacecrafts to Jupiter that would drop probes into the Jovian atmosphere. A detailed first-hand knowledge of the structure and composition of the Jovian atmosphere would be obtained as the probes descended into the thick cloud-cover, relaying data back to Earth via an orbiting satellite overhead. Plans for a Jupiter orbiter, a Jovian version of the successful Viking missions to Mars in 1976, are now being made for the 1980s.

# 12.4 Many Moons

JUPITER, the largest of all the planets, certainly has the lion's share of the moons in the solar system. Of the 34 known natural satellites in the solar system, 14 are in orbit about Jupiter. Four of Jupiter's moons are very large. These four Galilean satellites, discovered by Galileo in 1610, are comparable in size to our own moon and to the planet Mercury, as shown in Figure 12–22. The remaining ten Jovian satellites are very small, typically measuring less than 100 kilometers (60 miles) across. Indeed, some of these satellites are so very small and dim that they have escaped detection until recent years. For example, the thirteenth and fourteenth Jovian satellites, which may have diameters of less than 20 kilometers (12 miles), were discovered in the mid–1970s.

The four huge Galilean satellites are in nearly circular orbits comparatively close to Jupiter. As shown in Figure 12–23, their orbital radii range from about 422,000 kilometers (262,000 miles) to 1,880,000 kilometers (1,160,000 miles). This places

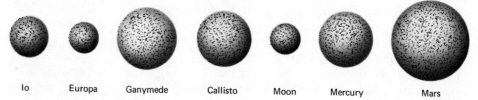

Io     Europa     Ganymede     Callisto     Moon     Mercury     Mars

**Figure 12–22.** *The Galilean Satellites.* Four of Jupiter's fourteen known satellites are very large. As shown in this scale drawing, they have sizes comparable to our own moon and the planet Mercury.

**Figure 12–23.** *Orbits of the Galilean Satellites.* The orbits of the Galilean satellites lie fairly close to Jupiter. In contrast, almost all of the smaller Jovian moons orbit the planet at much greater distances.

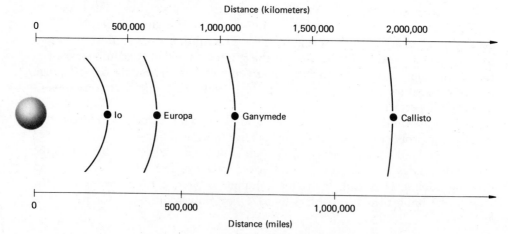

all four satellites well inside Jupiter's magnetosphere. Of the remaining ten smaller moons, only tiny Amalthea is located near the planet. Amalthea orbits Jupiter at a distance of only 180,000 kilometers (110,000 miles), which is even closer to the planet than any of the Galilean moons. The remaining nine small Jovian moons are always located very far from Jupiter. Their orbital radii range from 10 million kilometers to 24 million kilometers. Consequently, we note that the distance from Jupiter to Io, the innermost Galilean satellite, is roughly the same as the distance between Earth and our moon. As seen from the surface of Jupiter, Io would therefore appear roughly the same size as our own moon does to an Earth-based observer. In sharp contrast, the outer nine moons are so tiny and so far from Jupiter that they would be virtually invisible to the unaided eye of the Jupiter-based observer. Although the Galilean satellites superficially resemble Mercury and our own moon, data from Pioneers 10 and 11 revealed that each of these huge Jovian satellites is a unique world with its own distinctive properties.

Following Galileo's discovery of Jupiter's four largest moons in January of 1610, virtually no new astronomical data or discoveries concerning Jovian satellites were made for almost three centuries. Indeed, it was not until 1898 that Barnard discovered a fifth moon, Amalthea, orbiting the planet. The remaining nine satellites were all discovered during the twentieth century.

The first important earth-based observations of the Jovian satellites began in the mid-1920s at Lick Observatory. Using photometric techniques, it was shown that each of the Galilean moons varies in brightness by as much as 40 per cent. This strongly suggests that the surfaces of these satellites are not uniformly reflective. Instead, each of the satellites must have a glossy side and a dull side. Furthermore, the length of time over which a Galilean satellite varies its brightness is exactly equal to its orbital period. For example, Io orbits Jupiter once every $42\frac{1}{2}$ hours and varies its brightness once every $42\frac{1}{2}$ hours. The logical conclusion is that all of the Galilean satellites are in synchronous rotation about Jupiter. They all keep the same side facing Jupiter, just as our moon always keeps the same side facing Earth. As shown in Figure 12–24, for three of the Galilean satellites (Ganymede, Io, and Europa) the leading hemisphere is more reflective than the trailing hemisphere. For the outermost Galilean satellite, Callisto, exactly the opposite must be the case. Callisto's trailing hemisphere must be more reflective than its leading hemisphere.

In addition to revealing the synchronous rotation of the Galilean satellites, photometric observations also showed that the surfaces of these moons differ greatly in color and reflectivity. Io is very reddish and highly reflective, roughly equivalent to white sand. Europa is also highly reflective (both Io and Europa reflect about 62 per cent of the sunlight that falls on them) but has a more neutral color. Ganymede is somewhat less reflective; like dirty snow, Ganymede reflects 44 per cent of the sunlight. And finally Callisto, the outermost Galilean satellite, reflects only 19 per cent of the incident sunlight, similar to ordinary rocks. It therefore is entirely reasonable to suppose that different surface conditions exist on the various Galilean satellites.

Additional important information concerning surface conditions on the Galilean satellites comes from infrared observations. The first such observations were

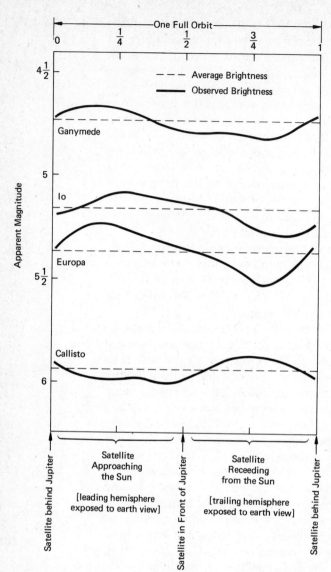

**Figure 12-24.** *Brightnesses of the Galilean Satellites.* As shown in this graph, the apparent magnitude of each Galilean satellite varies with a period equal to its orbital period. This means that the Galilean satellites are in synchronous orbit about Jupiter. (*Based on observations by Drs. D. P. Cruikshank and D. Morrison*)

made in the early 1960s at Hale Observatories and the University of Arizona. From infrared data it is possible to deduce the surface temperatures of the satellites. Since Jupiter is five times farther from the sun than Earth, the intensity of sunlight is only one twenty-fifth as strong as it is here at Earth. Consequently the surface temperatures of the satellites are quite low. Io and Europa, which reflect most of the infalling sunlight, have the lowest temperatures, about 140°K ( = −133°C = −206°F). Calisto, which absorbs a much larger percentage of the incident sunlight, is somewhat warmer. Its surface temperature is about 167°K ( = −106°C = −158°F). Ganymede, which has an intermediate albedo, also has an intermediate surface temperature of 154°K ( = −119°C = −182°F). These surface temperatures are for "high noon" on the satellites. On the dark sides of the satellites, or when the satellites pass through Jupiter's shadow, the surface temperatures are considerably lower. Temperatures on the satellites' night sides probably drop to about 80°K ( = −193°C = −316°F). At this temperature, the air here on Earth would liquify.

Another interesting property of the Galilean satellites becomes apparent when we examine their average densities. But in order to calculate their average densities, we must first know both the masses and sizes of each of the satellites. In this regard, the best available data came from the flights of Pioneers 10 and 11. The results are listed in Table 12–1, along with additional information.

**Table 12–1.** *Properties of the Galilean Satellites*

| | Io | Europa | Ganymede | Callisto |
|---|---|---|---|---|
| **Diameter** | | | | |
| (miles) | 2,260 | 1,890 | 3,260 | 3,040 |
| (kilometers) | 3,640 | 3,050 | 5,270 | 4,900 |
| **Mass** | | | | |
| (Jupiter = 1) | 0.000047 | 0.000025 | 0.000078 | 0.000057 |
| (Earth's moon = 1) | 0.81 | 0.65 | 2.18 | 1.13 |
| **Average Density** | | | | |
| (grams per cubic centimeter) | 3.5 | 3.1 | 1.9 | 1.6 |
| **Orbital Radius** | | | | |
| (miles) | 262,000 | 417,000 | 665,000 | 1,170,000 |
| (kilometers) | 421,600 | 670,900 | 1,070,000 | 1,880,000 |
| **Orbital Period** | | | | |
| (days) | 1.769 | 3.551 | 7.115 | 16.689 |
| (hours) | 42.46 | 85.22 | 170.76 | 400.54 |
| **Average Albedo** | | | | |
| (per cent reflectivity) | 63 | 64 | 43 | 17 |
| **Maximum Surface Temperature** | | | | |
| °K | 140 | 140 | 154 | 167 |
| °C | −133 | −133 | −119 | −106 |
| °F | −206 | −206 | −182 | −158 |
| **Average Apparent Magnitude** | | | | |
| (as seen from Earth) | 5.2 | 5.4 | 4.7 | 5.9 |

It is very striking to notice that the average density of the Galilean satellites decreases with increasing distance from the planet. The innermost satellite, Io, has the highest density while the outermost of the "big four," Callisto, has the lowest density. The average densities of Io and Europa are essentially the same as typical rocks. Thus, these two satellites must be composed primarily of rocks like our own moon, which has an average density of 3.34 grams per cubic centimeter. Both Ganymede and Callisto have significantly lower average densities. Consequently, each of these more distant satellites must contain a significant fraction of material whose density is much less than that of rocks. The best candidates for this low-density material are water and water-ice, which has a density of 1 gram per cubic centimeter.

The first evidence for water-ice on the Galilean satellites came in the mid-1950s as a result of the work of Gerard P. Kuiper at the University of Arizona. From observations between 1 and 3 microns, Kuiper noticed that Europa and Ganymede both reflected light in the same way as ordinary water-ice and frost. In the early 1970s, several astronomers at Kitt Peak and Catalina Observatories confirmed these observations. More than half (that is, between 50 and 100 per cent) of Europa's surface must be frost and ice while between 20 and 60 per cent of Ganymede's surface is covered with frost and ice.

Reflected sunlight from Io shows no evidence of water-ice and from this satellite's average density, it is reasonable to suppose that Io's surface is entirely rocky like our own moon. Reflected sunlight from Callisto also shows no evidence of water-ice or frost and Callisto's low albedo is consistent with rocks. The low albedo, low average density, and the absence of evidence for water-ice can all be reconciled by assuming that there is an icy crust that is very "dirty" and contains many rocks. This high rock content on Callisto's surface makes it impossible for earth-based astronomers to detect the underlying ice, yet a low-density substance must be there in order to explain the satellite's low average density. Since both

**Figure 12–25.** *Internal Structure of the Galilean Satellites.* Data from analyses of reflected sunlight and the average densities of the Galilean satellites can be combined to give a reasonable picture of the internal structures of the four largest Jovian moons.

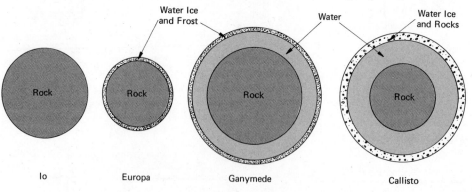

Europa and Ganymede show clear evidence of water-ice, it seems logical to suppose that this same substance is present in abundance on Callisto. As a result of these considerations, it is possible to make a reasonable guess at the internal structures of the Galilean satellites, as shown in Figure 12–25.

Of all of the Jovian satellites, Io is perhaps the most fascinating. First of all, its reddish color and high reflectivity cannot be easily explained in terms of ordinary rocks. Substances containing sulfur as well as salts such as sodium chloride and sodium sulfate seem much more likely candidates. In addition, it has been long known that Io influences radio emissions from Jupiter. When Io is at certain locations in its orbit, radio astronomers here on Earth detect an increase in the intensity of decametric radio noise from Jupiter. This effect of Io on Jupiter's magnetosphere was the first evidence that the innermost Galilean satellite possesses an ionosphere. As charged particles surrounding Jupiter encounter Io's ionosphere, they are accelerated thereby influencing Jupiter's radio emissions.

Additional evidence for an ionosphere around Io came in 1973 when Robert A. Brown at Harvard University discovered sodium lines in the emission spectrum of the satellite. Evidently, as Io orbits Jupiter deep inside the planet's magnetosphere, charged particles impinging on the surface of the satellite break apart salt molecules thereby liberating large quantities of sodium. It is believed that roughly 10 million sodium atoms are released from each square centimeter of Io's surface each second! The resulting cloud of sodium vapor continues to orbit Jupiter, thereby surrounding the planet with a torus or doughnut of sodium gas, as shown in Figure 12–26.

**Figure 12–26.  *Io's Sodium and Hydrogen Clouds.*** Io, which moves through the most intense regions of Jupiter's magnetosphere, is constantly emitting atoms from its surface. These atoms continue to orbit the planet. Jupiter is therefore surrounded by dough-nut-shaped rings of atoms and ions.

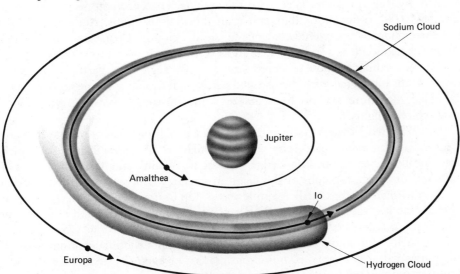

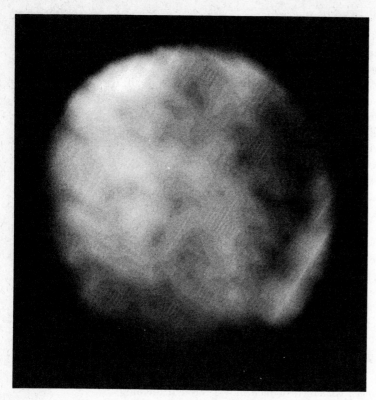

**Figure 12-27. *Ganymede.*** This view of Jupiter's largest moon was sent back from Pioneer 10 on December 3, 1973. The spacecraft was 750,000 kilometers (470,000 miles) from the satellite when the picture was taken. The surface resolution (that is, the size of the smallest feature that can be seen on Ganymede) in this view is about 400 kilometers (250 miles). Voyager spacecrafts in 1979 will be able to photograph features on Ganymede as small as a few hundred meters (roughly 1,000 feet) in size. (*NASA*)

While earth-based astonomers were discovering the extent of Io's sodium cloud around Jupiter, ultraviolet measurements from Pioneer 10 revealed an even larger cloud of hydrogen gas associated with the satellite. Hydrogen atoms in this cloud may be 10,000 times more abundant than the sodium atoms. And in the mid-1970s, astronomers discovered evidence for similar clouds of potassium and ionized sulfur. These clouds are so huge that an astronaut on Io would see the sky filled with aurora, similar to those observed near the polar regions on Earth, as the atoms in the clouds interacted with the charged particles in Jupiter's magnetosphere. The star-filled nighttime sky on Io is ablaze with constantly changing patterns of pastel-colored lights.

In 1979, two Voyager spacecrafts will coast past Jupiter on their way toward Saturn. Each of these spacecrafts will be similar to Mariners 9 and 10, which dramatically increased our knowledge of Mercury, Venus, and Mars. Hundreds of high-resolution photographs of the Galilean satellites will be transmitted back to Earth showing objects as small as a mile across. Up to now, the best view of one of Jupiter's moons was sent back by Pioneer 10 (see Figure 12-27) and reveals only vague surface markings. As a result of these new missions, four huge worlds will be made available for study. What we might find boggles the imagination; perhaps huge snowdrifts, vast canyons, numerous craters, or even volcanoes of water and liquid ammonia and methane.

# Questions and Exercises

1. How far is Jupiter from the sun?
2. How big is Jupiter?
3. How massive is Jupiter?
4. Describe the shape of Jupiter.
5. Compare the size and mass of Jupiter with Earth.
6. Why do scientists believe that any hydrogen or helium present on Jupiter billions of years ago is still on the planet?
7. Describe the appearance of the Great Red Spot.
8. What is the difference between belts and zones?
9. What are the Galilean satellites?
10. Briefly describe Jupiter's rotation.
11. What Earth-based observations led scientists to believe that Jupiter has a substantial magnetic field?
12. Why can't Earth-based astronomers ever see the dark side of Jupiter?
13. How much energy does Jupiter emit compared to the amount of energy it receives from the sun in sunlight?
14. Discuss one way in which Pioneers 10 and 11 differed from spacecrafts used in fly-bys of the inner planets.
15. Briefly describe how pictures of Jupiter were obtained by Pioneers' imaging photopolarimeters.
16. How are magnetic scattering regions produced in the solar wind and what effect do they have on cosmic rays?
17. What was learned about the distribution of small particles in the solar system as the Pioneers journeyed to Jupiter?
18. Briefly describe Jupiter's magnetosphere.
19. In what ways is Jupiter's magnetic field different from Earth's?
20. Compare and contrast the magnetospheres of Jupiter and Earth.
21. Why are charged particles in Jupiter's magnetosphere largely confined to a current sheet rather than something like the Van Allen belts surrounding Earth?
22. Briefly describe the effects of Jupiter's inner satellites on the planet's magnetosphere.
23. How thick is Jupiter's atmosphere?
24. How does the composition of Jupiter's atmosphere compare with that of the sun?
25. Describe the various cloud layers in Jupiter's atmosphere. What are they made of and where are they located?
26. Describe the variation of temperature in Jupiter's atmosphere.
27. What visible features on Jupiter correspond to regions of high atmospheric pressure? What visible features correspond to regions of low atmospheric pressure?
28. What is Coriolis force and what role does it play in air circulation patterns on Earth?

29. What is the difference between cyclones and anticyclones?
30. Compare and contrast circulation patterns on Earth and Jupiter.
31. Describe the wind flow patterns in the Great Red Spot.
32. Why do some scientists believe that Jupiter's atmosphere harbors primitive life forms?
33. How many satellites of planets are known to exist in the solar system? How many of these satellites are in orbit about Jupiter?
34. Briefly describe the observational evidence that supports the idea that the Galilean satellites always keep the same side facing Jupiter.
35. Briefly discuss the temperature ranges on the Galilean satellites. From what kind of observations do such data come?
36. How do the average densities of the Galilean satellites vary with distance from Jupiter?
37. What evidence is there that water-ice and frost exist on some of the Galilean satellites?
38. What evidence supports the idea that two of the Galilean satellites are primarily composed of rock while the other two satellites contain large quantities of a less dense substance such as water?
39. Briefly describe the ionosphere of Io.

# The Outer Planets 13

## Saturn—A Ringed Planet 13.1

SATURN, the sixth planet from the sun, is telescopically the most striking and beautiful object seen in the sky. Even a small, inexpensive amateur telescope often suffices to reveal the impressive system of rings that surround the planet. An excellent Earth-based view is shown in Figure 13-1.

Until the late 1700s, Saturn was the most distant planet known to astronomers. At an average distance of almost $1\frac{1}{2}$ billion kilometers (nearly 1 billion miles) from the sun, Saturn takes $29\frac{1}{2}$ years to complete one full orbit. Actually, because of the planet's elliptical orbit, the distance between the sun and Saturn varies from 1,350 million kilometers at perihelion to 1,440 million kilometers at aphelion (840 to 895 million miles) and the sidereal period is 29.458 years. In addition, Saturn's light-colored cloud-cover so efficiently reflects sunlight that near opposition the planet appears approximately as bright as any of the brightest stars in the sky. At opposition, Saturn's apparent magnitude reaches $-0.4$.

In many respects, Saturn takes "second place" to Jupiter. Saturn is the second largest and second most massive planet in the solar system. Saturn's mass is equal to

**Figure 13-1.** *Saturn.* Saturn, the sixth planet from the sun, is surrounded by a beautiful system of rings, which gives the planet its striking appearance. Saturn is one of the largest and most massive objects in the solar system, exceeded only by Jupiter and the sun. (*New Mexico State University Observatory*)

443

95 Earths (by comparison, recall that Jupiter's mass equals 318 Earth masses). While Saturn's mass is significantly smaller than Jupiter's, Saturn's diameter is only slightly less than that of its huge neighbor. Saturn's diameter is 120,000 kilometers (75,000 miles = 9.5 Earth diameters), which is only a little smaller than Jupiter's diameter of 143,000 kilometers (88,000 miles = 11.2 Earth diameters). Since Saturn has a moderate mass but a relatively large size, the planet's average density is unusually low. The average density of Saturn is only 0.7 grams per cubic centimeter. By comparison, the average density of Jupiter is 1.34 grams per cubic centimeter and that of Earth is 5.5 grams per cubic centimeter. Since the average density of water is 1 gram per cubic centimeter, Saturn is less dense than water. Saturn would literally float if placed in a huge tub of water.

Saturn also has the second fastest rotation rate of any of the planets. Saturn rotates about its axis once every $10\frac{1}{3}$ hours. A "day" on Saturn lasts a few minutes longer than a "day" on Jupiter. This rotation period of $10\frac{1}{3}$ hours is, however, an average. Like Jupiter, Saturn exhibits differential rotation. It rotates faster at the equator (10 hours, 14 minutes) than at the poles (10 hours, 38 minutes). This, of course, is possible because Saturn's surface is not solid. As in the case of Jupiter, when we look at Saturn through a telescope we are seeing only the outermost layers of a very thick cloud cover. This cloud cover is striped with belts and zones parallel to the planet's equator. Although these markings are somewhat less distinct than on Jupiter, they are probably regions of high and low atmospheric pressure that have been stretched all the way around the planet. The meteorologies of the two largest planets in the solar system might indeed be very similar, although data from future space flights will be needed in order to support or refute this speculation.

Since Saturn is so far from the sun, the planet's surface temperature is quite low, roughly $100°K$ ( $= -170°C = -280°F$), based on infrared observations. In view of this low temperature as well as the comparatively high escape velocity for the planet, Saturn has easily retained all the hydrogen and helium it had at the time of its formation. Saturn, like Jupiter, is composed mostly of hydrogen in both liquid and metallic form. Back in 1937, Rupert Wildt at Yale University first proposed that Saturn contains a small rocky core surrounded by a layer of ice buried deep inside the planet. This picture is still generally accepted today. A cross-section diagram of Saturn's interior is shown in Figure 13–2.

In addition to hydrogen and helium, other gases are known to exist in Saturn's atmosphere. Both methane and ammonia have been detected spectroscopically. A spectrum of reflected sunlight from Saturn is shown in Figure 13–3. For comparison, a spectrum of Jupiter is also included. Once again, the striking similarities between these two huge planets are evident.

While the structure and properties of the two largest planets in the solar system seem to be very similar, their telescopic appearance is not. A view of Saturn is, of course, dominated by its impressive system of rings. Although Galileo was the first person to see Saturn through a telescope, his instruments were so crude that he was unable to observe the planet's rings clearly. He noted, however, that the image of Saturn was very flattened as if the planet had "ears." Forty-five years later, in 1655, the optics of telescopes had improved to the extent that the Dutch astronomer

Christian Huygens could see the rings and correctly describe their appearance. It was at this time that astronomers realized there are actually three concentric rings around Saturn. The middle ring is by far the brightest and broadest. This *bright ring,* often called *Ring B,* has an inside diameter of 183,000 kilometers (113,000 miles) and an outside diameter of 233,000 kilometers (145,000 miles). The width of the bright ring is therefore 26,500 kilometers (16,000 miles). Two Earths could fit side by side between the inner and outer edges of the bright ring.

Beyond the bright ring is a fainter *outer ring,* often called *Ring A.* The inside diameter of this ring is 243,000 kilometers (15,000 miles), while the outside diameter is 274,000 kilometers (17,000 miles). The width of the outer ring is therefore 15,500 kilometers (9,600 miles).

Finally, inside the bright ring is the so-called *crepe ring,* or *Ring C.* This inner ring is very faint and extends from the inner edge of the bright ring down to a distance of only 12,000 kilometers (7,400 miles) above the planet's surface. The width of the crepe ring is therefore about 19,000 kilometers (12,000 miles). Earth could fit snugly between the inner edge of the crepe ring and Saturn's surface.

Some observers report seeing additional rings, perhaps one (Ring D) in the gap between Saturn and the inner edge of the crepe ring, or perhaps one (Ring D') beyond the outer edge of the outer ring. These additional rings must be extremely faint. They have never appeared on any earth-based photographs and are allegedly

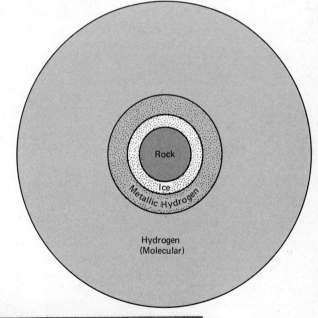

**Figure 13–2.** *Saturn's Interior.* Saturn's core probably consists of a rocky core surrounded by a layer of ice. Throughout most of the planet, hydrogen is probably in liquid form. Near the core, however, pressures are sufficiently high to convert the hydrogen to its metallic form.

**Figure 13–3.** *Spectra of Jupiter and Saturn.* Spectra of Jupiter and Saturn in the wavelength range 7,700–8,100 Å are shown. The atmospheres of both planets must have similar chemical compositions. (*Yerkes Observatory*)

Saturn ⟶

Jupiter ⟶

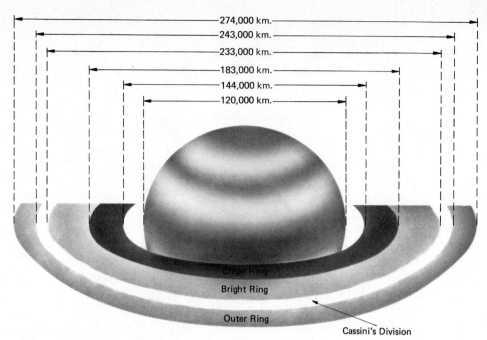

**Figure 13-4.** *Saturn's Rings.* Three major rings are seen around Saturn. This scale drawing gives the overall dimensions of the ring system.

observed only during brief moments of excellent "seeing" when the Earth's atmosphere is unusually steady and clear.

The structure of Saturn's rings is shown in Figure 13–4. Notice that there is a gap between Rings A and B. This gap was first observed by C. D. Cassini in the mid-1600s and is called *Cassini's division.* Cassini's division is about 5,000 kilometers (3,100 miles) wide. The planet Mercury could therefore easily fit in this gap between the bright and outer rings.

Saturn and its ring system are tilted through an angle of 27° from the plane of the planet's orbit. Recalling that an analogous tilting of Earth's equator from the plane of the ecliptic through an angle of $23\frac{1}{2}°$ is responsible for the seasons (see Section 2.1), we conclude that Saturn also must experience seasons. More importantly, however, this inclination of Saturn's rings means that the appearance of the ring system changes dramatically over the years. As shown in Figure 13–5, on certain occasions earth-based astronomers see the rings from the "top" while half a Saturnian year later we see the "bottom" side of the rings. During the spring and fall seasons on Saturn, earth-based observers see the rings edge on. A series of photographs spanning about 15 years (that is, half a Saturnian year) are shown in Figure 13–6. Notice that the rings seem to vanish when viewed edge on.

Since the rings of Saturn seem to disappear when viewed edge on, it is clear that the rings cannot be very thick. The thickness of the rings, which lie in the plane of

Saturn's equator, must be less than 20 kilometers (a dozen miles) and may be as thin as a few meters. Indeed, stars can be seen through the rings. In addition, by analyzing reflected sunlight from the rings, astronomers realize that the rings are composed of countless billions of tiny particles. These particles consist of rock and ice and are typically the size of pebbles. Each one of these rock and ice pebbles orbits Saturn like a tiny moon. As a result, the inner portions of the ring system revolve about Saturn much faster than the outer portions, just as Mercury goes about the sun much faster than Pluto.

The total mass of all the material in Saturn's rings is approximately equal to $1/23{,}000$ that of the planet itself. This means that if all the particles in the rings were lumped together, the resulting object would be roughly the size of our moon. This suggests that the rings are pebbles that did not manage to accrete to form a satellite. The rings are so close to Saturn that the tidal forces of the planet are greater than the mutual gravitational forces that the pebbles exert on each other. Therefore, the pebbles never managed to coalesce and form a satellite.

Beyond the outer edge of the rings, ten known satellites orbit Saturn. One of these satellites, Titan, is very large. With a diameter of 5,800 kilometers (3,600 miles), Titan is the second largest satellite in the solar system and orbits the planet once every 16 days. This satellite is so massive and so far from the sun that it has been able to retain an atmosphere. In 1944, Gerard P. Kuiper at the University of Chicago identified methane in the spectrum of reflected sunlight from Titan. More recently, Laurence M. Trafton at the University of Texas concluded that the

**Figure 13-5.** *The Inclination of Saturn's Rings.* Saturn's rings are tilted by 27° from the plane of the planet's orbit. As a result, Earth-based observers see the rings at various angles as Saturn orbits the sun. Once every 15 years, Earth-based observers view the rings edge on.

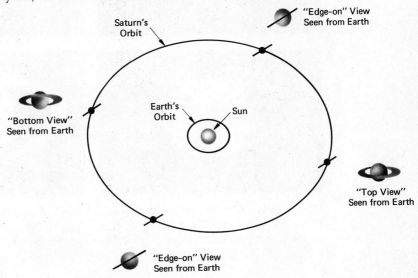

**Figure 13-6.** *The Changing Appearance of Saturn's Rings.* As Saturn orbits the sun, Earth-based observers see the rings at various angles. Notice that the rings seem to disappear when viewed edge on. (*Lowell Observatory photograph*)

atmospheric pressure of methane on Titan must be about four times greater than the atmospheric pressure on Mars. In view of Titan's mass and size, the satellite's average density is about $1\frac{1}{3}$ grams per cubic centimeter. As with the outer Galilean satellites of Jupiter, this means that a substantial fraction of Titan must be composed of water and ice. A reasonable interior structure of Titan is shown in Figure 13–7.

Titan was discovered by Huygens in 1655, the same year in which he first correctly described the rings. During the next few decades, from 1671 to 1684, Cassini discovered four more satellites. Three of Cassini's satellites (Tethys, Dione, and Rhea) are shown in Figure 13–8, along with Titan and an overexposed Saturn. The fourth satellite discovered by Cassini, Iapetus, is much farther from the planet than the other three. At an average distance of $3\frac{1}{2}$ million kilometers (about 2 million miles) from the planet, Iapetus is very peculiar. As the satellite rotates, its brightness changes dramatically. Evidently, one side of Iapetus is very shiny and highly reflective while the opposite side is very dull.

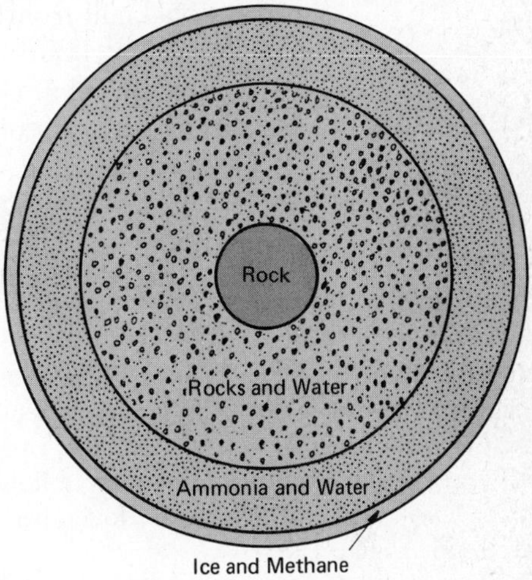

**Figure 13–7.** *The Structure of Titan.* Titan is the only satellite known to have an atmosphere. In addition, since Titan has a low average density, water and ice (rather than rock) must be important internal constituents.

Rock

Rocks and Water

Ammonia and Water

Ice and Methane

**Figure 13–8.** *Four of Saturn's Satellites.* Four of Saturn's brightest, inner satellites are seen in this overexposed view of the planet. Titan is the second largest satellite in the solar system. Rhea, Dione, and Tethys each have a diameter of roughly 1,000 kilometers (600 miles). (*Yerkes and McDonald Observatories photograph*)

Titan

Rhea
Dione

Saturn

Tethys

During the centuries following the work of Cassini and Huygens, five additional satellites have been discovered in orbit about Saturn. All of these satellites are very small, typically measuring only a few hundred kilometers across. The most recent satellite to be discovered is Janus, the innermost of all Saturn's moons. Janus orbits Saturn once every 18 hours only 20,000 kilometers (12,000 miles) beyond the outer edge of Saturn's outer ring. A. Dollfus first glimpsed Janus in December of 1966. At that time, the rings were exposed to Earth edge on and the resulting reduction in glare permitted Dollfus to discover Saturn's innermost moon.

The second inner moon of Saturn, Mimas, is also very small but has an important effect on the structure of Saturn's rings. Mimas orbits Saturn once every 22 hours, 25 minutes. *If* a particle were orbiting Saturn in Cassini's division, the orbital period would be 11 hours, $17\frac{1}{2}$ minutes. That is exactly half the orbital period of Mimas. Thus, every time this hypothetical particle orbits Saturn twice, Mimas orbits Saturn once. Every $22\frac{1}{2}$ hours, Mimas and the particle are lined up in exactly the same fashion. At each alignment, the particle would experience a tiny gravitational pull from Mimas. The net effect of these tiny "tugs" or perturbations once every $22\frac{1}{2}$ hours over millions of years is that the particle would be pulled out of its internal orbit. A gap in Saturn's rings would result. Cassini's division is therefore a direct result of gravitation perturbations by Mimas.

Over the years, several astronomers have reported seeing additional gaps in Saturn's rings. Presumably, these gaps would have been caused in the same way as Cassini's division. They would be located at distances from Saturn that correspond to orbital periods that happen to be convenient fractions of the orbital periods of

**Figure 13–9.** **The First Saturn Flyby.** Pioneer 11, which flew past Jupiter in December 1974, will arrive at Saturn in September 1979. This mission will give scientists their first close-up view of this ringed planet. Two more spacecrafts, Voyager 1 and Voyager 2, are now following in the footsteps of Pioneer 11. Voyager 1 will arrive at Saturn in November 1980. Voyager 2 will follow with a flyby in August 1981.

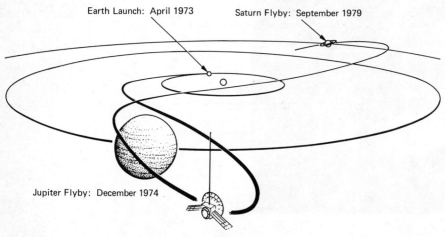

Saturn's satellites. Indeed, investigating the implications of one of these reported gaps led Dollfus to search for a satellite just beyond the outer ring in 1966.

Saturn's rings may indeed have numerous gaps in addition to Cassini's division. Unfortunately, the planet is so far away that Earth-based astronomers are often frustrated in their attempts to see fine details in the rings and on the planet's surface. The obvious remedy is to turn to unmanned spacecrafts. In September of 1979, Pioneer 11 will pass within a few hundred thousand kilometers of the planet. The trajectory of the spacecraft was chosen so that after encountering Jupiter in December 1974, Pioneer 11 would be directed toward Saturn, as shown in Figure 13–9. If all goes well, scientists will see views of Saturn far clearer and sharper than ever before.

While Pioneer 11 is silently gliding toward Saturn, scientists and engineers have been constructing two more spacecrafts that will visit the two largest planets in the solar system. Both vehicles in the Voyager Jupiter–Saturn series will use the enormous gravitational field of Jupiter like a slingshot to redirect the spacecrafts toward Saturn. The Jupiter fly-bys will occur in 1979 and the Saturn fly-bys in 1981.

In recent years, the space program has come under severe criticism from a variety of sources. For example, the entire Voyager Jupiter–Saturn project will cost $414 million. This sum is divided as follows: $316 million to the Jet Propulsion Laboratory for designing and building the two spacecrafts, $4 million for NASA administrative services, $72 million for the launch vehicles, and $22 million for flight support, tracking, and data reduction. It is often asked whether these funds could be better spent for social programs that would directly benefit disadvantaged citizens in the United States. Certainly, the welfare of all people in our society should be one of our primary concerns. Certainly $414 million sounds like a lot of money. But is it? In 1975, Americans spent $800 million on chewing gum. For approximately one half the amount of money Americans spend in one year on chewing gum, thousands of people are being employed in expanding the frontiers of human knowledge.

## *Uranus and Neptune—The Twin Planets* **13.2**

SATURN was the most distant planet known to ancient astronomers. Until the late 1700s, it constituted the outer boundary of our solar system. But unrecognized by astronomers of antiquity, a seventh more distant planet had often been observed. At opposition, Uranus reaches an apparent magnitude of $+5.6$, which means that it is just barely visible to the naked eye. Yet, because of its faintness and slowness, Uranus was invariably mistaken for a dim, uninteresting "star." It was plotted as a "star" on at least 20 star charts and sky surveys between 1690 and 1781.

On March 13, 1781, the German-born, English astronomer William Herschel

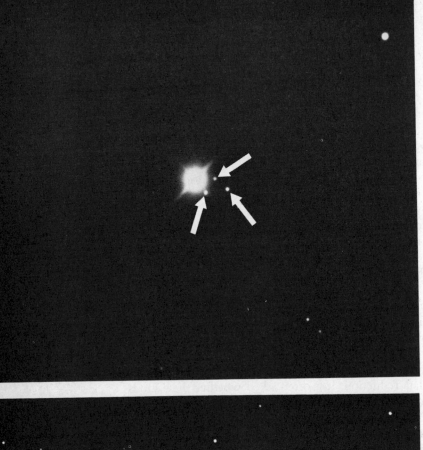

**Figure 13–10. _Uranus._** The seventh planet from the sun was discovered accidentally during a routine telescopic survey of the sky by William Herschel in 1781. Uranus is shown here along with three of its brightest satellites. (_Lick Observatory photograph_)

**Figure 13–11. _Neptune._** The existence of the eighth planet from the sun was inferred on the basis of gravitational perturbations of Uranus' orbit. Neptune was first seen in 1846 and is shown here with its largest moon, Triton. (_Lick Observatory photograph_)

was making a routine telescopic survey of the sky in the constellation of Gemini. During his observations, Herschel discovered a faint object that seemed to appear as a greenish-blue disk rather than a star. Thinking that it might be a distant comet, Herschel observed the object for several nights. Based on observations over several weeks, an orbit was calculated. The newly discovered object was in a nearly circular orbit 2,870,000,000 kilometers (1,780,000,000 miles) from the sun. At a distance of almost 3 billion kilometers, Uranus is twice as far from the sun as Saturn. It is unquestionably a planet and takes 84 years to complete one orbit.

By 1790, J. B. T. Delambre and other astronomers had calculated a more precise orbit for Uranus. However, it was soon found that this orbit, based on observations since Herschel's discovery, could not be reconciled with earlier records of the planet's position on star charts dating back to 1690. Something seemed to be wrong. By 1840, the discrepancy between the computed and observed positions of Uranus amounted to 2 minutes of arc. Although this is a tiny angle, hardly discernible to the naked eye, it was large enough to cause great concern among astronomers, whose calculations should have resulted in a much higher degree of accuracy. Even though astronomers had taken great care to include the gravitational perturbations of Jupiter and Saturn, it seemed as though Uranus was not moving along an orbit that could be predicted from Newtonian theory.

Beginning in 1841, a young undergraduate student at Cambridge University, John Couch Adams, decided to try a novel idea to explain the unusual motion of Uranus. Adams noted that during the first quarter of the nineteenth century, Uranus seemed to be accelerating slightly. During the next couple of decades, in the second quarter of the nineteenth century, Uranus seemed to be slowing down or decelerating slightly. Adams realized that this unusual motion of Uranus could be explained by the gravitational perturbations of an eighth, unknown planet still further from the sun. Specifically, Adams proposed that Uranus caught up with and passed a more slowly moving distant planet in 1822. In the years prior to 1822, as Uranus approached the line joining the sun and the unknown planet, Uranus experienced a small acceleration. But after 1822, as Uranus receded from the unknown planet, gravitational perturbations caused a small deceleration.

Adams proceeded to calculate the expected position of this unknown planet whose gravitational perturbations were deflecting Uranus slightly from its anticipated orbit. In October 1845, he sent the results of his computations to the Astronomer Royal of Great Britain, Sir George Airy. According to Adams, a new planet should be located in the constellation of Aquarius. Sir Airy had little faith in the work of such an undistinguished astronomer and the whole matter was dropped.

Meanwhile, in France, the astronomer Urbain Leverrier had a bright idea. Perhaps gravitational perturbations from an eighth, unknown planet were causing Uranus to be deflected slightly from its anticipated orbit. Leverrier published the results of his calculations in June of 1846, predicting the existence of a distant planet in the constellation of Aquarius. Noting that Leverrier's predicted position was less than 1° from Adams' predicted position, Sir Airy decided that somebody had better start examining the appropriate region of the sky in Aquarius. J. Challis, Director of the Cambridge Observatory, was given the task. Unfortunately, no up-to-date star charts of Aquarius were available to Challis and he started plotting all the faint stars in Aquarius. Challis' idea was to plot all the stars on two separate occasions to see if anything was moving. Unfortunately, it seems that Challis was not very careful in examining his observations and he failed to discover the new planet although he probably saw it.

Not wanting to bother with a bunch of uncouth Englishmen, Leverrier sent a letter to Johann Gottfried Galle at the Berlin Observatory. Galle received Leverrier's letter suggesting that he search Aquarius for a new planet on September 23, 1846. Fortunately, Galle possessed some fine star charts of the appropriate region of the sky and on that same night Neptune was discovered. It was less than 1° from Leverrier's predicted position. Needless to say, a hot debate arose between English and French astronomers, concerning who should be credited with discovering the eighth planet from the sun. It is today generally agreed that Adams and Leverrier share this honor equally.

The importance of the discovery of Neptune to the immediate history and philosophy of science cannot be overemphasized. In the years preceding Neptune's discovery, some scientists began to have serious doubts about classical mechanics and Newtonian gravitation. After all, they failed to predict Uranus' orbit correctly; perhaps Newton's theory of gravity simply doesn't work that far from the sun. These pessimistic speculations turned out to be totally wrong. Using Newtonian theory, both Adams and Leverrier found that they could account for Uranus' behavior and in doing so they both independently predicted the existence of yet another planet. Newton's ideas were obviously so universal and so powerful that they could be used to discover new planets. Neptune was literally discovered with paper and pencil. This triumph of classical mechanics permanently secured a position for Newton at the cornerstone of the very foundations of all physical science for decades to come.

While Uranus is nearly 3 billion kilometers from the sun, Neptune is considerably further away. The average distance between Neptune and the sun is $4\frac{1}{2}$ billion kilometers (about $2\frac{3}{4}$ billion miles). While the sidereal period of Uranus is 84 years, it takes 165 years for Neptune to circle the sun. Neptune is so very far away that it is never brighter than eighth magnitude and never larger than 2 seconds of arc across. By comparison, Uranus can appear twice as large, about 4 seconds of arc in diameter at opposition.

In many respects, Uranus and Neptune are very similar to each other. They both have nearly the same size and mass and probably have very similar chemical compositions and internal structures.

As you might expect, determining the diameter of planets billions of miles from Earth is a difficult task. In the case of Uranus, one of the best determination of the planet's diameter comes from balloon-borne telescopic observations. In 1970, a team of astronomers flew a remotely controlled telescope from a balloon at an altitude of 80,000 feet above the earth's surface. At such an altitude, their equipment was well above most of the earth's atmosphere and they were able to obtain many extraordinary sharp, clear images of Uranus. Figure 13–12 shows the results

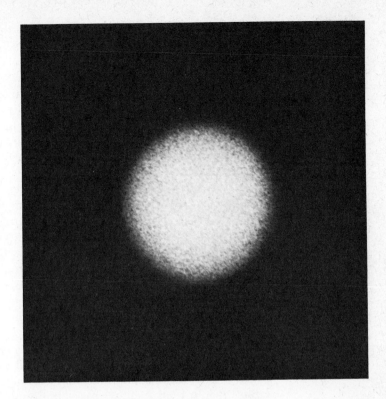

**Figure 13-12.** *The Best View of Uranus.* This extraordinary clear image of Uranus was obtained from a telescope carried by a balloon 80,000 feet above the Earth's surface. At this altitude, the remotely controlled telescope was above most of the Earth's atmosphere that adversely affects astronomical observation. Notice that no surface markings are seen on the planet. (*Project Stratoscope II, Princeton University supported by NSF and NASA*)

of their efforts. This photograph actually consists of a composite of over a dozen separate images and has a resolution of 0.15 seconds of arc. This is the best view of Uranus ever obtained by astronomers. From examining this photograph, Robert E. Danielson, Martin Tomasko, and Blair Savage conclude that Uranus' diameter is 51,800 kilometers (32,200 miles).

It is interesting to note that no discernible surface features are seen in Figure 13-12. Several Earth-based astronomers had reported seeing faint markings similar to belts and zones on Uranus. Yet, if such markings actually exist, they should have shown up in the photographs from Stratoscope II. With a resolution of 0.15 seconds of arc, continent-sized features as small as 2,000 kilometers (1,200 miles) across should have been detectable.

On March 10, 1977, a team of astronomers from Cornell University attempted to make an extremely accurate measurement of Uranus' diameter. On that date, Uranus was scheduled to pass in front of a star. This is called an *occultation.* By accurately measuring how long the star is hidden behind the planet, and from knowing how fast Uranus is moving along its orbit, the astronomers realized that they could easily calculate the planet's diameter. Preliminary investigations showed that this occultation would be visible only from the southern Indian Ocean. The Cornell astronomers therefore decided to make the observations from NASA's Kuiper Airborne Observatory (see Figure 4-17). It was all very straight-forward and no one expected anything dramatic to happen.

**455**

Flying at an altitude of 41,000 feet above the Indian Ocean, Dr. James Elliot, leader of the Cornell team, focused the airplane's telescope on the appropriate star. Much to everyone's surprise, however, the star disappeared from view 40 minutes *before* the occultation was due to begin. Seven seconds later, the star blinked back on. A few moments later, the star blinked out again. And then on again. And off again.

Over a period of 9 minutes, extending from 40 minutes to 30 minutes prior to the scheduled occultation. The star briefly blinked out *five* times. Then the occultation came. Half an hour after the end of the occultation, the star blinked on and off five more times.

The obvious interpretation of this unexpected result is that *Uranus is surrounded by five rings.* Saturn is no longer unique among the planets. Uranus also has rings.

Uranus' rings are very different from Saturn's. Saturn's rings are *very broad.* The total distance from the inner edge of the inner ('crepe) ring to the outer edge of the outer rings is a whopping 61,000 kilometers (38,000 miles). In sharp contrast, Uranus' rings seem to be *very thin.* The inner rings are only 5 to 10 kilometers wide and the outer ring is, at most, 100 kilometers wide. The innermost ring is about 18,000 kilometers (11,000 miles) above Uranus' cloud tops while the outer ring is at a distance of 25,000 kilometers (16,000 miles) from the planet. Of course, the thinness of Uranus' rings explains why no one has ever seen them before. The rings are so thin that they reflect very little light. As in the case of Saturn, Uranus' rings are probably composed of pieces of rock and ice.

The best determination of Neptune's diameter comes from the occultation of a star in 1968. In 1968, Neptune passed in front of a star and by observing how long the star was blocked out as Neptune moved along its orbit, astronomers in Japan, Australia, and New Zealand concluded that the planet's diameter is 49,500 kilometers (30,700 miles). Thus, Neptune is a little smaller than Uranus.

The masses of the planets are determined by observing the orbits of their moons. Uranus has five moons (see Figure 13–13) and Neptune has two. From knowing the sidereal periods and lengths of the semi-major axes of the orbits of these satellites, Kepler's third law can be used to give the masses of the planets. The mass of Uranus is 14.6 times that of Earth while the mass of Neptune is 17.2 times that of Earth. Thus, Neptune is slightly more massive than Uranus.

Since Neptune is more massive yet smaller than Uranus, Neptune's average density must be larger than that of Uranus. The average density of Uranus is 1.2 grams per cubic centimeter while the average density of Neptune is 1.7 grams per cubic centimeter. For comparison, recall that the average density of Jupiter is 1.3 grams per cubic centimeter. The higher average density of Neptune can be explained by assuming that the hypothetical rocky core of Neptune is slightly larger than the rock core of Uranus. Figure 13–14 shows a reasonable internal structure of the two planets.

As you might expect, the chemical compositions of Uranus and Neptune are arrived at by spectroscopic observations. Both hydrogen and methane have been detected in the reflected sunlight from these two planets. Of course, hydrogen is by far the most abundant substance. It is impossible to detect helium from earth-based observations but this gas is assumed to be present in roughly the same proportions as on the sun and Jupiter: one helium atom for every twenty hydrogen atoms. By analogy with the two inner Jovian planets, ammonia might be expected to exist in the atmospheres of Uranus and Neptune. No ammonia, however, has ever been detected. The probable explanation for this is that both planets are very cold. The surface temperature of Uranus is about 60°K ( = −210°C = −350°F) while that of Neptune is roughly 50°K ( = −220°C = −370°F). At such low temperatures, all the ammonia would be frozen out of the planets' atmospheres.

Since no surface features are seen on either Uranus or Neptune, determining their periods of rotation is also difficult. Nevertheless, it can be done spectroscopically. Recall from the discussion of the Doppler effect (see Section 4.3) that spectral

**Figure 13-13. *Uranus and its Five Satellites.*** All of Uranus' five known satellites are seen in this view. The image of Uranus is overexposed to reveal Miranda, the smallest and innermost moon, which had been discovered two weeks before this photograph was taken in 1948. (*Yerkes and McDonald Observatories photograph*)

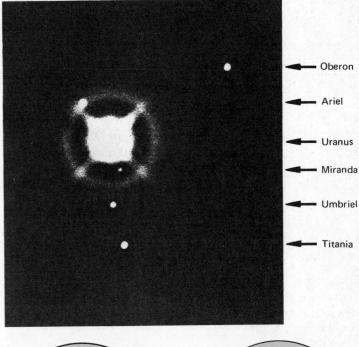

**Figure 13-14. *Probable Structure of Uranus and Neptune.*** It is reasonable to suppose that both Uranus and Neptune have rocky cores surrounded by a mantle of ice. The outer gaseous layers of these planets are composed almost entirely of hydrogen.

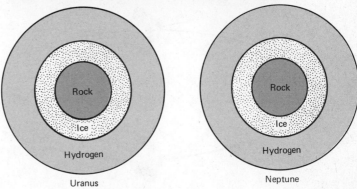

lines from an approaching source of light are blueshifted while spectral lines from a receding source of light are redshifted. If astronomers observe a rotating planet with a spectroscope, they will detect a slight blueshift from the side of the planet that is approaching Earth and a slight redshift from the side receding from Earth. Consequently, by measuring the Doppler shift across Uranus and Neptune, their rotation periods have been determined. The rotation period for Uranus is 10 hours, 45 minutes and the rotation period for Neptune is 15 hours, 50 minutes. This method is not as accurate as actually seeing features move across the planets' surfaces. Thus, these numbers may be wrong by as much as one hour.

In measuring the rotation periods of these planets, astronomers find themselves in a position to calculate the angle between the planet's equator and the plane of its orbit. Recall (see Section 2.1) in the case of our own planet, the angle between Earth's equator and the ecliptic is $23\frac{1}{2}°$. This inclination is, of course, responsible for the seasons. In the case of Neptune, its equatorial plane is inclined by 29° to its orbit. Thus, in principle, Neptune experiences seasons much like Earth. But, since the sun is so very far away from Neptune, the difference in the heating effect of the sun between summer and winter in a given hemisphere is much less pronounced than here on Earth.

While the inclination of Neptune's equator to its orbit is similar to the inclination between Earth's equator to its orbit, the case of Uranus is entirely different. Uranus' equator is tilted by 98° from its orbit. In other words, Uranus' axis of rotation lies very nearly in the plane of the planet's orbit. Thus, as shown in Figure 13–15, during summer in the northern hemisphere, Uranus' North Pole points almost directly at the sun. Similarly, the sun is almost directly overhead at the South Pole during summer in Uranus' southern hemisphere.

Actually, an inclination of 98° means that Uranus' axis of rotation is pointed in the opposite direction from all the other planets, except Venus. If we arbitrarily say that Earth's North Pole points "up" in the solar system, then Uranus' North Pole

**Figure 13–15.** *Seasons on Uranus.* Uranus' axis of rotation is tilted so far that it nearly lies in the plane of the planet's orbit. Thus, for example, during summer in Uranus' northern hemisphere, the sun is seen almost directly overhead at the planet's north pole.

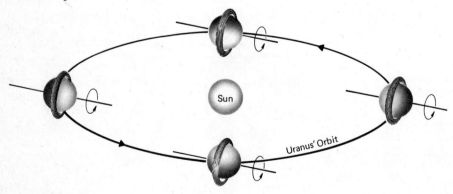

**Figure 13–16.** *Neptune and Triton.* With a diameter of 6,000 kilometers (almost 4,000 miles), Triton is the largest satellite in the solar system. Triton orbits Neptune backward or retrograde (that is, from east to west) once every six days. (*Yerkes and McDonald Observatories photograph*)

**Figure 13–17.** *The Satellites of Uranus and Neptune.* This scale drawing shows the relative sizes of the satellites of Uranus and Neptune. For comparison, the planet Mercury and our own moon are included. Triton, the largest satellite in the solar system, is almost as big as Mars.

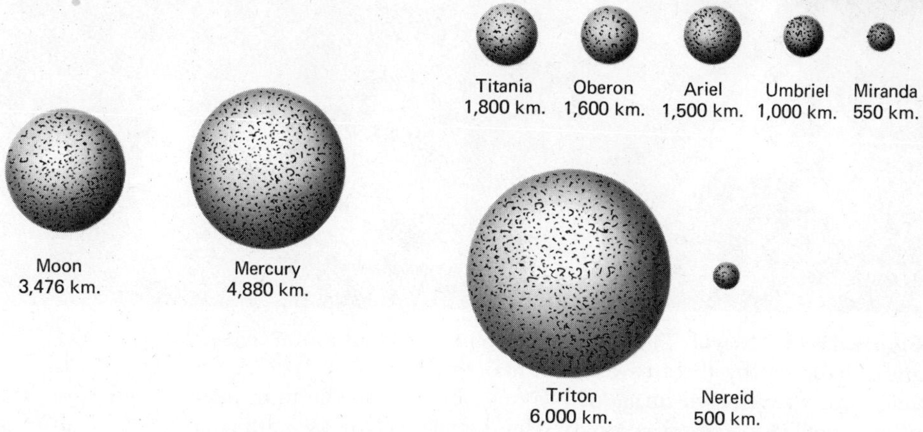

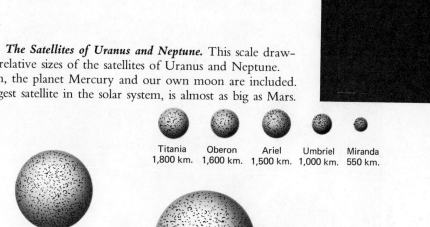

Titania 1,800 km.    Oberon 1,600 km.    Ariel 1,500 km.    Umbriel 1,000 km.    Miranda 550 km.

Moon 3,476 km.    Mercury 4,880 km.

Triton 6,000 km.    Nereid 500 km.

points "down." Alternatively, we could say that Uranus rotates *retrograde,* like Venus. As in the case of Venus, the sun appears to rise in the west and set in the east on Uranus. All of the orbits of Uranus' five moons lie nearly in the plane of the planet's equator. They all orbit the planet in the same direction as the plane rotates. Consequently, all of Uranus' satellites move retrograde. If we arbitrarily say that our moon goes around Earth "counterclockwise," then Uranus' satellites revolve about Uranus "clockwise."

Just as all of Uranus' satellites revolve retrograde about the planet, Neptune's largest satellite, Triton, also moves retrograde even though Neptune itself rotates in the usual, *direct* sense. Triton goes "backward" around Neptune once every 5 days, 21 hours. The average distance between Neptune and Triton is 353,400 kilometers (about 220,000 miles), which is very nearly the same distance as between our moon and Earth. Triton therefore never appears very far from Neptune, as shown in Figure 13–16.

With a diameter of 6,000 kilometers (3,700 miles) Triton is the largest satellite in the entire solar system. Triton is substantially larger than Mercury and is almost as big as Mars. A scale drawing showing the relative sizes of the satellites of Uranus and Neptune is shown in Figure 13–17.

**Figure 13–18.** *Nereid.* Tiny Nereid, Neptune's second moon, was discovered in 1949 and is indicated by the arrow in this photograph. Triton appears as a "lump" just below the overexposed image of Neptune. Nereid has the most highly elliptical orbit of any satellite in the solar system. The distance between Neptune and Nereid varies from 2 to 10 million kilometers ($1\frac{1}{4}$ to 6 million miles) as the satellite orbits the planet once every 360 days. (*Yerkes and McDonald Observatories photograph*)

Several Mariner and Pioneer missions to Uranus and Neptune are currently being planned for the late-1900s. For example, three separate spacecrafts may be launched in 1979 for Jupiter–Uranus fly-bys. This would be followed by a single Saturn–Uranus fly-by to be launched in 1981. Then, in 1986, two Mariner spacecrafts will be launched for Uranus–Neptune fly-bys. Thus, if all goes well, we may get our first close-up view of Triton just before the end of the century.

While the orbit of Triton is retrograde, the orbit of Neptune's other satellite, Nereid, is also unusual. Nereid is very, very small (its diameter is perhaps only 500 kilometers) and therefore escaped detection until 1949. A photograph of Nereid taken shortly after its discovery is shown in Figure 13–18. Nereid has the most highly eccentric orbit of any satellite in the solar system. The distance between Neptune and Nereid varies from 2 to 10 million kilometers ($1\frac{1}{4}$ to 6 million miles) during the 360 days it takes to go around the planet. As we shall see in the next section, the unusual orbits of Triton and Nereid may be directly related to the farthest planet from the sun, Pluto.

THE discovery of Neptune in the middle of the nineteenth century was one of the great triumphs of classical Newtonian mechanics. The existence of the eighth planet from the sun was inferred from anomalies in the observed orbit of Uranus. Uranus had not been exactly following its predicted orbit and the resulting deviations could be explained by the gravitational perturbations of a more distant, unknown planet. In this sense, Neptune was literally discovered with pencil and paper.

Shortly after the turn of the twentieth century, it was realized that there were some very tiny discrepancies between the observed and calculated orbits of Uranus and Neptune. For example, the existence of Neptune had explained about 98 per cent of the deviations of Uranus from its originally predicted orbit. But it seemed as though there were problems in arriving at an orbit that would be 100 per cent correct. There seemed to be similar problems with the orbit of Neptune, but since this planet moves so slowly and had been observed for only a few decades, no one could be quite certain if it were deviating from its predicted orbit. Nevertheless, these apparent deviations—which in reality were largely observational inaccuracies—led several astronomers to predict the existence of a ninth planet from the sun.

In 1909, A. Gaillot performed calculations predicting the existence of a 24-Earth-mass planet 66 AUs (ten billion kilometers or roughly six billion miles) from the sun. He was way off. These initial calculations were followed by similar computations in 1915 by Percival Lowell and in 1928 by William H. Pickering. Of these later calculations, Lowell's estimates turned out to be most nearly correct. Lowell concluded that the ninth planet from the sun should be seen in the constellation of Gemini and should have a mass almost 7 times larger than that of Earth. At a distance of roughly 40 AUs, this unknown planet was supposed to have an angular diameter of 1 second of arc. At his observatory in Arizona, Lowell searched the skies unsuccessfully for the planet until the time of his death.

The search for a ninth planet was facilitated by the donation of a wide-angle telescope to the Lowell Observatory by Lowell's brother. The central idea was to photograph the same portion of the sky on two separate occasions and then compare the views to see if anything had moved. Unfortunately, the suspected region of Gemini lies near the Milky Way and thousands upon thousands of stars appeared on each photographic plate. The task of comparing two plates simultaneously would seem hopeless if it were not for a device called a *blink microscope*. A blink microscope is a device in which two photographs of the same part of the sky can be placed. By means of a small mirror, the astronomer can rapidly shift his vision from one plate to the other. If the plates are properly aligned, all the star images remain stationary while the image of a moving object appears to jump back and forth.

In 1930, Clyde Tombaugh was examining two plates taken almost a week apart, on January 23 and 29, with his blink microscope. A tiny "star" appeared to jump back and forth by roughly the right amount for a trans-Neptunian planet. On March 13, 1930, Tombaugh announced the discovery of the ninth planet from the

**461**

sun only 6° from Lowell's predicted position. The planet was called Pluto, after the god of the underworld whose name prominently contains the initials of Percival Lowell.

With an apparent magnitude of only +15, Pluto always appears as a dim star on astronomical photographs, as shown in Figure 13–19. Although it is virtually impossible to measure the diameter of Pluto directly, astronomers can place reasonable limits on the size of the planet. For example, in the spring of 1965, Pluto was expected to occult a star. In spite of careful observation at several observatories in the United States at the predicted time of occultation, no dimming of the star was detected. Pluto obviously missed the star, which means that the planet must be less than 6,000 kilometers (3,800 miles) across. To obtain a lower limit on the size of the planet, assume that Pluto is 100 per cent reflective (that is, its albedo is 1.0) like a shiny mirror. If Pluto had a completely shiny surface and reflected all of the incident sunlight, the planet must be at least 2,000 kilometers (1,200 miles) in diameter in order to appear as a fifteenth-magnitude object in the sky. Thus, the diameter of Pluto must lie between 2,000 and 6,000 kilometers. It is perhaps about

**Figure 13–19. *Pluto.*** The ninth planet from the sun is easily identified by comparing these two photographs taken one day apart. Pluto takes 248½ years to complete one orbit of the sun. (*Lick Observatory photograph*)

the same size as Mercury, with a diameter in the range of 4,000 to 5,000 kilometers (roughly 3,000 miles).

At an average distance of almost 6 billion kilometers from the sun, the surface temperatures on Pluto would be extremely low, perhaps only 40°K ($-230$°C = $-390$°F). At such a temperature, all common gases would be frozen. For example, in 1976, Dale P. Cruikshank, Carl B. Pilcher, and David Morrison reported their discovery of methane ice on the planet. Consequently, Pluto does not have any appreciable atmosphere and the planet itself must consist entirely of ice and rock. Assuming a reasonable density for this ice–rock mixture, the planet's mass must be very low. Many astronomers estimate that Pluto's mass is somewhere in the range of $\frac{1}{10}$ to $\frac{1}{5}$ Earth-masses.

It is interesting to note that reasonable estimates of Pluto's mass are a far cry from Lowell's estimate of nearly 7 Earth-masses. This means that Pluto is far too tiny to cause any appreciable gravitational perturbations of either Neptune or Uranus. Indeed, it now seems clear that the alleged anomalies in the orbits of Uranus and Neptune, on the basis of which Lowell predicted the existence of Pluto, were observational inaccuracies. Pluto was therefore actually discovered by accident. The fact that Pluto has an orbit very similar to that predicted by Lowell is one of the remarkable coincidences of modern astronomy.

In many respects, Pluto seems to be a maverick among the outer planets. All of the outer planets, except for Pluto, are huge and massive with thick, gaseous atmospheres. Pluto is very small, has a low mass and a solid surface. In this regard alone, Pluto has characteristics quite similar to the satellites of the Jovian planets.

Pluto's orbit is also very different from the orbits of any of the Jovian planets. Pluto's orbit is steeply inclined to the plane of the ecliptic. The planes of the orbits of all the other outer planets lie very close to the ecliptic. The inclinations of the orbits of the Jovian planets from the plane of Earth's orbit are typically in the range of from 1 to 2 degrees. In sharp contrast, the plane of Pluto's orbit is tilted by 17° from the plane of the ecliptic, as shown in Figure 13–20. In addition, Pluto's orbit

**Figure 13–20.** *Pluto's Orbit.* Pluto's orbit is more steeply inclined and more eccentric than that of any of the other planets in the solar system. Indeed, Pluto is occasionally closer to the sun than Neptune.

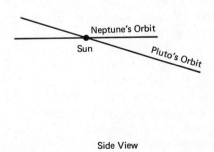

Side View

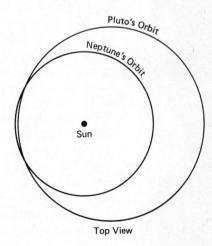

Top View

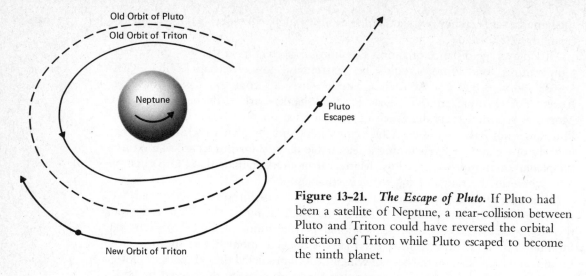

Old Orbit of Pluto
Old Orbit of Triton

Neptune

Pluto
Escapes

New Orbit of Triton

**Figure 13–21.** *The Escape of Pluto.* If Pluto had been a satellite of Neptune, a near-collision between Pluto and Triton could have reversed the orbital direction of Triton while Pluto escaped to become the ninth planet.

is very eccentric, much more elliptical than the orbits of any of the other planets in the solar system. As noted earlier, Pluto's average distance from the sun is almost 6 billion kilometers (about $3\frac{2}{3}$ billion miles). But the perihelion distance is only $4\frac{1}{2}$ billion kilometers ($2\frac{3}{4}$ billion miles) while the aphelion distance is slightly more than 7 billion kilometers ($4\frac{1}{2}$ billion miles). This means that Pluto is sometimes closer to the sun than Neptune, as shown in Figure 13–20. Pluto and Neptune are, however, in no danger of colliding. As a result of its high inclination, Pluto's orbit clears Neptune's orbit by 385 million kilometers (238 million miles).

The physical similarities between Pluto and the satellites of the Jovian planets, as well as Pluto's unusual orbit have led several astronomers to propose that long ago Pluto may have been a moon orbiting Neptune. This suspicion is heightened when we recall the unusual orbital properties of Neptune's remaining moons, Triton and Nereid. Triton's orbit is retrograde and Nereid's orbit is very highly eccentric. Specifically, both G. P. Kuiper and R. A. Lyttleton have argued that a close encounter between Triton and Pluto when it also was a satellite of Neptune could have caused Triton to reverse the direction of its orbit while Pluto escaped from the planet altogether. This near-collision and the resulting escape of Pluto are diagrammed in Figure 13–21. An often-cited objection to this hypothesis is the fact that Pluto's present orbit is relatively far from Neptune. With a simple two-body near-collision, the resulting orbit of Pluto would be expected to be closer to Neptune's orbit at one point. As noted earlier, the minimum distance between the orbits of Neptune and Pluto is about $\frac{1}{3}$ billion miles. Perhaps the answer lies in a three-body near-collision. Perhaps, when Pluto was a satellite of Neptune, all three satellites happened to have a near-collision at almost the same time. The same event that reversed the orbital direction of Triton also pushed Nereid into a highly eccentric path about the planet while Pluto escaped altogether.

The only remaining fact currently known about Pluto is its rotation rate. There is evidently a shiny area on the planet—perhaps a vast smooth expanse of ice—and thus, the brightness of Pluto increases by about 20 per cent when this shiny area is aimed at Earth. Based on this periodic variation in Pluto's brightness, J. S. Neff, W. A. Lane, and J. D. Fix have determined the rotation period of the planet to be 6 days, 9 hours, 18 minutes.

The details of what lies immediately beyond Pluto is largely a matter of speculation. Our star with all its planets is in orbit about the center of our galaxy along with billions of other stars. The sun moves at a speed of about 20 kilometers per second (45,000 miles per hour) with respect to the neighboring stars and takes 200 million years to go once around the Galaxy. Just as the planets orbiting the sun move through the solar wind, the sun orbiting the galactic center must be moving through an *interstellar wind*. The outer boundary where the interstellar wind first encounters the solar system results in a shock wave as the protons and electrons in the tenuous interstellar medium are slowed down to subsonic speeds. Still closer to the sun must be another boundary, called the *heliopause*, where the outward pressure from the solar wind counterbalances the flow of the interstellar wind. The exact dimensions of the resulting *heliosphere* are uncertain and therefore the scale used in drawing Figure 13–22 should not be taken too seriously. Perhaps data from Pioneer 10, currently on the way toward the star Aldebaran, will shed some insight on the size of the heliosphere.

To a certain degree, whether there might be planets farther from the sun than Pluto is anybody's guess. Nevertheless, we can be sure of one thing. *If* there is a tenth planet, it must be very tiny and very far away. Because the orbits of the other nine planets do not exhibit any unexpected perturbations, and therefore astronomers conclude that they have discovered all of the large planets orbiting the sun. In this regard, it is interesting to note that there are thousands upon thousands of very

**Figure 13–22.** *The Heliosphere.* It is reasonable to suppose that as the sun moves through the interstellar medium, interactions between the solar wind and the interstellar wind produce a heliosphere about the solar system. The dimensions of the heliosphere are uncertain.

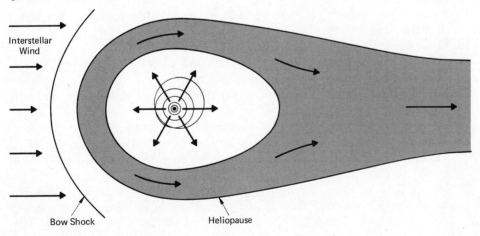

tiny objects orbiting the sun. But the vast majority of these boulder-sized objects, called *asteroids* or *minor planets,* are at distances from the sun between the orbits of Mars and Jupiter. Only *comets* with highly eccentric orbits pass beyond Pluto and perhaps escape into the reaches of interstellar space.

## Questions and Exercises

1. Describe the telescopic appearance of Saturn.
2. Contrast and compare Jupiter and Saturn, especially with regard to mass, size, density, rotation rate, and so forth.
3. What gases are known to exist in Saturn's atmosphere?
4. When and by whom were Saturn's rings discovered?
5. Describe Saturn's rings.
6. What is Cassini's division?
7. Describe the changes in the appearance of Saturn's rings over a $29\frac{1}{2}$ year period.
8. What are the rings made of?
9. Describe Titan and its atmosphere.
10. Describe one unusual property of Iapetus.
11. How many satellites does Saturn have?
12. Describe the effect which Mimas has on the structure of Saturn's rings.
13. How was Uranus discovered?
14. Describe the events leading up to the discovery of Neptune.
15. Compare and contrast Uranus and Neptune especially with regard to mass, size, density, chemical composition, rotation rate, number of satellites, and so on.
16. From what kinds of observations did the best determination of Uranus' diameter come?
17. From what kinds of observations did the best determination of Neptune's diameter come?
18. How were the rings around Uranus discovered?
19. Compare and contrast the rings around Uranus with the rings around Saturn. How are they similar? How are they different?
20. Describe the seasons on Uranus. Compared to the other planets, how and why are Uranus' seasons unusual?
21. In what ways are the orbits of Triton and Nereid unusual?
22. Why do astronomers feel that helium must be the second most abundant gas on Uranus and Neptune, even though helium has never been detected on either of these two planets?
23. How did astronomers determine the rotation rates of Uranus and Neptune, even though no surface features can be seen on either of these two planets?

24. Compare and contrast the prediction of Neptune by Adams and Leverrier with the prediction of Pluto by Lowell.
25. What is a blink microscope?
26. On what basis do astronomers conclude that Pluto's diameter must lie between 2,000 and 6,000 kilometers?
27. In what ways does Pluto's orbit differ from the orbits of the other outer planets?
28. Briefly describe the arguments for and against the idea that Pluto was once a satellite of Neptune.
29. Briefly describe the heliosphere.
30. Why do astronomers feel that there cannot be another large, massive, undiscovered planet just beyond the orbit of Pluto?

# 14 Vagabonds of Interplanetary Space

## 14.1 *The Asteroids*

In 1772, Johann Bode, director of the Berlin Observatory, popularized a little mathematical "game" for remembering the distances of the planets from the sun. Write down the sequence of numbers 0, 3, 6, 12, 24. . .doubling each time. Add 4 to each number. Then divide each of the resulting numbers by 10. The final result is a sequence of numbers that approximates the distances of the planets from the sun measured in astronomical units, as shown in Table 14–1.

Except for Neptune and Pluto, the agreement between the numbers in the sequence and the actual distances of the planets from the sun is quite remarkable. Indeed, in 1781, less than a decade after Bode began popularizing this sequence, William Herschel accidentally discovered Uranus. Uranus is located at 19.2 AU from the sun, very close to the "predicted" distance of 19.6 AU. As a result of this coincidence, the sequence of numbers became known as *Bode's law*. It is certainly not a "law" in the strict scientific sense; neither was it discovered by Bode. The sequence was first invented by the German astronomer J. D. Titius in 1766 in Wittenberg.

Bode's law seemed to work so well for giving planetary distances that many astronomers began wondering if there might be an undiscovered planet at 2.8 AU from the sun in the "gap" between Mars and Jupiter. This suspicion was so strong that by 1800, a team of six prominent German astronomers under the leadership of Johann Schröter resolved to search for this unknown planet. But before this team of "celestial police"—as they were soon nicknamed—could succeed, the announcements came from the Sicilian astronomer Giuseppe Piazzi. On January 1, 1801, Piazzi was mapping a region of the sky in the constellation of Taurus when he noticed an uncharted "star." Over the next several nights, this "star" gradually shifted its position and by mid-January, Piazzi sent a letter describing his discovery to Bode at the Berlin Observatory. Unfortunately, the letter did not arrive until late March when the object was too near the sun for observation. Equally unfortunately, Piazzi became ill in mid-February and was forced to discontinue his observations of this object, which Bode suspected was indeed the "missing planet." With only six weeks of observation, astronomers were unable to calculate an orbit

**Table 14-1.** *Bode's Law*

| Titius' Progression | Planet | Actual Distance (AU) |
|---|---|---|
| (0+4)/10 = 0.4 | Mercury | 0.39 |
| (3+4)/10 = 0.7 | Venus | 0.72 |
| (6+4)/10 = 1.0 | Earth | 1.00 |
| (12+4)/10 = 1.6 | Mars | 1.52 |
| (24+4)/10 = 2.8 | ? | |
| (48+4)/10 = 5.2 | Jupiter | 5.20 |
| (96+4)/10 = 10.0 | Saturn | 9.54 |
| (192+4)/10 = 19.6 | Uranus | 19.18 |
| (384+4)/10 = 38.8 | Neptune | 30.6 |
| (768+4)/10 = 77.2 | Pluto | 39.4 |

and it was feared that Piazzi's object was lost. It seemed that the only hope was to wait until another astronomer stumbled upon it during some future survey of the sky.

Upon hearing of these difficulties, the young brilliant German mathematician Karl Friedrich Gauss decided to try his hand at the problem of calculating the orbit of Piazzi's object. Prior to the work of Gauss, it was necessary to have several observations of an object over a long period of time in order to compute a reliable orbit. Only if months separated the observations—during which the object in question moved substantial distances across the sky—could astronomers come up with an accurate orbit. Fortunately, Gauss had recently invented a mathematical method whereby a number of observations or measurements could be combined to give the *most probable answer,* even though each individual measurement contained a large margin of error. His method was ideally suited for the observations of Piazzi's object, which had been seen for only six weeks.

By November of 1801, Gauss had succeeded in calculating the most probable orbit of Piazzi's object from the short series of observations and predicted that it should be in the constellation of Virgo. On December 31, 1801, F. X. von Zach rediscovered Piazzi's object very near the position predicted by Gauss. At Piazzi's request, the object was named *Ceres,* after the protecting goddess of Sicily. Ceres' average distance from the sun (or, more precisely, the length of the semimajor axis of Ceres' orbit) is 2.77 AU—another remarkable agreement with Bode's law! Ceres orbits the sun once every 4.6 years.

Since Ceres is always very faint, even at opposition when it is nearest Earth, it must be quite small—perhaps only 1,000 kilometers (600 miles) across. Ceres was therefore too tiny to be the "missing planet" and the team of German astronomers continued searching. In March of 1802, one of them, Heinrich Oblers, discovered a second faint starlike object, which he named *Pallas.* Like Ceres, the average distance between Pallas and the sun is 2.77 AU. Astronomers began to suspect that the "missing planet" had somehow blown up or fragmented and they were now in the process of discovering the pieces. It seemed entirely reasonable to suppose that

**Table 14–2.** *Selected Belt Asteroids*

| Number | Name | Year Discovered | Semi-major axis (AU) | Sidereal Period (years) | Size (kilometers) |
|--------|------|-----------------|----------------------|-------------------------|-------------------|
| 1 | Ceres | 1801 | 2.77 | 4.6 | 1,000 |
| 2 | Pallas | 1802 | 2.77 | 4.6 | 590 |
| 3 | Juno | 1804 | 2.67 | 4.4 | 300 |
| 4 | Vesta | 1807 | 2.36 | 3.6 | 550 |
| 5 | Astraea | 1845 | 2.56 | 4.1 | 130 |

**Figure 14–1.** *Two Asteroids.* Asteroids are most often discovered by blurred "trails" on time-exposure photographs. The images of two asteroids appear on this photograph. (*Yerkes Observatory photograph*)

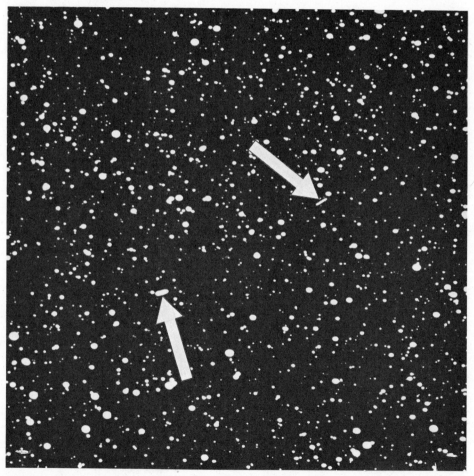

additional pieces were floating around, so the search continued. In 1804, *Juno* was discovered and *Vesta* was sighted in 1807. The length of the semimajor axis of Juno's orbit is 2.67 AU, while that of Vesta is 2.36 AU. No more of these *asteroids* had been found by 1815 and the search was abandoned.

In 1830, the Prussian amateur astronomer Karl Hencke resumed the search and 15 years later, in 1845, succeeded in discovering the fifth asteroid, which he named Astraea. All of these five asteroids have orbits in the "gap" between Mars and Jupiter. Relevant data about each are contained in Table 14–2.

By 1890, a total of 300 asteroids had been discovered, the vast majority of which had orbits between the orbits of Mars and Jupiter. This region of the solar system therefore became known as the *asteroid belt* and the tiny objects that orbit the sun in this region are called *belt asteroids*.

In 1891, the German astronomer Max Wolf began the first photographic search for asteroids. Rather than painstakingly scrutinizing the skies for uncharted "stars" whose positions gradually change from night to night, Wolf looked for asteroids by taking long-exposure photographs. If an asteroid happened to be in the field of view, it would leave a blurred "trail." The image of the asteroid would therefore be easily recognized provided care was taken to insure that the telescope accurately tracked the stars so that the star images appear as sharp pinpoints of light. A good example of this technique is shown in Figure 14–1.

At the present time, almost 2,000 asteroids have been identified and catalogued. Their orbits are listed in the famous Soviet *Ephemerides of Minor Planets*. Hundreds of other asteroids have been observed once or twice but "lost" before reliable orbits could be determined. It has been estimated that there are 100,000 asteroids that appear bright enough to leave trails on photographs taken with the 200-inch Palomar telescope.

All of the large asteroids have been discovered. With a diameter of about 1,000 kilometers (600 miles), Ceres is the largest. Only five asteroids have diameters greater than about 300 kilometers (200 miles), but almost 250 asteroids are bigger than 100 kilometers (60 miles) across. Most observable asteroids must have diameters of roughly 1 kilometer (about ½ mile).

The astronomer deduces the size of an asteroid by assuming an albedo and measuring the asteroid's apparent magnitude. It is estimated that a typical asteroid has an albedo of roughly 0.1 (that is, it reflects only 10 per cent of the sunlight). From knowing the distance to the asteroid, and from measurements of its apparent magnitude, the astronomer can calculate the asteroid's size. In this regard, it should be noted that the apparent magnitudes of many asteroids vary in a regular and periodic fashion. It is entirely reasonable to suppose that this is caused by the fact that such asteroids are irregularly shaped and tumbling. By measuring the way in which the apparent magnitude of an asteroid varies, astronomers can deduce the asteroid's rotation period and make reasonable estimates of its overall shape.

An asteroid is said to be officially "discovered" once a reliable orbit has been determined. The accuracy of the orbit must be verified by resighting the asteroid on at least one succeeding opposition. The asteroid is then given a number and a name. The number is simply a running index which indicates the order of discovery (for example, 1 Ceres, 2 Pallas, 433 Eros, 1566 Icarus). The astronomer who discovered the asteroid is then given the option of choosing a name. Since astronomers have long since run out of names of Greek and Roman gods and heroes, and since most astronomers (for some unknown sexist reason) are men, asteroids are usually named after girl friends, wives, and lovers.

As mentioned earlier, many astronomers supposed that the asteroids came from a planet that mysteriously blew up or fragmented. Indeed, this idea can still be found in some science fiction stories today. Modern astronomers, however, realize that this once-popular shattered-planet hypothesis is physically unreasonable. First of all, it has been estimated that the total mass of all the asteroids—both known and unknown—adds up to only $\frac{1}{50}$ of an earth mass. Thus, even if all the asteroids were lumped together, they would not produce a good-sized planet. The second major problem has to do with how the planets were created. It is believed that planets formed by a process called *accretion* whereby small chunks of rock that condensed out of the primordial solar nebula 4.6 billion years ago managed to stick together. As the resulting blob of matter, or *protoplanet,* grows in size, its increasing gravitational field gradually attracts more particles until finally a planet is formed. But recall that Jupiter lies just beyond the asteroid belt. As Jupiter orbits the sun, the gravitational field of this enormous planet perturbs the asteroids, probably increas-

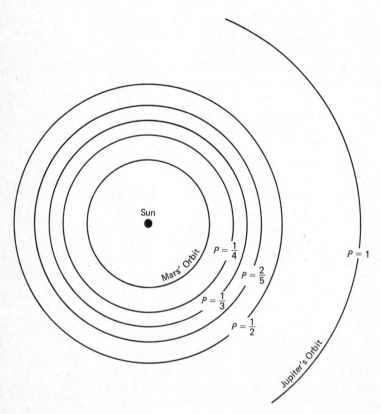

**Figure 14–2. *Kirkwood's Gaps.***
Asteroids orbiting the sun with periods that are simple fractions of Jupiter's period (for example, $\frac{1}{2}$, $\frac{3}{7}$, $\frac{2}{5}$, $\frac{1}{3}$, $\frac{1}{4}$, and so forth) would regularly line up with Jupiter. Such asteroids would therefore regularly experience gravitational perturbations that would cause the asteroids to deviate from their original orbits. The final result is "gaps" in the orbits of particles in the asteroid belt. Four typical gaps are shown in this scale drawing.

ing their relative speeds with respect to one another. Indeed, today the relative speeds between asteroids is typically 5 kilometers per second (1,100 miles per hour). Thus, when two asteroids come together, they would collide violently rather than stick to each other. This process would tend to produce smaller fragments rather than a larger planet. V. S. Safronov and other astronomers therefore conclude that the total amount of material accreted in the asteroid belt 4.6 billion years ago must have been very small.

The influence of Jupiter's gravitational field on the asteroids can be seen today in a phenomenon called *Kirkwood's gaps*. Jupiter orbits the sun once every 11.86 years. Half of Jupiter's sidereal period is 5.93 years, which corresponds to an orbit (according to Kepler's third law) whose semimajor axis is 3.28 AU. There are no asteroids with orbits at 3.28 AU from the sun. Similarly, one third of Jupiter's sidereal period is 3.95 years, which corresponds to an orbit at 2.50 AU from the sun. No asteroids have orbits whose semimajor axes equal 2.50 AU.

The reason for these and similar gaps in the asteroid belt was first explained by Daniel Kirkwood in 1866. Suppose, for example, there were an asteroid at 3.28 AU from the sun. Every time Jupiter orbited the sun once, this asteroid would orbit the sun twice. Thus, after every two revolutions about the sun, this asteroid would be lined up with Jupiter in exactly the same fashion. After every two orbits of the sun, this asteroid would experience the same pull or perturbation from Jupiter's enormous gravitational field. The net effect of these perturbations over millions of years would cause the asteroid to deviate from its original orbit. A gap would eventually result at 3.28 AU. For the same reason, there are no asteroids at other distances that correspond to simple fractions of Jupiter's orbital period. If there had been any asteroids at these distances, they would have long ago perturbed from their original orbits so that by today these asteroids are circling the sun in different orbits that do not regularly line up with Jupiter.

Figure 14–2 shows some typical orbits about the sun in which no asteroids are found. More precisely, no asteroids orbit the sun with periods that are simple fractions of Jupiter's period. This set of "forbidden" sidereal periods corresponds to a set of "forbidden" semimajor axes, according to Kepler's third law. This set of "forbidden" semimajor axes are called Kirkwood's gaps, four of which are shown in Figure 14–2.

While Jupiter's gravitational field produces gaps in the asteroid belt, the combined action of Jupiter and the sun works together to trap asteroids at two specific locations in Jupiter's orbit. This remarkable situation occurs because there are two places in Jupiter's orbit, 60° in front of the planet and 60° behind the planet, where particles could stay forever. As proved in the late 1700s by the French mathematician J. L. Lagrange, there should be two locations where the combined gravitational forces of Jupiter and the sun can trap particles. These two locations are called *Lagrangian points*.

Between 1906 and 1908, four asteroids were found at the Lagrangian points. By 1959, the total number had increased to 14. In 1970, C. J. van Houten, I. van Houten-Groenveld, and T. Gehrels searched for very faint objects and found more than 45 asteroids at one of the Lagrangian points alone. It is estimated that there may be as many as 700 asteroids at each Lagrangian point. These asteroids are called the *Trojan asteroids* and are named after Homeric heroes such as 588 Achilles, 624 Hektor, and 911 Agamemnon. They lie in Jupiter's orbit, as shown in Figure 14–3, and orbit the sun along with Jupiter every 11.86 years. In reality, the number of Trojan asteroids is constantly changing. Occasionally, perhaps as a result of a collision, an asteroid may escape from a Lagrangian point and new asteroids may be captured.

Although the vast majority of asteroids have orbits in the asteroid belt at distances between 2.1 and 3.3 AUs from the sun, some asteroids have highly eccentric orbits that take them into the inner regions of the solar system. Specifically, there are more than a dozen asteroids whose orbits take them inside Earth's orbit. They are called the *Apollo asteroids*. Of the Apollo asteroids, 1566 Icarus has the shortest perihelion distance. Icarus comes within 0.19 AU of the sun, which is about half the average distance between Mercury and the sun. Adonis, Hermes, Apollo, Toro, and Geographos are additional well-known Apollo asteroids with perihelion distances between 0.6 and 0.8 AU. Orbits of several belt and Apollo asteroids are shown in Figure 14–4. By 1975, 19 Apollo asteroids have been listed and range in diameter from 200 meters (600 feet) to about 6 kilometers (4 miles).

One remarkable feature of the Apollo asteroids is that they potentially could strike Earth. Indeed, according to E. J. Öpik, most Apollo asteroids will someday end their journeys about the sun with such catastrophic collisions with our planet. Fortunately, it is estimated that collisions between Earth and a sizable asteroid occur only once every 10 to 100 million years. A collision with an asteroid 1 kilometer (3,000 feet) in diameter would create a crater 13 kilometers (8 miles) across with a force equivalent to 20,000 megaton hydrogen bombs. The remarkably circular shapes of the Gulf of Mexico and Hudson Bay suggest to some scientists that enormously catastrophic collisions have occurred in the past.

While actual collisions are rare, some near-misses have occurred in recent years. In 1931, Eros passed within 23 million kilometers (14 million miles) of Earth. This was so close that astronomers looking through large telescopes could see the irregularly shaped rock (see Figure 14–5) tumbling end over end every 5.27 hours. On January 23, 1975, Eros eclipsed a third-magnitude star in the constellation of Gemini and observations of this occultation permitted astronomers to calculate the size of the brick-shaped asteroid to be 7 by 19 by 30 kilometers (4 by 12 by 19 miles).

On June 14, 1968, Icarus passed within (only!) 6.4 million kilometers (4 million miles) of our planet. Icarus is nearly spherical. It has a diameter of about 1.1 kilometers (0.7 miles) and rotates once every 2.3 hours.

On April 10, 1972, Air Force satellites and people on the ground observed a small asteroid pass over Montana at an altitude of only 60 kilometers (40 miles). Although this object is perhaps more properly called a meteoroid (see Section 14.3), it may have weighed as much as 1,000 tons. Fortunately, this object skipped off of the earth's atmosphere and returned to outer space. If it had approached earth a few kilometers lower, it would have produced an enormous explosion and crater 100 meters (300 feet) wide in Alberta, Canada.

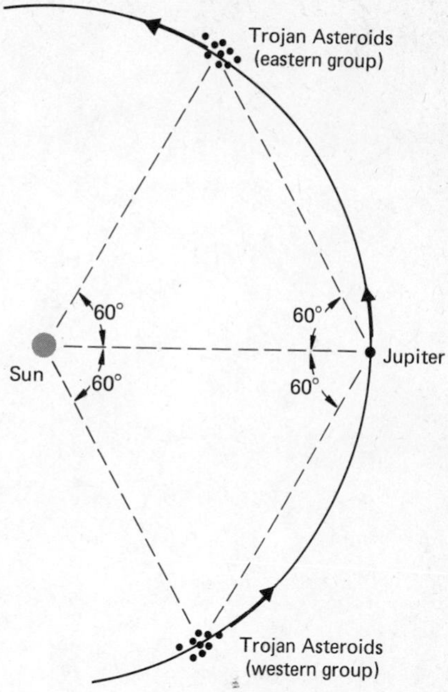

**Figure 14–3.** *The Trojan Asteroids.* There are two locations along Jupiter's orbit where the combined gravitational forces of the sun and Jupiter can trap asteroids. Asteroids orbiting the sun at either one of these stable Lagrangian points are called Trojan asteroids.

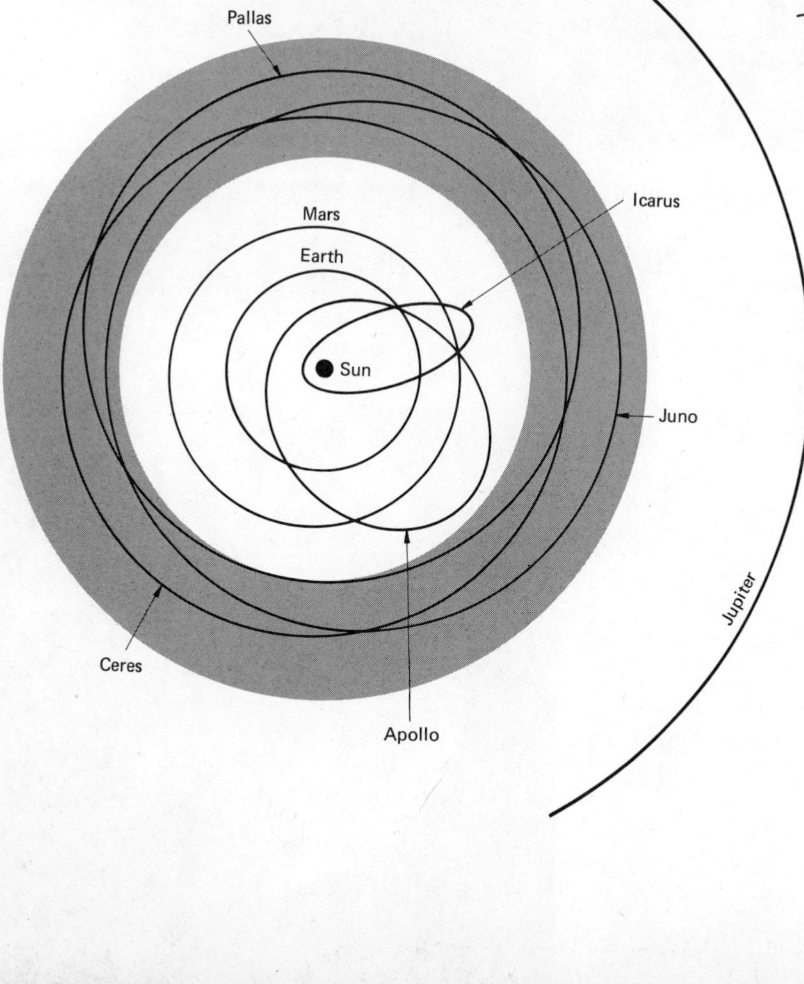

**Figure 14–4.** *Asteroid Orbits.* Most asteroids orbit the sun in the asteroid belt (shaded gray region). Ceres, Pallas, and Juno are typical belt asteroids. A few asteroids have perihelions of less than 1 AU and are called Apollo asteroids. The orbits of two such asteroids, Icarus and Apollo, are shown here.

**Figure 14-5.** *Eros.* In 1931, Eros passed within 23 million kilometers (14 million miles) of Earth. Eros measures 7 by 19 by 30 kilometers (4 by 12 by 19 miles) and rotates once every 5.27 hours. This photograph was taken on February 16, 1931. (*Yerkes Observatory photograph*)

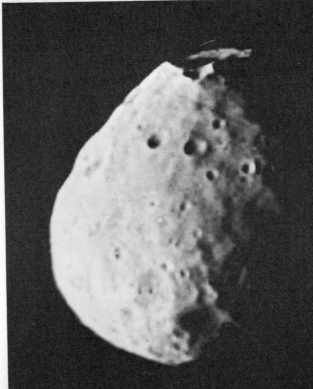

**Figure 14-6.** *Phobos.* This photograph of Mars' larger satellite was taken in July 1976, by Viking orbiter 1. The large crater at the top (Phobos' north pole) is approximately 5 kilometers (3 miles) across. (*NASA*)

No one is sure what an asteroid would look like close up. Nevertheless, many scientists suppose that it would look similar to Phobos, one of the satellites of Mars. As shown in Figures 14–6 and 14–7, Phobos is heavily cratered. Numerous collisions over the ages between asteroids would probably have resulted in similar heavily cratered surfaces. Indeed, as early as 1918, the Japanese astronomer K. Hirayama noticed that various groups of asteroids have very similar orbits. These groupings, called *Hirayama families,* are probably fragments of larger asteroids that suffered collisions. And today we see the resulting pieces all orbiting the sun along very similar paths. Many tiny fragments of asteroids regularly collide with earth, producing meteors and meteorites. Meteorites can be collected by a person who knows how to recognize them.

One of the most promising areas of current research with asteroids also involves meteorites. Since meteorites are very common here on earth, their minerology and chemical composition can be easily measured in the laboratory. Recently, Thomas B. McCord and Clark R. Chapman have recorded the spectra of reflected sunlight from about 100 asteroids. By comparing the spectra of the asteroids with spectra of reflected light from known rocks and meteoroids in the laboratory, they have been able to deduce the minerology of some of these asteroids. In this way scientists are beginning to discover exactly what asteroids are made of. In many cases, they are very similar to meteorites. It is reasonable to suppose that asteroids and meteoroids are among the most ancient objects in the solar system. By examining the chemical composition of asteroids, scientists should be able to gain deep insight into the details of how the solar system formed. Of course, the best data would come from a rendezvous with an asteroid. Two such unmanned missions are tentatively planned for launch in 1986.

**Figure 14–7.** *Close-up of Phobos.* This incredible view of Phobos' surface was photographed in September 1976 by Viking orbiter 2 at a range of only 880 kilometers (545 miles). The smallest feature visible is 40 meters (130 feet) across. The area shown measures approximately 14 by 7 kilometers (9 by 4½ miles). Many scientists feel that the surface of an asteroid would look like Phobos. (*NASA*)

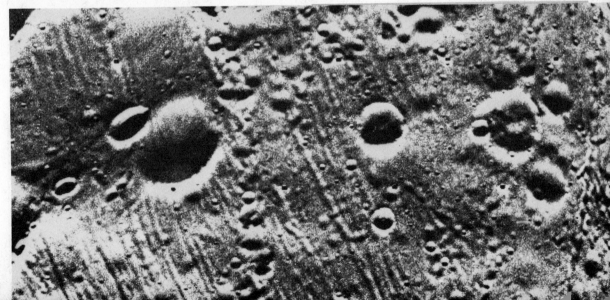

COMETS have been seen and recorded for thousands of years. A bright comet slowly moving across the sky from night to night is surely among the most spectacular celestial events visible to the unaided eye. The ghostly apparition of a comet in the sky was usually regarded with fear and terror by our superstitious ancestors. During medieval times in Europe, for example, it was believed that comets were poisonous gases in the earth's atmosphere. It is sad to note that, at least among some people, things have not changed much. In 1910, earth passed through the tail of Halley's comet and some unscrupulous but enterprising people made a fortune selling "comet pills," which were supposed to protect the buyer against the poisonous vapors. Even as late as 1973, when Comet Kohoutek passed near the sun, equally unscrupulous publishers produced a series of paperbacks predicting havoc and disaster.

The first serious scientific observations of comets date back to Tycho Brahe in 1577. At issue was the Aristotlian concept that comets were atmosphere phenomena

**Figure 14–8.** *Encke's Comet.* Orbiting the sun once every 3.3 years, Encke's comet has the shortest period of all known periodic comets. Its orbit lies entirely inside the orbit of Jupiter. Usually, when a comet is first sighted, it has an amorphous fuzzy appearance like this view of Encke's comet. In this photograph, the images of stars are blurred because the telescope was tracking the comet during the time exposure. (*Yerkes Observatory photograph*)

here on earth. Tycho Brahe reasoned that if comets are really located in the earth's atmosphere, then it should be possible to detect a parallactic shift in the comet's position from various places on earth. In other words, for example, astronomers in Germany and France observing the same comet should see it in different parts of the sky since the observers themselves are widely separated. Of course, Tycho Brahe failed to detect any such parallax and therefore concluded that comets must be far away, at least three times more distant than the moon.

The next major advance came in 1705 when Edmond Halley, a personal friend of Isaac Newton, published calculations dealing with the orbits of two dozen comets observed during the previous centuries. In particular, he noted that bright comets seen in 1531, 1607, and 1682 had very similar orbits. Indeed, Halley speculated that they could be the *same* comet returning every 76 years and therefore predicted that the comet should return in 1758.

Halley did not live long enough to see his predictions borne out. Following up on Halley's calculations, Alexis Clairaut noted that the comet should experience perturbations by both Jupiter and Saturn and concluded that the comet should be visible late in 1758. It was sighted on the evening of December 25, 1758. A new permanent member of the solar system had been found. It was promptly christened *Halley's comet*.

Halley's comet is not an especially spectacular comet. Bigger, brighter comets have often been seen. Halley's comet is of interest from historical aspects and also because it is one of the brighter *periodic comets*. It was last seen in 1910 and is due back in 1986. Most comets travel along nearly parabolic orbits that take them far beyond the outermost planets. Thus, the periods of most comets (if, indeed, they *ever* return) are measured in thousands of years. Some comets may take a million years to orbit the sun once. By contrast, the elliptical orbit of Halley's comet extends only slightly beyond the orbit of Neptune. Consequently, Halley's comet returns to the inner regions of the solar system at regular intervals of only 76 years. Historical records show that Halley's comet has been observed since 239 B.C.

On the average, a dozen comets are discovered each year, often by accident. For example, an astronomer photographing some interesting galaxy or nebula might notice that an unexpected faint, fuzzy object was also recorded on his photographic plate. Further observations then reveal that this object is approaching the sun along a highly eccentric elliptical or nearly parabolic orbit. At this point, the astronomer realizes that he has discovered a comet. The comet is then named after the astronomer and receives an official designation indicating the year and order of the discovery. For example, late in 1975, the Danish astronomer Richard West found a new comet. Since it was the fourteenth comet discovered in that year, Comet West is also called 1975n. Although over a dozen comets may be seen in a given year, very few of them (perhaps only 1 in 20) ever become bright enough to be visible to the naked eye.

Comets are usually discovered at distances of several astronomical units on their inward journeys toward the sun. At such large distances from the sun, an inbound comet appears as a small fuzzy blob through a telescope, similar to the view of Encke's comet seen in Figure 14–8. Only after the comet reaches the inner portions of the solar system does the usual characteristic *tail* begin to develop. Radiation from the sun vaporizes the ices in the comet's *nucleus* and the solar wind pushes the resulting gas away from the sun, thereby producing a long flowing tail. After the comet swings around the sun and returns to the depths of space, the tail begins decreasing in size and brightness. These changes in the appearance of a comet are easily seen in the series of photographs of Halley's Comet in Figure 14–9. Since the

**Figure 14–9.** *Halley's Comet.* As a comet approaches the sun, solar radiation vaporizes the ices in the comet's nucleus. The resulting gases are pushed outward by the solar wind producing the comet's tail. The tail is long and bright only when the comet is near the sun. As the comet recedes from the sun, the tail gradually fades, as shown in this series of photographs of Halley's comet. (*Hale Observatories*)

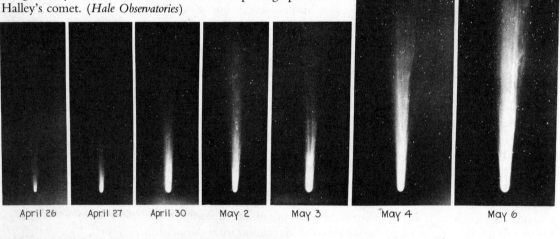

April 26    April 27    April 30    May 2    May 3    May 4    May 6

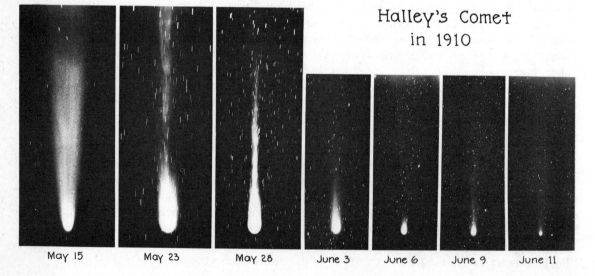

Halley's Comet
in 1910

May 15    May 23    May 28    June 3    June 6    June 9    June 11

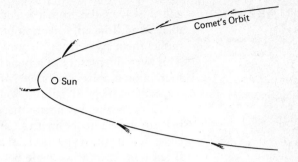

**Figure 14–10.** *A Comet's Tail.* Particles in the solar wind push the comet's gases radially away from the sun, thereby producing a long flowing tail. The comet's tail always points directly away from the sun, regardless of the direction in which the comet is moving.

solar wind consists of particles moving radially outward from the sun, the comet's tail always points directly away from the sun, as shown in Figure 14–10.

The shapes and sizes of comets vary widely. Some comets have a bright head and a short stubby tail while others may have a tiny, faint head and a very long flowing tail. In spite of these differences, all comets have a number of properties in common. For example, the only solid part of a comet is its nucleus. The nature of comets can be understood by assuming that this nucleus consists of ices (methane ice, ammonia ice, and ordinary water ice) along with bits of rock and dust. This is the famous "dirty iceberg" model first proposed by Fred Whipple in 1950. A cometary nucleus is typically 10 to 15 kilometers in diameter.

As a comet comes within a few astronomical units of the sun, solar ultraviolet radiation begins vaporizing the ices in the comet's nucleus. The nucleus therefore becomes surrounded by a round, hazy, glowing ball of gases called the *coma*. The coma increases in size as the comet approaches the sun and reaches a maximum diameter of roughly a million kilometers (several hundred thousand miles) while still at distances of $1\frac{1}{2}$ to 2 astronomical units. The glowing ball of gas exhibits spectral lines of molecules and radicals such as $C_2$, OH, NH, CN, and $NH_2$. The vaporization of water-ice and the subsequent dissociation of $H_2O$ molecules liberate large numbers of hydrogen atoms, which results in the formation of a huge *hydrogen envelope* surrounding the coma. The existence of hydrogen envelopes around the heads of comets was unknown until the 1970s simply because this hydrogen emits radiation only at ultraviolet wavelengths, which do not penetrate the earth's atmosphere. In 1970, however, ultraviolet observations of Comet Bennett from the Orbiting Astronomical Observatory revealed a hydrogen envelope roughly ten million kilometers (six million miles) across. Similar space observations of Comet Kohoutek in 1973 and 1974 showed that the nucleus was surrounded by a hydrogen envelope 40 million kilometers (about 25 million miles) in diameter.

As the comet moves still closer to the sun, both radiation pressure from sunlight as well as the particles in the solar wind push the material in the coma away from the sun to produce the comet's tail. In general, a comet actually develops *two* tails. The low-mass ionized molecules from the coma (for example, $CO^+$, $N_2^+$, $CO_2^+$, and $H_2O^+$) are easily pushed directly away from the sun by the solar wind. This results in a straight *ion tail* or *Type I tail,* which may exhibit knots and lumps whose position and shape may vary from day to day. In contrast, more massive bits of rock

**481**

and dust liberated from the cometary nucleus are not so easily influenced by the solar wind. Instead, radiation pressure from sunlight striking minute dust particles pushes them away from the sun, thereby producing a curved *dust tail* or *Type II tail*. This overall structure of a typical comet is sketched in Figure 14–11. The series of four photographs of Comet Mrkos shown in Figure 14–12 clearly display both Type I and Type II tails.

Occasionally, because of the angle at which a comet is observed, the Type II dust tail can appear to be in front of the comet. This is, however, only a projection effect. While the Type I ion tail always points directly away from the sun, the Type II tail is inclined at an angle between the direction to the sun and the direction in which the comet is moving. Under certain circumstances, the directions of the sun, the earth, and comet's motion can combine so that a portion of the dust tail seems to be in front of the comet. The result is a so-called *anti-tail*. In the spring of 1957, Comet Arend–Roland displayed a fine anti-tail, as shown in Figure 14–13.

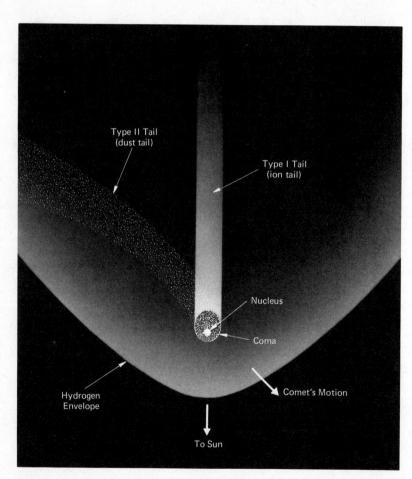

**Figure 14–11.** *The Structure of a Comet.* The only solid part of a comet is its nucleus, which measures from 2 to 15 kilometers across. Vaporization of the ices in the cometary nucleus by sunlight liberates gas and dust, which are pushed away from the sun by the solar wind and radiation pressure.

# COMET MRKOS

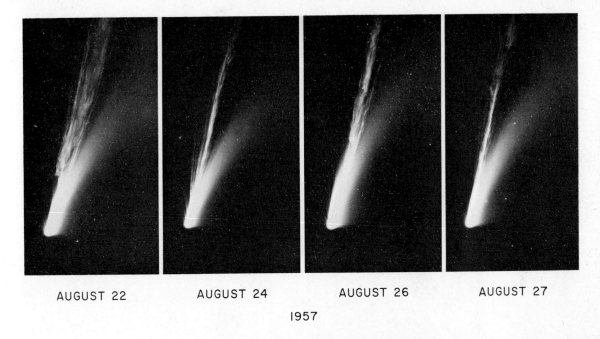

AUGUST 22       AUGUST 24       AUGUST 26       AUGUST 27

1957

Photographed with the 48-inch schmidt telescope.

**Figure 14-12.  *Comet Mrkos.*** This bright comet seen in 1957 clearly displayed both Type I and Type II tails. A Type I tail consists of ionized atoms pushed directly away from the sun by the solar wind. The curved Type II tail consists of tiny dust particles pushed away from the nucleus by radiation pressure. (*Hale Observatories*)

**Figure 14-13.  *Comet Arend-Roland.*** This bright comet seen in the spring of 1957 displayed a fine anti-tail. Actually, this anti-tail is not really in front of the comet. Instead, because of the angle and orientation at which the comet was observed, the curved Type II dust tail is seen in projection. (*Lick Observatory photograph*)

**Figure 14-14.** *Comet Ikeya-Seki.* This bright comet graced the predawn sky during October 1965. It had a very tiny nucleus and a huge tail, which measured nearly one astronomical unit from end to end. (*Lick Observatory photograph*)

The tail of a bright comet can grow to enormous lengths. Comet Ikeya-Seki shown in Figure 14-14, was easily visible to the unaided eye during the predawn hours of October 1965. The comet's tail was roughly one astronomical unit long. Cometary tails measuring 150 million kilometers (about 100 million miles) have occasionally been observed. Yet, in spite of their lengths, their tails do not contain very much matter. Figuratively speaking, you could pack a comet's tail in a suitcase and walk off with it. Earth passed through the tail of Halley's comet in 1910 with no observable effects. The tails of comets are almost perfect vacuums.

Since roughly a dozen comets are sighted each year and since the orbital periods of most comets are measured in thousands or tens of thousands of years, there must be a vast number of comets very far from the sun. Unlike the planets and asteroids, which are confined to orbits near the plane of the ecliptic, comets approach the sun from every conceivable angle. In addition, about half of the comets have retrograde orbits and revolve about the sun in a direction opposite to that of all the planets. Based on these orbital characteristics of comets, the Dutch astronomer Jan H. Oort in 1950 proposed that there must be a swarm of comets surrounding the solar system out to distances of 50,000 astronomical units (almost one light year, or approximately one quarter of the distance to the nearest star). Since comets approach the sun at all angles, this so-called *Oort cloud* must be spherical and may contain as many as 100 billion comets. The combined total mass of all these comets must be approximately as large as the mass of Earth.

As comets approach the inner portions of the solar system, their orbits can be dramatically affected by the planets, especially by Jupiter. For example, Jupiter's gravitational field may slow down the comet. In this way, a long-period comet orbiting the sun once every million years can become a short-period comet. There are about four dozen comets, called *Jupiter's family of comets,* which orbit the sun with periods of five to ten years that may have been affected in this fashion. Alternatively, a comet may be sped up by gravitational perturbations with a planet and escape from the solar system. E. J. Opik estimates that about 96 per cent of the short period comets might be ejected from the solar system in this manner. The remaining 4 per cent will suffer collisions with the planets. Such a collision with earth probably occurred on June 30, 1908, when a violent explosion took place in the atmosphere over Siberia. This explosion was equivalent to a nuclear bomb of several hundred kilotons and devastated a hundred square miles surrounding the impact site.

Occasionally a comet may plunge toward the solar system along an orbit that comes very close to the sun. As the nucleus of one of these *sun-grazing* comets passes within a million kilometers of the sun's surface, enormous tidal forces are experienced that try to break up the cometary nucleus. For example, during perihelion passage in 1846, Comet Biela broke into two pieces. Comet Biela was a periodic comet that orbited the sun once every 7 years. When it returned in 1852, two comets were observed. These pieces of Comet Biela failed to return on subsequent scheduled visits. Instead, on November 27, 1872 and November 27, 1885, when earth passed through the orbit of Comet Biela, spectacular meteor showers occurred. Evidently, all that remained of Comet Biela was a swarm of dust particles.

More recently, the sun-grazing comet Ikeya–Seki broke into two pieces after perihelion passage and the nucleus of Comet West, shown in Figure 14–15, split into four fragments in March of 1976. The nuclei of comets must be very fragile.

After a comet has disintegrated, either as a result of tidal disruption or simply by the vaporization of all its ices, all that remains is a swarm of dust particles. If earth should happen to pass through one of these swarms, a *meteor shower* is observed as

**Figure 14–15.** *Comet West.* Following perihelion passage, the nucleus of this bright comet broke into four pieces in March of 1976. ( *Joint Observatory for Cometary Research*)

8°

**Figure 14–16.** *The Breakup of Comet West.* Shortly after passing near the sun, the nucleus of Comet West broke into four fragments. This remarkable series of photographs clearly shows the disintegration of the comet's nucleus. (*New Mexico State University Observatory*)

| 8 MAR 76 | 12 MAR 76 | 14 MAR 76 | 18 MAR 76 | 24 MAR 76 |

the high-speed dust particles burn up in the earth's atmosphere. In the mid–1800s, G. V. Schiaparelli showed that the Perseid meteors (so-named because they seem to radiate from the constellation of Perseus) are clustered along the orbit of a comet seen in 1862. Similarly, the Draconids are associated with the periodic comet Giacobini–Zinner and the Geminids probably come from a burned-out prehistoric comet. Most of the dust particles or *cometary meteoroids* that produce these showers are extremely tiny. No cometary meteoroid has ever survived passage through the earth's atmosphere. However, during rocket flights, meteorites have been collected above the atmosphere. These micrometeorites are literally the sizes of pieces of dust and can be seen clearly only with the aid of a microscope.

ALTHOUGH it might sound absolutely incredible at first, it has been estimated that 3,000 tons of material falls on earth each day! Most of this material striking our planet is in the form of tiny dust particles, which, as we saw in the previous section, come from comets.

The fact that interplanetary space contains abundant quantities of dust is apparent to the naked-eye observer. Just after sunset or just before sunrise on an especially clear, dark, moonless night far from any city lights, it is possible to see a faint hazy glow in the sky called the *zodiacal light*. This zodiacal light is widest at the horizon nearest the sun (in the west after sunset and in the east before sunrise) and slants upward along the ecliptic. This faint glow, seen in Figure 14–17, is caused by sunlight reflecting off of tiny dust particles orbiting the sun along with the planets. A typical grain of interplanetary dust is only 1 micron ($\frac{1}{10,000}$ centimeter, or about $\frac{1}{25,000}$ inch) in size.

It is believed that most of the fine-grain material that gives rise to the zodiacal light is of cometary origin. But because of the action of the solar wind, radiation

**Figure 14–17.** *The Zodiacal Light.* After sunset or before sunrise, a faint glow can be seen in the night sky. This glow, called the zodiacal light, is caused by sunlight reflecting off of tiny grains of dust in interplanetary space. (*Yerkes Observatory photograph*)

pressure from sunlight, and collisions with planets, which sweep up this material, approximately eight tons of interplanetary dust is lost every second. In order to maintain the zodiacal light, therefore, approximately eight tons of dust must be added to interplanetary space each second. Most of this new material probably comes from old comets. In addition, some of this dust comes from collisions between asteroids. During such collisions, however, much larger fragments of rock are also produced. These fragments—too large to be called dust, yet too small to be asteroids—are *meteoroids*.

When a meteoroid or a dust particle strikes the earth's atmosphere at a high speed, the particle is vaporized in one or two seconds thereby producing a *meteor*. A meteor, sometimes called a "shooting star," is simply the brief flash of light seen by a ground-based observer who happens to be looking up in the sky as a small rock or piece of dust from interplanetary space collides with earth. Meteors are typically seen at distances of less than 160 kilometers (100 miles) and it is estimated that 25 million meteors bright enough to be seen occur around the globe each day. A photograph of a meteor is shown in Figure 14–18.

**Figure 14–18.  *A Meteor.*** As an astronomer was taking a photograph of this region of the sky, a meteor flashed across the field of view. Meteors result when a tiny rock, called a meteoroid, strikes the earth's atmosphere at a high speed. (*Yerkes Observatory photograph*)

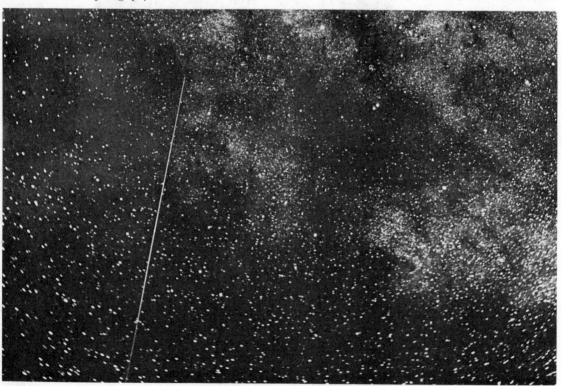

Although the estimated number of meteors over the entire earth per day is large, on a given night an individual observer gazing up at the sky may see very few—if any—"shooting stars." The flash of a meteor is so brief that the observer must be looking at exactly the right portion of the sky at the very moment that a grain of interplanetary matter strikes the atmosphere less than a hundred miles away. Nevertheless, there are selected times during the year when a person stands a very good chance of seeing a substantial number of bright meteors. These are the times of *meteor showers.* When earth happens to pass through a swarm of particles in space, a meteor shower is produced as the particles collide with the atmosphere. One well-known shower is the Perseid shower, so named because the *radiant* or location from which the meteors appear to originate lies in the constellation of Perseus. Other major showers that often produce fine displays are listed in Table 14–3. It is customary to name a meteor shower after the constellation from which the meteors appear to come. Thus, for example, the radiant of the Geminids is located in the constellation of Gemini.

A person who wishes to observe a dramatic meteor shower usually has more "luck" during the early morning hours rather than in the late evening. This is because from midnight until sunrise, the observer is standing on the side of the earth that faces the direction in which the earth is moving around the sun, as shown in Figure 14–19. Most meteoroids have speeds between 10 and 70 kilometers per second (22,000 and 150,000 miles per hour). As these particles make head-on collisions with our planet, even the slowest meteoroids are capable of producing

**Table 14–3.** *Meteor Showers*

| Name | Date of Maximum Display |
|------|-------------------------|
| Lyrids | April 21 |
| Perseids | August 12 |
| Draconids | October 10 |
| Orionids | October 21 |
| Taurids | November 7 |
| Leonids | November 16 |
| Geminids | December 12 |

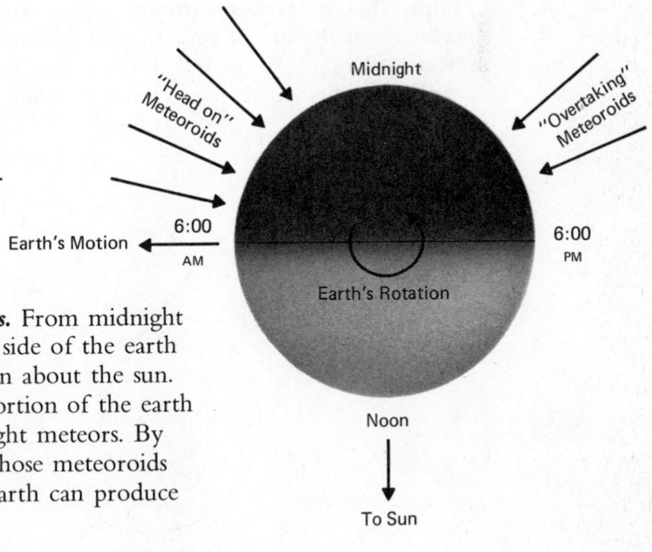

**Figure 14–19.** *Morning vs. Evening Meteors.* From midnight until dawn, the observer is standing on the side of the earth that faces the direction of the earth's motion about the sun. Meteoroids and cometary dust strike this portion of the earth "head on" thereby producing frequent, bright meteors. By contrast, from sunset until midnight, only those meteoroids moving fast enough to catch up with the earth can produce meteors.

meteors. But Earth's orbital velocity around the sun is 30 kilometers per second (67,000 miles per hour). Therefore, during the evening hours from sunset until midnight only those meteoroids moving fast enough to catch up with Earth can produce meteors. The situation is analogous to driving down a freeway during a rainstorm. Numerous raindrops vigorously pelt the front windshield while very few raindrops strike the rear window.

During a good meteor shower, an observer may see as many as one meteor per second. But on rare occasions, the number of meteors is much more frequent. For example, during the predawn hours of November 17, 1966, over 2,000 Leonid meteors were sighted each minute!

The frequency of meteors depends on how the meteoroids are distributed in the *meteoroid swarm* orbiting the sun. If, as shown in Figure 14–20, the meteoroids are uniformly scattered around the orbit of an old comet, observers can consistently count on a meteor shower each time Earth passes through the orbit of the meteoroid swarm. Such showers are very reliable and predictable but do not necessarily produce overwhelming numbers of meteors. On the other hand, if the meteoroid swarm is concentrated or bunched up in one location along its orbit, a meteor shower occurs only when Earth happens to intersect the swarm. Such showers are much less reliable because Earth may simply miss the swarm. But when Earth passes directly through the swarm, enormous numbers of meteors can be sighted.

Most meteoroids are small, perhaps the size of grains of sand or pebbles, and burn up upon striking the atmosphere. Occasionally, however, larger meteoroids collide with Earth. In such cases, air friction does not succeed in vaporizing the entire object. Only the surface layers are worn away or *ablated* and the object survives the journey down to the ground. Such objects as are found on the ground are called *meteorites*.

**Figure 14–20.** *Meteoroid Swarms.* Meteoroids may be strewn uniformly around their orbit about the sun or they may be bunched up. Reliable meteor showers result from swarms that are distributed all the way around their orbit. The most spectacular showers occur when Earth passes through a tightly concentrated swarm.

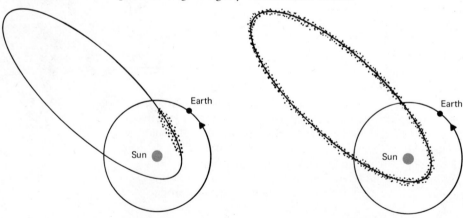

On very rare occasions, a huge meteoroid may collide with Earth. Not only does the object easily survive the journey through the atmosphere, but impact with the ground can be so violent as to produce a huge crater. More than a dozen *meteorite craters* have been identified around the globe. One of the most famous of these craters is the Barringer Meteorite Crater near Winslow, Arizona. This crater, shown in Figure 14–21, is 1,300 meters (almost one mile) across and 180 meters (600 feet) deep. It is believed that the meteoroid which produced this crater collided with Earth 22,000 years ago.

Although meteorites are commonly seen in museums today, the idea that "stones fall from the sky" was very unpopular only two centuries ago. Indeed, one early American president ridiculed "Yankee professors" for suggesting such an absurd idea. Nevertheless, ancient writings describe "celestial stones" and our ancestors deeply revered these objects. The Ka'ba, a sacred Muslim stone enshrined in Mecca, is one example. In spite of these ancient records, however, the first

**Figure 14–21.** *The Barringer Meteorite Crater.* This huge crater, measuring almost one mile across, was created by the impact of a huge meteoroid some 22,000 years ago. Over 25 tons of meteoritic material have been found in the surrounding regions. (*Yerkes Observatory photograph*)

**Table 14–4.** *Meteorite Classification*

| | | |
|---|---|---|
| Stones | Carbonaceous Chondrites | 5.7% |
| (Total = 92.8%) | Chondrites | 80.0% |
| | Achondrites | 7.1% |
| Stony Irons | Mesosiderites | 0.9% |
| (Total = 1.5%) | Pallasites | 0.5% |
| | Others | 0.1% |
| Irons | Hexahedrites | 0.6% |
| (Total = 5.7%) | Octahedrites | 4.3% |
| | Ataxites | 0.8% |

scientific work on meteorites did not occur until the 1790s. In 1794, the German physicist E. F. F. Chladni published a study of meteorites. He even suggested that meteoroids might have come from a planet that had fragmented for some unknown reason, an idea that was destined to become very popular a few years later with the discovery of asteroids.

Chladni's work was severely criticized, especially in France, until a large meteorite fell in the town of L'Aigle on April 26, 1803. The French Academy sent a famous physicist, J. B. Biot, to investigate and only after his exhaustive report were scientists convinced that stones fall from the sky.

All meteorites fall into one of three broad categories based on their iron content: stones, stony irons, and irons. Of course, it is easiest to find meteorites which have a high iron content simply because they stick to magnets and respond to metal detectors. Thus, irons constitute a large percentage of "finds." In spite of the relative ease with which iron meteorites are found, the vast majority of meteoroids that strike Earth are stones. More than 90 per cent of the "falls" are stones. These meteorites are, however, very hard to find simply because it is usually difficult to distinguish them from ordinary terrestrial rocks. This is especially true if the stone meteorite has lain on the ground and been weathered for a considerable period of time. Thus, most of the *falls* are stones while most of the *finds* are irons. Table 14–4 lists the major different types of meteorites along with the measured percentage of falls in each category.

By far, *chondrites* constitute the highest percentage of all falls. Chondrites are stony meteorites that contain small round bodies called *chondrules* (from the Greek word for "seed"). Chondrules are usually composed of minerals such as olivine and orthopyroxene and their structure reveals that they were formed by rapid cooling of molten droplets. Typical chondrites are shown in Figures 14–22 and 14–23. In both cases, the meteoritic nature of these specimens is clear from the fact that their surfaces are covered with a *fusion crust,* which formed as they passed through the earth's atmosphere. All stony meteorites contain some (10 to 15 per cent) nickel and iron. A slice of a chondrite is shown in Figure 14–24 and clearly reveals tiny flecks of metal embedded in the meteorite.

**Figure 14–22.** *A Chondrite.* This stony meteorite fell in Plainview, Texas. It is easily recognized as a meteorite because its surface is covered with a fusion crust formed during the high-speed firey descent through the atmosphere. (*Collection of Ronald A. Oriti, photograph by Paul Roques; Griffith Observatory*)

**Figure 14–23.** *A Chondrite.* This stony meteorite fell near Bruderheim, Alberta in Canada. Notice that there are two chips in the fusion crust, exposing the light-colored interior of the meteorite. (*Photograph—Paul Roques; Griffith Observatory*)

**Figure 14–24.** *A Slice of a Chondrite.* This slice of a chondrite reveals numerous tiny flecks of nickel–iron metal embedded in the meteorite. This chondrite fell near Neenach, California. (*Collection of Ronald A. Oriti, photograph by Paul Roques; Griffith Observatory*)

**Figure 14-25.** *A Carbonaceous Chondrite.* Carbonaceous chondrites are the most ancient of all meteorites. Their chemical composition reveals that they have experienced little or no changes since they were formed. This carbonaceous chondrite fell in Mexico. A chondrule measuring about 5 millimeters ($\frac{1}{5}$ inch) across is seen protruding from the meteorite. (*Collection of Ronald A. Oriti, photograph by Paul Roques; Griffith Observatory*)

**Figure 14-26.** *An Achondrite.* Achondrites are stony meteorites without chondrules. Their properties strongly suggest that they have experienced some melting. This achondrite fell in Kansas. Some of its grey fusion crust can be seen. (*Collection of Ronald A. Oriti, photograph by Paul Roques; Griffith Observatory*)

**Figure 14-27.** *An Achondrite.* This achondrite fell near Pasamonte, New Mexico. Its dark fusion crust is heavily cracked. Achondrites may have come from the melting and subsequent cooling of chondrites. During such a process, chondrules would have been destroyed. (*Collection of Ronald A. Oriti, photograph by Paul Roques; Griffith Observatory*)

Carbonaceous chondrites, so-named because of their high carbon content, are the most ancient of all meteorites. It is believed that carbonaceous chondrites come very close to being the unaltered, unprocessed primordial material out of which the solar system was created. They have a high water content (roughly 20 per cent) and a low density (2.2 grams per cubic centimeter as opposed to 3.5 grams per cubic centimeter for ordinary chondrites). From their high content of water and other volatiles, it is apparent that carbonaceous chondrites have never been subjected to high temperatures or pressures that would have driven off these substances.

A typical carbonaceous chondrite is shown in Figure 14–25 (note the chondrule protruding from the meteorite's surface). The carbon in these meteorites is not free carbon but rather is contained in complex organic molecules. The fact that these ancient meteorites contain organic molecules has some fascinating implications. Some scientists have even suggested that these carbon compounds could have had a biological origin. In fact, a carbonaceous chondrite that fell near Murchison, Australia in 1969 was shown to contain several amino acids, which are the chemical building blocks of proteins. The idea that living organisms could have been present billions of years ago at the time of the formation of the solar system is, of course, highly speculative and controversial.

Achondrites constitute the third and final type of stony meteorites. Achondrites are stony meteorites without chondrules and their crystal structure is coarser than that of chondrites. Achondrites closely resemble terrestrial igneous rocks. Achondrites, such as those shown in Figure 14–26 and 14–27, may have originated from a melting and subsequent cooling of chondrites.

Stony iron meteorites are divided into two main classes depending on the minerals of the stony part. *Mesosiderites* (see Figure 14–28) contain minerals such as plagioclase and pyroxene. *Pallasites,* (see Figures 14–29 and 14–30) contain olivine. When a large asteroid cools down from a molten state, nickel and iron sink toward the center of the object while the silicate minerals rise toward the surface. It is

**Figure 14–28.** *A Mesosiderite.* This stony-iron meteorite fell near Vaca Muerta in Chile. It has been sliced and polished to show the mixture of minerals and nickel–iron metal. (*Collection of Ronald A. Oriti, photograph by Paul Roques; Griffith Observatory*)

**Figure 14-29.** *A Pallasite.* Pallasites are stony-iron meteorites composed of olivine crystal inclusions in nickel–iron metal. This pallasite fell near Antofagasta in Chile. (*Collection of Ronald A. Oriti, photograph by Paul Roques; Griffith Observatory*)

**Figure 14-30.** *A Slice of a Pallasite.* This pallasite, which fell in New Mexico, has been sliced and polished to show the arrangement of olivine and nickel–iron metal. (*Collection of Ronald A. Oriti, photograph by Paul Roques; Griffith Observatory*)

believed that stony irons represent material from the interfaces between the nickel-iron cores and stony surfaces of asteroids that have been broken apart by collisions over the ages.

Iron meteorites are composed almost entirely of metal and can be divided into three categories based on their nickel content. *Hexahedrites* have the lowest nickel content (10 to 11 per cent). Octahedrites have a moderate nickel content (13 to 15 per cent) and *ataxites* have a very high nickel content, typically more than 20 per cent. As mentioned earlier, these meteorites are the easiest to find even though they represent a relatively small fraction of all falls. When cut, polished, and etched with acid, the internal structure of an iron meteorite often reveals information about its history. For example, hexahedrites often exhibit a series of fine lines, called *Neumann bands,* caused by shock waves passing through the meteorite during a violent collision in interplanetary space. An excellent example of Neumann bands is shown in Figure 14–31.

In 1808, Count Alois von Widmanstätten, director of the Imperial Porcelain Works in Vienna, discovered that when certain iron meteorites are cut, polished, and then dipped in acid, remarkable patterns of interlocking metallic crystals become visible. These so-called *Widmanstätten figures* occur only with octahedrites and are believed to have resulted from the slow cooling of the meteoroid or asteroid in space. Widmanstätten figures are seen in Figure 14–32.

Finally, ataxites contain more nickel than either hexadrites or octahedrites. A slice of an ataxite is shown in Figure 14–33.

**Figure 14–31.** *A Hexahedrite.* When cut, polished, and etched with acid, hexahedrites often exhibit striations called Neumann bands. It is believed that Neumann bands are evidence of violent collisions suffered by the meteorite while it was still in space. This hexahedrite was found in Arkansas. (*Collection of Ronald A. Oriti, photograph by Paul Roques; Griffith Observatory*)

**Figure 14–32.** *An Octahedrite.* When cut, polished, and etched with acid, octahedrites exhibit intricate interlocking patterns of metallic crystals called Widmanstätten figures. This octahedrite was found near Henbury, Australia. (*Collection of Ronald A. Oriti, photograph by Paul Roques; Griffith Observatory*)

**Figure 14–33.** *An Ataxite.* Ataxites are iron meteorites with a very high nickel abundance. This particular ataxite was found in New Mexico. (*Photograph—Paul Roques; Griffith Observatory*)

All of these various types of irons are believed to have been the cores of shattered asteroids and large meteoroids. When an asteroid formed billions of years ago, short-lived radioactive isotopes would have provided enough heat to melt the inner parts. While still molten, the heavier iron atoms would gradually sink toward the center of the asteroid while lighter elements are pushed toward the surface. As a result of subsequent collisions and shattering, a wide variety of meteoroids are produced: stones from the surface layers, irons from the interior, and stony irons from the regions in between.

Meteorites are found all around the world. Their extraterrestrial origin is undebatable and the study of these objects provides important clues concerning the early history of the solar system. But in addition to meteorites, a second type of objects is found in only certain locations that—at first glance—appears to have an extraterrestrial origin. They are called *tektites*. Tektites are small glassy objects with a high silica ($SiO_2$) content. They are typically a few centimeters in size and their heavily ablated surfaces are evidence of the fact that they entered (or re-entered) the atmosphere at a high speed. Several tektites are shown in Figure 14–34.

**Figure 14–34.** *Tektites.* Tektites are small glassy objects with severely ablated surfaces. They are typically found in Australia, Southeast Asia, Western Africa, Czechoslovakia, Texas, and Georgia. (*Collection of Ronald A. Oriti, photograph by Paul Roques; Griffith Observatory*)

From radioisotope dating it is clear that tektites have not been in space nearly as long as ordinary meteorites. While meteorites have been in space for millions of years since they broke off from their parent bodies, tektites have been exposed to space only for a few years or minutes. This is far too short for tektites to be of cometary or asteroidal origin. Most scientists believe that tektites come from Earth. If a large meteoroid collides with our planet, great quantities of rock are blown out of the impact crater. Some of these rocks might be ejected upward with such high speeds that they momentarily rise above the atmosphere. Upon falling back to Earth, the rocks are melted and ablated by air friction and fall to the ground as tektites. Meteorites and tektites are clear evidence of the fact that collisions with cosmic objects over the ages have been an important factor in shaping the face of our planet.

## Questions and Exercises

1. What is Bode's law?
2. Why did astronomers begin searching for a "missing planet" between the orbits of Mars and Jupiter?
3. How was the first asteroid discovered?
4. What is the asteroid belt?
5. Approximately how many asteroids have been discovered and listed in catalogues?
6. Give two objections to the once-popular shattered-planet hypothesis that stated that asteroids are fragments of a planet that blew up.
7. What are Kirkwood's gaps?
8. What are the Trojan asteroids?
9. What are the Apollo asteroids?
10. Briefly discuss the probability of a collision between Earth and an asteroid.
11. What is a Hirayama family?
12. Briefly describe a method by which an astronomer can deduce the mineralogy of an asteroid.
13. What is a comet?
14. How are most comets discovered?
15. Roughly, how many comets are discovered each year?
16. Briefly describe the structure of a comet.
17. Contrast and compare Type I and Type II tails.
18. What is the size of a typical cometary nucleus and what is it made of?
19. What is the Oort cloud?
20. What is a sun-grazing comet?
21. How are comets and some meteor showers related?

22. Describe the changes in the appearance of a comet as it moves through the inner regions of the solar system.
23. How much material (cometary dust, meteorites, and so forth) falls on earth each day?
24. What is the zodiacal light?
25. What is the difference between a meteoroid, a meteor, and a meteorite?
26. What is a meteor shower?
27. Why are more meteors usually seen during predawn hours rather than in the late evening?
28. What are the three main categories into which all meteorites can be classified?
29. What is the difference between a fall and a find?
30. What is a chondrule? What kind of meteorites contain chondrules?
31. What are Neumann bands?
32. What are Widmanstätten figures?
33. What is a tektite?

# Evolution of the Solar System

## *The Life History of Stars*

To ancient man the heavens must have seemed eternal, permanent, and unchanging. The stars and constellations that greet us when we walk outside on a clear night are virtually identical to what our most ancient ancestors saw as they gazed up into the star-filled sky. Yet, in spite of all appearances, we realize that the heavens *must* be changing. Stars can be seen in the sky because they emit light. In order to give off light, stars must use up energy. And, as the sources of this energy are depleted, the stars must change; they must evolve. The reason why stars give the illusion of being permanent and unchanging is simply because this evolution is very slow. Milestones marking major changes in the structure of a star during its life cycle are separated by hundreds of millions or even billions of years, far longer than the memory of men or the span of recorded history.

In order to understand how stars evolve, the astrophysicist must first have information about the physical properties of stars. He must know exactly what a star is before he can make intelligent statements about how a star changes over the years. It is therefore necessary to have data concerning stellar brightnesses, stellar masses, surface temperatures, chemical compositions, and so forth. Without this basic data about the true nature of stars, it would be absurd to try to construct a reasonable theory of stellar evolution.

As a result of laborious work by countless astronomers over the past century, we now have a good understanding of the physical properties of the stars we see in the sky. As mentioned in Section 5.3, we know that the absolute magnitudes (that is, the "real brightnesses") of stars cover a wide range. Expressing stellar luminosities in solar units (that is, the sun's real brightness is "1 sun"), we find that the dimmest stars typically have brightnesses of only $1/10,000$ sun, while the most luminous stars shine with a brightness of up to 1,000,000 suns. In addition, the surface temperatures of stars range from about 3,000°K for the coolest, up to about 30,000°K for the very hottest stars. And finally, it was noted that the range of stellar masses is not exceptionally large. Expressing the masses of stars in solar units (that is, the sun's mass is "1 solar mass"), we found that the range extends from $1/10$ solar mass up to about 60 solar masses.

**501**

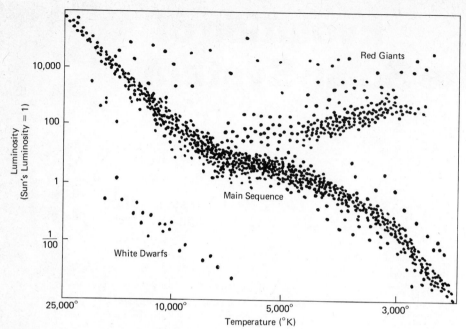

**Figure 15-1.** *The Hertzsprung–Russell Diagram.* In plotting the luminosities of stars against their surface temperatures, the dots on the resulting graph cluster into three main groups: main-sequence stars, red giant stars, and white dwarf stars.

In the early part of the twentieth century, a remarkable discovery was made concerning some of this data about the nature of stars. It was found that when a graph is drawn on which the luminosities of stars are plotted against their surface temperatures, all the dots (each dot represents a star of known brightness and temperature) fall into one of three broad regions. As discussed in Section 5.3, the data for most stars seen in the sky fall along a band on the graph called the *main sequence.* Data for most reddish-appearing stars (called *red giants*) seen in the night sky occupy a second well-defined region of the graph to one side of the main sequence. And finally, data for an important class of dim, hot stars (called *white dwarfs*) occupy a third distinct region on the opposite side of the main sequence. This graph, called a *Hertzsprung–Russell diagram,* is shown in Figure 15-1.

As stars live out their lives, their structures and properties must change in response to the gradual depletion of nuclear fuels in their interiors. As a result, the luminosities and surface temperatures of stars must change. On the Hertzsprung–Russell diagram, we see where visible stars spend *most* of their time. They spend a great deal of time as main-sequence stars, as red giants and as white dwarfs, and very little time as anything else in between. Therefore, by understanding how dots representing individual stars move around on the Hertzsprung–Russell diagram, we would know how stars evolve. An understanding of why the data for stars cluster into three distinct regions on the H–R diagram and how these regions are related to each other is the same as knowing how stars are born, how they change as they mature, and perhaps what happens to them as they die.

As astronomers look out into space, they often find huge clouds of gas called *nebulae*. The famous Orion nebula, shown in Figure 15–2, is a fine example. Some of these nebulae contain dark regions that were originally believed to be "holes" in the gas clouds. Astronomers today realize that these dark regions (such as those in the Trifid nebula shown in Figure 15–3) are not holes at all, but rather consist of large foreground clouds of very cool gas and dust obscuring the brighter nebulosity in the background. It is believed that stars are born in these enormous dark clouds of cool gas.

Consider one of these large, cold clouds of gas floating in space. As you might expect, the gas in the cloud would not be uniformly distributed everywhere. Instead, at various locations, there might be some inhomogeneities or "lumps." One of these lumps contains a little more matter than the typical surrounding regions. But because it contains more matter, it has a slightly stronger gravitational field and therefore will attract some of the nearby gas. As it attracts nearby gas, the lump grows in size and mass. As the mass of the lump increases, the strength of its gravitational field goes up and more surrounding gas is attracted. This is obviously a self-escalating process: the bigger it gets, the more it attracts—the more it attracts, the bigger it gets. In this way, a small inhomogeneity in a cold interstellar gas cloud can grow by *gravitational accretion* into a huge ball of gas.

Calculations reveal that one of these huge spheres of gas is gravitationally unstable. There is simply nothing to hold up the trillions of tons of gas pressing inward from all sides. The ball of gas therefore begins to contract. As the *protostar* contracts, the pressures, densities, and temperatures inside the ball of gas start to rise. In a very short time, temperatures reach a few thousand degrees and the protostar starts to shine. This is the first step in the birth of a star.

Using the laws of physics and high-speed computers, astrophysicists such as L. Henyey and C. Hayashi have shown how a hypothetical protostar must change. In a very short period of time, a protostar contracts under the influence of gravity until temperatures at its core reach a few million degrees. At this stage, however, the core is so hot that hydrogen nuclei collide with sufficient violence that they fuse to form helium nuclei. This thermonuclear reaction, in which hydrogen is fused into helium, releases enormous quantities of energy. The outpouring of energy from the star's core dramatically changes the structure of the rest of the star. For the first time, pressures become sufficient to hold up the outer layers of the sphere of gas. As a result, a star is born.

The ignition of *hydrogen burning* at the center of a contracting protostar marks the first important milestone in stellar evolution. Prior to this stage, the dot representing the contracting protostar had been moving very rapidly across the H–R diagram simply because the structure of the star itself was rapidly changing. But with the ignition of hydrogen burning, the dot on the H–R diagram comes to rest on the main sequence because the star itself becomes stable. This reveals the true meaning of the main sequence. Main-sequence stars are new-born stars in which hydrogen burning is occurring at their cores. The exact position of a star on the

**Figure 15–2 (*opposite*).** *The Orion Nebula.* This beautiful nebula in the constellation of Orion is just barely visible to the naked eye as a fuzzy "star." (*Lick Observatory photograph*)

**Figure 15–3.** *The Trifid Nebula.* The dark regions in this nebula are caused by large foreground clouds of very cool gas that are obscuring the bright nebulosity in the background. It is believed that stars are born in these enormous cold clouds of interstellar gas. (*Lick Observatory photograph*)

main sequence depends on the star's mass. Low-mass protostars become dim, cool main sequence stars while high-mass protostars become very bright, hot main-sequence stars. Our own star, the sun, has an intermediate mass and therefore lies in the middle portion of the main sequence.

Most stars we see in the sky, including our sun, are main-sequence stars. Astronomers have discovered stars that are just in the process of igniting hydrogen burning at their cores. These include so-called *T Tauri stars,* many of which can be seen in the nebula known as NGC 2264 in Figure 15–4. Dots representing these stars on the H–R diagram are just about to settle down on the main sequence.

**Figure 15–4.** *The Birth of Stars.* This nebula, called NGC 2264, contains many very young stars that are now in the process of igniting hydrogen burning at their cores. (*Hale Observatories*)

**Figure 15–5.** *Young Main-Sequence Stars.* This cluster of stars, called the Pleiades, is visible to the naked eye. Very recently hydrogen burning began at the cores of these stars. (*Lick Observatory photograph*)

Figure 15–5 shows a cluster of stars called the Pleiades. All of the stars in this cluster have only recently ignited hydrogen burning at their cores thereby becoming full-fledged main-sequence stars. In the case of the sun, hydrogen burning at its core began some 4½ billion years ago. The dot on the H–R diagram that represents the sun has been sitting on the main sequence ever since.

Hydrogen burning at the core of a main-sequence star cannot go on forever. Quite simply, after a period of time (depending on the mass of the star) all the hydrogen at the star's center gets used up and converted into helium. In the sun, 600 million tons of hydrogen are converted into helium every second. While this may sound like an incredible rate, there is so much hydrogen in the sun's interior that hydrogen burning can go on for a total of about 10 billion years. Our sun therefore has a little more than 5 billion years left on the main sequence.

**507**

The depletion of hydrogen at the center of a main-sequence star signals the beginning of some major changes in the star's structure. Recall that the ignition of hydrogen burning stopped the gravitational contraction of a protostar. Consequently, when hydrogen burning finally shuts off, the star once again becomes gravitationally unstable. As a result of calculations by astrophysicists like I. Iben and R. Kippenhahn, it is now realized that the hydrogen-depleted core of the star begins to contract rapidly. In doing so, the star's core becomes very hot as gravitational energy from the collapsing core is converted into thermal energy. As this outpouring of thermal energy spreads outward from the center of the star, the star's outer layers get pushed outward. Thus, as the core contracts, the surface of the star begins to expand. Since the star gets bigger, it also gets brighter. But any time a gas expands, it cools. Consequently, the temperature of the star's expanding surface begins to decrease.

Of course, astronomers cannot actually see what is going on deep inside a star. Through their telescopes they can only observe a star's outer layers. The contraction of a star's hydrogen-depleted core causes the outer layers to expand. Observationally, astronomers would see the star getting brighter and cooler. The dot representing such a star on the H–R diagram therefore begins to move toward the region occupied by red giants.

As the temperature of the contracting core increases, the temperatures immediately above the core also start to rise. But between the star's core and surface, there is still some fresh hydrogen that had not been used up while the star was on the main sequence. As a result, hydrogen burning can be ignited in a *shell* surrounding the core. This ignition of hydrogen shell burning accelerates the expansion of the star's outer layers. As the star expands, its total luminosity increases while its surface temperature decreases. The final result is a full-fledged red giant.

In about 5 billion years, our sun will become a red giant. It will become several hundred times more luminous than it is today and its surface temperature will fall to about 4,500°K from its present value of 6,000°K. In doing so, the red giant sun will become so huge (roughly 350 million kilometers = 200 million miles in diameter) that Earth's orbit will lie *inside* the red giant's surface. Of course our planet will be vaporized!

In the later stages of stellar evolution, the temperature of the contracting red giant's core can become so high that *helium burning* is ignited deep inside the star. This ignition, which is gradual in the case of massive stars but sudden in the case of low-mass stars, typically occurs when the star's helium-rich core reaches 100 million degrees Kelvin. With two thermonuclear processes occurring inside the star, the star's structure begins to change and the dot representing the star on the H–R diagram starts to wander around in the red giant region.

Eventually all the helium at the star's center is used up and the core of the star again begins to contract. In the case of a sufficiently massive star, the temperature at its center can become so incredibly high (billions of degrees) that *carbon burning* is ignited. Indeed, a host of bewildering thermonuclear reactions become possible. The end result is that the star becomes unstable and blows up. For low-mass stars, such as the sun, the resulting explosion is fairly modest. The outer layers of the star are cast off into space. Low-mass stars that have ended their life cycles in this fashion produce beautiful *planetary nebulae* such as the Ring Nebula seen in Figure 15-6.

Massive stars end their lives in a spectacular detonation called a *supernova*. During a supernova explosion, runaway nuclear reactions tear the star apart. In doing so, the star suddenly and dramatically increases its luminosity so that it shines with the brightness of a billion suns. For a couple of days, a supernova can outshine

**Figure 15-6** *A Planetary Nebula.* The Ring Nebula in the constellation of Lyra is an excellent example of a star that gently cast off its outer atmosphere near the end of its life cycle. (*Lick Observatory photograph*)

the entire galaxy in which it resides. The remains of these cataclysmic events are seen as *supernova remnants* such as the Veil Nebula shown in Figure 15–7.

In looking at planetary nebulae or supernova remnants, astronomers see only the outer portions of dying stars cast off during explosive events at the end of their life cycles. But what happened to the star's burned-out core? Astrophysicists have diligently examined this question in the recent past and found that three remarkably different kinds of objects can be created. The nature of these dead stars depends entirely on the mass of the burned-out stellar core.

A "dead" star is one in which all possible sources of nuclear fuel have been used up. It is the final stage of stellar evolution. It is what (if anything) is left behind after the creation of a planetary nebula or supernova explosion. Since all the nuclear fuel has been used up, a dead star can leave a "corpse" only if some internal pressures develop inside the burned-out star that resist further compression by gravity.

The first important discoveries concerning dead stars date back to the 1930s and the pioneering work of S. Chandrasekhar. Once all the nuclear fuel in a dying star

**Figure 15–7.** *A Supernova Remnant.* This delicate nebula in the constellation of Cygnus is an excellent example of a massive star that ended its life with a supernova explosion several tens of thousands of years ago. (*Hale Observatories*)

has been exhausted, the star begins to contract under the influence of gravity. As a result of trillions upon trillions of tons of stellar material pressing inward from all sides, the star simply shrinks in size. Dr. Chandrasekhar discovered that if the total mass of the dying star is not too large (specifically, less than 1.44 solar masses), then the electrons inside the contracting star can develop a powerful pressure, called *degenerate electron pressure,* which is sufficiently strong enough to resist any further contraction. Once one of these low-mass dying stars has contracted down to a sphere about 10,000 kilometers in diameter (that is, roughly the size of earth), a degenerate electron pressure develops that is strong enough to support the star. This is a stable configuration and the dead star stops contracting. Such a star is small and therefore dim. It is also very hot, having a surface temperature typically in the range of 20,000 to 50,000°K. These dim, hot stars are called white dwarfs.

Astronomers have discovered many white dwarfs in the sky. A photograph of one is shown in Figure 15–8. Indeed, our own star will someday become a white dwarf. Billions of years in the future, after the sun has exhausted all its possible sources of nuclear fuel, it will contract under the influence of gravity. When it has finally shrunk down to an object about the size of earth, internal pressures from the

**Figure 15–8. *A White Dwarf.*** Sirius, the brightest appearing star in the sky, has a white dwarf companion. White dwarfs, such as the one seen here alongside an overexposed Sirius, are dead stars. (*Courtesy of R. B. Minton*)

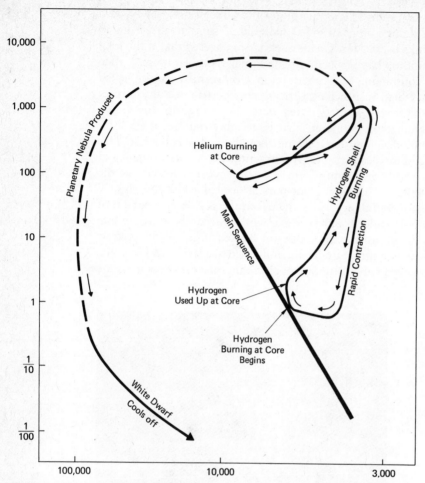

**Figure 15-9.** *Evolution of the Sun.* The complete life history of the sun can be represented as a dot moving around on the Hertzsprung–Russell diagram. The sun spends billions of years, first as a main-sequence star, then as a red giant, and finally as a white dwarf.

electrons in the burned-out sun will become strong enough to resist any further contraction. Like all white dwarfs, the sun will simply cool off as it radiates light into space.

At this point, we now have a complete understanding of the Hertzsprung–Russell diagram. We know the true meaning of the various regions of the H–R diagram: main sequence, red giants, and white dwarfs. Based on this understanding, we can draw the *evolutionary track* of the sun on the H–R diagram. This evolutionary track is the path followed by the dot representing the sun. As shown in Figure 15–9, the path begins with the initial, rapid gravitational contraction of the

**512**

protosun. As soon as hydrogen burning is ignited at the sun's core, the dot settles down on the main sequence (where it is today). Eventually the sun will evolve into a red giant and the dot representing our star moves into the upper right-hand portion of the H–R diagram. Finally, after all the nuclear fuel is used up, the sun will contract to become a white dwarf and the dot rapidly moves from the red giant region across the H–R diagram to the lower left-hand corner of the graph. There it remains forever, gradually shifting to lower temperatures and luminosities as the dead sun slowly cools off.

An important property of white dwarfs is that their total masses must be less than the critical value of 1.44 solar masses, called the *Chandrasekhar limit.* If a dying star has a mass greater than 1.44 solar mass, it could never become a white dwarf simply because degenerate electron pressure is not strong enough to support a larger amount of dead stellar material. What, then, might become of a massive dying star?

Late in 1967, radio astronomers discovered objects in space that emit rapid pulses of radio noise at incredibly precise intervals. It is today realized that these objects, called *pulsars,* constitute a second type of dead star. If a dying star has a mass slightly greater than the Chandraeskhar limit, it will not stop contracting when it has shrunk down to the size of earth. The contraction continues until the density of the dead stellar material is so great that negatively charged electrons and positively charged protons get squeezed together. In this process, the electrons and protons combine to form neutrons. When, finally, the dying star consists almost entirely of neutrons, a *degenerate neutron pressure* develops that prevents further contraction. The result is a *neutron star* roughly 20 kilometers (a dozen miles) in diameter. It is believed that pulsars are rapidly rotating neutron stars with intense magnetic fields.

A pulsar, called NP 0532, is located at the center of a famous supernova remnant known as the Crab Nebula. Evidently, the mass of the dying star which produced this supernova remnant must have been slightly greater than the Chandrasekhar limit. The dead star therefore collapsed down into a neutron star. The arrow in Figure 15–10 identifies the neutron star.

**Figure 15–10.** *A Neutron Star.* A neutron star is located at the center of this famous supernova remnant called the Crab Nebula in the constellation of Taurus. (*Lick Observatory photograph*)

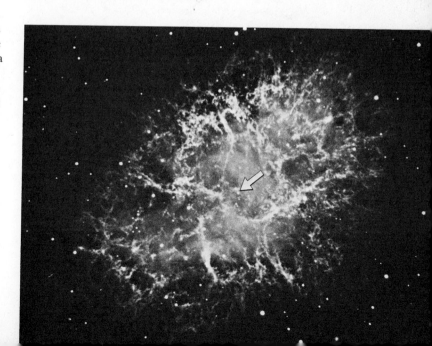

Just as there is an upper limit to the mass of a white dwarf, there is also a maximum mass for neutron stars. Dead stars whose masses exceed 3 solar masses cannot become neutron stars simply because degenerate neutron pressure is not strong enough to support the enormous weight of such a huge amount of stellar material.

By the early 1970s, it became clear that if the mass of a dead star exceeds 3 solar masses, then *no* physical forces can ever develop to prevent catastrophic gravitational collapse. Very massive dying stars therefore get smaller and smaller and smaller. Finally, the entire star gets crushed out of existence at a single point. In doing so, the intensity of the gravitational field above the collapsing massive star becomes so incredibly enormous that space and time fold in over themselves and the star disappears from the universe! Such an object is called a *black hole*. A diagram showing the curvature of space around a black hole is given in Figure 15–11.

**Figure 15–11.** *A Black Hole.* No physical forces can ever support very massive dying stars. As massive dying stars catastrophically implode, space and time fold in over themselves and the star disappears from the universe. The highly warped space–time that remains is called a black hole.

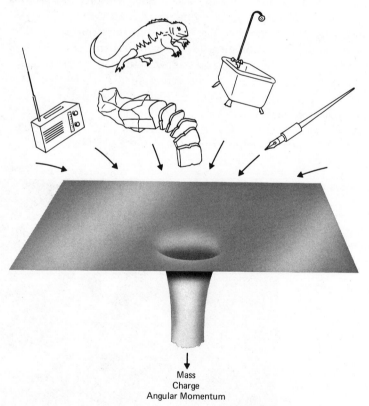

Mass
Charge
Angular Momentum

Using Earth-orbiting X-ray telescopes, such as that shown in Figure 4–22, astronomers have discovered a black hole orbiting an ordinary star in the constellation of Cygnus. As matter from a nearby normal star is sucked into the black hole, the gas is compressed and heated to temperatures of several million degrees, thereby emitting enormous quantities of X rays. Of course, nothing ever gets out from inside the black hole itself. The intensity of gravity inside a black hole is so great that even light cannot escape.*

Regardless of what form dying stars might take, large quantities of matter are ejected into space as stars near the end of their life cycles. Due to nuclear processes that have occurred in the stars, this matter has been enriched with numerous heavy elements. This ejected matter therefore enriches the *interstellar medium* with all sorts of heavy elements and the entire process of stellar formation can begin anew. But since the clouds of interstellar gas contain heavy elements, solid objects can also be created along with these later-generation stars. These objects can become planets, satellites, asteroids, meteoroids, and comets. Thus, the deaths of ancient stars have made it possible for new stars to have planetary systems as does our sun.

## *The Creation of the Solar System* 15.2

FOR THOUSANDS of years, people have looked up toward the heavens and seen what we see in the starfilled night sky. But few human beings are satisfied with an endless list of observations alone. Instead, we want to know why things are the way they are. Questions concerning the origin of the sun, the moon, the planets, and Earth have therefore been with us since before the dawn of recorded history.

Every civilization and every religion has had some sort of answers to these fundamental astronomical questions. In ancient times, mystical and mythical explanations were sufficient. The classic story of how "in the beginning, God created heaven and earth" in seven days is an excellent example. But a few hundred years ago, some people began yearning for a physical explanation for the formation of the physical world. Having discovered some of the physical laws that describe how the solar system works, they wanted to know if these laws could be used to explain how the solar system was formed. A collection of ideas that gives such an explanation is called a *cosmogony.*

The first physical cosmogony was proposed by the famous French philosopher René Descartes in 1644. Descartes envisioned a universe initially filled with gas. He believed that this gas would contain *vortices,* like eddies in a stream of flowing water.

---

*For further reading about black holes, the following two books by the author of this text are recommended: *The Cosmic Frontiers of General Relativity* (Boston: Little, Brown & Co., 1977) and *Relativity and Cosmology, Second Edition* (New York: Harper & Row, 1977).

Perhaps the sun condensed out of a large vortex while the planets and their satellites condensed out of much smaller vortices in this swirling primordial gas.

Descartes' idea was the first of several *evolutionary theories* in which the planets form as a result of ordinary processes that would naturally happen anytime a star is created. In sharp contrast, *catastrophic theories* occasionally became popular. In these theories, the planets are formed as a result of some unusual event or accident.

In 1745, a century after Descartes' work, the French scientist Georges Louis Leclerc de Buffon proposed the first catastrophic theory. Buffon suggested that a massive object (he called it a "comet") collided with the sun and knocked out enough matter to form the planets. A glancing blow by the "comet" would explain the sun's rotation as well as the rotation of the planets.

During the next few years in the eighteenth century, several scientists and philosophers elaborated on Descartes' work to show how the solar system could have condensed out of a primordial nebula. In 1755, the great German philosopher Immanuel Kant applied newly invented Newtonian mechanics to Descartes' proposal. Kant argued that rotation would have caused the primordial cloud of gas and dust to become very flattened as the sun and planets started to form by gravitational accretion.

A few years later, in 1796, the French mathematician Pierre Simon de Laplace independently proposed a similar theory but added a few interesting details. Laplace's *nebular hypothesis* also begins with a slowly rotating, contracting primordial cloud of gas and dust. Laplace argued that as the rotating cloud contracted under the influence of gravity, its rate of rotation would speed up just as an ice skater doing a pirouette spins faster as she pulls in her arms. But as the rate of rotation of the contracting nebula increases, centrifugal forces at the edge of the nebula also rise. Thus, the contracting nebula would periodically shed rings of matter. Planets then condense out of the material in these *Laplacian rings*. In this way, Laplace apparently explained the flattened appearance of the solar system, the nearly circular orbits of the planets, as well as the fact that the sun rotates in the same direction that the planets move along their orbits. These two eighteenth-century evolutionary cosmogonies are depicted in Figures 15–12 and 15–13.

As the years passed by, it became increasingly clear that there were some severe problems with the ideas of Kant and Laplace. First of all, there was the difficulty that the contracting nebula should produce a sun that is rotating extremely rapidly, far more rapidly than we actually observe. Secondly, it was not at all easy to explain how the Laplacian rings could be generated—and if they could, the gas would simply float off into space without forming planets. As a result of these seemingly insurmountable difficulties, astronomers revived Buffon's ideas in the early 1900s.

Between 1901 and 1905, T. C. Chamberlin and F. R. Moulton proposed a catastrophic theory in which a star passed very near the sun. During the near miss, the resulting high tides caused huge filaments of gas to erupt from the star and the sun. As shown in Figure 15–14, the matter from these eruptions later condensed into planets.

This interesting theory soon came under severe criticism from the British

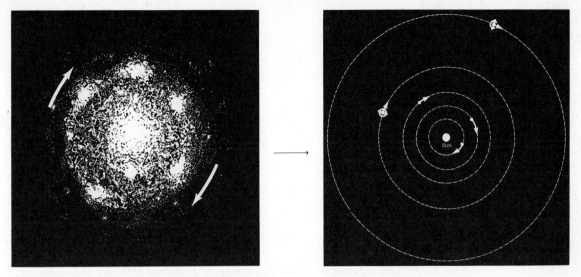

**Figure 15–12.** *Kant's Cosmogony (1755).* Kant envisioned the solar system forming from a clotting mass of gas and dust. Lumps grow by accretion to form planets and satellites. Most of the nebula contracts to form the sun. (*Adapted from Yerkes Observatory photographs*)

**Figure 15–13.** *Laplace's Cosmogony (1796).* As the rotating primordial nebula contracts, its rate of rotation increases. As a result of centrifugal forces, rings of matter are shed. These rings condense into planets. Most of the nebula contracts to form the sun. (*Adapted from Yerkes Observatory photographs*)

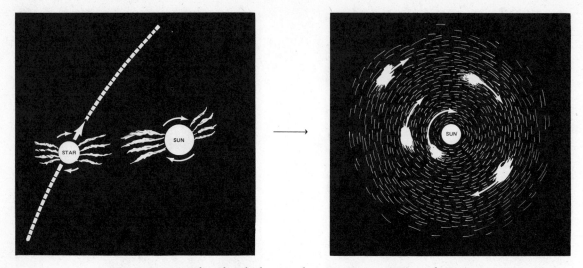

**Figure 15-14.** *The Chamberlin–Moulton Cosmogony (1901 and 1905).* A passing star narrowly misses the sun. As a result of high tides, huge filaments of gas erupt during the close encounter. The matter in these filaments later condenses into planets and satellites. (*Adapted from Yerkes Observatory photographs*)

**Figure 15-15.** *Jeffreys–Jeans Cosmogony (1917).* A passing star sideswipes the sun tearing out a long cigar-shaped filament of matter. The matter in the filament later condenses into planets, with the largest ones in the middle and the smaller ones at either end. (*Adapted from Yerkes Observatory photographs*)

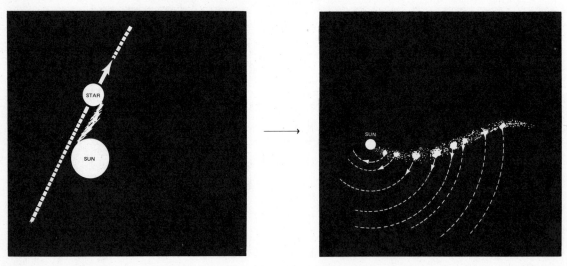

astronomer H. Jeffreys who argued that collisions between gaseous filaments would heat up the matter rather than allowing it to cool down to condense into planets. Jeffreys, along with J. H. Jeans in 1917, therefore proposed an alternative catastrophic theory in which a passing star sideswiped the sun. This grazing collision drew out a massive cigar-shaped filament of matter, as shown in Figure 15–15, which eventually cooled and condensed into planets. Since the cloud was fatter in the middle than at the ends, this theory allegedly explained why the most massive planets (Jupiter and Saturn) are located at intermediate distances from the sun.

All of these early cosmogonies contained a lot of assumptions, difficulties, and improbabilities. For example, in our Galaxy, the stars are separated by huge distances usually measured in dozens of light years. The probability of two stars colliding is therefore extremely small, so small that it is very unlikely that a near miss or grazing collision would ever take place. Some astronomers, such as H. N. Russell, R. A. Lyttleton, and F. Hoyle, tried to cope with this high degree of improbability by assuming that the sun was once a member of a binary or triple-star system. But their ideas also met with insurmountable difficulties and during the early 1940s it became clear that astronomers did not possess any reasonable cosmogony at all. To appreciate what has happened since then, perhaps it would be a good idea to re-examine some of those well-known facts that a reliable cosmogony *must* explain.

The first thing that is obvious when we look at the solar system is that it is very flattened. Specifically, the orbits of all the planets lie nearly in the same plane. Furthermore, the plane of the planetary orbits is nearly the same as the plane of the sun's equator.

In addition to the *coplanar* properties of the solar system, the direction of rotation of objects in the solar system is remarkably well defined. For example, all the planets orbit the sun in the same direction. This direction also happens to be the same direction in which the sun itself rotates. And with only two exceptions (Venus and Uranus), the planets themselves rotate about their axes in the same sense that they revolve about the sun. Even the overwhelming majority of the 34 satellites revolve about their planets in the same direction that the planets themselves rotate about their axes and revolve about the sun.

Of course there are numerous additional properties of the solar system. For example, the orbits of the planets are nearly circular and the spacing of these orbits are given by relations such as Bode's law. In addition, there is a remarkable difference in the chemical composition of the inner (terrestrial) and outer (Jovian) planets. But nevertheless, any successful cosmogony must begin by accounting for the fact that the plane of the solar system is very well defined, as is the direction of rotation and revolution of the major objects in the solar system. These basic properties strongly suggest an evolutionary cosmogony. They are the sort of properties one might naturally expect to result from processes of gravitational contraction, accretion, and condensation in a primordial nebula. Consequently, after World War II, the majority opinion among astronomers returned to a nebula hypothesis similar to that of Descartes three centuries earlier.

**Figure 15–16.** *Von Weisäcker's Cosmogony (1946).* Vortices or eddies form in the equatorial plane of the primordial nebula. Accretion of the gas and dust into planets takes place along the boundaries between the various zones of eddies. (*Adapted from Yerkes Observatory photographs*)

**Figure 15–17.** *Kuiper's Cosmogony (1951).* Vortices are randomly distributed in the primordial nebula. Small retrograde-rotating eddies act like ball bearings between larger primary eddies. Accretion into protoplanets occurs in the smaller eddies and tidal forces from the sun eventually get the contracting planets rotating in the correct direction. (*Adapted from Yerkes Observatory diagram*)

The first of the contemporary cosmogonies was proposed by the German physicist C. F. von Weizsäcker in the mid-1940s. Von Weizsäcker argued that whirlpool-like eddies, or vortices, would have formed during the initial stages in the contraction of the primordial nebula. As depicted in Figure 15–16, the largest of these vortices would arrange themselves into specific zones, the spacing of which could be in agreement with Bode's law. Planets then form along the interfaces between the zones of eddies.

A few years later, in 1951, the American astronomer G. P. Kuiper elaborated on von Weizsäcker's cosmogony by arguing that the eddies would be much more chaotically distributed in the primordial nebula, as shown in Figure 15–17. Kuiper assumed that grains of matter would condense from the turbulent primordial nebula and grow into larger objects by accretion. But he found that collisions between dust grains at the boundaries of large eddies would be far too violent. He was therefore forced to accept the idea that accretion into protoplanets occurs in *retrograde-rotating eddies,* which act like ball bearings between the larger primary eddies. Unfortunately, all the protoplanets are then rotating backward. To overcome this difficulty, Kuiper argued that tidal forces caused by the newly formed sun on the protoplanets would raise tides that eventually would reverse the direction of rotation of the contracting protoplanets.

The currently accepted cosmogony takes some of the best ideas from these earlier theories along with what is now known about stellar formation and the condensation and accretion of dust grains in a primordial nebula. It is commonly known as the *inhomogeneous accretion model.* The story begins with the contraction of an interstellar cloud of gas and dust such as that shown in Figure 15–18. In earlier evolutionary cosmogonies, it was believed that most of the primordial nebula contracted to form the sun. It is now realized that this could not be correct. The reason is that the sun is rotating fairly slowly. If *most* of the matter in the primordial nebula contracted into the protosun, then the sun should be rotating very rapidly. Indeed, this was a primary objection to the ideas of Kant and Laplace. A fundamental law of physics, called the *conservation of angular momentum,* dictates that anytime a rotating object contracts, its rate of rotation speeds up. In order *not* to form a rapidly rotating sun, only a portion of the primordial nebula could have contracted to form our star. A large amount of matter must have remained at substantial distances from the new-born sun. In this way it is possible to form a slowly rotating sun because much of the angular momentum in the early solar system remains behind in a huge cloud of gas and dust spread out over a large volume of space. Recent calculations by M. R. Pine and A. G. W. Cameron suggest that the primordial nebula contained at least three solar masses of gas and dust. One solar mass went into forming the sun while two solar masses of matter remained behind at distances up to 100 AU (roughly ten billion miles) from the newly formed sun.

The processes leading up to the creation of planets probably began as the primordial nebula first began contracting. Tiny grains of interstellar dust would collide and stick to each other. By the time the sun started to form, these amalgamated clumps would be roughly a few millimeters in diameter, like small fluffy

**Figure 15-18.** *A Nursery of Stars and Planets.* Planets form by a combination of condensation and accretion of grains in a contracting interstellar cloud of gas and dust. These processes are now occurring in this beautiful object called the Eagle Nebula (also known as M16 or NGC 6611) in the constellation of Serpens. (*Hale Observatories*)

pebbles. In only a few hundred years, these clumps of matter would settle toward the mid-plane of the nebula.

The next important stage in the formation of the planets depends on a recently discovered mechanism proposed by Peter Goldreich at Caltech and William R. Ward of Harvard. Goldreich and Ward showed that in a disk of small particles, gravitational instabilities will break up the sheet into much larger clumps. In this way, the millimeter-sized fluffy pebbles stick together to form kilometer-sized asteroid-like objects.

The next step involves the fact that the mutual gravitational attraction between the asteroid-sized objects will cause them to aggregate into clusters. As these swarms of objects collide and intermingle, they gradually condense into large solid bodies. Through this continued process of accretion and consolidation, planets and their satellites eventually form.

The chemical composition of a particular accreting mass depends on where it is in the primordial solar system. In the inner portions of the solar system, the interstellar grains and resulting clumps are composed of substances that are not easily vaporized, such as iron and silicates. Due to the proximity of the sun, all volatile substances remain in gaseous form. As a result, planets forming near the sun will contain large percentages of metals and silicates. At somewhat larger distances from the sun, the temperature is lower and thus interstellar grains would contain a covering of water-ice, which condenses at a temperature of $160°K$ ( $-110°C = -170°F$). Ammonia and methane condense at still lower temperatures, and thus interstellar grains fairly far from the sun would also have coverings of these frozen gases. In this way, we understand why the inner (terrestrial) planets are rocky objects primarily composed of iron and silicates while the outer (Jovian) planets contain enormous quantities of volatile gases like methane and ammonia.

After all the planets and their satellites had formed through this process of accretion, there still must have been large quantities of extra gas and dust in orbit about the sun. It is well known that very young stars usually pass through a phase, called the *T Tauri stage,* when they eject large quantities of matter in a very short period of time. Indeed, T Tauri stars may eject one solar mass in *only* a million years. It is entirely reasonable that the new-born sun passed through such a phase early in its history. When thermonuclear reactions were finally ignited at the sun's core, the outpouring of energy produced a *T Tauri wind,* which blew all the excess material out of the solar system. In this way, the T Tauri wind (a very intense version of today's solar wind) cleaned up the solar system.

The general picture presented here depicts the formation of the solar system as a natural consequence of the creation of the sun. No extraordinary, accidental events—such as a grazing collision with a passing star—are employed. It therefore seems entirely reasonable to suppose that *any* star like the sun might be surrounded by a system of planets. For this reason, it is generally agreed that there must be numerous planetary systems scattered around our Galaxy.

A few billion years after the formation of the solar system, a remarkable chemical event occurred on at least one of the planets. Conditions of temperature and pressure on the third planet from the sun were such that atoms eventually combined into complex molecules that have the incredible ability to make exact replicas of themselves. As a result, living organisms appeared on Earth.

While the process of planetary formation is now viewed as the consequence of a natural sequence of events, it is *not* clear that the appearance of life would also occur any time appropriate conditions exist. The science of biology simply has not yet advanced to the point that the necessary and sufficient conditions for the formation of living organisms can be specified with a reliable degree of certainty. Although a lot of speculation exists, we still do not know if the appearance of life on Earth was a rare and unique event, or if such biochemical phenomenon would be commonplace on any hospitable planet. Are human beings just one of the countless living creatures scattered around the Galaxy, or are we terrifyingly alone in an otherwise lifeless universe? This is certainly one of the most profound questions modern scientists can ask.

## Are We Alone?

THE POSSIBILITY of extraterrestrial life involves some of the most fascinating speculation in all of modern science. Are we alone? Or are there alien races of intelligent creatures who gaze up at the stars asking the same questions we ask?

If extraterrestrial life exists elsewhere in the universe, it seems entirely reasonable that these alien life forms evolved on "habitable" planets. It is quite difficult to imagine how life could evolve in the sparce interstellar medium. Speculations concerning extraterrestrial life therefore must begin with questions concerning the possibility of other planetary systems.

It is flatly impossible for an astronomer sitting at a telescope to see a planet orbiting a distant star in the night sky. Planets are very small and since they shine by reflected light from their parent star, they must be exceedingly dim. Direct observation of distant planetary systems is therefore completely beyond the reach of modern astronomers. The best they can hope for is indirect evidence.

One important piece of indirect evidence involves the speed at which stars rotate. More precisely, it involves the *angular momentum* of stars. Using spectroscopic techniques, astronomers have been measuring rotation rates of stars with reasonable accuracy since the 1950s. In particular, it was found that massive stars rotate fairly rapidly, while low-mass stars (like the sun) rotate slowly. Specifically, the angular momentum of massive stars is high, while low-mass stars possess a much smaller amount of angular momentum. In the case of our own solar system, the sun is spinning relatively slowly and most of the angular momentum in the solar system is possessed by the planets. In general, stars with masses greater than about $1\frac{1}{2}$ solar masses are spinning rapidly while with less than $1\frac{1}{2}$ solar masses are spinning slowly.

It is interesting to note that if the angular momentum of the planets is added to the angular momentum of the sun, a total value is obtained for the solar system that is nearly equal to the angular momentum of a rapidly spinning high-mass star. Some astronomers have therefore noted that the small amount of angular momentum possessed by low-mass stars can be explained by assuming that these stars are orbited by planets. Presumably, these planets orbiting low-mass stars possess the "missing" angular momentum, as first suggested by S. S. Huang in 1965. In addition, low-mass stars are much more common than high-mass stars. A large fraction of the billions of stars in our Galaxy have masses less than $1\frac{1}{2}$ solar masses. In view of these considerations, it seems entirely reasonable to suppose that planetary systems are *very* common. Few astronomers would therefore argue with the conjecture that there are billions of planets in our Galaxy.

Making intelligent guesses about planets orbiting distant stars is easy. Finding these planets is a nearly impossible task. Nevertheless, Peter van de Kamp of the Sproul Observatory at Swarthmore College has reported some suggestive data concerning a nearby star called Barnard's star. Many years of careful observations reveal that Barnard's star seems to be wiggling back and forth. In the 1960s, Dr. van de Kamp pointed out that this periodic oscillation of Barnard's star about its normal path in the sky could be explained by the gravitational influence of two large planets orbiting this low-mass star. Although there are large uncertainties in the observations, van de Kamp suggested that the two unseen planets have masses equal to 1.1 and 0.8 times Jupiter's mass. There are several other low-mass stars which exhibit similar tiny "wiggles" over the years (technically they are called *astrometric binaries*) and may be orbited by comparatively large planets. Unfortunately, the observations are so very difficult that it is impossible to draw any firm conclusions. Suffice it to say that the probability of extrasolar planets seems to be very high. With increased precision and years of painstaking observation, we might someday be able to infer conclusively the existence of a large planet about some nearby low-mass star.

Of all the possible extrasolar planets in the Galaxy, only a few will possess conditions suitable for the appearance and evolution for life. For example, planets orbiting very near their parent stars will be too hot for the formation of complex organic molecules. Similarly, planets very far from their stars will be too cold. There is, therefore, an optimum range of distances from stars where habitable planets can exist. In addition, a habitable planet must have a sufficiently high surface gravity so that it can retain an atmosphere.

In 1965, the American physicist Stephen Dole presented arguments that stars whose masses range from 0.9 to 1.0 solar masses are most likely to possess habitable planets. This means that there may be as many as a billion such planets in our galaxy. Of this total number, Dole further estimates that roughly 6 per cent could support human-like creatures.

Having a habitable planet does not necessarily mean that life *must* appear on that planet. Unfortunately, the biological sciences have not advanced to the point that they can tell us the probability of the appearance of living organisms given suitable initial conditions. We know of only one planet where initial conditions were suitable and life did develop: Earth. Extrapolation based on one example is indeed pure speculation. Nevertheless, one of the important lessons of the history of science over the past several centuries is that we do not occupy some special, unique place in the universe. We live on a typical planet, orbiting a typical star in a typical galaxy. In so far as mankind's biological place in the universe is as unspecialized as his physical place, we are perhaps a typical race of intelligent creatures. For this reason, many scientists believe that extraterrestrial life must be a common phenomenon.

Surely, it would be nice to have some data to back up (or invalidate) the belief that life is a common phenomenon. During the early 1970s, radio astronomers around the world were discovering all sorts of organic molecules scattered through the interstellar medium. Obviously the building blocks for life are present floating between the stars even before planets are formed. Four and a half billion years ago, the atmosphere of primordial Earth was rich in ammonia and methane, with only a small amount of water vapor. In a classic experiment in 1955, H. C. Urey and S. L. Miller passed electric sparks (simulating lightning) through a mixture of these gases and found that *amino acids* were formed. Amino acids are a class of molecules that constitute the building blocks of *proteins*. It is clear that as soon as primordial Earth had cooled and volcanic outgassing had produced an abundant supply of water, conditions were ripe for the appearance of life.

Careful examination of certain types of sedimentary rocks with ages between 2.7 and 3.5 billion years reveals numerous tiny inclusions only a hundredth of a millimeter in diameter. These inclusions have recently been identified as the remnants of bacteria and blue-green algae by E. S. Barghoorn at Harvard and J. W. Schopf at UCLA. But bacteria and algae are fairly complicated and therefore must themselves have evolved from more primitive organisms. In view of the ages of these rocks, it is clear that life appeared on Earth only a few hundred million years after our planet became habitable.

While life made its appearance fairly early in Earth's history, nothing dramatic happened for the next two billion years. Up until a billion years ago, multi-celled animals were simple sack-like creatures. It was not until the beginning of the so-called Cambrian period some 500 million years ago that creatures finally evolved that could leave fossils. The dominant fossils of that time were left by trilobites, hard-shelled sea creatures that inhabited the ocean floors. All of the higher life forms—fishes, reptiles, birds, and mammals—followed in the remaining few hundred million years.

Scientists investigating earlier life forms on Earth recognize certain *periods* that mark various stages in the development of life on our planet. These periods comprise the *geologic time scale* listed in Table 15–1. Two things are very striking. First of all, life forms appeared early in the history of our planet. And secondly, there was a comparatively rapid succession of evolutionary changes culminating in the appearance of man. In other words, once things got started, intelligent creatures appeared in a relatively short period of time. This suggests more than the hypothesis that life is a common phenomenon. If Earth is indeed typical, it means that once life gains a foothold, there is an inevitable and rapid evolution toward intelligence.

This seems to make sense from the viewpoint of natural selection and "survival of the fittest." If two creatures appear on a planet, and if one is stupid while the other is smart, the more intelligent creature will be better able to cope with its environment and insure the survival of its offspring. The natural process of evolution therefore seems to favor the development of intelligence.

Based on this viewpoint many scientists believe that intelligent technologically advanced creatures inhabit planets in our Galaxy. And indeed the search has begun both in the United States and the Soviet Union. Using giant radio telescopes, American and Russian scientists have been listening for signals sent by intelligent creatures. In 1960, Dr. Frank Drake of Cornell used a radio telescope at the National Radio Astronomy Observatory to initiate "Project Ozma." For four weeks he collected data from two nearby stars: epsilon Eridani and tau Ceti. The results were negative. This listening technique was repeated in 1968 by Dr. V. S. Troitsky, who examined 12 nearby stars using a 45-foot antenna at Gorky University and again at NRAO by Dr. G. Verschuur who observed 10 more nearby stars in 1972. All results were negative.

Undaunted, scientists began three more systematic searches for alien signals in 1972. These on-going programs were joined by similar continuing searches in Canada and Puerto Rico in 1974 and 1975, respectively. Most recently, in 1977, a

**Table 15-1.** *Geologic Time Scale*

| Period | Date (millions of years ago) | Dominant Life Form |
|---|---|---|
| | 0 | |
| Quarternary | | Man |
| | 2 | |
| Tertiary | | Mammals |
| | 70 | |
| Cretaceous | | |
| | 130 | |
| Jurassic | | Dinosaurs |
| | 180 | |
| Triassic | | Reptiles |
| | 225 | |
| Permian | | Conifers |
| | 260 | |
| Pennsylvanian | | Ferns |
| | 300 | |
| Mississippian | | |
| | 340 | |
| Devonian | | Fishes |
| | 405 | |
| Silurian | | Early land plants |
| | 435 | |
| Ordovician | | |
| | 480 | |
| Cambrian | | Trilobites |
| | 560 | |
| Precambrian | | Soft, small life forms |
| | 4500 | |

project called SETI (Search for Extraterrestrial Intelligence) was begun by JPL scientists using the radio antennas originally built for tracking NASA spacecrafts.

The negative results from these searches to date are not surprising. If there are one million advanced civilizations scattered among 100 billion stars in our Galaxy, we must observe 100,000 stars before we have a good statistical chance of detecting an alien message. So far we have listened to only 200 stars.

In 1974, humanity announced its presence to the universe. Using the 1,000-foot radio telescope at Arecibo, astronomers beamed a coded message toward the Great Cluster in the constellation of Hercules. This cluster, called M13, is shown in

**Figure 15-19.** *The Great Cluster in Hercules.* Using the 1,000-foot radio telescope at Arecibo, Puerto Rico, astronomers have sent a coded message toward this globular cluster (called M13) in the constellation of Hercules. (*Hale Observatories*)

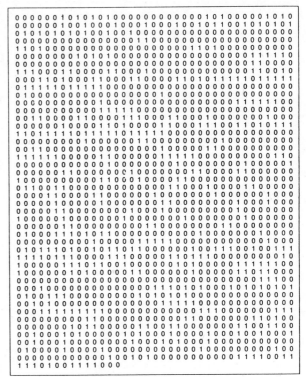

**Figure 15-20.** *The Arecibo Message in Binary Code.* It will take 24,000 years for this message from the human race to reach the stars in the globular cluster M13. Any message of this type, consisting of dots and dashes, is clearly the work of intelligent creatures. (*From "The Search for Extraterrestrial Life" by Carl Sagan and Frank Drake, Copyright © 1975 by Scientific American, Inc. All rights reserved*)

Figure 15–19. The entire message in raw form is shown in Figure 15–20 and may be thought of as a series of "dots" and "dashes" (or, alternatively, as "0s" and "1s" in binary code). The message is decoded by arranging the characters into 73 groups of 23 each. If each "1" is represented with a black square and each "0" is represented with a white square, then an interesting picture emerges, as shown in Figure 15–21. Obviously, the alien recipients of this message must be clever enough to arrange the data in this fashion. They must realize that the total number of characters (1679) is the product of two prime numbers (73 and 23). Once this is done, the message begins defining the numbers 1 through 10. This is followed by the atomic weights of the most important elements in terrestrial life. Then comes the formulas for the sugars and bases in the DNA molecule and finally the DNA molecule itself. The message ends with pictures of a human being, the solar system, and a diagram of the Arecibo telescope.

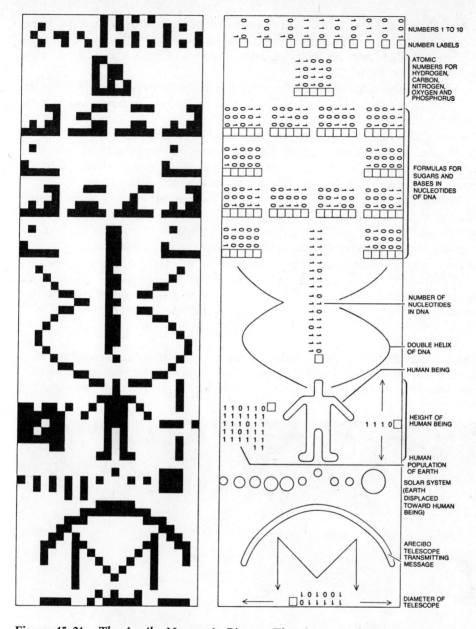

**Figure 15–21.** *The Arecibo Message in Pictures.* The message is decoded by arranging it into 73 consecutive groups of 23 characters each. If each 0 (each "dot") is given a white square and each 1 (each "dash") is given a black square, then a picture emerges. This message includes information about our solar system, about human beings, about the DNA molecule, and the chemistry of terrestrial life. (*From "The Search for Extraterrestrial Life" by Carl Sagan and Frank Drake, Copyright © 1975 by Scientific American, Inc. All rights reserved*)

It will take 24,000 years for the message to reach the 30,000 stars in M13. *If any* intelligent creatures in this cluster possess radio telescopes like the one at Arecibo, and *if* they just happen to be observing in the direction of Earth, they *will* detect the message with no trouble.

While mankind has deliberately beamed only one message toward the stars, radio waves have been "leaking" off Earth since the 1930s. For five decades, weak signals from ordinary radio and television programs have been leaving our planet. The probability that an alien race just happens to be observing toward Earth when the Arecibo message arrives is small. But in contrast, the probability of these same creatures discovering our existence by "eavesdropping" on soap operas, football, games, quiz shows, and the Five O'clock News is much higher.

Perhaps this works the other way around. Perhaps our chances of discovering extraterrestrial intelligence would be greatly increased if we could eavesdrop on *their* TV and radio shows. Our present radio telescopes are capable of sending and receiving *beamed* messages almost anywhere in our Galaxy. But if we want to pick up interalien communications, we must build much more sensitive equipment.

A system for effective eavesdropping was devised by Bernard M. Oliver at the Hewlett-Packard Company. Known as Project Cyclops, this system requires the building of 1,500 radio telescopes, each having a diameter of 100 meters (300 feet).

**Figure 15–22.** *Project Cyclops.* In order to eavesdrop on interalien communications (for example, on alien radio or TV shows), a large array of radio telescopes must be built. One proposal, called Project Cyclops, involves building 1,500 radio telescopes each having a diameter of 100 meters (300 feet). (*NASA/Ames*)

When linked by computers and operated in unison, they would act like one huge radio telescope with incredible sensitivity. The overall system is shown in Figures 15–22 and 15–23. Project Cyclops would cost $10 billion.

Detecting communications from or among intelligent alien creatures would clearly be one of the most monumental discoveries of modern science. What they might have to say to us or to each other is beyond our imagination. But the probability of success of Project Cyclops critically depends on the lifetimes of alien civilizations. We have been transmitting weak signals for five decades; in another five decades we may be extinct. If technologically advanced civilizations typically last for only a hundred years, then the probability of communicating with one is very small.

Along with the ability to build sensitive radio receivers comes the ability to produce the hydrogen bomb. In discovering the laws of physics, all intelligent races of creatures must, at some point, realize that they can construct powerful weapons. In learning about their own biology, they will develop the means (perhaps through vaccination) of coping with diseases. The resulting longevity of these creatures may result in an overpopulation problem that taxes the natural resources of their planet.

Mankind has only recently been confronted with these problems. In learning about thermonuclear processes in the sun, we learned how to construct thermonu-

**Figure 15–23.** *The Complete Array.* When operated in unison, the 1,500 radio telescopes of Project Cyclops would greatly increase our chances of detecting intelligent communications of alien creatures. (*NASA/Ames*)

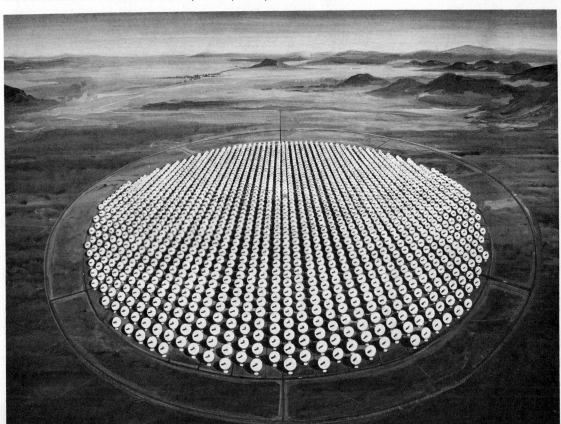

clear weaponry. In learning how to control disease, we have created terrifying conditions of overpopulation. As a result, there is a constantly increasing number of nations that simultaneously possess nuclear weaponry along with a populace largely suffering from malnutrition or even starvation. Clearly, this is a highly unstable situation. Major changes must occur in the social, political, and economic structure of the human race in order to insure the survival of our species. Antiquated methods that had insured the survival of primitive tribes and nations (when spears and clubs were the most advanced forms of weaponry) are obviously suicidal in a civilization that can detonate thousands of megatons of destructive power with the flick of a finger.

It is perhaps reasonable to suppose that as intelligent civilizations evolve, they all sooner or later must confront the crises that humanity now faces. As the laws of physics are revealed, advanced creatures must rapidly abandon antiquated modes of behavior and adopt new social structure to insure their survival. The magnitude of these revolutionary changes is enormous and the time span during which these changes must be implemented is terrifyingly short. For example, in the case of humanity, reliable computer models involving population, pollution, industrial and agricultural production, and natural resources clearly indicate that we shall be extinct within 50 years if such changes are not made.

It is obviously impossible to estimate the survival probability of technologically advanced civilizations. Perhaps the best we can do is look to nature for some clues. Thousands of sperm die so that one egg is fertilized. Thousands of seeds rot on the ground so that one germinates and grows into a tree. Perhaps thousands of intelligent civilizations perish so that one can survive with the awesome knowledge and power that necessarily accompanies unlocking the secrets of physical reality.

## Questions and Exercises

1. Briefly discuss the range of masses, luminosities, and surface temperatures of stars in the sky.
2. Draw a Hertzsprung–Russell diagram and indicate which portions of the graph are occupied by main-sequence stars, red giants and white dwarfs.
3. What is meant by gravitational accretion?
4. Briefly describe how stars are born.
5. When is hydrogen burning ignited at the core of a star? Where is it located on the H–R diagram during this stage of its evolution?
6. When is hydrogen shell burning ignited inside a star? How is it moving on the H–R diagram at this stage in its evolution?
7. Briefly describe what is going on inside a star as it evolves off of the main sequence toward the red giant region of the H–R diagram.

8. When is helium burning ignited at the core of a star? Where is it located on the H–R diagram during this stage of its evolution?

9. What is a planetary nebula? How do planetary nebulae fit into the story of stellar evolution?

10. What is a supernova?

11. What is a supernova remnant? How do supernova remnants fit into the story of stellar evolution?

12. What is a white dwarf?

13. Briefly describe the three possible final products of stellar evolution.

14. On a H–R diagram, sketch the evolutionary track of the sun.

15. What is a cosmogony?

16. What is the basic difference between evolutionary and catastrophic cosmogonies?

17. Briefly describe the cosmogonies of Descartes and Buffon.

18. Contrast and compare the evolutionary cosmogonies of Kant and Laplace.

19. Why did astronomers return to the notion of a catastrophic cosmogony around 1900?

20. Contrast and compare the Chamberlin–Moulton and Jeffrey–Jeans catastrophic cosmogonies.

21. Contrast and compare the evolutionary cosmogonies of von Weizsäcker and Kuiper.

22. Briefly describe the inhomogeneous accretion model for the formation of the solar system.

23. What is the Goldreich–Ward instability mechanism and what role did it play in the formation of planets?

24. What is the T Tauri wind and what role did it play in the formation of the solar system?

25. Briefly describe why the rotation rates of stars suggest that low-mass stars may have planetary systems.

26. Assuming that the terrestrial events leading up to the appearance of humanity are indeed typical, present arguments that there must be alien civilizations of intelligent creatures.

27. What is the purpose of Project Cyclops?

28. Even if a million technologically advanced alien civilizations have appeared or will appear in our Galaxy, communication with these creatures would be possible only if their civilizations can survive for a long enough time to be detected. In your own words, briefly discuss what advanced civilizations must do to insure their survival.

29. One of the important questions that faces astronomers searching for intelligent alien messages is "Would we recognize an alien message?" One simple message we would recognize is a sequence of prime numbers (that is, 1, 3, 5, 7, 11, 13, 17 . . . ). What other messages can you think of that we might recognize?

# Glossary

**Absolute magnitude:** A measure of the real brightness or luminosity of a celestial object.

**Absolute zero:** The lowest possible temperature; a temperature of $0°K$ or $-273°C$.

**Absorption spectrum:** Dark lines superimposed on a continuous spectrum of electromagnetic radiation.

**Acceleration:** A change in velocity; speeding up, slowing down, or changing direction.

**Accretion:** The process whereby small rocks or gas accumulate under the influence of gravity.

**Activity center (solar):** A region on the sun where sunspots, flares, and prominences occur.

**Albedo:** The percentage of sunlight reflected by a planet, a satellite, or an asteroid.

**Almanac:** A book or table listing astronomical events.

**Altitude:** Angular distance above or below the horizon, measured along a vertical circle, to a celestial object.

**Angstrom (Å):** A unit of length equal to one hundred millionth of a centimeter.

**Angular size:** The angle subtended by an object.

**Annular eclipse:** An eclipse of the sun in which the moon is too distant to cover the sun completely, so that a ring of sunlight shows around the moon.

**Antipodal:** On the opposite side.

**Aphelion:** The point in a planet's orbit at which it is farthest from the sun.

**Apogee:** The point in the orbit of a satellite at which it is farthest from the earth.

**Apollo asteroids:** Asteroids whose orbits take them closer to the sun than one Astronomical Unit.

**Apparent magnitude:** A measure of the brightness of a star or another celestial object viewed from the earth.

**Apparent solar day:** The interval between two successive transits of the sun's center across the meridian.

**Apparent solar time:** Time reckoned by the actual position of the sun in the sky.

**Asteroid:** One of several tens of thousands of small planets, ranging in size from a few hundred kilometers to less than one kilometer in size.

**Asteroid belt:** That part of the solar system, between the orbits of Mars and Jupiter, where most of the asteroids are found.

**Asthenosphere:** A plastic layer of dense rock immediately below the lithosphere on which the lithosphere floats.

**Astronautics:** The science of laws and methods of space flight.

**Astronomical unit (AU):** The average distance between the earth and the sun; a unit of length equal to approximately 93 million miles.

**Astronomy:** The study of the universe beyond the earth's atmosphere.

**Astrophysics:** The study of the physical properties and phenomena associated with planets, stars, and galaxies.

**Atom:** The smallest particle of an element; it retains the properties that characterize that element.

**Atomic mass unit:** A unit of mass approximately equal to the mass of a hydrogen atom.

**Atomic number:** The number of protons in each of the nuclei of the atoms of a particular element.

**Atomic weight:** The average mass of the atom of a particular element measured in atomic mass units.

**Aurora:** Light radiated by atoms and ions high in the earth's atmosphere, sometimes called the northern lights or southern lights.

**Auroral zone:** That part of the earth's surface, near the magnetic poles, where aurora are commonly seen.

**Autumnal equinox:** The point at which the sun crosses the celestial equator moving from north to south.

**Bar:** A unit of atmospheric pressure approximately equal to the air pressure on Earth at sea level, or about 15 pounds per square inch.

**Barred spiral:** A spiral galaxy in which the arms project from the ends of a bar that runs through the nucleus.

**Belts (Jovian):** Dark colored bands in the temporate regions of Jupiter's atmosphere parallel to Jupiter's equator.

**Beta decay:** The emission of an electron by a radioactive nucleus.

**Big-bang theory:** A cosmological theory in which the expansion of the universe is presumed to have begun with a primeval explosion.

**Binary star:** A double star; two stars revolving about each other.

**Bipolar group:** A large sunspot that consists of two pronounced groups of smaller spots having opposite magnetic polarity.

**Blackbody:** A hypothetical, perfect radiator; a body that absorbs and reemits all the radiation falling on it.

**Black hole:** A region of highly warped space–time caused by an intense gravitational field.

**Bode's law:** A technique for obtaining a sequence of numbers that gives the approximate distances of the planets from the sun in astronomical units.

**Bohr atom:** A model of the atom developed by Niels Bohr in which the electrons revolve about the nucleus in circular orbits.

**Breccia:** A rock that is formed from smaller jaggered rock fragments that have been cemented together.

**Bright ring:** The brightest of the rings surrounding Saturn.

**Caldera:** The central crater at the summit of a volcano.

**Caloris Basin:** A large, relatively flat area on Mercury.

**Carbonate:** A rock-forming mineral involving $CO_3$.

**Cassini's division:** A 5,000-kilometer-wide gap between the bright and outer rings around Saturn.

**Celestial equator:** A great circle on the celestial sphere, 90° from the celestial poles.

**Celestial mechanics:** A branch of astronomy dealing with the motions of the members of the solar system.

**Celestial poles:** The points about which the celestial sphere appears to rotate; the points of intersections of the celestial sphere with the extension of the earth's polar axis.

**Celestial sphere:** The apparent sphere of the sky; a sphere of large radius centered on the earth.

**Cepheid variable:** A type of pulsating star.

**Ceres:** The largest of the minor planets and the first to be discovered.

**Chromosphere:** The part of the solar atmosphere immediately above the photosphere.

**Cleavage plane:** A plane of weakness along which a crystal breaks when struck with a sharp blow.

**Cluster of galaxies:** A system of galaxies with at least several, and in some cases thousands, of members.

**CNO Cycle:** A series of thermonuclear reactions occurring in certain stars that convert hydrogen into helium using carbon as a catalyst.

**Coma (cometary):** The diffuse, gaseous component of the head of a comet.

**Comet:** A swarm of solid particles and gases that revolves about the sun, usually in a highly eccentric orbit.

**Comet head:** The main part of a comet, consisting of a nucleus and a coma.

**Condensation temperature:** The temperature at which substances condense out of a gaseous state.

**Conglomerate:** A rock that is formed from smaller rounded, weathered rock fragments that have been cemented together.

**Conic section:** The curve of intersection between a circular cone and a plane—an ellipse, circle, parabola, or hyperbola.

**Conjunction:** The configuration in which a planet appears nearest to the sun or some other planet.

**Constellation:** A configuration of stars named for a particular object, person, or animal.

**Continental drift:** The tendency for continents to move slowly across the earth's surface.

**Continental shelf:** That portion of a continent that is submerged in relatively shallow water.

**Continuous spectrum:** A spectrum of light comprised of radiation of a continuous range of wavelengths, or colors.

**Convection:** The transport of energy by the motions of hot gases.

**Convective zone:** The outer portions of the

solar interior where energy is transported outward by convection.

**Coriolis effect:** The deflection of a moving object caused by the fact that the object is moving in a rotating system.

**Corona:** The outer atmosphere of the sun.

**Coronagraph:** An instrument for photographing the chromosphere and corona of the sun when it is not eclipsed.

**Cosmic rays:** Atomic nuclei (mostly protons) that strike the earth's atmosphere at exceedingly high speeds.

**Cosmogony:** The study of the origin of the solar system.

**Cosmological model:** A specific theory of the organization and evolution of the universe.

**Cosmology:** The study of the organization and evolution of the universe.

**Crater:** A circular depression on the surface of a planet or a moon usually caused by the impact of a meteoroid.

**Crepe ring:** A faint inner ring surrounding Saturn.

**Crescent moon:** A phase of the moon during which the moon appears less than half full.

**Crust (earth):** The outermost layers of the earth's surface.

**Daylight savings time:** A time one hour ahead of standard time, usually adopted in spring and summer to take advantage of long evening twilights.

**Decametric radiation:** A type of radio radiation coming from Jupiter and having wavelengths of roughly ten meters.

**Decimetric radiation:** A type of radio radiation emitted by Jupiter's magnetosphere and having wavelengths around ten centimeters.

**Declination:** Angular distance north or south of the celestial equator.

**Deferent:** A stationary circle in the Ptolemaic system along which the center of another circle (an epicycle) moves.

**Density:** The ratio of the mass of an object to its volume.

**Deuterium:** An isotope of hydrogen whose nuclei contain one proton and one neutron.

**Differential gravitational force:** The difference in the strength of the gravitational force across an extended object that tends to deform the object.

**Differential rotation:** Rotation in which all the parts of an object do not behave like a solid. The sun and galaxies exhibit differential rotation.

**Direct motion:** The apparent eastward motion of a planet on the celestial sphere or with respect to the stars.

**Diurnal motion:** Motion repeated or recurring every day.

**Doppler shift:** An apparent change in wavelength of the radiation from a given source caused by its relative motion in the line of sight.

**Dust tail (cometary):** The curved tail of a comet composed mostly of dust and tiny fragments of rock.

**Earthquake:** A sudden shift in the earth's crust.

**Eccentricity:** A measure of how "flattened" an ellipse or an elliptical orbit is.

**Eclipse:** The cutting off of all or part of the light of one body by another body passing in front of it.

**Eclipse path:** The track along the earth's surface swept out by the tip of the shadow of the moon during a solar eclipse.

**Eclipse season:** A period during the year when an eclipse of the sun or moon is possible.

**Eclipsing binary star:** A binary star in which the plane of revolution of the two components is nearly edge on to our line of sight, so that one star periodically passes in front of the other.

**Ecliptic:** The apparent annual path of the sun on the celestial sphere.

**Ejecta:** Material dug up and tossed out during crater formation on the moon or a planet.

**Electromagnetic radiation:** Radiation consisting of waves propagated with the speed of light. Radio waves; infrared, visible, and ultraviolet light; X rays; and gamma rays are all forms of electromagnetic radiation.

**Electromagnetic spectrum:** The entire range of electromagnetic waves.

**Electron:** A negatively charged subatomic particle that normally moves about the nucleus of an atom.

**Ellipse:** A conic section; the curve of intersection of a circular cone and a plane cutting completely through it.

**Elliptical galaxy:** A galaxy whose apparent contours are ellipses and that contains no conspicuous interstellar material.

**Elongation:** The angular distance between two objects in the sky such as the sun and a planet.

**Emission line:** A discrete, bright spectral line.

**Emission nebula:** A gaseous nebula that derives its visible light from the flourescence of ultraviolet light from a star.

**Emission spectrum:** A spectrum of emission lines.

**Energy level:** A particular level, or amount, of energy possessed by an atom or ion above that which it possesses in its ground state.

**Ephemeris:** A table giving the positions of a celestial body at various times or other astronomical data.

**Ephemeris time:** Time that passes at a strictly uniform rate, used to compute the instant of various astronomical events.

**Epicenter:** The location on the earth's surface directly above the focus of an earthquake.

**Epicycle:** A circular orbit of a body in the Ptolemaic system, the center of which revolves about another circle (the deferent).

**Equator:** A great circle on the earth, 90° from the North and South Poles.

**Equinox:** One of the intersections of the ecliptic and celestial equator.

**Escape velocity:** The speed needed to escape from the gravitational field of an object.

**Excitation:** The imparting of energy to an atom or an ion.

**Extragalactic:** Beyond the galaxy.

**Eyepiece:** A magnifying lens used to view the image produced by the objective lens of a telescope.

**Faculus (faculae):** Bright region near the limb of the sun.

**Filtergram:** A photograph of the sun taken through a special narrow band-pass filter.

**Fireball:** A spectacular meteor.

**First quarter:** A phase of the moon (approximately one week after new moon) during which half of the illuminated side of the moon can be seen as seen from Earth.

**Fission:** The breakup of a heavy atomic nucleus into two or more lighter ones.

**Flare:** A sudden and temporary outburst of light from an extended region of the solar surface.

**Floccules (flocculi):** A bright region of the solar surface observed in the monochromatic light of a spectral line; usually called a plage.

**Focal length:** The distance from a lens, or mirror, to the point where light is focused.

**Focal point:** The point where light is focused by a lens or mirror.

**Focus (earthquake):** The place inside the earth's crust where an earthquake originates.

**Focus (optical):** The point where the rays of light converged by a mirror or lens meet.

**Fraunhofer line:** An absorption line in the spectrum of the sun or a star.

**Fraunhofer spectrum:** The array of absorption lines in the spectrum of the sun or of a star.

**Frequency:** The number of vibrations in a unit of time; the number of waves that cross a given point in a unit of time.

**Front:** The boundary between warm and cold air masses in the earth's atmosphere.

**Full moon:** A phase of the moon during which the entire daylight hemisphere is visible from the earth.

**Fusion:** The building up of heavier atomic nuclei from lighter ones.

**Galactic rotation:** The rotation of the galaxy.

**Galaxy:** A large assemblage of stars. A typical galaxy contains millions to hundreds of billion of stars.

**Galilean satellites:** The four largest satellites of Jupiter.

**Gamma rays:** Photons of electromagnetic radiation with wavelengths shorter than those of X rays. Gamma rays have the highest frequency of any form of electromagnetic radiation.

**General relativity:** A theory of gravitation proposed by Dr. Einstein in which gravity is expressed as the warping of space–time.

**Geocentric:** Centered on the earth.

**Giant (star):** A highly luminous star with a large radius.

**Gibbous moon:** A phase of the moon during which more than half, but not all, of the moon's sunlit hemisphere is visible from the earth.

**Globular cluster:** A large, spherical cluster of stars.

**Gondwanaland:** One of the two supercontinents into which Pangaea originally divided.

**Granulation (solar):** The ricelike pattern evident in photographs of the solar photosphere.

**Gravitation:** The attraction of matter for matter.

**Great Red Spot:** A large, reddish semipermanent feature in Jupiter's atmosphere which has been observed for centuries.

**Greenhouse effect:** That phenomenon whereby incoming sunlight is trapped in an environment causing the temperature of that environment to increase.

**Greenwich meridian:** The meridian of longitude passing through the site of the old Royal Greenwich Observatory, near London.

**Gregorian calendar:** A calendar (now in common use) introduced by Pope Gregory XIII in 1582.

**HI region:** A region of neutral hydrogen in interstellar space.

**HII region:** A region of ionized hydrogen in interstellar space.

**Half-life:** The time it takes for half of the radioactive nuclei in a particular sample to decay.

**Hardness:** A measure of the ability of a mineral to scratch or be scratched by other minerals.

**Heliocentric:** Centered on the sun.

**Helium burning:** Thermonuclear reactions occurring in stars in which helium is converted into carbon.

**Hertzsprung–Russell (H-R) diagram:** A diagram showing the relationship of the absolute magnitudes of a group of stars to their temperature, spectral class, or color index.

**High:** A region of higher-than-average atmospheric pressure in a planetary atmosphere.

**Hirayama family:** A class of asteroids that have very similar orbits.

**Horizon (astronomical):** A great circle on the celestial sphere, 90° from the zenith.

**Hubble law:** The linear relationship between the distances and velocities of remote galaxies.

**Hubble constant:** The constant of proportionality in the relationship between the velocities and distances of remote galaxies.

**Hydrogen burning:** Thermonuclear reactions occurring in stars in which hydrogen is converted into helium.

**Hyperbola:** A curve of intersection between a circular cone and a plane that is at too small an angle with the axis of the cone to cut all the way through it and is not parallel to a line in the face of the cone.

**Ice age:** One of several brief periods in the earth's history when a comparatively large fraction of the earth's surface was covered with ice.

**Igneous rock:** A rock that has cooled from a molten state.

**Image:** The optical representation of an object produced by the refraction or reflection of light rays from the object by a lens or mirror.

**Image tube:** A device in which electrons emitted from a photocathode surface exposed to light are focused electronically.

**Inclination (orbital):** The angle between the orbital plane of a revolving body and a fundamental plane—usually the plane of the celestial equator or the ecliptic.

**Inertia:** The property of matter that makes the action of a force necessary to change the state of motion of an object.

**Inferior conjunction:** The configuration of an inferior planet when it is between the sun and the earth.

**Inferior planet:** A planet whose distance from the sun is less than the earth's.

**Infrared radiation:** Electromagnetic radiation with a wavelength longer than the longest visible wavelength (red) but shorter than wavelengths in the radio range.

**Intercalate:** To insert, as a day, in a calendar.

**Interferometer (stellar):** An optical device that uses light-interference phenomena to measure small angles.

**International Date Line:** An arbitrary line on the surface of the earth, near a longitude of 180°, on either side of which the date changes by one day.

**Interplanetary medium:** Gas and solid particles in interplanetary space.

**Interstellar dust:** Microscopic solid grains in interstellar space.

**Interstellar gas:** Sparse gas in interstellar space.

**Interstellar matter:** Interstellar gas and dust.

**Interstellar wind:** The hypothetical flow of particles in interstellar space.

**Inverse beta decay:** The emission of a positron by a radioactive nucleus.

**Ion:** An atom that has become electrically charged through the addition or loss of one or more electrons.

**Ionization:** The process by which an atom gains or loses electrons.

**Ionosphere:** The upper region of the earth's atmosphere, in which many of the atoms are ionized.

**Ion tail (cometary):** The nearly straight tail of a comet that points directly away from the sun and is composed mostly of ionized molecules; Type I tail.

**Irregular galaxy:** A galaxy without rotational symmetry (that is, neither spiral nor elliptical).

**Isotope:** Any of two or more forms of an element having the same number of protons but different numbers of neutrons in their nuclei.

**Jovian planet:** Jupiter, Saturn, Uranus, or Neptune.

**Julian calendar:** A calandar introduced by Julius Caesar in 45 B.C.

**Jupiter:** The fifth planet from the sun in the solar system.

**Kepler's laws:** Three laws developed by Johannes Kepler to describe the motions of the planets.

**Kirkwood's gaps:** Gaps in the spacing of the minor planets caused by perturbations produced by the major planets.

**KREEP norite:** A type of lunar rock containing unusually high concentrations of potassium (K), rare earth elements (REE), and phosphorous (P).

**Lagrangian points:** Places where the combined gravitational fields of two massive objects can capture and trap smaller objects and particles.

**Last quarter:** A phase of the moon (approximately one week before new moon) during which half of the illuminated side of the moon can be seen from Earth.

**Latitude:** A north-south coordinate on the surface of the earth; the angular distance north or south of the equator.

**Laurasia:** One of the two supercontinents into which Pangaea originally divided.

**Law of polarity:** A statement of the relationship between the magnetic polarity of the components of a bipolar group of sunspots and the solar hemisphere in which the group is located during the sunspot cycle.

**Leap year:** A year with 366 days, to make the average length of the calendar year as nearly equal as possible to the tropical year.

**Libration:** A change in the visible hemisphere of the moon viewed from earth.

**Light:** Electromagnetic radiation visible to the human eye.

**Light curve:** A graph showing the variation in light, or magnitude, of a variable or an astronomical object.

**Light year:** The distance light travels in a vacuum in 1 year; 6 trillion miles.

**Limb:** The apparent edge of a celestial body.

**Limiting magnitude:** The faintest magnitude that can be observed with a given instrument or under given conditions.

**Line of nodes:** The line of intersection between the plane of the orbit of an object (such as the moon) and another plane in space (such as the ecliptic).

**Lithosphere:** A shell of hard rocky material covering the earth's surface.

**Lobate scarp:** A long, low cliff.

**Local Group:** The cluster of galaxies to which our galaxy belongs.

**Longitude:** An east–west coordinate on the earth's surface; the angular distance along the equator east or west of the Greenwich meridian to another meridian.

**Low:** A region of lower-than-average atmospheric pressure in a planetary atmosphere.

**Luminosity:** The rate of radiation of energy into space by a celestial object.

**Lunar eclipse:** An eclipse of the moon.

**Magellanic Clouds:** Two neighboring galaxies visible to the naked eye in southern latitudes.

**Magma:** Molten rock beneath or within the earth's crust.

**Magnetic field:** The region of space near a magnetized body within which magnetic forces can be detected.

**Magnetic scattering:** The collision and rebounding of streams in the solar wind.

**Magnetopause:** The outer boundary of a planet's magnetosphere where the magnetosphere meets the solar wind.

**Magnetosheath:** A turbulent region in between the bow shock and magnetosphere of a planet where the solar wind flows around the planet's magnetosphere.

**Magnetosphere:** That region surrounding a planet where effects are dominated by the planet's magnetic field.

**Magnifying power:** A measure of the strength of a telescope based on the increase in angular diameter of an object viewed through it.

**Magnitude:** A measure of the amount of light received from a star or another luminous object.

**Main sequence:** The largest sequence of stars on the H–R diagram.

**Major axis (of an ellipse):** The maximum diameter of an ellipse.

**Mantle:** That portion of the earth between the crust and the core.

**Mare:** Latin for sea; the name applies to many sealike flat areas on the moon and Mars.

**Mars:** The fourth planet from the sun in the solar system.

**Mascon:** A region of higher-than-average gravity on the moon; short for mass concentration.

**Mass:** A measure of the total amount of material in a body.

**Mass–luminosity relationship:** An empirical relationship between the masses and luminosities of many stars.

**Mass number:** The total number of protons and neutrons in the nuclei of an isotope.

**Massif:** A rugged, compact portion of a mountain range.

**Maunder butterfly diagram:** A graph showing the migration of sunspots towards the solar equator during the sunspot cycle.

**Maximum elongation:** The point in the orbit of an inferior planet at which the planet appears farthest from the sun.

**Mean solar day:** The interval between successive passages of the mean sun across the meridian; the average length of the apparent solar day.

**Mean solar time:** Time reckoned from the position of the mean sun.

**Mean sun:** A fictitious body that moves eastward with constant speed along the celestial equator.

**Mechanics:** The branch of physics that deals with the behavior of material bodies.

**Mercury:** The nearest planet to the sun in the solar system. It is also the smallest, both in size and mass.

**Meridian (celestial):** A great circle on the celestial sphere that passes through an observer's zenith and the celestial poles.

**Meridian (terrestrial):** A great circle on the surface of the earth that passes through a particular place and the North and South Poles of the earth.

**Metamorphic rock:** A rock that has been transformed by heat and pressure deep inside the crust of a planet.

**Meteor:** The flash of light observed when a meteoroid enters the earth's atmosphere and burns up; sometimes called a "shooting star."

**Meteor shower:** The apparent descent of a large number of meteors radiating from a common point in the sky, caused by the collision of the earth with a swarm of meteoritic particles.

**Meteorite:** A portion of a meteoroid that survives passage through the atmosphere and strikes the ground.

**Meteoroid:** A meteoritic particle in space.

**Micrometeorites:** Interplanetary dust.

**Micron:** A unit of length equal to ten thousand angstroms.

**Milky Way:** A band of light encircling the sky, caused by the many stars lying near the plane of the galaxy; also used as synonym for the galaxy to which the sun belongs.

**Millibar:** One thousandth of a bar.

**Mineral:** A crystalline substance out of which rocks are formed.

**Minor axis (of an ellipse):** The smallest diameter of an ellipse.

**Minor planet:** One of several tens of thousands of small planets, ranging from a few hundred kilometers to less than 1 kilometer in diameter.

**Mohs scale:** A standardized scheme of denoting the hardness of a mineral.

**Molecule:** Two or more atoms bound together.

**Monochromatic:** Of one wavelength, or color.

**Moraine:** A pile of debris in front of a glacier.

**Nadir:** A point on the celestial sphere 180° from the zenith.

**Neap tides:** The lowest tides in the month, which occur when the moon is near the first or third quarter phase.

**Nebula (nebulae):** A cloud of interstellar gas or dust.

**Neptune:** The eighth planet from the sun in the solar system.

**Neutrino:** A subatomic particle that travels at the speed of light.

**Neutron:** A subatomic particle with no charge and with a mass approximately equal to that of the proton.

**New moon:** A phase of the moon during which Earth-based observers see only the unilluminated side of the moon.

**Newton's laws:** Laws of mechanics and gravitation formulated by Isaac Newton.

**Newtonian focus:** An optical arrangement in a reflecting telescope whereby light is reflected by a flat mirror to a focus at the side of the telescope tube just before it reaches the focus of the objective lens.

**Node:** An intersection of the orbit of a celestial body with a fundamental plane, usually the plane of the celestial equator or the ecliptic.

**North point:** That intersection of the celestial meridian and the astronomical horizon lying nearest the north celestial sphere.

**Nova:** A star that experiences a sudden outburst of radiant energy, temporarily increasing its luminosity hundreds or thousands of times.

**Nuclear transformation:** The transformation of one atomic nucleus into another.

**Nucleus (atomic):** The heavy part of an atom, composed of protons and neutrons, about which the electrons revolve.

**Nucleus (cometary):** A swarm of solid particles in the head of a comet.

**Nucleus (galactic):** A concentration of stars and gas at the center of a galaxy.

**Objective lens:** The principal image-forming component of a telescope or another optical instrument.

**Obliquity:** The angle between the plane of a planet's equator and the plane of its orbit.

**Occultation:** An eclipse of a star or planet by the moon or another planet.

**Olympus Mons:** The largest volcano on Mars.

**Oort cloud:** A hypothetical spherical distribution of billions of comets surrounding the solar system and at a very great distance from the sun.

**Opposition:** The configuration of a planet when it is directly opposite the sun as seen from earth.

**Optics:** The branch of physics that deals with the properties of light.

**Orbit:** The path of a body revolving about another body or point.

**Orbital elements:** Mathematical quantities which specify an orbit.

**Orbital inclination:** The angle between the

orbital plane of a planet and the plane of the ecliptic.

**Pangaea:** The original large continent which broke up into several smaller continents some 200 million years ago.

**Panthalassa:** A vast primordial ocean on the earth which surrounded Pangaea.

**Parabola:** The curve of intersection between a circular cone and a plane parallel to one side of the cone.

**Parallax:** An apparent displacement of an object caused by a motion of the observer.

**Parsec:** The distance of an object with a stellar parallax of 1 second of arc. A parsec equals 3.26 light years.

**Partial eclipse:** An eclipse in which the concealed body is not completely obscured.

**Penumbra:** The portion of a shadow from which only part of the light source is occulted by an opaque body.

**Penumbra (sunspot):** The lighter regions of a sunspot.

**Penumbral eclipse:** A lunar eclipse in which the moon passes through the penumbra, but not the umbra, of the earth's shadow.

**Perigee:** The place in the orbit of a satellite where it is closest to the center of the earth.

**Perihelion:** The place in the orbit of an object revolving about the sun where it is closest to the sun.

**Period:** Generally the interval of time required for a celestial body to rotate once on its axis, revolve once about a primary, or return to its original state after an increase in luminosity.

**Period-luminosity relationship:** An empirical relationship between the periods and luminosities of Cepheid variable stars.

**Perturbation:** A small disturbance in the motion of a celestial body produced by the gravitational field of another nearby body.

**Phases (lunar):** The changes in the moon's appearance as different portions of its illuminated hemisphere become visible from the earth.

**Photon:** A discrete unit of electromagnetic energy.

**Photosphere:** The region of the solar (or stellar) atmosphere from which radiation escapes into space.

**Plage:** A bright region of the solar surface observed in the monochromatic light of a particular spectral line; a flocculus.

**Planck's law of radiation:** A formula for calculating the intensity of radiation at various wavelengths emitted by a black body.

**Planet:** Any of nine solid, nonluminous bodies revolving about the sun.

**Planetary nebula:** A shell of gas ejected from, and enlarging about, an extremely hot star.

**Planetoid:** A minor planet.

**Plate tectonics:** The idea that the earth's crust consists of large moving plates.

**Pluto:** The ninth planet from the sun in the solar system.

**Pore (solar):** A small blemish on the sun which eventually grows into a sunspot.

**Positron:** An anti-electron.

**Precession (of the equinoxes):** The slow westward motion of the equinoxes along the ecliptic as a result of precession.

**Precession (of earth):** A slow, conical motion of the earth's axis of rotation, caused by the gravitational torque of the moon and sun on the earth's equatorial bulge.

**Prime focus:** The point in a telescope where the objective lens or mirror focuses light.

**Prime meridian:** The meridian of longitude passing through the site of the old Royal Greenwich Observatory, near London; the great circle from which terrestrial longitude is measured.

**Primordial fireball:** The extremely hot opaque gas presumed to have contained the entire mass of the universe at the time of, or immediately following, the explosion of the primeval atom.

**Prism:** A wedge-shaped piece of glass used to disperse white light into a spectrum.

**Prominence:** A flamelike phenomenon in the solar corona.

**Proton-proton chain:** A thermonuclear process occurring in stars in which protons are fused together to produce helium.

**Protoplanet:** A newly formed object that will eventually evolve into a planet.

**Pulsar:** A small but powerful celestial radio source that emits short, regular bursts of radio noise.

**P-wave:** A longitudinal wave produced by an earthquake.

**Quantum mechanics:** That field of physics that deals with the structure and properties of atoms and subatomic particles.

**Radiant (of a meteor shower):** The point in the sky from which the meteors belonging to a shower seem to radiate.

**Radiation:** The emission and transmission of energy in the form of waves or particles.

**Radiative transport:** The radiative flow of energy outward from the center of a star.

**Radiative zone:** The inner portions of the sun where energy flows outward by radiative transport.

**Radioactive decay:** The process whereby unstable nuclei emit particles.

**Radio astronomy:** The use of radio wavelengths to make astronomical observations.

**Radio telescope:** A telescope designed to make observations in radio wavelengths.

**Ray (lunar):** Any of a system of bright, elongated streaks, sometimes associated with a crater.

**Red giant:** A large, cool star of high luminosity in the upper-right portion of the H–R diagram.

**Reflecting telescope:** A telescope in which the principal optical component is a concave mirror.

**Refracting telescope:** A telescope in which the principal optical component (objective) is a lens or system of lenses.

**Refraction:** The bending of light rays passing from one transparent medium to another.

**Relative orbit:** The orbit of one of two mutually revolving bodies about the other.

**Resolution:** The degree to which fine details in an image are separated or resolved.

**Resolving power:** A measure of the ability of an optical system to resolve, or separate, fine details in the image it produces.

**Retrograde motion:** The apparent westward motion of a planet on the celestial sphere or with respect to the stars.

**Revolution:** The motion of one body around another.

**Richter scale:** A standardized scheme for denoting the intensity of earthquakes.

**Rift:** The boundary between adjacent plates on the earth's surface where the plates are separating and hot magma surges up to form new crust.

**Right ascension:** A coordinate, measured eastward along the celestial equator, for denoting the east–west positions of celestial bodies.

**Rille (lunar):** A crevasse, or trenchlike depression, in the moon's surface.

**Rock:** A piece of the solid material out of which planets such as the earth are made.

**Rotation:** The turning of a body about an axis running through it.

**Saturn:** The sixth planet from the sun in the solar system.

**Sedimentary rock:** A type of rock usually formed by precipitation.

**Seismograph:** An instrument that records movements in the crust of a planet.

**Seleno-:** A prefix referring to the moon.

**Semimajor axis:** One half of the major axis of an ellipse.

**Semiminor axis:** One half of the minor axis of an ellipse.

**Shadow zone:** A region of the earth opposite the location of the epicenter of an earthquake where no P-waves or S-waves are felt.

**Shock waves:** A region of abrupt change in pressure and density caused by supersonic flow around a body.

**Sidereal day:** The interval between two successive meridian transits of the vernal equinox.

**Sidereal month:** The period of the moon's revolution about the earth with respect to the stars.

**Sidereal period:** The period of revolution of one body about another with respect to the stars.

**Sidereal time:** Time reckoned according to the position of the vernal equinox in the sky.

**Sidereal year:** The period of the earth's revolution about the sun with respect to the stars.

**Silicate:** A common rock-forming mineral based on the element silicon.

**Solar atmosphere:** The outer layers of the sun, including the photosphere, chromosphere, and corona.

**Solar system:** The system of the sun and the planets, satellites, asteroids, comets, meteoroids, and other objects revolving around it.

**Solar time:** Time measured according to the sun.

**Solar wind:** A radial flow of particle radiation leaving the sun.

**Solstice:** Either of two points on the celestial sphere where the sun reaches its maximum distances north or south of the celestial equator.

**Specific gravity:** The ratio of the mass of an object to the mass of an equal volume of water.

**Spectrogram:** A photograph of a spectrum.

**Spectrograph:** An instrument for photographing a spectrum; usually attached to a telescope to photograph the spectrum of a star.

**Spectroheliogram:** A photograph of the sun obtained with a spectroheliograph.

**Spectroheliograph:** An instrument for photographing the sun, or part of the sun, in the monochromatic light of a particular spectral line.

**Spectroscope:** An instrument for viewing the spectrum of a light source directly.

**Spectroscopic binary:** A double star whose binary nature is deduced from the periodic shifting of spectral lines in its spectrum.

**Spectroscopy:** The study of spectra.

**Spectrum:** The array of colors, or wavelengths, obtained when light from a source is dispersed by passing it through a prism or grating.

**Spectrum analysis:** The study and analysis of spectra, especially stellar spectra.

**Spicule:** A narrow jet of material rising in the solar chromosphere.

**Spin–orbit coupling:** That phenomena whereby the periods of rotation and revolution of a planet or satellite are related by whole numbers.

**Spiral arms:** Armlike areas of interstellar material and young stars that wind out in a plane from the central nucleus of a spiral galaxy.

**Spiral galaxy:** A flattened, rotating galaxy with wheel-like arms of interstellar material and young stars winding out from its nucleus.

**Spring tide:** The highest tide of the month, produced near the time of full moon or new moon.

**Standard time:** Local mean solar time of a standard meridian, adopted over a large region to avoid the inconvenience of continuous time changes around the earth.

**Star:** A self-luminous sphere of gas.

**Star cluster:** An assemblage of stars held together by their mutual gravitation.

**Stefan's law:** A relationship between the temperature of a black body and the total amount of radiation it emits.

**Stellar evolution:** The changes that take place in the size, luminosity, structure, and other characteristics of a star as it ages.

**Stratosphere:** The second lowest layer in the earth's atmosphere, between the troposphere and the mesosphere.

**Subduction zone:** The boundary between two colliding plates on the earth's surface where the crust is being forced downward into the asthenosphere.

**Summer solstice:** The point on the celestial sphere where the sun is farthest north of the celestial equator.

**Sun:** The star about which the earth and other planets revolve.

**Sunspot:** A temporary cool region in the solar photosphere that appears dark in contrast to the surrounding hotter photosphere.

**Sunspot cycle:** A semiregular 11-year period during which the number of sunspots fluctuates.

**Superior conjunction:** The configuration of a planet in which the planet and earth are on opposite sides of the sun.

**Superior planet:** A planet farther from the sun than the earth.

**Supernova:** A stellar outburst, or explosion, in which a star suddenly increases in luminosity.

**Supernova remnant:** A nebula resulting from the ejection of matter from a star that had undergone a supernova explosion.

**Surface gravity:** A measure of the strength of gravity at the surface of a planetary body.

**S-wave:** A transverse wave produced by an earthquake.

**Synchrotron radiation:** Radiation emitted by charged particles moving in a magnetic field.

**Synodic month:** The period of revolution of the moon with respect to the sun; or the period of the cycle of lunar phases.

**Synodic period:** The interval between successive occurrences of the same configuration of a planet (for example, between successive oppositions or successive superior conjunctions).

**Tail (cometary):** Gases and solid particles ejected from the head of a comet and forced away from the sun by radiation pressure and the solar wind.

**Tektites:** Rounded glassy bodies suspected to be of meteoritic origin.

**Telescope:** An instrument used to view, measure, or photograph distant objects.

**Telluric:** Of terrestrial origin.

**Temperature (absolute):** Temperature measured in centigrade degrees from absolute zero.

**Temperature (centigrade):** Temperature measured on a scale calibrated so that water freezes at $0°$ and boils at $100°$.

**Temperature (Fahrenheit):** Temperature measured on a scale calibrated so that water freezes at $32°$ and boils at $212°$.

**Temperature (Kelvin):** Absolute temperature measured in centigrade degrees ($°K = °C + 273°$).

**Terminator:** The line of sunrise or sunset on a celestial body such as the moon.

**Terrestrial planet:** Mercury, Venus, Earth, Mars, and sometimes Pluto.

**Tethys Sea:** The body of water that separated Laurasia from Gondwanaland; the ancestor of the Mediterranean Sea.

**Tharsis Plateau:** That region on Mars containing four of the largest and most recent volcanoes on the planet.

**Thermal radiation:** Radiation emitted by an object because the object has a temperature above absolute zero.

**Thermonuclear energy:** Energy associated with thermonuclear reactions.

**Thermonuclear reaction:** A nuclear reaction or transformation that results from encounters between high-velocity nuclear particles.

**Tidal force:** A differential gravitational force that tends to deform a body.

**Tide:** A deformation of a body caused by the differential gravitational force exerted on it by another body.

**Titan:** Saturn's largest satellite.

**Total eclipse:** An eclipse of the sun in which the photosphere is hidden entirely by the moon; a lunar eclipse in which the moon passes completely into the umbra of the earth's shadow.

**Train (of a meteor):** A temporarily luminous trail in the wake of a meteor.

**Transit (of the meridian):** The passage of an object across the celestial meridian.

**Transit (of a planet):** The passage of an inferior planet across the disc of the sun.

**Trench (oceanic):** A deep depression in the ocean floor at a subduction zone.

**Triple-alpha process:** A thermonuclear reaction in which helium nuclei are converted into carbon nuclei.

**Triton:** Neptune's largest satellite.

**Trojan asteroid:** One of several minor planets that share Jupiter's orbit around the sun but are located approximately 60° around the orbit from Jupiter.

**Tropical year:** The period of revolution of the earth about the sun with respect to the vernal equinox.

**Troposphere:** The lowest layer in the earth's atmosphere, from sea level to an altitude of about 11 kilometers (7 miles).

**Ultraviolet radiation:** Electromagnetic radiation whose wavelength is shorter than the shortest wavelengths to which the eye is sensitive (violet); radiation whose wavelength ranges from approximately 100 to 4,000 Å.

**Umbra:** The central, completely dark part of a shadow.

**Umbra (sunspot):** The darkest portions of a sunspot.

**Universal time:** The local mean time of the prime meridian.

**Universe:** The totality of matter, radiation, and space.

**Uranus:** The seventh planet from the sun in the solar system.

**Valles Marineris:** A vast canyon on Mars almost 4,000 kilometers in length.

**Van Allen belt:** One of two large radiation belts surrounding the earth where electrons and protons are trapped by the earth's magnetic field.

**Venus:** The second planet from the sun in the solar system.

**Vernal equinox:** The point on the celestial sphere where the sun crosses from the south to the north of the celestial equator.

**Vertical circle:** Any great circle passing through the zenith.

**Visual binary star:** A binary star in which the two components can be resolved telescopically.

**Volume:** A measure of the total space occupied by a body.

**Wavelength:** The spacing of the crests or troughs in a wave train.

**Weight:** A measure of the force of gravitational attraction.

**Wein's law:** A relationship between the temperature of a blackbody and the wavelength at which it emits the greatest intensity of radiation.

**White dwarf:** A star that has exhausted most or all of its nuclear fuel and has collapsed to a very small size.

**Widmanstätten figures:** Crystalline structures observable in cut and polished meteorites.

**Winter solstice:** The point on the celestial sphere where the sun is farthest south of the celestial equator.

**X rays:** Photons whose wavelengths are shorter than ultraviolet wavelengths and longer than gamma wavelengths.

**Zap crater:** A tiny crater found on lunar rocks thought to be caused by the impact of micrometeorites.

**Zeeman effect:** A splitting or broadening of spectral lines as a result of magnetic fields.

**Zenith:** The point on the celestial sphere that is directly overhead of an observer.

**Zodiac:** A belt around the sky centered on the ecliptic.

**Zodiacal light:** A faint illumination along the zodiac, believed to be caused by sunlight reflected and scattered by interplanetary dust.

**Zones (Jovian):** Light-colored bands in the temperate regions of Jupiter's atmosphere parallel to Jupiter's equator.

# Appendix

**Table A-1.** *The Planets (Physical Data)*

| Planet | Diameter (kilometers) | Diameter (Earth = 1) | Mass (Earth = 1) | Surface Gravity (Earth = 1) | Period of Rotation | Number of Moons |
|---|---|---|---|---|---|---|
| Mercury | 4,880 | 0.38 | 0.06 | 0.38 | 58.65 days | 0 |
| Venus | 12,100 | 0.95 | 0.82 | 0.90 | 243 days | 0 |
| Earth | 12,760 | 1.00 | 1.00 | 1.00 | 23 hrs. 56 min. | 1 |
| Mars | 6,790 | 0.53 | 0.11 | 0.38 | 24 hrs. 37 min. | 2 |
| Jupiter | 142,800 | 11.19 | 318.0 | 2.64 | 9 hrs. 50 min. | 14 |
| Saturn | 121,000 | 9.47 | 95.2 | 1.13 | 10 hrs. 14 min. | 10 |
| Uranus | 51,800 | 3.69 | 14.6 | 1.07 | 10 hrs. 49 min. | 5 |
| Neptune | 49,500 | 3.50 | 17.3 | 1.08 | 16 hrs. | 2 |
| Pluto | 6,000? | 0.5? | 0.1? | 0.3? | 6.39 days | 0 |

**Table A-2.** *The Planets (Orbital Data)*

| Planet | Average Distance from the Sun (in AUs) | Average Distance from the Sun (in millions of kilometers) | Orbital Period (in years) | Orbital Period (in days) | Average Orbital Speed (in kilometers per second) | Orbital Inclination (in degrees) |
|---|---|---|---|---|---|---|
| Mercury | 0.387 | 57.9 | 0.241 | 88.0 | 47.8 | 7.0 |
| Venus | 0.723 | 108.2 | 0.615 | 224.7 | 35.0 | 3.4 |
| Earth | 1.000 | 149.6 | 1.000 | 365.3 | 29.8 | 0.0 |
| Mars | 1.524 | 227.9 | 1.881 | 687.0 | 24.2 | 1.8 |
| Jupiter | 5.203 | 778.3 | 11.862 | | 13.1 | 1.3 |
| Saturn | 9.539 | 1,427 | 29.458 | | 9.7 | 2.5 |
| Uranus | 19.18 | 2,870 | 84.013 | | 6.8 | 0.8 |
| Neptune | 30.06 | 4,497 | 164.793 | | 5.4 | 1.8 |
| Pluto | 39.44 | 5,900 | 247.686 | | 4.7 | 17.2 |

**Table A-3.** *Satellites of Planets*

| Name | Maximum Magnitude | Diameter (in kilometers) | Average Distance from Planet (in kilometers) | Period of Revolution | Discoverer |
|---|---|---|---|---|---|
| *Satellite of Earth* | | | | | |
| Moon | −12.7 | 3,480 | 384,400 | 27$^d$ 07$^h$ 43$^m$ | |
| *Satellites of Mars* | | | | | |
| Phobos | 11.6 | 25 | 9,400 | 0$^d$ 07$^h$ 39$^m$ | Hall, 1877 |
| Deimos | 12.8 | 13 | 23,500 | 1$^d$ 06$^h$ 18$^m$ | Hall, 1877 |
| *Satellites of Jupiter* † | | | | | |
| V | 13.0 | (150)* | 180,500 | 0$^d$ 11$^h$ 57$^m$ | Barnard, 1892 |
| Io | 4.8 | 3,640 | 421,600 | 1$^d$ 18$^h$ 28$^m$ | Galileo, 1610 |
| Europa | 5.2 | 3,050 | 670,800 | 3$^d$ 13$^h$ 14$^m$ | Galileo, 1610 |
| Ganymede | 4.5 | 5,270 | 1,070,000 | 7$^d$ 03$^h$ 43$^m$ | Galileo, 1610 |
| Callisto | 5.5 | 4,900 | 1,882,000 | 16$^d$ 16$^h$ 32$^m$ | Galileo, 1610 |
| XIII | (20) | (<10) | 4,356,000 | 239$^d$ 06$^h$ | Kowal, 1974 |
| VI | 13.7 | (120) | 11,470,000 | 250$^d$ 14$^h$ | Perrine, 1904 |
| VII | 16 | (40) | 11,800,000 | 259$^d$ 16$^h$ | Perrine, 1905 |
| X | 18.6 | (10) | 11,850,000 | 263$^d$ 13$^h$ | Nicholson, 1938 |
| XII | 18.8 | (10) | 21,200,000 | 631$^d$ 02$^h$ | Nicholson, 1951 |
| XI | 18.1 | (20) | 22,600,000 | 692$^d$ 12$^h$ | Nicholson, 1938 |
| VIII | 18.8 | (20) | 23,500,000 | 738$^d$ 22$^h$ | Melotte, 1908 |
| IX | 18.3 | (15) | 23,700,000 | 758$^d$ | Nicholson, 1914 |
| *Satellites of Saturn* | | | | | |
| Janus | (14) | (300) | 157,500 | 0$^d$ 17$^h$ 59$^m$ | Dollfus, 1966 |
| Mimas | 12.1 | (400) | 185,400 | 0$^d$ 22$^h$ 37$^m$ | Herschel, 1789 |
| Enceladus | 11.8 | (400) | 237,900 | 1$^d$ 08$^h$ 53$^m$ | Herschel, 1789 |
| Tethys | 10.3 | 1,000 | 294,500 | 1$^d$ 21$^h$ 18$^m$ | Cassini, 1684 |
| Dione | 10.4 | 1,000 | 377,200 | 2$^d$ 17$^h$ 41$^m$ | Cassini, 1684 |
| Rhea | 9.8 | 1,600 | 526,700 | 4$^d$ 12$^h$ 25$^m$ | Cassini, 1672 |
| Titan | 8.4 | 5,800 | 1,221,000 | 15$^d$ 22$^h$ 41$^m$ | Huygens, 1655 |
| Hyperion | 14.2 | (400) | 1,479,300 | 21$^d$ 06$^h$ 38$^m$ | Bond, 1848 |
| Iapetus | 11.0 | (1,200) | 3,558,400 | 79$^d$ 07$^h$ 56$^m$ | Cassini, 1671 |
| Phoebe | (14) | (300) | 12,945,500 | 550$^d$ 11$^h$ | Pickering, 1898 |
| *Satellites of Uranus* | | | | | |
| Miranda | 16.5 | (600) | 123,000 | 1$^d$ 09$^h$ 56$^m$ | Kuiper, 1948 |
| Ariel | 14.4 | 1,500 | 191,700 | 2$^d$ 12$^h$ 29$^m$ | Lassell, 1851 |
| Umbriel | 15.3 | 1,000 | 267,000 | 4$^d$ 03$^h$ 38$^m$ | Lassell, 1851 |
| Titania | 14.0 | 1,800 | 438,000 | 8$^d$ 16$^h$ 56$^m$ | Herschel, 1787 |
| Oberon | 14.2 | 1,600 | 586,000 | 13$^d$ 11$^h$ 07$^m$ | Herschel, 1787 |
| *Satellites of Neptune* | | | | | |
| Triton | 13.6 | 6,000 | 353,400 | 5$^d$ 21$^h$ 03$^m$ | Lassell, 1846 |
| Nereid | 18.7 | (500) | 5,560,000 | 359$^d$ 10$^h$ | Kuiper, 1949 |

* Parentheses indicate that numbers are estimated.

† Precise data are not yet available for a fourteenth satellite recently discovered in orbit about Jupiter.

**Table A–4.** *Significant Events in Lunar Exploration*

| Spacecraft Name | Launch Date | Landing Site |
|---|---|---|
| *Ranger Program* | | |
| Ranger 7 | July 28, 1964 | Mare Nubium |
| Ranger 8 | February 17, 1965 | Mare Tranquilitatis |
| Ranger 9 | March 21, 1965 | Alphonsus |
| *Surveyor Program* | | |
| Surveyor 1 | May 30, 1966 | Oceanus Procellarum |
| Surveyor 3 | April 17, 1967 | Oceanus Procellarum |
| Surveyor 5 | September 8, 1967 | Mare Tranquilitatis |
| Surveyor 6 | November 7, 1967 | Sinus Medii |
| Surveyor 7 | January 7, 1968 | Tycho |
| *Orbiter Program* | | |
| Lunar Orbiter 1 | August 10, 1966 | |
| Lunar Orbiter 2 | November 6, 1966 | |
| Lunar Orbiter 3 | February 5, 1967 | |
| Lunar Orbiter 4 | May 4, 1967 | |
| Lunar Orbiter 5 | August 1, 1967 | |
| *Apollo Program* | | |
| Apollo 11 | July 16, 1969 | Mare Tranquilitatis |
| Apollo 12 | November 14, 1969 | Oceanus Procellarum |
| Apollo 14 | January 31, 1971 | Fra Mauro |
| Apollo 15 | July 26, 1971 | Hadley–Apennine |
| Apollo 16 | April 16, 1972 | Descartes |
| Apollo 17 | December 7, 1972 | Taurus–Littrow |
| *Luna Program* | | |
| Luna 3 | October 3, 1959 | |
| Luna 9 | January 31, 1966 | Oceanus Procellarum |
| Luna 13 | December 21, 1966 | Oceanus Procellarum |
| Luna 16 | September 12, 1970 | Mare Fecunditatis |
| Luna 17 | November 10, 1970 | Mare Imbrium |
| Luna 20 | February 14, 1972 | Mare Fecunditatis |
| Luna 21 | January 8, 1973 | Mare Serenitatis |
| Luna 24 | August 9, 1976 | Mare Crisium |

**Table A–5.** *Significant Events in Planetary Exploration*

| Spacecraft Name | Launch Date | Destination | Date of Landing or Encounter |
|---|---|---|---|
| Mariner 2 | August 26, 1962 | Venus | December 14, 1962 |
| Mariner 4 | November 28, 1964 | Mars | July 14, 1965 |
| Venera 4 | June 12, 1967 | Venus | October 18, 1967 |
| Mariner 5 | June 14, 1967 | Venus | October 19, 1967 |
| Venera 5 | January 5, 1969 | Venus | May 16, 1969 |
| Venera 6 | January 10, 1969 | Venus | May 17, 1969 |
| Mariner 6 | February 25, 1969 | Mars | July 31, 1969 |
| Mariner 7 | March 27, 1969 | Mars | August 5, 1969 |
| Venera 7 | August 17, 1970 | Venus | December 15, 1970 |
| Mariner 9 | May 30, 1971 | Mars | November 13, 1971 |
| Pioneer 10 | March 3, 1972 | Jupiter | December 4, 1973 |
| Venera 8 | March 26, 1972 | Venus | July 22, 1972 |
| Pioneer 11 | April 6, 1973 | Jupiter | December 3, 1974 |
|  |  | Saturn | September 1979 |
| Mariner 10 | November 3, 1973 | Venus | February 5, 1974 |
|  |  | Mercury | March 29, 1974 |
| Viking 1 | August 20, 1975 | Mars | July 20, 1976 |
| Viking 2 | September 9, 1975 | Mars | September 3, 1976 |
| Voyager 2 | August 20, 1977 | Jupiter | July 1979 |
|  |  | Saturn | August 1981 |
|  |  | Uranus | January 1986 |
|  |  | Neptune | September 1989? |
| Voyager 1 | September 5, 1977 | Jupiter | March 1979 |
|  |  | Saturn | November 1980 |

# Index

**557**